Lecture Notes in Computer Science 8393

William G. Kennedy Nitin Agarwal
Shanchieh Jay Yang (Eds.)

Social Computing, Behavioral-Cultural Modeling, and Prediction

7th International Conference, SBP 2014
Washington, DC, USA, April 1-4, 2014
Proceedings

Springer

Volume Editors

William G. Kennedy
George Mason University
Krasnow Institute for Advanced Study
Center for Social Complexity
and Department of Computational Social Science
Fairfax, VA, USA
E-mail: wkennedy@gmu.edu

Nitin Agarwal
University of Arkansas at Little Rock
George W. Donaghey College
of Engineering and Information Technology
Little Rock, AR, USA
E-mail: nxagarwal@ualr.edu

Shanchieh Jay Yang
Rochester Institute of Technology
Department of Computer Engineering
Rochester, NY, USA
E-mail: jay.yang@rit.edu

ISSN 0302-9743 e-ISSN 1611-3349
ISBN 978-3-319-05578-7 e-ISBN 978-3-319-05579-4
DOI 10.1007/978-3-319-05579-4
Springer Cham Heidelberg New York Dordrecht London

Library of Congress Control Number: 2014933219

LNCS Sublibrary: SL 3 – Information Systems and Application, incl. Internet/Web and HCI

Typesetting: Camera-ready by author, data conversion by Scientific Publishing Services, Chennai, India

Printed on acid-free paper

Springer is part of Springer Science+Business Media (www.springer.com)

Preface

This volume contains the accepted papers presented orally and as posters at the 2014 International Conference on Social Computing, Behavioral-Cultural Modeling, and Prediction (SBP 2014). This was the seventh year of the SBP conference, and the fourth since it merged with the International Conference on Computational Cultural Dynamics (ICCCD). This year the SBP 2014 conference was co-located with the Behavioral Representation in Modeling and Simulation (BRiMS) conference, which had a separate submission, review, and acceptance process and produces a separate record of its proceedings.

In 2014 the SBP conference continued its traditions. We received a set of 101 submissions (from 17 countries), down from last year's 137 submissions but better than the prior highest number of 88 submissions. SBP continued to be a selective, single-track conference. Twenty-four submissions were accepted as oral presentations, a 24% acceptance rate. We also accepted 31 posters, more than we had previously, for an overall 54% acceptance rate. Three papers were withdrawn and one paper did not meet the length requirements to be published. This resulted in 51 papers included in this volume. Finally, continuing our tradition, a bound copy of these proceedings was distributed to participant attendees at the conference and made available electronically as well as part of Springer's *Lecture Notes in Computer Science* series.

This conference is strongly committed to fostering multidisciplinary research, consistent with recent trends in computational social science and related fields. Authors were asked to indicate from a checklist which topics fitted the papers they were submitting. Many papers covered multiple topics across categories, reflecting the multidisciplinary nature of the submissions and the accepted papers. Consequently, the papers in this volume are presented in alphabetical order by the surname of the first author, instead of placed into preset categories. The topic areas that formed the core of previous SBP conferences were all represented: social and behavioral sciences, health sciences, military science, methodology, and information science.

Across the five topics, the overall number of submissions varied somewhat with the most papers in the category of information sciences (91 submissions, 30%), followed closely by methodology (87 submissions, 28%) and the topic of behavioral, social sciences, and economics (77 submissions, 25%). Health sciences and military science the least represented topics (29 and 24 submissions, respectively, and just less than 10% each). However, the acceptance rate by topic area varied very little across the topics, demonstrating that the quality of the submissions did not vary by topic. Although we cannot compare the quality of this years' submissions with last year, we can say that they were practically uniform across this wide range of topics.

There were a number of events that took place at the conference that are not contained in these proceedings, but they added greatly to the scientific and collegial value of the experience for participants. The first day of the conference included several tutorials on topics ranging from social media to computational models, among others. In the main conference program, each day was marked by a scientific lecture offered by a distinguished keynote speaker. These and other activities are documented on the SBP 2014 conference website, which also contains additional information (http://sbp-conference.org).

Conference activities such as SBP only succeed with assistance from many contributors, which involves a year-long and in many cases a multi-year effort on the part of many volunteers whose names are acknowledged in the next section. Following last year's success, the conference was held once again in downtown Washington, DC, at the University of California's DC Center on Rhode Island Avenue, during April 1–4. We are very grateful for their hospitality and many forms of logistical support. The 2014 Conference Committee met frequently during the year preceding the event, making numerous decisions concerning papers, posters, speakers, sessions, and all other components of a complex event such as this. We are grateful to the program co-chairs and reviewers, to each of the special activities chairs (for diverse activities such as sponsorships, registration, publicity, website, tutorials, challenge problems, and others), as well as the Steering Committee of founding members and the Advisory Committee of funding members, all of whose names are gratefully acknowledged in the next section. Finally, a special thanks to Dorothy Kondal, Administrative Assistant at the Center for Social Complexity, George Mason University, for her indispensable help in executing tasks, keeping us on schedule and organized. We thank all for their kind help, dedication, and collegiality. Funding for the SBP 2014 Conference was also made available by the Mason Center for Social Complexity.

January 2014

Claudio Cioffi-Revilla
Jeffrey C. Johnson

Organization

Conference Co-chairs

Claudio Cioffi-Revilla	George Mason University
Jeffrey Johnson	East Carolina University

Program Co-chairs

William G. Kennedy	George Mason University
Nitin Agarwal	University of Arkansas at Little Rock
Shanchieh Jay Yang	Rochester Institute of Technology

Steering Committee

John Salerno	Air Force Research Lab
Huan Liu	Arizona State University
Sun Ki Chai	University of Hawaii
Patricia Mabry	National Institutes of Health
Dana Nau	University of Maryland
V.S. Subrahmanian	University of Maryland

Conference Committee

Claudio Cioffi-Revilla	George Mason University
Jeffrey Johnson	East Carolina University
William G. Kennedy	George Mason University
Nitin Agarwal	University of Arkansas at Little Rock
Shanchieh Jay Yang	Rochester Institute of Technology
Nathan D. Bos	Johns Hopkins University/Applied Physics Lab
Ariel M. Greenberg	Johns Hopkins University/Applied Physics Lab
John Salerno	Air Force Research Lab
Huan Liu	Arizona State University
Donald Adjeroh	West Virginia University
Katherine Chuang	Soostone, Inc.

Advisory Committee

Fahmida N. Chowdhury	National Science Foundation
Rebecca Goolsby	Office of Naval Research
John Lavery	Army Research Lab/Army Research Office
Patricia Mabry	National Institutes of Health
Tisha Wiley	National Institutes of Health

Sponsorship Committee Chair

Huan Liu	Arizona State University

Publicity Chair

Donald Adjeroh	West Virginia University

Web Chair

Katherine Chuang	Soostone, Inc.

Tutorial Chair

Dongwon Lee	Penn. State University

Challenge Problem Co-chairs

Wen Dong	Massachusetts Institute of Technology, Media Lab
Kevin S. Xu	3M Company
Fred Morstatter	Arizona State University
Katherine Chuang	Soostone, Inc.

Workshop Co-chairs

Fahmida N. Chowdhury	National Science Foundation
Tisha Wiley	National Institutes of Health

Topic Area Chairs

Edward Ip	Wake Forest School of Medicine, for Health Sciences
Chandler Johnson	Stanford, for the Behavioral & Social Sciences & Economics
Amy Silva	Charles River Associates, for Military & Security
Gita Sukthankar	University of Central Florida, for Methodology
Kevin S. Xu	3M Company, for Information, Systems & Network Science

Technical Program Committee

Mohammad Ali Abbasi
Myriam Abramson
Donald Adjeroh
Nitin Agarwal
Kalin Agrawal
Muhammad Ahmad
Yaniv Altshuler
Chitta Baral
Geoffrey Barbier
Asmeret Bier
Lashon Booker
Nathan Bos
David Bracewell
Erica Briscoe
George Aaron Broadwell
Jiangzhuo Chen
Alvin Chin
David Chin
Peng Dai
Yves-Alexandre de Montjoye
Bethany Deeds
Wen Dong
Koji Eguchi
Jeffrey Ellen
Yuval Elovici
Zeki Erdem
Ma Regina Justina E. Estuar
Laurie Fenstermacher
William Ferng
Michael Fire
Anthony Ford
Wai-Tat Fu
Armando Geller
Matthew Gerber
Brian Gurbaxani
Shuguang Han
Soyeon Han
Daqing He
Walter Hill
Michael Hinman
Shen-Shyang Ho
Robert Hubal
Samuel Huddleston
Edward Ip
Terresa Jackson
Lei Jiang
Kenneth Joseph
Ruben Juarez
Byeong-Ho Kang
Jeon-Hyung Kang
William G. Kennedy
Halimahtun Khalid
Masahiro Kimura
Shamanth Kumar
Kiran Lakkaraju
John Lavery
Jiexun Li
Zhuoshu Li
Huan Liu
Ting Liu
Xiong Liu
Jonathas Magalhães
Matteo Magnani
Juan F. Mancilla-Caceres
Achla Marathe
Stephen Marcus
Mathew Mccubbins
Michael Mitchell
Sai Moturu
Marlon Mundt
Keisuke Nakao
Radoslaw Nielek
Kouzou Ohara
Alexander Outkin
Wei Pan
Wen Pu
S.S. Ravi
Nasim S. Sabounchi
Tanwistha Saha
Kazumi Saito
David L. Sallach
Kaushik Sarkar
Philip Schrodt
Arun Sen
Samira Shaikh
Amy Silva

Samarth Swarup
George Tadda
Sarah Taylor
Gaurav Tuli
Trevor Van Mierlo
Craig Vineyard
Anil Kumar Vullikanti
Changzhou Wang
Haiqin Wang
Xiaofeng Wang
Zhijian Wang
Rik Warren
Wei Wei
Nicholas Weller
Paul Whitney
Kevin Xu
Christopher Yang
Laurence T. Yang
Shanchieh Jay Yang
John Yen
Bei Yu
Lei Yu
Mo Yu
Serpil Yuce
Mi Zhang
Qingpeng Zhang
Kang Zhao
Yanping Zhao
Inon Zuckerman
Fatih Özgül

Table of Contents

Oral Presentations

Poster Presentations

Oral Presentations

Human Development Dynamics: An Agent Based Simulation of Adaptive Heterogeneous Games and Social Systems

Mark Abdollahian, Zining Yang, and Patrick deWerk Neal

School of Politics and Economics, Claremont Graduate University, Claremont, USA
{mark.abdollahian,zining.yang,patrick.neal}@cgu.edu

Abstract. In the context of modernization and development, the complex adaptive systems framework can help address the coupling of macro social constraint and opportunity with individual agency. Combining system dynamics and agent based modeling, we formalize the Human Development (HD) perspective with a system of asymmetric, coupled nonlinear equations empirically validated from World Values Survey (WVS) data, capturing the core qualitative logic of HD theory. Using a simple evolutionary game approach, we fuse endogenously derived individual socio-economic attribute changes with Prisoner's Dilemma spatial intra-societal economic transactions. We then explore a new human development dynamics (HDD) model behavior via quasi-global simulation methods to explore economic development, cultural plasticity, social and political change.

Keywords: economic development, modernization, cultural shift, democratization, co-evolution, game theory, agent based model, techno-social simulation, complex adaptive systems.

1 Introduction

Rooted in comparative political economy, the HD perspective is a qualitative, trans-disciplinary approach to understanding modernization and development through the lens of interdependent economic, cultural, social and political forces across individual, institutional and societal scales. Here we extend Abdollahian et al.'s [1] novel, quantitative systems dynamic representation of HD theory at the societal level towards integrated macro-micro scales in an agent based framework. Quek et al [26] also design an interactive macro-micro agent based framework, which they call a spatial Evolutionary Multi-Agent Social Network (EMAS), on the dynamics of civil violence. We posit a new, Human Development Dynamics (HDD) approach where agency matters.

In order to create a robust techno-social simulation [32], we instantiate a system of asymmetric, coupled nonlinear difference equations that are then empirically validated with five waves of data from the World Values Survey (2009). We then fuse this system to agent attribute changes with a generalizable, non-cooperative Prisoner's Dilemma game following Axelrod [3-5] and Nowak and Sigmund [24, 25] to simulate intra-societal, spatial economic transactions where agents are capable of Robust Adaptive Planning (RAP). Understanding the interactive political-cultural effects of macro-socio dynamics and individual agency in intra-societal

W.G. Kennedy, N. Agarwal, and S.J. Yang (Eds.): SBP 2014, LNCS 8393, pp. 3–10, 2014.

transactions are key elements of a complex adaptive systems (CAS) approach. We find strong epistatic interactions, where strategies are interdependent, and local social co-evolution [20] help determine global-macro development outcomes in a particular society.

2 HD Dynamics Background

HD postulates a complex modernization process where value orientations drive an individual's level of existential security and change in predictable ways given shifts in existential security. HD theory provides a framework in which economic development, societal wealth and human needs create generalizable shifts in cultural predispositions and political behavior [17-19] [33].

HD theory expands upon economic drivers from neoclassical growth theory [30, 31] [6] commonly attributed to high growth paths and convergence [21] [27]. Such approaches specify detailed and interactive vectors of economic determinants, country and time-specific effects separately [10]; HD theory fuses cultural, social and political development process into economic growth (*Y*) dynamics.

Rational-secular (*RS*) *cultural* values correspond to individuals' growing emphasis on technical, mechanical, rational, and bureaucratic views of the world. During economic industrialization phases, cultural dispositions tend to progress from an emphasis on traditional pre-industrial values—often measured in terms of religious ceremony attendance—to secular world views, transferring authority from traditional religious figures to technological progress and bureaucratic political life.

Self-expressive (*SE*) *social* values corresponds to the post-industrial phase of economic development where the wealth and advanced welfare system generated by education, increased productivity and service-related economic activities provides individuals with an overwhelming sense of existential security [7] and the freedom to seek self-expression and demand political participation. Self-expression values promote liberal political institutions through two mechanisms. First, to the extent that there is incongruence between cultural demand for, and political supply of, liberal institutions, individuals are more or less prone to elite-challenging activity [16] [13]. Second, self-expression values support the social acceptance of basic democratic norms such as trust and political participation. The end result is a gradual transition toward democratization in autocratic nations and more effective political representation in democratic nations [19].

Lastly, HD theory expects democratic (*D*) *political* values to exhibit positive feedbacks with economic progress, based on previous work on liberal institutions and economic development [12] [9] [14] [2]. Declining economic conditions reintroduce the primacy of basic needs, fueling conditions for more traditional value orientations and less self-expression. Disequilibrium between culturally defined political expectations and political realities promotes and provides motivation for revolutionary change.

The HD perspective suggests a staged process in which rising level of existential security via economic development leads to an increased emphasis on rational-secular and self-expression values. However, these effects are neither linear nor monotonic, as we see strong reversion towards autocratic institutional preferences in survival-minded societies. Democratic norms and institutions that outpace economic progress are inherently unstable with a persistent, turbulent reversion processes, even at high levels of democratic norms and existential security. This suggests that societies experiencing democratization can frequently expect punctuated reversals and revolutions towards more autocratic institutions until more sustainable economic growth and democratic institutions re-emerge.

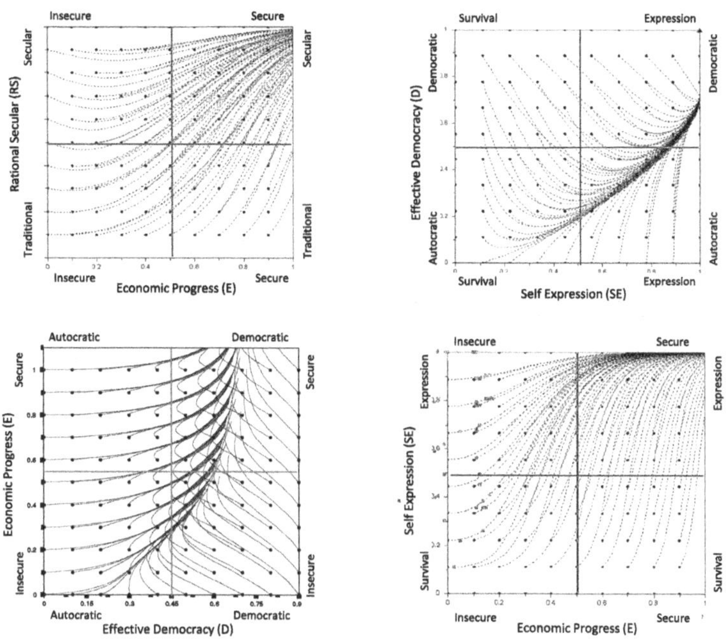

Fig. 1. HD Phase Portraits (Source: Abdollahian et al 2012)

3 A Human Development Dynamics Model

We maintain individual agent attribute relationships and postulated changes of *RS*, *SE*, *D* and *Y* in keeping with HD theory. These endogenously derived, individual agent attributes (RS^i, SE^i, D^i and Y^i) impact how economic transaction games occur, either increasing or decreasing individual wealth and, at increasing scales, determining societal productivity [8]. Geography and proximity are allowed to play a role by instantiating in random two-dimensional lattice worlds.

Social co-evolutionary systems allow each individual to either influence or be influenced by all other individuals as well as macro society [29] [35], perhaps eventually becoming coupled and quasi-path interdependent. Accordingly, we instantiate non-cooperative, socio-economic Prisoner's Dilemma (PD) transaction games given the similarity of agent *i*'s attribute vector (A^i) of social, cultural, political and economic preference (RS^i, SE^i, D^i and Y^i) to agent *j*'s attribute vector (A^j) for selected A^{ij} pairs. Here, symmetric preference rankings and asymmetric neighborhood proximity distributions allows "talk-span," a Euclidean radius measure, to proxy for communications reach, social connectivity and technology diffusion constraining the potential set of A^{ij} game pairs. Low talk-span values restrict games to local neighborhoods among spatially proximate agents, while higher talk-span values expand potential A^{ij} pairs globally, modeling socially compressed space.

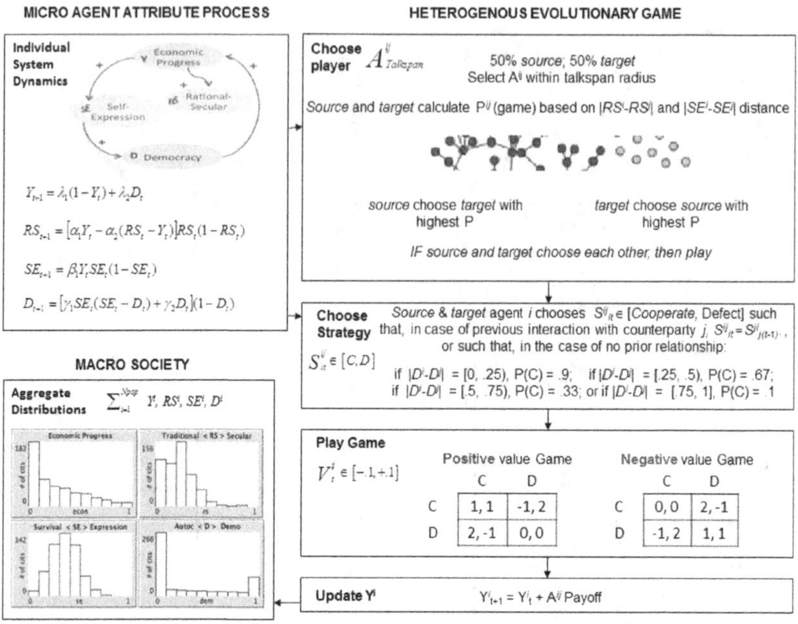

Fig. 2. HDD Architecture (implemented in NetLogo [34])

Following Social Judgment Theory, the attribute positions of two agents are conceived as a Downsian continuum [11] [15] where distance between these positions symmetrically affects the likelihood of one accepting the other's position. Agent i evaluates the likelihood of conducting a transaction with agent j based on similarity of socio-cultural preferences $|RS^i\text{-}RS^j|$ and $|SE^i\text{-}SE^j|$ within the given neighborhood. This captures communications and technology diffusion for frequency and social tie formation [22].

After transaction counterparties are identified, similarity is measured against an exogenous threshold to gauge compatibility. If both parties are satisfied, compatible agents, endowed with RAP cognition, enter into an engagement and search their memory for prior transactions with their period t counterparty. In the case of no prior transaction experience, agents individually each select strategy $S^{ij}_{it} \in$ [*Cooperate, Defect*] probabilistically based on similarity of political preferences as expressed by $|D^i\text{-}D^j|$ [28].

In repeat transactions, agents have perfect memory of $t\text{-}n$ and will predicate their strategy in period t transactions on their counterparties' $t\text{-}1$ behavior such that $S^{ij}_{it} = S^{ji}_{j(t-1)}$. Agents are unaware of counterparties' strategy rule at any point in time. This can lead to the emergence of stable productive relationships, bad relationships featuring pure defection strategies over repeated interactions, and tit-for-tat relationships, where agents alternate between strategies and never sync into a stable productive transactional relationship. This reflects recent work on the affects on co-evolution of both dynamic strategies and updating rules based on agent attributes [20] [23].

Following Nowak and Sigmund [24], we randomly assign game transaction values. However, we do not asymmetrically constrain such values; any particular game transaction value between pairs, V^{ij}, lies in between [-.1, .1]. This instantiation allows for different potential deal sizes, costs, or benefits. We specifically model socio-economic transaction games as producing either positive or negative values as we want to capture behavioral outcomes from games with both upside gains or downside losses.

In our HDD framework, A_i strategies are adaptive, which affect A_{ij} pairs locally within a proximate radius as first order effects. Other agents, within the system but outside the talk-span radius, are impacted through cascading higher orders. Agents simultaneously co-evolve as strategy pair outcomes CC, DC/CD or DD at t affect Y^i at t+1, thus driving both positive and negative RS, SE and D feedback process through $t+n$ iterations. These shape A^i attributes which spur adaptation to a changing environment, summing Y^i, RS^i, SE^i and D^i vector values. Feedback into subsequent A^{ij} game selection networks and strategy choice yields a CAS representation across multiple scales.

4 Sensitivity Analysis

In order to make more generalizable model inferences, Table 1 details the interactive parameter effects[1] on economic prosperity Y, as well as strategy choice pairs CC, CD/DC and DD. As all variables are relatively scaled, we can interpret magnitude and substantive effects across OLS β coefficients. The results reflect a quasi-global sensitivity analysis with 500 agents in180 runs and 700 iterations in each run, randomly down-sampled for pooled OLS tractability.

Table 1. Impact on Economic Development and Strategy Pairs

Model	Economic	CC	CD	DD
Economic		1.099*	0.666*	0.498*
		(0.000)	(0.000)	(0.000)
Rational Secular	0.492*	-0.354*	-0.186*	-0.137*
	(0.000)	(0.000)	(0.000)	(0.000)
Self Expression	-0.128*	0.156*	0.071*	0.411*
	(0.000)	(0.000)	(0.000)	(0.000)
Democracy	0.262*	0.028*	-0.209*	-0.392*
	(0.000)	(0.000)	(0.000)	(0.000)
Cooperate	0.354*			
	(0.000)			
Defect	-0.080*			
	(0.000)			
Talk-span	0.255*	-0.199*	-0.051*	-0.068*
	(0.000)	(0.000)	(0.000)	(0.000)
Time	-0.111*	-0.176*	-0.065*	-0.334*
	(0.000)	(0.000)	(0.000)	(0.000)
Threshold	-0.063*	-0.204*	-0.318*	-0.365*
	(0.000)	(0.000)	(0.000)	(0.000)
RAP	0.024*	-0.020*	-0.289*	-0.135*
	(0.000)	(0.000)	(0.000)	(0.000)
N	78591	81982	73499	61877
Prob > F	0.000	0.000	0.000	0.000
R-squared	0.946	0.809	0.472	0.412
Root MSE	0.041	0.795	0.978	0.877

Numbers in parentheses are corresponding robust standard errors.
* Significance at 1% level.

[1] Parameter setting: talk-span = 0, 1, 4, 7, 10; threshold = 0, 0.04, 0.09, 0.16, 0.25, 0.36; RAP = true, false.

Our first model on mean societal economic development Y confirms HD theory that positive values of mean societal RS and D values significantly speed the pace of economic development, although SE is significant and slightly negative; this may relate to a loss of productivity when efforts in isolation are directed away from production and towards self-expression. Looking at the impact of evolutionary games, we see that cooperation has a stronger positive impact than defection or mixed strategies in increasing transaction value to society. Talk-span spatial proximity is positive and significant, confirming priors that increasing technology and compressing potential social space also speed development processes. Time is slightly negative, indicating that economic prosperity is not endogenous to the model. Threshold, agent willingness to engage in transactions, is slightly negative, implying that reduced trust has a slightly negative impact on growth. Lastly, RAP is slightly positive, suggesting increased cognition is beneficial in our simulated environment. Future research will investigate to what extent the RAP coefficient increases with agent analytical sophistication, and may include an endogenous "education" component.

5 Conclusions

Consistent with qualitative HD theory and empirical reality, our HDD model finds complexity and nonlinear path dependence in three areas: adaptive development processes, social co-evolutionary transactions and near equilibrium development trajectories. From a complex adaptive system perspective on HD theoretical processes, economic progress is a necessary condition for successful secularization and expressive political behavior, which are antecedents for lasting democratic institutions. While modernization is not inevitable, our results support empirical observations for a staged process where increasing existential security via economic development leads to increased emphasis on rational-secular and self-expressive values that results in societal development. Here we find that rational-secular norms strongly impact economic growth and speed up the pace of development more than self-expressive societal values alone. Beyond supporting HD theory, agents do adapt interactively with their environments as mutual cooperation does result in higher societal wealth than defection alone and is self reinforcing over time.

While only an initial, rough approximation at the truly complex, interdependent and highly nonlinear nature of modernization, our HDD approach provides insights into the interactivity of individual agency and societal outcomes seen through the lens of evolutionary games. Perhaps techno-social simulations like HDD can assist policy makers and scholar alike, to better understand, anticipate and shape positive social outcomes for all.

References

1. Abdollahian, M.A., Coan, T., Oh, H.N., Yesilda, B.: Dynamics of cultural change: the human development perspective. International Studies Quarterly, 1–17 (2012)
2. Acemoglu, D., Robinson, J.: Why Nations Fail: The Origins of Power, Prosperity, and Poverty. Crown Business, New York (2012)
3. Axelrod, R.: The evolution of strategies in the iterated Prisoner's Dilemma. In: Davis, L. (ed.) Genetic Algorithms and Simulated Annealing, pp. 32–41. Morgan Kaufman, Los Altos (1987)
4. Axelrod, R.: The Complexity of Cooperation: Agent-Based Models of Competition and Collaboration. Princeton University Press, Princeton (1997a)

5. Axelrod, R.: The dissemination of culture: a model with local convergence and global polarization. Journal of Conflict Resolution 41, 203–226 (1997b)
6. Barro, R.: Economic growth in a cross section of countries. Quarterly Journal of Economics 106(2), 407–444 (1991)
7. Bell, D.: The Coming of Postindustrial Society. Basic Books, New York (1973)
8. Binmore, K.G.: Game Theory and the Social Contract. MIT Press, Cambridge (1994)
9. Boix, C., Stoke, S.: Endogenous democratization. World Politics 55, 517–549 (2003)
10. Caselli, F., Esquivel, G., Lefort, F.: Reopening the convergence debate: a new look at cross-country growth empirics. Journal of Economic Growth 1(3), 363–389 (1996)
11. Darity, W.: Social Judgment Theory. Macmillan Reference USA, Detroit (2008)
12. Diamond, L.: Economic development and democracy reconsidered. In: Diamond, L., Marks, G. (eds.) Reexamining Democracy. Sage, London (1992)
13. Eckstein, H., Gurr, T.R.: Patterns of Authority: A Structural Basis for Political Inquiry. John Wiley & Sons, New York (1975)
14. Feng, Y.: Democracy, Governance, and Economic Performance: Theory and Evidence. The MIT Press, Cambridge (2003)
15. Griffin, E.: A First Look at Communication Theory. McGraw-Hill Higher Education, Boston (2009)
16. Gurr, T.R.: Why Men Rebel. Princeton University Press, Princeton (1970)
17. Inglehart, R.: Modernization and Postmodernization: Cultural, Economic and Political Change in 43 Societies. Princeton University Press, Princeton (1997)
18. Inglehart, R., Baker, W.E.: Modernization, cultural change, and the persistence of traditional values. American Sociological Review 65, 19–51 (2000)
19. Inglehart, R., Welzel, C.: Modernization, Cultural Change, and Democracy: The Human Development Sequence. Cambridge University Press, New York (2005)
20. Kauffman, S.A.: The Origins of Order: Self-Organization and Selection in Evolution. Oxford University Press, Oxford (1993)
21. Mankiw, N.G., Romer, D., Weil, D.N.: A contribution to the empirics of economic growth. Quarterly Journal of Economics 107, 407–437 (1992)
22. McPherson, M., Smith-Lovin, L., Cook, J.: Birds of a feather: homophily in social networks. Annual Review of Sociology 27, 415–444 (2001)
23. Moyano, L.G., Sanchez, A.: Spatial Prisoner's Dilemma with Heterogeneous Agents. Elsevier Manuscript (2013)
24. Nowak, M.A., Sigmund, K.A.: Strategy of win-stay, lose-shift that outperforms tit-for-tat in the prisoner's dilemma game. Nature 364, 56–58 (1993)
25. Nowak, M.A., Sigmund, K.A.: Evolution of indirect reciprocity by image scoring. Nature 393, 573–577 (1998)
26. Quek, H.Y., Tan, K.C., Abbass, H.A.: Evolutionary game theoretic approach for modeling civil violence. IEEE Transactions on Evolutionary Computation 13, 1–21 (2009)
27. Sala-i-Martin, X.X.: Regional cohesion: evidence and theories of regional growth and convergence. European Economic Review 40, 1325–1352 (1996)
28. Siero, F.W., Doosje, B.J.: Attitude change following persuasive communication: Integrating social judgment theory and the elaboration likelihood model. Journal of Social Psychology 23, 541–554 (1993)
29. Snijders, T.A., Steglich, C.E., Schweinberger, M.: Modeling the co-evolution of networks and behavior. Longitudinal Models in the Behavioral and Related Sciences, 41–71 (2007)
30. Solow, R.: A contribution to the theory of economic growth. The Quarterly Journal of Economics 70, 65–94 (1956)

31. Swan, T.: Economic growth and capital accumulation. Economic Record 32, 334–361 (1956)
32. Vespignani, A.: Predicting the behavior of techno-social systems. Science 325, 425–428 (2009)
33. Welzel, C., Inglehart, R., Klingemann, H.: The theory of human development: a cross-cultural development. European Journal of Political Science 42, 341–380 (2003)
34. Wilensky, U.: NetLogo. Center for Connected Learning and Computer-Based Modeling, Northwestern University, Evanston, IL (1999), http://ccl.northwestern.edu/netlogo/
35. Zheleva, E., Sharara, H., Getoor, L.: Co-evolution of social and affiliation networks. In: Proceedings of the 15th ACM SIGKDD International Conference on Knowledge Discovery and Data Mining. ACM, New York (2009)

Mobility Patterns and User Dynamics in Racially Segregated Geographies of US Cities

Nibir Bora, Yu-Han Chang, and Rajiv Maheswaran

Information Sciences Institute,
University of Southern California, Marina del Rey, CA 90292
nbora@usc.edu,{ychang,maheswar}@isi.edu

Abstract. In this paper we try to understand how racial segregation of the geographic spaces of three major US cities (New York, Los Angeles and Chicago) affect the mobility patterns of people living in them. Collecting over 75 million geo-tagged tweets from these cities during a period of one year beginning October 2012 we identified home locations for over 30,000 distinct users, and prepared models of travel patterns for each of them. Dividing the cities' geographic boundary into census tracts and grouping them according to racial segregation information we try to understand how the mobility of users living within an area of a particular predominant race correlate to those living in areas of similar race, and to those of a different race. While these cities still remain to be vastly segregated in the 2010 census data, we observe a compelling amount of deviation in travel patterns when compared to artificially generated ideal mobility. A common trend for all races is to visit areas populated by similar race more often. Also, blacks, Asians and Hispanics tend to travel less often to predominantly white census tracts, and similarly predominantly black tracts are less visited by other races.

Keywords: Mobility patterns, racial segregation, Twitter.

1 Introduction

Sociologists and economists have long been trying to understand the influence of racial segregation in the United States, on various social aspects like income, education, employment and so on. Every decennial census has hinted the gradually decreasing residential racial segregation in many major metropolitans, but, it is important to continually analyze the effects and how they change over time to have a better understanding of today's social environment.

In this paper, we try to understand if, and how, racial segregation affect the way people move around in large metropolitans. Ubiquitously available data from geo-location based sharing services like Twitter poses a prudent source of real-time spatial movement information. Coalescing users belonging to racially predominant geographic areas with their mobility patterns, we analyze to find variations in travel to areas of similar and dissimilar races. We also build generalized models of ideal human mobility and create a corpus of travel activity

W.G. Kennedy, N. Agarwal, and S.J. Yang (Eds.): SBP 2014, LNCS 8393, pp. 11–18, 2014.

analogous to the actual data. Comparing the actual mobility of users to the ideal models, we look for bias and interesting behavior patterns and dynamics in three U.S. cities- New York, Los Angeles and Chicago.

2 Related Work

A number of studies ([1], [2], [3]) have accounted for the qualitative statistics of racial segregation in major U.S. metropolitans, and how its extent has changed over time. Most of these studies use one or more of the five indexes explained by Massey and Denton [4] to compare the magnitude of segregation between two racial groups. Analyzing data from the 1980 census in 60 U.S. metropolitans, Denton and Massey's [1] findings indicated that blacks were highly segregated from white in all socioeconomic levels, relative to Hispanics or Asians. Although the levels of segregation has declined modestly during the 1980s [5], through the 1990s [2], and up to 2000 [3], blacks still remain more residentially segregated than Hispanics and Asians. Clark [6] points out that although a certain degree of racial integration is acceptable it is unrealistic to expect large levels of integration across neighborhoods, because there exists a tendency for households of a given race to cluster with others of similar race [7].

Veering from the conventional studies that measure the extent of segregation, few researchers have tried to identify social problems arising as a result of it. Peterson and Krivo [8] studied the effect of racial segregation on violent crime. Card and Rothstein [9] found that black-white SAT test score gaps during 1998-2001 were much higher in more segregated cities compared to nearly integrated cities. In our study, we consider another interesting effect- biases in mobility patterns as a result of segregation, and at the same time shed some light on the extent of segregation in the 2010 census data.

Spatiotemporal models of human mobility have been studied on various datasets, such as circulation of US bank notes [10] and cell phone logs [11]. Temporal human activities like replying to emails, placing phone calls, etc. are known to occur in rapid successions of short duration followed by long inactive separations, resembling a Pareto distribution. The truncated power law distribution characterizing heavy-tailed behavior for both distance and time duration of hops between subsequent events in a trajectory of normal travel pattern has been established by many studies [10], [11]. Although geo-location based data from Twitter has been used in several applications like spotting and tracking earthquake shakes [12] and street-gang behavior [13], it has not been used to model effects of racial segregation on mobility patterns and behavior.

3 Data Description

Census Tracts and Racial Segregation Data. To build a geographic scaffolding for our experiments, we use census tract polygons defined by the U.S. Census Bureau[1] in three large metropolitans- New York City, Los Angeles and

[1] `http://www.census.gov/`

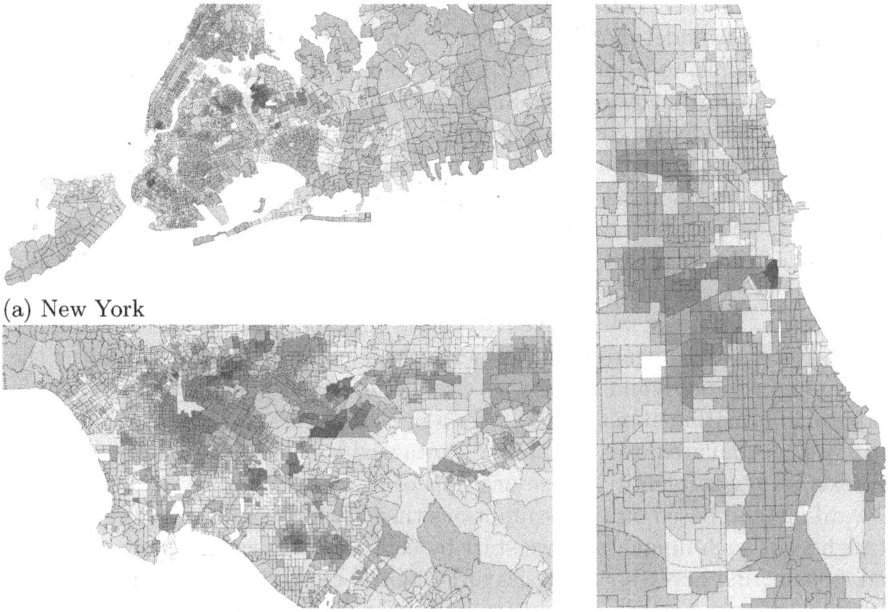

(a) New York

(b) Los Angeles

(c) Chicago

Fig. 1. Figure shows racial segregation maps for (a) New York City, (b) Los Angeles, and (c) Chicago. Colors- blue:white, green:black, red:Asian, orange:Hispanic, brown:others. *Note: The maps show only a portion of the entire city.*

Chicago. Census tracts are small, relatively permanent statistical subdivisions of a county or equivalent entity, typically having a population between 1,200 and 8,000 people. Census information is also available for city blocks, however, using such small geographic entities would account for scarce user movement data. Census tract shapefiles can be downloaded from the National Historical Geographic Information System (NHGIS) website[2].

To find the predominant race in each census tracts we use table P5 (Hispanic or Latino Origin by Race) in the 2010 Summary File 1 (SF1), also available on the NHGIS website. Table P5 enumerates the number of people in each census tract belonging to the racial groups- non-Hispanic White, non-Hispanic Black, non-Hispanic Asian, Hispanic or Latino, and other. The population of these five categories sums up to the total population of the tract. The predominant race in a tract is the one having a majority (50% or more) population. Figure 1 shows census tract boundaries color coded by the majority race. The tracts with a lighter shade of color represents the ones where the race with maximum population did not have a 50% majority.

---

[2] `https://www.nhgis.org/`

Users and Twitter Data. Once the geographic canvas is ready, we need human entities to model mobility patterns. An ideal dataset would consist of a large set of people, location of their homes and their daily movement traces on a geographic coordinate system. Since such a corpus is difficult to build and acquire, we consider an alternative source- Twitter. A geo-tagged tweet is up to 140 characters of text and is associated with a user id, timestamp, latitude and longitude. Frequent Twitter users who use its location sharing service, would produce a close representation of their daily movement in their tweeting activity.

We collected strictly geo-tagged tweets using Twitter's Streaming API[3] and limited them to the polygon bounding the three cities New York, Los Angeles and Chicago. This way, we received all tweets with location information and not just a subset. We disregard all other fields obtained from Twitter, including the user's twitter handle, and in no way use the information we retain to identify personal information about a user. Over a period of one year, beginning October 2012, we accumulated over 75 million geo-tagged tweets.

Next, we try to identify home locations of users by following a very straight forward method based on the assumptions that users generally tweet from home at night. For each unique user we start by collecting all tweets between 7:00pm and 4:00am, and apply a single pass of DBSCAN clustering algorithm [14]. The largest cluster produced by the cluster analysis is chosen as the one corresponding to the user's home, and its centroid is used as the exact coordinates. Skipping users with very few tweets, and ones for whom a cluster could not be formed, we were able to identify home locations for over 30,000 unique users. A user is assigned to a particular census tract if the coordinates of his/her home lies within its geographic bounds, and is assumed to belong to the race of majority population in that tract.

Once these preprocessing steps have been carried out, we are left with a rich set of data comprising of users, their race, their home location, and their movement activity on a geographic space. This data will act as the seed for all our following experiments.

4 Experimental Setup

Keeping in line with our primary objective of identifying effects of racial segregation on movement behavior, we calculate number of visits among users living in tracts of different races. Human mobility, however, tends to follow uniform patterns and can be simulated by parameterized models. The question that arises is whether or not by visiting a tract of similar or dissimilar race, a user is simply adhering to the ideal movement pattern he/she is supposed to follow, or is there a bias due to the presence of a particular race. To answer this question, we build models of movement patterns for each of the three cities and generate synthetic datasets. Measuring the variation of actual mobility data from the ideal (simulated) movement patterns would indicate the presence of any inter-race bias. This steps involved in this process are explained next.

[3] `https://dev.twitter.com/`

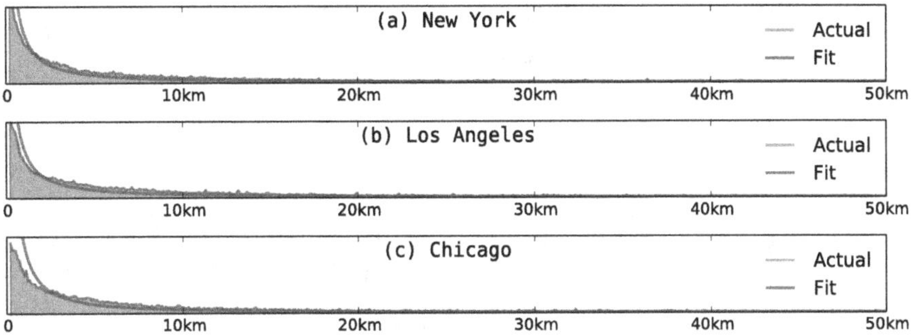

Fig. 2. Figure shows displacement from home while tweeting in (a) New York City, (b) Los Angeles, and (c) Chicago

4.1 Models of Movement Pattern

An established characteristic of human mobility is its Levy flight and random walk properties. In essence, the trajectory of movement follows a sequence of random steps where the step size belongs to a power law distribution (probability distribution function (PDF): $f(x) = Cx^{\alpha}$), meaning, there are a large number of short hops and fewer long hops. As shown in [13] the distance from home while tweeting also follows a power law distribution.

Figure 2 shows the distance from home distribution over the range 100m to 50km, and the corresponding least square fit for power law in the three cities. As a test of correctness we use a two-sample Kolmogorov-Smirnov test, where the null hypothesis states that the two samples are drawn from a continuous power law distribution. In each case, the null hypothesis was accepted with significance ($p < 0.05$), hence verifying the correctness of the parameter fits. Tweets within 100m from the home location of users were removed as such small shifts in distance may occur due to GPS noise even when the user is stationary.

As shown in [13], the direction of travel from home also follows a uniform distribution, only to be skewed by physical and geographic barriers like freeways, oceans etc. For computational simplicity we disregard any such skew and assume that the distribution follows a perfect uniform distribution, i.e. equally likely to travel in any direction. The resulting PDF, shown below, is the product of two probabilities- one for distance and the other for direction of travel θ (θ is constant).

$$f(x) = Cx^{\alpha} \times \frac{1}{\theta} \tag{1}$$

Artificial location data for a user is generated by creating a random sample from the distribution in Equation 1. Keeping the number of simulated tweet locations equal to the number of actual tweets, a synthetic dataset is created by sampling for each user.

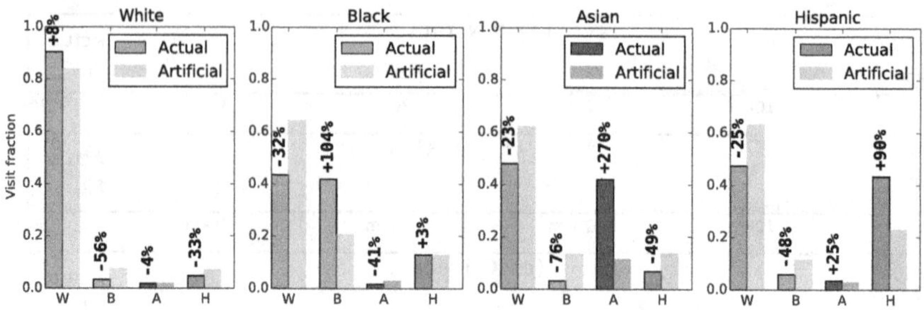

(a) New York

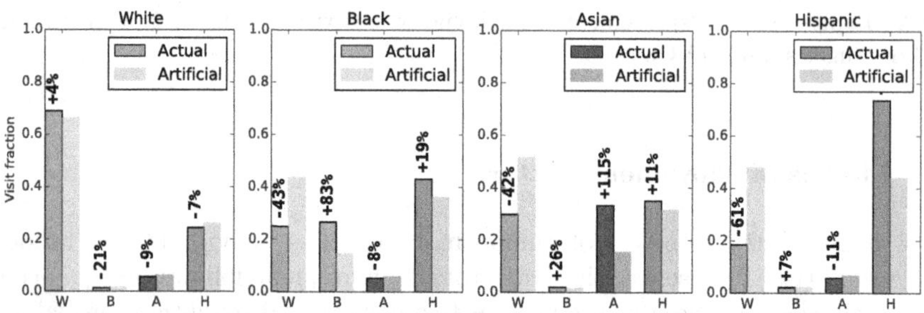

(b) Los Angeles

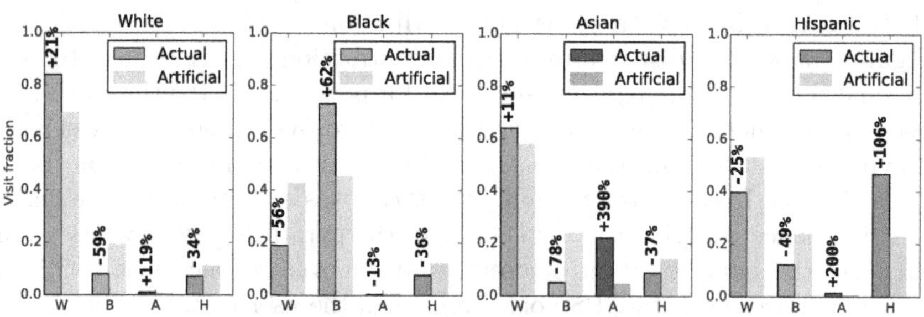

(c) Chicago

Fig. 3. Figure shows fraction of visits between each of the four race in the three cities. Colors- blue:white, green:black, red:Asian, orange:Hispanic.

5 Results and Discussion

With the artificial data being a representation of ideal movement pattern one should follow, the comparison with actual movements yield a number of interesting results. Figure 3 shows, for each of the three cities, fraction of visits by people living in white, black, Asian and Hispanic tracts, alongside the simulated ideal fractions. As clearly visible, the actual movement patterns of users deviate significantly from the expected behavior.

A noticeable trend for any given race is that visits are always higher to tracts of similar race, with the difference from simulated visits for blacks, Asians and Hispanics being very high, going up to four times for Asians. The difference is not as high for whites, however, both the actual and simulated fractions are noticeably larger than visits to any other races. This is explained by the fact that there are far too many white tracts compared to other races, and white segregation clusters are large as well. While visiting tracts of other races the actual and simulated data are very close except in New York and Chicago where the actual visits by people living in white tracts to black tracts are over 50% less than what ideally should have been. Likewise, people living in black tracts in all three cities would visit white tracts less often, while their visits to Asian and Hispanic tracts does not skew much from the artificial data. Hispanics in New York and Chicago visits blacks less often than expected, but it is just the opposite in Los Angeles.

In general, visits to white tracts by blacks, Asians and Hispanics are always much lower than simulated. The only exception to this trend is for Asians in Chicago, where there are very few tracts with a majority Asian population, meaning, it could simply be a bias in sample size. The visits to predominantly black tracts by other races is also lower than the artificial data, although there is an interesting exception in Los Angeles where Asians and Hispanics visit blacks more often than expected.

It is fascinating to see that all races are biased towards areas of identical race, and tend to keep away from others. It is also interesting to note that these trends do not resonate equally in all cities. For example, blacks in New York and Los Angeles, would visit Hispanics close to or even more than expected, but in Chicago the fraction of visits is less.

6 Conclusion

In this paper we try to understand the effects of racial segregation on mobility patterns of people living in three major U.S. metropolitans- New York City, Los Angele and Chicago. We assembled a dataset comprising of human entities, their home locations and daily movement data by accumulating geo-tagged tweets from these cities and performing simple preprocessing steps. The human entities were combined with geographic entities, in this case census tract polygons, and each user was associated with a particular race homologous to the race with majority population in that tract, as calculated from the 2010 census data. Building parameterized models of human mobility for these cities we generated synthetic data to compare with the actual movement of people. We observed significant effects of racial segregation on people's mobility, leading to some interesting observations.

Although racial segregation in the U.S. has been decreasing in the past few decades the major metropolitans are still vastly segregated. People living within tracts of any particular race are biased towards other races and tend to visit tracts of similar race more often. However, the difference in visits to other races is not evenly distribute. Blacks, Asians and Hispanics usually have a higher percentage of difference in their visits to white tracts, and similarly, black tracts are less visited by other races. Within these patterns we also observed some variations among the three cities, for example, the higher than expected visits to black tracts in Los Angeles.

Our approach allows to use readily available geo-location based data from Twitter to model human mobility and investigate effects of geographic and sociological constraints. However, this approach is far from being perfect and opens up numerous avenues for future research. For instance, census tracts have people from different races living in them, but human entities are designated to the race with majority population, when in reality they may belong to a different race. Another assumption we made was the uniform distribution of direction of travel. It would be interesting to introduce skews in the distribution according to the presence of geographic barriers.

References

1. Denton, N.A., Massey, D.S.: Residential segregation of blacks, hispanics, and asians by socioeconomic status and generation. Social Science Quarterly 69(4), 797–817 (1988)
2. Glaeser, E.L., Vigdor, J.L.: Racial segregation in the 2000 census: Promising news. Brookings Institution, Center on Urban and Metropolitan Policy (2001)
3. Logan, J.R., Stults, B.J., Farley, R.: Segregation of minorities in the metropolis: Two decades of change. Demography 41(1), 1–22 (2004)
4. Massey, D.S., Denton, N.A.: The dimensions of residential segregation. Social Forces 67(2), 281–315 (1988)
5. Massey, D.S., Denton, N.A.: Trends in the residential segregation of blacks, hispanics, and asians: 1970-1980. American Sociological Review, 802–825 (1987)
6. Clark, W.A.: Residential preferences and neighborhood racial segregation: A test of the schelling segregation model. Demography 28(1), 1–19 (1991)
7. Bayer, P., McMillan, R., Rueben, K.S.: What drives racial segregation? new evidence using census microdata. Journal of Urban Economics 56(3), 514–535 (2004)
8. Peterson, R.D., Krivo, L.J.: Racial segregation and black urban homicide. Social Forces 71(4), 1001–1026 (1993)
9. Card, D., Rothstein, J.: Racial segregation and the black–white test score gap. Journal of Public Economics 91(11), 2158–2184 (2007)
10. Brockmann, D., Hufnagel, L., Geisel, T.: The scaling laws of human travel. Nature 439(7075), 462–465 (2006)
11. Gonzalez, M.C., Hidalgo, C.A., Barabasi, A.L.: Understanding individual human mobility patterns. Nature 453(7196), 779–782 (2008)
12. Sakaki, T., Okazaki, M., Matsuo, Y.: Earthquake shakes twitter users: real-time event detection by social sensors. In: Proceedings of the 19th International Conference on World Wide Web, pp. 851–860. ACM (2010)
13. Bora, N., Zaytsev, V., Chang, Y.H., Maheswaran, R.: Gang networks, neighborhoods and holidays: Spatiotemporal patterns in social media. In: 2013 ASE/IEEE International Conference on Social Computing (2013)
14. Ester, M., Kriegel, H.P., Sander, J., Xu, X.: A density-based algorithm for discovering clusters in large spatial databases with noise, Kdd (1996)

Incorporating Interpretation into Risky Decision-Making

A Computational Model

David A. Broniatowski[1] and Valerie F. Reyna[2]

[1] The George Washington University, Department of Engineering Management and Systems Engineering, Washington, DC, USA
broniatowski@gwu.edu
[2] Cornell University, Departments of Psychology and Human Development, Ithaca, NY, USA
vr53@cornell.edu

Abstract. Most leading computational theories of decision-making under risk do not have mechanisms to account for the incorporation of cultural factors. Therefore, they are of limited utility to scholars and practitioners who wish to model, and predict, how culture influences decision outcomes. Fuzzy Trace Theory (FTT) posits that people encode risk information at multiple levels of representation – namely, gist, which captures the culturally contingent meaning, or interpretation, of a stimulus, and verbatim, which is a detailed symbolic representation of the stimulus. Decision-makers prefer to rely on gist representations, although conflicts between gist and verbatim can attenuate this reliance. In this paper, we present a computational model of Fuzzy Trace Theory, which is able to successfully predict 14 experimental effects using a small number of assumptions. This technique may ultimately form the basis for an agent-based model, whose rule sets incorporate cultural and other psychosocial factors.

Keywords: framing, gist, verbatim, cultural modeling.

1 Introduction

Anthropologists, such as Mary Douglas (e.g., [1]) argue that a group member's perception of risk is driven by cultural norms that define that group's identity. Similarly, Jasanoff (e.g., [2]) and others in the field of Science and Technology Studies, argue that risk is a social construct that is group-based. In contrast, scholars such as Sunstein (e.g., [3]) argue that risks are objective, and must therefore be addressed in a manner consistent with known statistics. In this paper, we draw upon Fuzzy Trace Theory (FTT; e.g., [4]), a theory of decision-making under risk, which explicitly acknowledges that risk perception contains elements of subjective perception that are shaped by culture, emotion, and prior experience. Our goal here is to formalize FTT, thereby generating a computational theory that can be used to predict the outcomes of risky decisions given the gists that are held by a member of a given group. We will therefore specify the theory to such an extent that it may be used as a set of rules, e.g., for a population of agents within an agent-based model. To that end, we have generated a novel computational representation of FTT, which is described in Section 2.

W.G. Kennedy, N. Agarwal, and S.J. Yang (Eds.): SBP 2014, LNCS 8393, pp. 19–26, 2014.

Adherents of the expected utility paradigm claim that a rational decision-maker would choose between two risky options based upon which option yielded the largest expected payoff. Nevertheless, previous studies, most notably those of Tversky and Kahneman ([5]), have demonstrated the existence of consistent heuristics and biases in human decision making, using scenarios such as what has become known as the Disease Problem (DP) in the decision-making literature:

> "Imagine that the U.S. is preparing for the outbreak of an unusual Asian disease, which is expected to kill 600 people. Two alternative programs to combat the disease have been proposed. Assume that the exact scientific estimates of the consequences of the program are as follows:
> If Program A is adopted, 200 people will be saved
> If Program B is adopted, there is a 1/3 probability that 600 people will be saved and a 2/3 probability that no people will be saved." [5]

A second framing contains the same preamble, but presents the following options:

> "If Program C is adopted 400 people will die.
> If Program D is adopted there is a 1/3 probability that nobody will die, and a 2/3 probability that 600 people will die."

Although subjects are more likely to choose option A than option B, they are more likely to choose option D than option C, even though these problems are mathematically identical. Adherents of the heuristics and biases approach point to Prospect Theory (PT; [6]) which weights losses and gains relative to an individual's reference point differently. PT and its successor, Cumulative Prospect Theory [7], both retained utility theory's assumption of a continuous and monotonic utility function. Such continuous models tend to be favored because they are computationally tractable.

According to Fuzzy Trace Theory (FTT), these, and similar, decision options are stored at a categorical, qualitative level of processing – known as gist – which is encoded simultaneously with the detailed verbatim numbers. "The gist of a decision varies with age, education, culture, stereotypes worldview, and other factors that affect the meaning or interpretation of information…" [8]. According to FTT, the majority of subjects in the gain frame of the DP prefer option A (D) because they interpret its decision options as:

A) Some live (die) vs.

B) Some live (die) OR none live (die)

A central tenet of FTT is that detailed representations of these numerical options are recorded, but that the gist representation is used preferentially, especially when verbatim decision-making is insufficient to distinguish between decision options. Finally, when a decision between gist categories is made, it is made on the basis of simple binary valenced affect. (e.g., good vs. bad, approach vs. avoid, etc.).

We have implemented our theory in a computer program that makes predictions regarding outcomes in risky-decision problems. The theory and its computer implementation successfully predict the outcome of 14 different experimental effects reported in the literature. We further test our theory against the outcomes of a meta-analysis of 13 studies reported in the literature. We begin with a description of the theory.

2 A Theory and Model of Risky Decision-Making

Fuzzy Trace Theory makes four main predictions about risky decision problems:

1. Decision options are encoded at a detailed verbatim level and a high level of abstraction – or gist – simultaneously. Effects that are not consistent with expected utility, such as framing, may arise when decision options encode different gists.
2. Choices between gist categories are made upon the basis of binary valenced affect. Decision-makers prefer the positive-affect option.
3. When categorical contrasts are not possible, decision-makers will revert to more precise gists, e.g., ordinal representations, and framing effects will be attenuated.
4. Gist categories are encoded based upon a subject's prior experience (e.g., cultural norms). These take the form of categorical contrasts, such as psychologically special representations of numbers (e.g., all, none, certainty, etc) and may be culturally contingent. Changing these will change what gists are encoded.

Rivers et al., [9] illustrate FTT with the example of an underage adolescent who must decide between two options for how to spend her evening. She can go to a fun party where alcohol is served, but which might be broken up by parents (and therefore take a risk of being culturally sanctioned); or she can go to a sleepover that is not as much fun, but is also not culturally sanctioned (and therefore not take a risk). Suppose the adolescent knows that the sleepover will be fun with certainty, whereas there is a 90% chance that the party will be twice as much fun as the sleepover, but there is a 10% chance that the parents will shut the party down, which is no fun. A classically rational decision-maker would play the odds – i.e., they would attend the risky party because, on average, the party is likely to be 1.8 times as fun as the sleepover. In contrast, a decision-maker who relies only on gist interpretations would perceive the choice as between some fun with certainty (thc sleepover) and maybe some fun but maybe no fun (the party). This gist-based decision-maker would choose the sleepover.

This gist hierarchy has been implemented as part of a computational model of decision-making under risk, which makes predictions for outcomes. We next describe our theory and illustrate its workings with examples from the computer program.

We begin by representing our decision problem in a decision space, where each axis corresponds to a variable of interest. Our computer program first prompts the user to indicate the number of axes in the decision space, and the names of these axes. In the adolescent decision problem, one axis would represent some quantity of fun, whereas a second axis would represent the probability with which that fun is attained. Each point in our space is a complement in a decision option.

The gist routine partitions this space into categories, each representing a set of points in the space. According to the theory, these categories may arise from the decision-maker's prior knowledge and expectations, as well as psychological regularities. For example, there is strong evidence to indicate that some of a quantity is psychologically different than none of a quantity [10-13], leading most decision-makers to distinguish between these categories. Similarly, "all" of a quantity is treated qualitatively distinct manner [14]. Analogously, no chance (probability = 0%) and certainty (probability = 100%) are psychologically distinct values on the probability axis [7].

After collecting information about the decision space (i.e., what are its axes/major variables), our program prompts the user to enter constraints on the decision space. These constraints take the form of an algebraic expression that is set equal to zero.

2.1 Categorical Routine

Once the decision-space has been established, the user enters the decision options describing the problem. For each axis in the space (i.e., each variable), the user enters the numerical data as it appears in the decision problem. An element of the problem may not be explicitly specified (for example, in the DP, where the probabilities in options A and C are not given, even though they are implicitly assumed to be 100%). In such cases the user leaves the field blank, and is subsequently prompted to enter the implicit value. The computer implementation of the theory uses an internal representation corresponding to the categories of the space in which each decision complement is located. These categories are then compared in a pairwise fashion – for each pair of different decision complements that are in different decision options, the user is prompted to indicate which complement is preferred, corresponding to the decision-maker's valenced affect; values are required to choose between categories.

2.2 Ordinal Routine

If two decision-options fall into the same category the decision-maker is indifferent between these complements at the categorical level. Our theory predicts that the decision-maker will descend the gist hierarchy, and revert to a more precise ordinal representation of the decision options. When this happens in our computer implementation, the user is asked to indicate the preferred direction along each of the axes in our decision space. The decision-options are then compared independently along each axis. For example, if our hypothetical adolescent knew for certain that the parents would not break up the party (thereby removing the categorical possibility of no fun), she would face a choice between two options, which advertise "some fun with certainty." She would choose the party because it would be "more fun" in contrast to the sleepover, which would be "less fun." If one of the options is preferred along at least one axis, and equal to or preferred along each of the other axes, then this option is chosen.

It may occur that the options are equal along all dimensions, or preferred along one dimension, but not preferred along at least one other dimension. For example, if, instead of parental intervention, our hypothetical adolescent suspected that there was a 10% chance that adult supervision would be present at the party to ensure that it remained dry – an option that would be half as much fun as the sleepover – then the adolescent would have to choose between an option that is interpreted as more fun with some chance and less fun with some chance, and an option that is some fun with certainty. Here the decision-maker is also indifferent at the ordinal level.

2.3 Interval Routine

Our theory predicts that all decision-makers also use yet a more precise interval representation. Here, the user is asked to indicate if there were any implicit decision options that had been left out in earlier routines. For example, truncated complements or implicit probabilities might be added. In this case, our computer implementation calculates the expected utility value of each of the decision options, and chooses the options with the highest such value. In the adolescent example, the adolescent would choose to go to the party because it would be 1.85 times as fun (that is 2 * 0.9 + 0.1 * 0.5), on average, than the sleepover. If one option's expected utility dominates, then this option is chosen; otherwise the decision-maker is indifferent at the interval level.

3 Testing Our Model

This model does not incorporate an error term, and is therefore deterministic. It successfully predicts the modal outcome of each experiment in Table 1.

Table 1. Overview of the 14 effects replicated by our model (Sources: [15-18] & Table 2)

Experimental Effect		Outcome
A: 200 live	B: 1/3 * 600 live or 2/3 * none live	A
C: 400 die	D: 1/3 * 600 die or 2/3 * none die	D
A: 200 live	B: 1/3 * 600 live	Indifference
C: 400 die	D: 2/3* 600 die	Indifference
A: 200 live	B: 2/3 * none live	Strong A
C: 400 die	D: 1/3 * none die	Strong D
A: 200 live	B: 1/3* all live or 2/3 * none live	Attenuated
C: 400 die	D: 1/3 * none die or 2/3 * all die	Attenuated
A: 200 live and 400 don't live	B: 1/3 * 600 live or 2/3 * none live	Indifference
C: 400 die and 200 don't die	D: 1/3 * none die or 2/3 * 600 die	Indifference
A: 400 do not live	B: 1/3 * 600 live or 2/3 * none live	B
A: 200 do not die	B: 1/3 * none die or 2/3 * 600 die	C
A: $1m with certainty	B: 0.89*$1m or 0.1*$5m or 0.01*$0	Weak A
C: 0.89 * $0 or 0.11*$1m	D: 0.90*$0 or 0.10*$5m	Strong D

3.1 Meta-analysis

We next performed a meta-analysis of 13 studies to test our prediction that changing the interpretation of an existing problem, without changing the numerical information, would change what gists are encoded. We examined studies in the literature that deviated from Tversky & Kahneman's original experimental protocol by substituting the word "all" to options B and D in the DP. The new options B & D read:

B) Some chance that all live and some chance that none live

D) Some chance that all die and some chance that none die

The effect sizes for these studies were calculated using Cramer's V – the difference between the proportions of the samples that chose each option in each frame (thus, a perfect framing effect in which every subject chose option A would have an effect size of 1.0, whereas no framing effect, in which subjects were equally likely to choose options A and B, would have an effect size of 0) – and then compared to the corresponding effect sizes for those studies in the literature that replicated Tversky & Kahneman's original protocol. Results are shown in Table 2. We analyzed only between-subjects comparisons so as to avoid the heuristic override effect [19].

Table 2. Results of our meta-analysis showing the effect of including the word "All"

Hypothetical Population Size	Wording	Reference	N	p-value (Chi Square)	Effect Size
60	Standard	[20]	213	<0.001 (12.27)	0.24
	Includes "All"	**[21]**	**80**	**0.813 (0.07)**	**0.03**
		[22]	**63**	**0.458 (0.51)**	**0.09**
600	Standard	[5]	307	<0.001 (76.75)	0.50
		[23]	90	<0.001 (21.61)	0.49
		[24]	244	<0.001 (49.41)	0.45
		[25]	292	<0.001 (31.80)	0.33
		[26]	148	<0.001 (29.97)	0.45
		[27]	105	<0.001 (20.33)	0.44
	Includes "All"	[21]	100	0.005 (7.84)	0.28
		[22]	65	0.005 (7.96)	0.35
6000	Standard	[28]	46	0.003 (8.91)	0.44
	Includes "All"	**[21]**	**88**	**0.055 (3.52)**	**0.20**
		[22]	61	0.029 (4.78)	0.28

For hypothetical population sizes of 60 and 6000, at least one of the tests including the word "all" did not have a significant framing effect. Furthermore, when combining across all three population sizes, effect sizes for tests using the word "all" are significantly smaller than they are for tests that follow the standard DP formulation (Mann-Whitney U test; $U = 44.0$; $p=0.004$). Note that overall effect sizes for populations of size 60 are smaller than those for larger populations. We explain this by the fact that 20 is interpreted as essentially nil [29], attenuating the framing effect.

3.2 Discussion

As the combined results of Tables 1 and 2 suggest, our theory successfully predicted the outcomes of the fourteen effects studied. These results support our theory and its implementation. These studies showed that individuals tend to exhibit framing effects when decision options encode different gists (prediction 1), as in the standard, as opposed to non-zero-truncated, DP, and the first decision of the Allais Paradox. Furthermore these choices are based on valenced affect (prediction 2) – i.e., some money is better than no money; some lives saved is better than no lives saved, but some die is worse than none die, etc. When categorical contrasts are not possible, decision-makers will revert to more precise representations (prediction 3). Thus, we see indifference in the zero-complement-truncated DP, and maximization of expected utility in the second gamble of the Allais paradox because both options have the gist of "some money with some chance, or no money with some chance." Finally, gist categories are encoded based upon interpretations which, when changed, can change subjects' behaviors (prediction 4). Thus, adding the word "all" to the standard DP attenuates the framing effect because a decision-maker must choose between an option that advertises "some saved with some chance" and an option that advertises "none saved with some chance OR all saved with some chance." To our knowledge, no rival theories are able to explain these several different effects within one unified framework. Furthermore, our approach is the only one to explain the circumstances under which subjects will use qualitative and categorical (i.e., discrete) representations of numbers, rather than continuous and quantitative representations. Critics might take issue with our notion of constraints, because it is in principle possible to find some constraint that will yield results consistent with data after the fact. Our reaction to this claim is that we only use constraints that are grounded in empirically validated patterns – i.e., some-none or all-some distinctions. For these, our model predicts an attenuated framing effect, and the data from several studies bears out this prediction. The effects explained by our model are therefore robust and novel, lending support to our hypothesis that risky decisions are made upon information that is encoded simultaneously at multiple levels of abstraction, and that resulting categories are culturally contingent.

References

1. Douglas, M., Wildavsky, A.B.: Risk and culture: An essay on the selection of technological and environmental dangers. University of California Pr. (1983)
2. Jasanoff, S. (ed.): Handbook of science and technology studies. Sage (1995)
3. Sunstein, C.R.: Risk and reason: Safety, law, and the environment. Cambridge University Press (2002)
4. Reyna, V.F.: A new intuitionism: Meaning, memory, and development in Fuzzy-Trace Theory. Judgment and Decision Making 7(3), 332–359 (2012)
5. Tversky, A., Kahneman, D.: The framing of decisions. Science 211, 453–458 (1981)
6. Kahneman, D., Tversky, A.: Prospect theory: An analysis of decision under risk. Econometrica: Journal of the Econometric Society, 263–291 (1979)
7. Tversky, A., Kahneman, D.: Advances in prospect theory: Cumulative representation of uncertainty. Journal of Risk and Uncertainty 5(4), 297–323 (1992)
8. Reyna, V.F., Adam, M.B.: Fuzzy-Trace Theory, Risk Communication, and Product Labeling in Sexually Transmitted Diseases. Risk Analysis 23(2), 325–342 (2003)

9. Rivers, S.E., Reyna, V.F., Mills, B.: Risk taking under the influence: A fuzzy-trace theory of emotion in adolescence. Developmental Review 28(1), 107–144 (2008)
10. Frank, M.C., Fedorenko, E., Lai, P., Saxe, R., Gibson, E.: Verbal interference suppresses exact numerical representation. Cognitive Psychology 64(1), 74–92 (2012)
11. Venkatraman, V., Huettel, S.A.: Strategic control in decision-making under uncertainty. European Journal of Neuroscience 35(7), 1075–1082 (2012)
12. Venkatraman, V., Payne, J.W., Bettman, J.R., Luce, M.F., Huettel, S.A.: Separate neural mechanisms underlie choices and strategic preferences in risky decision making. Neuron 62(4), 593–602 (2009)
13. Schurr, A., Ritov, I.: The Effect of Giving It All Up on Valuation: A New Look at the Endowment Effect. Management Science (2013)
14. Reyna, V.F., Chick, C.F., Corbin, J.C., Hsia, A.N.: Developmental Reversals in Risky Decision Making Intelligence Agents Show Larger Decision Biases Than College Students. Psychological Science (2013) 0956797613497022
15. Kühberger, A., Tanner, C.: Risky choice framing: Task versions and a comparison of prospect theory and fuzzy-trace theory. Journal of Behavioral Decision Making 23(3), 314–329 (2010)
16. Conlisk, J.: Three variants on the Allais example. The American Economic Review, 392–407 (1989)
17. Carlin, P.S.: Is the Allais paradox robust to a seemingly trivial change of frame? Economics Letters 34(3), 241–244 (1990)
18. Allais, M.: Fondements théoriques, perspectives et conditions d'un Marché commun effectif: extrait de la Revue d'économie politique numéro spécial" Le Marché commun et ses problèmes". CNRS, Janvier-Février (1958)
19. Stanovich, K.E., West, R.F.: On the relative independence of thinking biases and cognitive ability. Journal of Personality and Social Psychology 94(4), 672 (2008)
20. Bohm, P., Lind, H.: A note on the robustness of a classical framing result. Journal of Economic Psychology 13(2), 355–361 (1992)
21. Wang, X.T.: Framing effects: Dynamics and task domains. Organizational Behavior and Human Decision Processes 68(2), 145–157 (1996)
22. Wang, X.T., Johnston, V.S.: Perceived social context and risk preference: A re - examination of framing effects in a life - death decision problem. Journal of Behavioral Decision Making 8(4), 279–293 (1995)
23. Takemura, K.: Influence of elaboration on the framing of decision. The Journal of Psychology 128(1), 33–39 (1994)
24. Highhouse, S., Yüce, P.: Perspectives, perceptions, and risk-taking behavior. Organizational Behavior and Human Decision Processes 65(2), 159–167 (1996)
25. Stanovich, K.E., West, R.F.: Individual differences in framing and conjunction effects. Thinking & Reasoning 4(4), 289–317 (1998)
26. Druckman, J.N.: The implications of framing effects for citizen competence. Political Behavior 23(3), 225–256 (2001)
27. Druckman, J.N.: Evaluating framing effects. Journal of Economic Psychology 22(1), 91–101 (2001)
28. Levin, I.P., Chapman, D.P.: Risk taking, frame of reference, and characterization of victim groups in AIDS treatment decisions. Journal of Experimental Social Psychology 26(5), 421–434 (1990)
29. Stone, E.R., Yates, J.F., Parker, A.M.: Risk communication: Absolute versus relative expressions of low-probability risks. Organizational Behavior and Human Decision Processes 60(3), 387–408 (1994)

Labeling Actors in Social Networks Using a Heterogeneous Graph Kernel

Ngot Bui and Vasant Honavar

College of Information Sciences and Technology
The Pennsylvania State University
University Park, PA 16802, USA
{npb123,vhonavar}@ist.psu.edu

Abstract. We consider the problem of labeling actors in social networks where the labels correspond to membership in specific interest groups, or other attributes of the actors. Actors in a social network are linked to not only other actors but also items (e.g., video and photo) which in turn can be linked to other items or actors. Given a social network in which only some of the actors are labeled, our goal is to predict the labels of the remaining actors. We introduce a variant of the random walk graph kernel to deal with the heterogeneous nature of the network (i.e., presence of a large number of node and link types). We show that the resulting heterogeneous graph kernel (HGK) can be used to build accurate classifiers for labeling actors in social networks. Specifically, we describe results of experiments on two real-world data sets that show HGK classifiers often significantly outperform or are competitive with the state-of-the-art methods for labeling actors in social networks.

1 Introduction

Social networks (e.g. Facebook) and social media (e.g. Youtube) have provided large amounts of network data that link actors (individuals) with other actors, as well as diverse types of digital objects or items e.g., photos, videos, articles, etc. Such data are naturally represented using *heterogeneous* networks with multiple types of nodes and links. We consider the problem of labeling actors in social networks where the labels correspond to membership in specific interest groups, participation in specific activities, or other attributes of the actors. However, in many real-world social networks, labels are available for only a subset of the actors. Given a social network in which only some of the actors are labeled, our goal is to predict the labels of the remaining actors.

Accurate prediction of actor labels is important for many applications, e.g., recommending specific items (e.g., movies, musics) to actors. A variety of approaches to labeling nodes in social networks have been explored in the literature including methods that develop a relational learner to classify an actor by iteratively labeling an actor to the majority class of its neighbors [1, 2]; methods that effectively exploit correlations among the labels and attributes of objects [3–5]; semi-supervised learning or transductive learning methods [6, 7] such as random-walk based methods [8, 9] that assign a label to an actor based on the known label(s) of objects represented by node(s) reachable via random walk(s) originating at the node representing the actor. However, with

W.G. Kennedy, N. Agarwal, and S.J. Yang (Eds.): SBP 2014, LNCS 8393, pp. 27–34, 2014.

the exception of RankClass [10], Graffiti [8], EdgeCluster [11], and Assort [12, 13], most of the current approaches to labeling actors in social networks focus on *homogeneous* networks, i.e., networks that consist of a single type of nodes and/or links. RankClass and Graffiti offer probabilistic models for labeling actors in heterogeneous social networks. EdgeCluster mines the latent multi-relational information of a social network and convert it into useful features which can be used in constructing a classifier. Assort augments network data by combining explicit links with links mined from the nodes' local attributes to increase the amount of the information in the network and hence improve the performance of the network classifier [2]. Against this background, we introduce a heterogeneous graph kernel (HGK), a variant of the random walk graph kernel for labeling actors in a heterogeneous social network.

HGK is based on the following intuition: Two actors can be considered "similar" if they occur in the similar *contexts;* and "similar" actors are likely to have similar labels. We define the *context* of an object to include its direct and indirect neighbors and links between those neighbors. The similarity of two actors is defined in terms of the similarity of the corresponding contexts. We extend the random walk graph kernel [14–16] which has been previously used for labeling nodes in homogeneous networks to the setting of heterogeneous networks. The resulting HGK is able to exploit the information provided by the multiple types of links and objects in a social networks to accurately label actors in such networks. Results of experiments on two real-world data sets show that HGK classifiers often significantly outperform or are competitive with the state-of-the-art methods for labeling actors in social networks.

2 Preliminaries

A social network can be considered as a heterogeneous network of multiple types of objects and links. Formally, we define a social network as follows.

Definition 1. ***Heterogeneous Social Network.*** *A heterogeneous social network with multiple types of objects and links is represented by a graph* $G = (V, E)$ *in which* $V = \{V_1 \cup V_2 \cup \ldots \cup V_m\}$ *is a vertex set where* V_p *denotes the set of vertices of type* p*; and* $E = \{\cup E_{pq} | 1 \le p, q \le m\}$ *is an edge set where* $E_{pq} = \{(x, y) \,| x \in V_p, y \in V_q\}$ *is a set of edges between the two objects of types p and q, respectively.*

We define the transition probability between a node x and its neighbors as being inversely proportional to the number of x's neighbors and the probability of remaining at node x with a certain stopping probability. Let T_{pq} be the transition probability matrix between nodes in V_p and V_q (note that $T_{pq} \neq T_{qp}$) and S_{pq} be the stopping probability matrix of nodes in V_q. Let $A \in \{V_1, V_2, \cdots, V_m\}$ be a set of actors in a network, we formally describe our problem as follows. Consider a social network in which each actor $x \in A$ belongs to one category in $C = \{c_1, c_2, \cdots, c_N\}$ and a subset of labeled actors A^L, our task is to complete the labeling for unlabeled actors in the subset $A^U = A - A^L$.

3 Heterogeneous Graph Kernel

3.1 Kernel Function

We consider two objects to be similar if they occur in similar contexts. We generalize the marginalized kernel [14] that has been used for computing similarity between two objects in a homogeneous networks to the heterogeneous network setting as follows.

Let x be a node in a heterogeneous network G, and let $h_x = x - x_1 - x_2 - \cdots - x_l$ be a random walk starting from x with the length of l. The probability of h_x is defined as: $P(h_x) = P_t(x_1|x) P_t(x_2|x_1) \cdots P_t(x_l|x_{l-1}) P_s(x_l)$ where P_s is stopping probability and P_t is transition probability.

We assume that the stopping probability for all nodes of all types is $P_s = \rho$ $(0 < \rho < 1)$. Let $\mathcal{N}_q(x_{i-1})$ be the neighbors of type q of x_{i-1}, then we have:

$$\sum_{q=1}^{m} \sum_{x_i \in \mathcal{N}_q(x_{i-1})} P_t(x_i|x_{i-1}) + P_s(x_{i-1}) = 1 \tag{1}$$

Suppose that x_{i-1} is an object of type p. Let w_{pq} be the *transition probability weight* (TPW) from type p to type q where $\sum_q w_{pq} = 1$. Then, $\sum_{x_i \in \mathcal{N}_q(x_{i-1})} P_t(x_i|x_{i-1}) = (1-\rho) w_{pq}$. Transition probabilities from x_{i-1} to neighbors x_i of type q are assumed to be equal, i.e., $P_t(x_i|x_{i-1}) = \frac{(1-\rho)w_{pq}}{|\mathcal{N}_q(x_{i-1})|}$. We define the *linking preferences* of type p to be proportional to the TPWs of type p, i.e., $w_{p1} : w_{p2} : \ldots : w_{pm}$. In the absence of prior knowledge, we assume that $\forall q : 1 \le q \le m$, w_{pq} are equal.

We define the *kernel* induced *similarity* between two objects x and y of type p as follows.

$$K^p(x,y) = \sum_{h_x} \sum_{h_y} R_h(h_x, h_y) P(h_x) P(h_y) \tag{2}$$

where $K^p(x,y)$ is a kernel function; $R_h(h_x, h_y)$, the similarity between two paths h_x and h_y, is equal to 0 if they are of different lengths; otherwise, $R_h = \prod_{i=1}^{l} R_0(x_i, y_i)$. If x_i and y_i are of the same type (say p), then $R_0(x_i, y_i) = R_0^p(x_i, y_i)$ where $R_0^p(x_i, y_i)$ is defined using *Jaccard similarity* coefficient on the sets of directed neighbors of x_i and y_i as follows.

$$R_0^p(x_i, y_i) = \frac{\sum_{q=1}^{m} |\mathcal{N}_q(x_i) \cap \mathcal{N}_q(y_i)|}{\sum_{q=1}^{m} |\mathcal{N}_q(x_i) \cup \mathcal{N}_q(y_i)|} \tag{3}$$

where $\mathcal{N}_q(x_i)$ and $\mathcal{N}_q(y_i)$ are neighbors of type q of objects x_i and y_i, respectively. Otherwise, $R_0(x_i, y_i) = 0$.

3.2 Efficient Computation

Computing the kernel value between two nodes by generating random walks starting from the two nodes is computationally expensive. We adapt a technique introduced in [14, 17] for efficient computation of the random walk graph kernel in the case of homogeneous networks to the heterogeneous network setting as follows.

$$K^p(x,y) = \sum_{l=1}^{\infty} R_l^p(x,y) \tag{4}$$

where $R_l^p(x,y)$ is recursively defined using matrix form as follows: $R_l^p = \sum_{q=1}^{m} T_{pq} \left(R_0^q \circ R_{l-1}^q\right) T_{pq}^t$ and $R_1^p = \sum_{q=1}^{m} (T_{pq} \circ S_{pq}) R_0^q (T_{pq} \circ S_{pq})^t$, T_{pq}^t is transpose of T_{pq} and "$\circ$" is Hadamard product (see the Appendix A and B for the formation and convergence proof of (4), respectively).

Let $d = \max(|V_1|, \ldots, |V_m|)$, then the time for computing kernel matrix corresponding to a random walk of length of 1 (i.e., R_1) is $O\left(|V_p|d^2\right)$. The time for computing kernel matrix R_l $(l > 1)$ is $O\left(d^3\right)$. As a result, time complexity for computing kernel matrix for type p is $O\left(ld^3\right)$.

3.3 Learning to Label Actors in Social Network

We first compute the kernel matrix that captures the pair-wise similarity between actors and normalize it to obtain: $\hat{K}^p(x,y) = K^p(x,y)/\sqrt{K^p(x,x)K^p(y,y)}$. We train a support vector machine (SVM[1]) using HGK for labeling actors in social networks[2].

4 Experimental Settings and Results

We describe results of experiments that compare the performance of HGK with that of several baseline classifiers. We also study the sensitivity of the performance of HGK to length of random walk and to linking preferences of type actor.

4.1 Social Media Data

We crawled two real-world heterogeneous social networks. The first data set is from Last.fm music network. We manually identified 11 disjoint groups (categories of users who share similar interests in music e.g., `http://www.last.fm/group/Metal` in the case of users who enjoy Heavy Metal) that contain approximately equal number of users in the network; we then crawled users and items and the links that denote the relations among the objects in the network. In particular, the subset of the Last.fm data that we use consists of 1612019 links that connect 25471 nodes. The 25471 nodes belong to one of 4 types: 10197 users (actors), 8188 tracks, 1651 artists and 5435 tags; The 1612019 links belong to one of 5 types: 38743 user-user, 765961 user-track, 8672 track-artist, 702696 track-tag, and 95947 artist-tag links.

Our second data set is from Flickr. We manually identified 10 disjoint groups (communities of users who share the same taste in pictures e.g., `http://www.flickr.com/groups/iowa/` in the case of users who share an interest in pictures that relate to the state of Iowa) of approximately equal numbers of users. The data set constructed by crawling the Flickr network contains 361787 links that connect 22347 nodes. The nodes are of one of three types: 6163 users, 14481 photos, and 1703 tags; and the links are of one of three types: 88052 user-user, 144627 user-photo, and 129108 photo-tag. In both data sets, we use the group memberships of users as class labels to train and test all models.

[1] `http://www.csie.ntu.edu.tw/~cjlin/libsvm/`

[2] The method can be applied more generally, e.g., for labeling *any* type of nodes in social networks.

4.2 Methods

We compare the performance of SVM trained using HGK with several state-of-the-art methods for labeling actors in social networks:

1. Weighted-Vote Relational Neighbor Classifier with network data augmentation (wvRN-Assort) [2, 12, 13]: a method that first augments networked data by combining explicit links with links mined from the nodes' local attributes and then uses the augmented network as input to a Weighted-Vote Relational Neighbor Classifier.
2. Network-Only Link-Based Classification with network data augmentation (nLB-Assort) [2, 12, 13]: a method which is similar to wvRN-Assort but uses the augmented network as input to a Network-Only Link-Based [4] classifier that constructs a relational feature vector for each node by aggregating the labels of its neighbors which is used to train a logistic regression model.
3. EdgeCluster [11]: a method which extracts the social dimensions of each actor, i.e., the affiliations of the actor in a number of latent social groups and uses the resulting features to generate a discriminative model to classify actors.
4. EdgeCluster-Cont [11]: a method that combines both social dimensions and features extracted from user profiles to build predictive models. For both data sets, we report results obtained using a subset of user profile features (e.g., artists, tags) that yield the best performance.
5. Augmented-Graph Kernel (AGK): a method that uses a homogeneous graph kernel [14, 17]. We augment the network data by adding an edge between two actors if they share links to a specified number (n) of items (When $n = \infty$, the method defaults to the use of homogeneous graph kernel on the unaugmented network data). For both data sets, we report results for a choice of n that yields the best performance.

4.3 Experimental Design and Results

In the first set of experiments, we compare the performance of different methods as a function of the percentage of actors in the network with known labels. For each choice of the percentage of labeled actors, we randomly select the corresponding fraction of labeled data for each node label for training and the rest for testing. We repeat this process 10 times and report the average accuracy. The length of the random walk was set equal to 1 and the linking preferences were set to be equal (i.e., TPWs of a type were set to be equal).

Figure 1 shows the results of the first set of experiments. In particular, in Last.fm data set, HGK significantly ($p < 0.05$) outperforms all other methods when at least 4% of the actors are labeled. nLB-Assort does not work well when the fraction of actors with known labels is less than 10%; This can be explained by the fact that it relies on the statistics of labels aggregated from the neighbors of an actor to label an actor. Not surprisingly, EdgeCluster-Cont which uses more information than EdgeCluster outperforms EdgeCluster. In the Flickr data set, AGK and HGK outperform other methods when the fraction of actors with known labels is less than 10%. Furthermore, AGK significantly outperforms HGK when the fraction of actors with known labels is less than 7% and both HGK and AGK significantly outperform other methods when the fraction

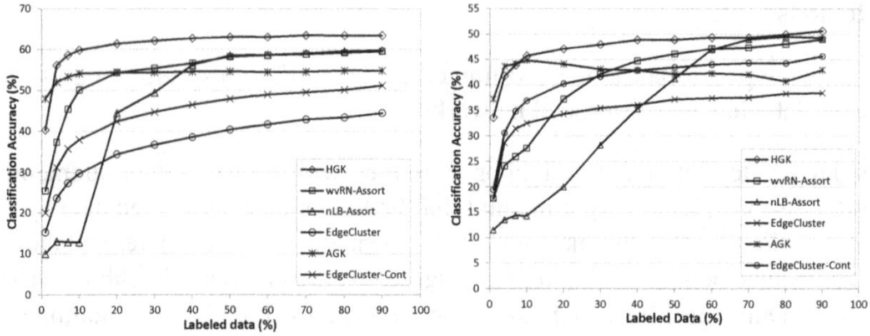

Fig. 1. Accuracies of six methods on Last.fm (left) and on Flickr (right)

of actors with known labels is between 7% to 10%. HGK significantly outperforms all other methods with labeled data when the fraction of actors with known labels ranges between 10% and 60%. Both HGK and nLB-Assort outperform other methods when the fraction of actors with known labels is between 70% and 80%. On both data sets, HGK often significantly outperforms, or is at least competitive with all other methods. This can be explained by the fact that HGK is able to exploit information provided by multiple node and link types to uncover multi-relational latent information to reliably discriminate between different actor labels.

The second set of experiments explores the sensitivity of kernel methods (HGK and AGK) as a function of the length of the of random walk. The length l of the random walk is varied from 0 to 10 with the linking preferences to be equal (across all the links from an actor). We report results averaged over 10-fold cross validation runs.

Table 1. Accuracies (%) of kernel methods with different lengths of walk. Bold numbers represent best results based on paired t-test ($p < 0.05$) on 10-fold cross validation.

	l	0	1	2	3	4	5	6	7	8	9	10
Last.fm	HGK	**61.4**	**63.8**	**62.9**	**61.0**	**59.9**	**58.8**	**58.3**	56.7	56.6	55.4	55.5
	AGK	39.0	43.8	51.2	53.1	54.5	55.1	55.0	55.1	54.9	54.2	54.0
Flickr	HGK	**49.8**	**49.7**	**46.1**	**46.0**	**46.5**	**44.8**	**44.6**	**43.2**	42.8	41.8	41.3
	AGK	33.9	38.1	42.1	42.2	41.2	41.6	40.9	41.1	41.1	41.1	41.2

Table 1 shows that kernel methods work well at some shorter walks (e.g., $l = 1, 2$). As the walk becomes longer, the performances of HGK and AGK decrease or remain the same. This indicates that the further the neighbor is from the node to be classified, the less informative it is for prediction. HGK significantly outperforms AGK with $l \leq 6$ (Results for $l = 0$ correspond to simply using the similarity values given by R_0^p).

The last set of experiments examines the performance of the learned model by fixing the length of random walk (i.e., $l = 1$) and varying the linking preferences of actors. Specifically, in Last.fm, let $w_1 = w_{useruser}$ and $w_2 = w_{usertrack}$ be the TPWs from type user. We examine the performance of the model by changing ratio $w_1 : w_2$ (1:5, 1:4, 1:3, 1:2, 1:1, 2:1, ..., 9:1). We do the same for Flickr with $w_1 = w_{useruser}$ and $w_2 = w_{userphoto}$. We report classification accuracy averaged over a 10-fold cross-validation runs.

Figure 2 shows the results of the last experiment set which investigates the influence of *linking preferences* on the performance of the HGK. In the case of Last.fm, the model performs better when the linking preference between a user and a track is higher than that between users whereas in the case of Flickr, the situation is reversed. Our results suggest that in the case of Last.fm, a surfer (exploring music) is likely to move from a user to a track more often than from a user to another user, with the opposite being true in the case of a surfer exploring pictures in Flickr.

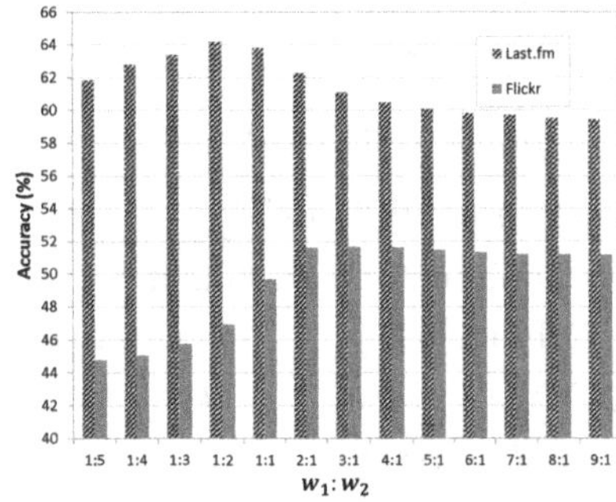

Fig. 2. Accuracies when turning $w_1 : w_2$

5 Summary

We introduced a generalization of random walk graph kernel from the setting of homogeneous networks, i.e., networks consisting of only one type of nodes and one type of links, to the setting of heterogeneous networks, i.e., networks consisting of multiple types of nodes and links. We used the resulting kernel, HGK, to train an SVM classifier for labeling actors in heterogeneous social networks. The results of our experiments show that HGK classifiers often significantly outperform or are competitive with the state-of-the-art methods for labeling actors in social networks. Some promising directions for further research include: (i) combining multiple kernels [18] that capture different notions of similarity between nodes in heterogeneous networks; and (ii) using *linking preferences* directly estimated from the data to improve the accuracy of predicted labels.

Appendix A. Based on (2), $K^p(x,y)$ can be regrouped [14] as follows: $K^p(x,y) = \sum_{l=1}^{\infty}\Big(\sum_{x_1,y_1}\Big(P_t(x_1|x)P_t(y_1|y)R_e(x_1,y_1)\Big(\sum_{x_2,y_2}\Big(P_t(x_2|x_1)P_t(y_2|y_1)\times R_e(x_2,y_2)\Big(\cdots\Big(\sum_{x_l,y_l}\Big(P_t(x_l|x_{l-1})P_t(y_l|y_{l-1})R_e(x_l,y_l)P_s(x_l)P_s(y_l)\Big)\Big)\cdots\Big)\Big)\Big)\Big)\Big)$. Now let $R_1^p(x,y) = \sum_{x_1,y_1}\Big(P_t(x_1|x)P_t(y_1|y)R_0(x_1,y_1)P_s(x_1)P_s(y_1)\Big)$. After some derivation steps, we have $R_l^p(x,y) = \sum_{x_1,y_1}\Big(P_t(x_1|x)\,P_t(y_1|y)\,R_0(x_1,y_1)\,R_{l-1}(x_1,y_1)\Big)$. So, $K^p(x,y) = \sum_{l=1}^{\infty} R_l^p(x,y)$.

We have $R_1^p = \sum_{x_1,y_1}\Big(P_t(x_1|x)P_t(y_1|y)R_0(x_1,y_1)P_s(x_1)P_s(y_1)\Big)$. Since $R_0(x_1,y_1)=0$ for all pair (x_1,y_1) when x_1 and y_1 are not the same type, so $R_1^p = \sum_{q=1}^{m}\Big(\sum_{x_1,y_1\in V_q}\Big(P_t(x_1|x)P_s(x_1)R_0^q(x_1,y_1)P_t(y_1|y)P_s(y_1)\Big)\Big) = \sum_{p=1}^{m}(T_{pq}\circ S_{pq})R_0^q(T_{pq}\circ S_{pq})^t$. Using the same derivation method, we have $R_l^p = \sum_{q=1}^{n} T_{pq}\left(R_0^q \circ R_{l-1}^q\right)T_{pq}^t$.

Appendix B. We prove that $K^p(x, y)$ is converged when $l \to \infty$. Applying the ratio test, (4) converges when $\lim_{l\to\infty} \frac{R^p_{l+1}(x,y)}{R^p_l(x,y)} < 1$ (5). We have $R^p_l(x, y) - R^p_{l+1}(x, y) = \sum_{x_1,y_1}\Big(P_t(x_1|x)P_t(y_1|y)R_0(x_1, y_1)\Big(\sum_{x_2,y_2}\Big(P_t(x_2|x_1)P_t(y_2|y_1)R_0(x_2, y_2)\times \Big(\cdots P_s(x_l)P_s(y_l)\Big(1-\sum_{x_{l+1},y_{l+1}}P_t(x_{l+1}|x_l)P_t(y_{l+1}|y_l)R_0(x_{l+1}, y_{l+1})\Big)\cdots\Big)\Big)\Big)\Big)$ (6). It is sufficient for (6) to hold if $\sum_{x_{l+1},y_{l+1}} P_t(x_{l+1}|x_l) P_t(y_{l+1}|y_l) R_0(x_{l+1}, y_{l+1}) < 1$ (7) From (1), $\sum_{x_{l+1}} P_t(x_{l+1}|x_l) = \sum_{p=1}^{m} \sum_{x_{l+1}\in\mathcal{N}_p(x_l)} P_t(x_{l+1}|x_l) = 1 - \rho$. So, $\sum_{x_{l+1},y_{l+1}} P(x_{l+1}|x_l)P_t(y_{l+1}|y_l) = \sum_{x_{l+1}} P_t(x_{l+1}|x_l)\sum_{y_{l+1}} P_t(y_{l+1}|y_l) \leq (1-\rho)^2$. Since $R_0(.,.) \leq 1$ and $\rho \in (0, 1)$, $\sum_{x_{l+1},y_{l+1}} P_t(x_{l+1}|x_l)P_t(y_{l+1}|y_l)R_0(x_{l+1}, y_{l+1}) < 1$. □

References

1. Macskassy, S.A., Provost, F.: A simple relational classifier. In: MRDM Workshop at KDD 2003, pp. 64–76 (2003)
2. Macskassy, S.A., Provost, F.: Classification in networked data: A toolkit and a univariate case study. J. Mach. Learn. Res. 8, 935–983 (2007)
3. Sen, P., Namata, G., Bilgic, M., Getoor, L., Gallagher, B., Eliassi-Rad, T.: Collective classification in network data. AI Magazine, 93–106 (2008)
4. Lu, Q., Getoor, L.: Link-based classification. In: ICML, pp. 496–503 (2003)
5. Eldardiry, H., Neville, J.: Across-model collective ensemble classification. In: AAAI (2011)
6. Zhou, D., Bousquet, O., Lal, T.N., Weston, J., Schölkopf, B.: Learning with local and global consistency. In: NIPS, pp. 321–328 (2004)
7. Wang, J., Jebara, T.: Graph transduction via alternating minimization. In: ICML (2008)
8. Angelova, R., Kasneci, G., Weikum, G.: Graffiti: graph-based classification in heterogeneous networks. World Wide Web Journal (2011)
9. Lin, F., Cohen, W.W.: Semi-supervised classification of network data using very few labels. In: Proceedings of the 2010 International Conference on Advances in Social Networks Analysis and Mining, pp. 192–199 (2010)
10. Ji, M., Han, J., Danilevsky, M.: Ranking-based classification of heterogeneous information networks. In: KDD, pp. 1298–1306 (2011)
11. Tang, L., Liu, H.: Scalable learning of collective behavior based on sparse social dimensions. In: CIKM, pp. 1107–1116 (2009)
12. Macskassy, S.A.: Improving learning in networked data by combining explicit and mined links. In: AAAI, pp. 590–595 (2007)
13. Macskassy, S.A.: Relational classifiers in a non-relational world: Using homophily to create relations. ICMLA (1), 406–411 (2011)
14. Kashima, H., Tsuda, K., Inokuchi, A.: Marginalized kernels between labeled graphs. In: ICML, pp. 321–328 (2003)
15. Vishwanathan, S.V.N., Schraudolph, N.N., Kondor, R., Borgwardt, K.M.: Graph kernels. Journal of Machine Learning Research 11, 1201–1242 (2010)
16. Kang, U., Tong, H., Sun, J.: Fast random walk graph kernel. In: SDM, pp. 828–838 (2012)
17. Li, X., Chen, H., Li, J., Zhang, Z.: Gene function prediction with gene interaction networks: a context graph kernel approach. Trans. Info. Tech. Biomed. 14(1), 119–128 (2010)
18. Cortes, C., Mohri, M., Rostamizadeh, A.: Multi-class classification with maximum margin multiple kernel. In: ICML 2013, vol. 28, pp. 46–54 (May 2013)

Sample Size Determination to Detect Cusp Catastrophe in Stochastic Cusp Catastrophe Model: A Monte-Carlo Simulation-Based Approach

Ding-Geng(Din) Chen[1,2], Xinguang (Jim) Chen[3], Wan Tang[2], and Feng Lin[1]

[1] School of Nursing, University of Rochester Medical Center, Rochester
(din_chen,vankee_lin)@urmc.rochester.edu
[2] Department of Biostatistics and Computational Science,
University of Rochester Medical Center, Rochester, USA
(din_chen,wan_tang)@urmc.rochester.edu
[3] Wayne State University School of Medicine,
Pediatric Prevention Research Center, Detroit, MI, USA
Jimax168chen@gmail.com

Abstract. Stochastic cusp catastrophe model has been utilized extensively to model the nonlinear social and behavioral outcomes to detect the exisitance of cusp catastrophe. However the foundamental question on sample size needed to detect the cusp catastrophe from the study design point of view has never been investigated. This is probably due to the complexity of the cusp model. This paper is aimed at filling the gap. In this paper, we propose a novel Monte-Carlo simulation-based approach to calculate the statistical power for stochastic cusp catastrophe model so the sample size can be determined. With this approach, a power curve can be produced to depict the relationship between its statistical power and samples size under different specifications. With this power curve, researchers can estimate sample size required for specified power in design and analysis data from stochastic cusp catastrophe model. The implementation of this novel approach is illustrated with data from Zeeman's cusp machine.

Keywords: Stochastic cusp catastrophe model, power analysis, sample size de-termination, Monte-Carlo simulations.

1 Introduction

Study should be well-designed. An important aspect of good design of study is to determine the number of study subjects (i.e. sample size) required to adequately statistically power the study to address the research questions or objectives. From this perspective, statistical power analysis is in fact an essential component of any valid study. By definition, statistical power analysis is to calculate the (frequentist) statistical power which is the probability of failing to reject the null hypothesis when it is false. The formal statistical basis for sample size determination requires: (i) the questions or objectives of the study to be defined; (ii) the most relevant outcome measures reflecting the objective to be identified; (iii) the specification of the effect

W.G. Kennedy, N. Agarwal, and S.J. Yang (Eds.): SBP 2014, LNCS 8393, pp. 35–41, 2014.

size (which embodies the research question) in the study that can be detected; (iv) specification of the magnitudes of the Type-I and Type-II decision errors; and (v) estimates of the mean and variability of the endpoint. Statistical power analysis is specific to different data type and dependent on the distribution as well as the statistical model.

Recently, cusp catastrophe model has been used extensively [1]. However, to the best of our knowledge, there is no research and publications on determining the sample size and calculating the statistical power for this model. This paper is then aimed to investigate this gap.

2 Overview of the Cusp Catastrophe Model

2.1 Deterministic Cusp Catastrophe Model

To overcome the limitation of linear analytical approach, cusp catastrophe model is proposed to model any system outcomes which can incorporate both linear and nonlinear along with discontinuous transitions in equilibrium states as control variables are varied. According to the catastrophe systems theory [2], the dynamics for system outcome is modeled as $-V(y;\alpha,\beta) = \alpha y + \frac{1}{2}\beta y^2 - \frac{1}{4}y^4$ with dynamical system in the form of $\frac{\partial y}{\partial t} = -\frac{\partial V(y,\alpha,\beta)}{\partial y}$, where α is called asymmetry control variable and β is called bifurcation control variable which are linked to determine the health outcome variable y.

2.2 Stochastic Cusp Catastrophe Model

This cusp catastrophe model is fundamentally a deterministic. In order to use this cusp model for real-life applications which are stochastic nature, Cobb and his colleagues [2-3] casted this model into a stochastic differential equation as follows $dy = \frac{\partial V(y,\alpha,\beta)}{\partial y}dt + dW(t)$, where dW(t) is a white noise Wiener process with variance σ^2. With this SDE, the probability distribution of the health behavior measure (y) under equilibrium can be expressed as $f(y) = \frac{\Psi}{\sigma^2} exp\left[\frac{\alpha(y-\lambda)+\frac{1}{2}\beta(y-\lambda)^2-\frac{1}{4}(y-\lambda)^4}{\sigma^2}\right]$ where ψ is a normalizing constant and λ is to determine the origin of y. With this density function, the theory of maximum likelihood can be employed for estimating parameters and statistical inference which is implemented in *R* Package "Cusp" [1], 2009). Specifically for data from *n* study subjects, we denote the observed *p* dependent variables as $Y_i = (Y_{i1}, Y_{i2},\ldots,Y_{ip})$, *q* predictor variables $X_i = (X_{i1}, X_{i2},\ldots, X_{iq})$, the behavior measure y_i and the control variables α_i and β_i are modeled as linear combinations of the *X* and *Z* as follows:

$$\begin{aligned} y_i &= w_0 + w_1 Y_{i1} + w_2 Y_{i2} + \ldots + w_p Y_{ip} = w \times Y_i \\ \alpha_i &= \alpha_0 + \alpha_1 X_{i1} + \alpha_2 X_{i2} + \ldots + \alpha_q X_{iq} = \alpha \times X_i \\ \beta_i &= \beta_0 + \beta_1 X_{i1} + \beta_2 X_{i2} + \ldots + \beta_q X_{iq} = \beta \times X_i \end{aligned} \quad (1)$$

Then the log-likelihood function for these *n* observations is as follows:

$$l(w,\alpha,\beta \mid X,Y) = \Sigma_{i=1}^{n} \log \psi_i - \Sigma_{i=1}^{n}(\alpha_i y_i + \frac{1}{2}\beta_i y_i^2 - \frac{1}{4}y_i^4) \quad (2)$$

which is then maximized to estimate model parameters for associated statistical inference. A likelihood-ratio Chi-square test is used for model comparison and to test the goodness-of-fit between cusp model to the linear regression model. Other model selection criteria are also calculated, such as the R^2 =1-(error variance/variance of y) as well as the model selection information indices of *AIC* and *BIC* [5-6].

2.3 Detection of Cusp Catastrophe

In order to establish the presence of a cusp catastrophe, three guidelines are proposed by Cobb [5]. First, the cusp model should be substantially better than linear model which can be evaluated by the likelihood ratio test. Second, any of the coefficients w_1, ..., w_p should be statistically significant (w_0 does not have to) and at least one of the α_i's or β_i's should be statistically significant. Thirdly, at least 10% of the (α_i, β_i) pairs should lie within the bifurcation region. This 10% guideline of Cobb was modified by Hartelman [6] to compare the cusp model to the non-linear logistic regression: $y_i = 1/\left(1 + e^{-\frac{\alpha_i}{\beta_i^2}}\right)+\epsilon_i \ (i = 1, \ldots, n)$ with better AIC and BIC for cusp model than for the logistic regression.

3 Monte-Carlo Simulation-Based Power Analysis Approach

3.1 Statistical Power and Sample Size Determination

Sample size determination is a essential step in planning and designing a study. Sample size is usually associated with statistical power. In statistics, power is the probability of correctly rejecting the null hypothesis (i.e. no cusp catastrophe) which in common sense is the fraction of the times that the specified null-hypothesis will be rejected from statistical tests. The sample size and power calculation for the stochastic cusp catastrophe model has never been done to the best of our knowledge which probably due to the theoretical complexity of this model from the high-order density function as well as to comply with the three guidelines from Section 2.3. This theoretical derivation of a power function might be impossible.

3.2 Monte-Carlo Simulation-Based Approach

To overcome this difficult, we propose a Monte-Carlo simulation-based approach to calculate the statistical power for a series of specified sample size (*n*, i.e. the number of observations for the stochastic cusp catastrophe modeling) to generate a sample size-power curve. With this curve, the sample size can be then reverse-determined for specific statistical power (say 80% or 85% as typically chosen). Specifically, for a pre-specified sample size (n), the following steps are needed to calculate the statistical power:

1) Specify α, β and w from Equation (1) to be detected based on prior knowledge;
2) Simulate data from q predictor variables $X_i = (X_{i1}, X_{i2},..., X_{iq})$ with pre-specified distributions and then calculate the corresponding asymmetry (α_i) and bifurcation (β_i) variables from the last two equations of Equation (1);
3) Simulate the p dependent variables $Y_i = (Y_{i1}, Y_{i2},..., Y_{ip})$ and calculate the corresponding state measure (y_i) from the first equation of Equation (1);
4) Fit the stochastic cusp catastrophe model using the maximum-likelihood as outlined in Equation (2) with the data generated from Steps 2) and 3); and make conclusion on whether there is a significant cusp catastrophe based on the guidelines in Section 2.3;
5) Repeat Steps 2) to 4) a large number of time (say 1,000 times) and calculate the proportion of simulations which satisfy the decision rules. This proportion is then the statistical power for the pre-specified n and the cusp parameters given in Step 1);
6) Sample size determination can be carried out by running Steps 1) to 5) with a series of ns to produce a power curve and then back-calculate the sample size required for pre-specified power, such as power at 0.8(or 0.85) in typical study design.

4 Zeeman's Data Analysis

4.1 Zeeman's Data

We make use of the Zeeman's data built in the R "Cusp" package [1]. There are three datasets obtained from three different settings of a Zeeman catastrophe machine. This machine was an architecture made for demonstration of the deterministic cusp catastrophe model: $x + yz - z^3 = 0$ with different settings of (x, y). Notice that we changed our coordinate system from (y, α, β) to (x, y, z) here to be consistent to Zeemen's notations.

To demonstrate the proposed approach in Section 3, we utilize the dataset 1 which consists of 150 observations generated from experiments from this Zeeman's machine. The data have three columns of (x, y, z) which are the asymmetry (x) that is orthogonal to the central axis, the bifurcation (y) that is parallel to the central axis and state variable (z) that is the shortest distance from the wheel strap point to the central axis.

4.2 Stochastic Cusp Catastrophe Model Fitting

Similar to the analysis in [1], we fit a series cusp catastrophe models and the best cusp catastrophe model with corresponding equations in equation (4) is as follows:

$$y_i = w_1 Z_i, \quad \alpha_i = \alpha_0 + \alpha_1 X_i, \quad \beta_i = \beta_0 + \beta_1 y_i$$

with estimated parameters of $\hat{\alpha}_0 = 0.45$, $\hat{\alpha}_1 = 1.15$, $\hat{\beta}_0 = 0.99$, $\hat{\beta}_1 = -1.48$ and $\hat{w}_1 = 0.90$ which are all statistically significant. With this model, the value of the log-likelihood function is 74.7 and the corresponding value for the linear model is -170.4 which yielded a likelihood-ratio $\chi^2 = 490.3$ with degree of freedom of 1 leading to a very small p-value < 0.00001 indicating the cusp model fitted the data better than the linear model. Other model goodness-of-fit statistics, such as R^2, AIC and BIC, gives the same conclusion. The Zeeman data (points) and the best fitted response surface of the stochastic cusp catastrophe model are illustrated in Figure 1.

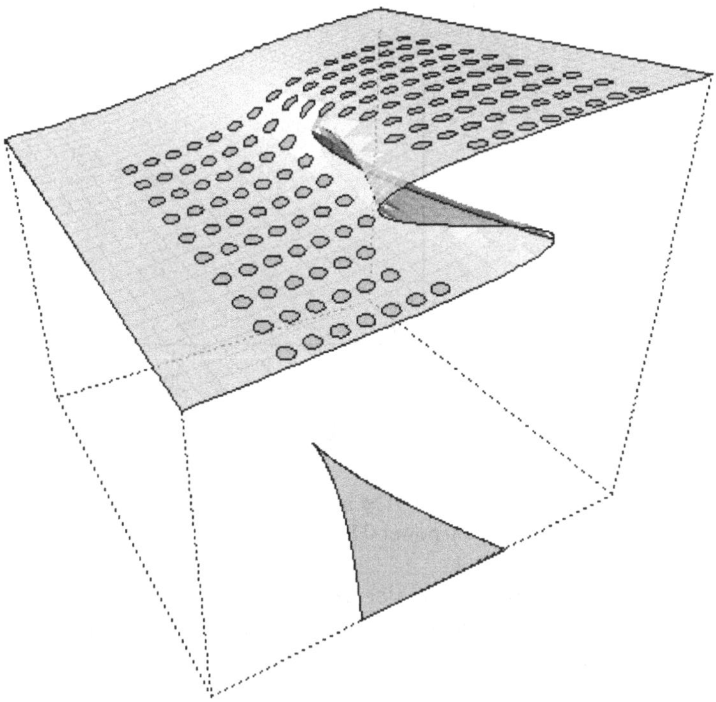

Fig. 1. Zeeman's Data (points) and the fitted cusp catastrophe model (surface)

4.3 Power Calculation and Sample Size Determination

To implement the simulation-based approach in Section 3, we use these estimated parameters with different sample sizes from 10 to 30 by 5, and run the Steps 2 to 6 for 1,000 simulations for each sample size of 10, 15, 20, 25 and 30. The estimated values of statistical power are 0.26, 0.83, 0.96, 0.97 and 0.98, respectively. Figure 2 graphically illustrates the relationship between the sample size and the corresponding statistical power. It can be seen that as sample size increases from 10 to 30, the statistical power increases from 0.26 to 0.98.

In order to get the required sample size for specific statistical power, we can make use of this estimated power curve as illustrated in Figure 2 and back-calculate the

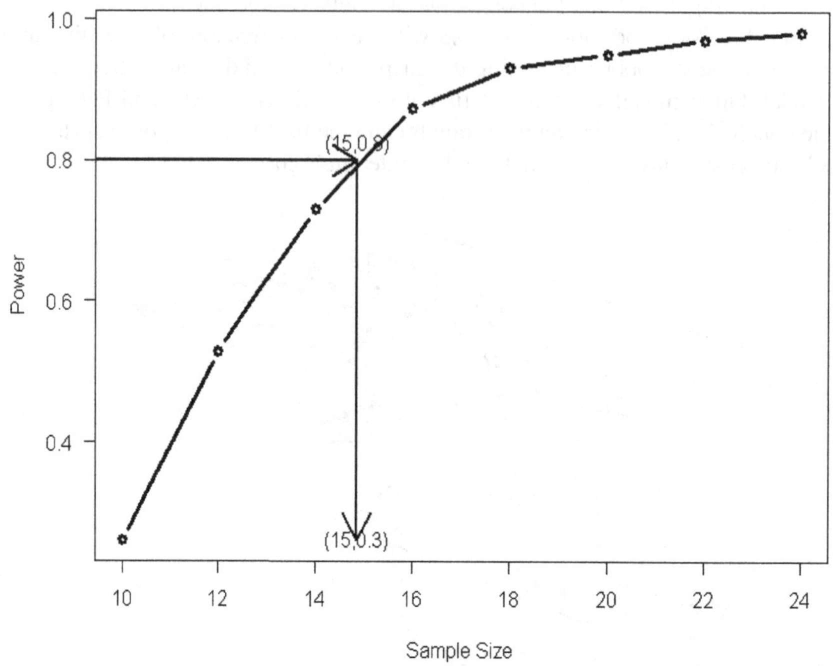

Fig. 2. Power curve to Zeeman's data 1. The arrow-headed line segments indicated the back-calculation of sample size of 15 from power 0.8.

sample size for the specific power. For example in Figure 2, we illustrate this approach for power at 0.8. For power of 0.8, we can back-calculate the sample size based on this estimated power curve to estimate the required sample size which is 15.

4.4 Reverse-Verification

If this novel simulation-based approach is valid, the sample size estimate of 16 described in Section 4.3 would allow approximately 80% chance to detect the underlying cusp catastrophe. Therefore, we took a reverse approach to compute statistical power by repeatedly sample 16 data points from Zeeman's dataset and fitting the stochastic cusp catastrophe model to detect cusp catastrophe. Among 1,000 repeats of the Monte-Carlo simulations with sample size n = 16, we found 851 times (85.1%) significant. This result indicates that the power analysis of the novel simulation-based method is very close to 85%. In another word, the method we proposed is valid.

5 Discussions

In this paper, we proposed a novel Monte-Carlo simulation-based approach to estimate the statistical power for the stochastic cusp catastrophe model. With is approach, we could produce the power curve which depicts the relationship between its statistical power and samples size under different specifications. This power curve can then be used to estimate the sample size required for specified power in design and analysis data from cusp catastrophe model.

We validated this proposed approach by a reverse-verification and demonstrated it is very reasonable. This validation can be further verified by empirically examining Zeeman's data generation. The Zeeman's data were generated for each y with exactly 15 observations for x from -7 to 7 by unit 1 for each y from -2.5 to 2 by 0.5 to produce 150 (= 15×10) observations. Therefore, for each value of y, the 15 data points should be able to generate a cusp catastrophe model.

Acknowledgements. This research was supported in part by three NIH grants from the Eunice Kennedy Shriver National Institute of Child Health and Human Development (NICHD, R01HD075635, PIs: Chen X and Chen D), the National Institute On Drug Abuse (NIDA, R01 DA022730, PI: Chen X) and University of Rochester CTSA award number KL2 TR000095 from the National Center for Advancing Translational Sciences (PI: Lin).

References

1. Grasman, R.P., van der Mass, H.L., Wagenmakers, E.: Fitting the cusp catastrophe in R: A cusp package primer. Journal of Statistical Software 32(8), 1–27 (2009)
2. Thom, R.: Structural stability and morphogenesis. Benjamin-Addison-Wesley, New York (1975)
3. Cobb, L., Ragade, R.K.: Applications of Catastrophe Theory in the Behavioral and Life Sciences. Behavioral Science 23, 291–419 (1978)
4. Cobb, L., Zacks, S.: Applications of Catastrophe Theory for Statistical Modeling in the Biosciences. Journal of the American Statistical Association 80(392), 793–802 (1985)
5. Cobb, L.: An Introduction to cusp surface analysis. Technical report. Aetheling Consultants, Louisville (1998)
6. Hartelman, A.I.: Stochastic catastrophe theory. University of Amsterdam, Amsterdam (1997)

Who Will Follow a New Topic Tomorrow?

Biru Cui and Shanchieh Jay Yang

Department of Computer Engineering, Rochester Institute of Technology,
83 Lomb Memorial Dr., Rochester, NY 14623, United States

Abstract. When a novel research topic emerges, we are interested in discovering how the topic will propagate over the bibliography network, *i.e.*, which author will research and publish about this topic. Inferring the underlying influence network among authors is the basis of predicting such topic adoption. Existing works infer the influence network based on *past* adoption cascades, which is limited by the amount and relevance of cascades collected. This work hypothesizes that the influence network structure and probabilities are the results of many factors including the social relationships and topic popularity. These heterogeneous information shall be optimized to learn the parameters that define the homogeneous influence network that can be used to predict *future* cascade. Experiments using DBLP data demonstrate that the proposed method outperforms the algorithm based on traditional cascade network inference in predicting novel topic adoption.

1 Introduction

Information cascade has been well studied to explain how individuals adopted information, but not as much to predict future cascades especially when the information is new and possibly not quite relevant to past cascades. An individual adopts the information when she receives enough influence from her infected neighbors [4], one who already adopted the information, or randomly if the underlying decision making process is unclear [4]. A set of work was developed to infer the inherent influence network [2] [5] [1] based on a number of cascades; the influence network inferred is the one which best fits all cascades. In general, the more cascades the more accurate influence network one can recover.

The inferred influence network can recover the most likely influence flows based on the distribution of collected past topic adoption cascades. However, it is unclear whether the future topic cascade can be explained by the distribution of past ones, due to two reasons. First, past cascades can be limited to cover the relationships among all actors. Second, the new topic propagation could be irrelevant to how past topics propagated. Though the topic cascade can be volatile and in many cases there are not sufficient cascades available, the *way* an author being influenced to work on a new topic is relative stable; an author is likely to be influenced by her colleagues or other researchers with social connections. These social connections contain rich information describing different relationships between authors and can be used to infer the inherent influence flows.

Existing research on heterogeneous information networks mostly focus on ranking and clustering [3], similar objects searching [8] and link prediction [6]. Differently, this

W.G. Kennedy, N. Agarwal, and S.J. Yang (Eds.): SBP 2014, LNCS 8393, pp. 43–50, 2014.

work tries to predict how a new topic is being adopted by authors as a cascade without knowing the influence network structure, but not to predict an additional new link over the existing network. This work proposes to leverage the rich heterogeneous bibliography network information to complement the past topic cascades for determining the inherent influence network. Besides the social connections, the popularity of the topic itself also affects the adoption process. In general, authors are more likely to follow popular topics than less widely accepted ones. To this end, this paper aims at developing an algorithm that finds an influence network by optimizing over the social connections and topic popularity subject to past cascades. The influence networks will then be used to predict new topic adoption. DBLP data is used to demonstrate the performance improvements of the proposed method in Mean Average Precision (MAP) and Area under ROC Curve (AUC).

2 Influence Factors for Research Topic Adoption

This work concerns the adoption of research topics as evidenced by authors' publications. A "topic" is defined as a popular term that represents a specific scientific concept. Adopting a topic means an author has published at least one paper that contains the term in either the title or the abstract.[1] An author X "follows" another author Y if X adopts a topic after Y adopts the same topic. This following behavior is also interpreted as an infection process [2]. The following lists the factors that potentially play a role for the influence one author has on the others to adopt a research topic.

2.1 Direct Observation

The direct observation of the adoption is the time when an author adopts the topic, and a cascade can be generated based on the chronological ordered observations. Each topic has its own cascade. According to Gomez-Rodriguez *et al.* [2], recovering an influence network with K edges will require $2K$~$5K$ cascades. This inferred influence network could be used to predict new topic adoption based on the assumption that if author X followed Y in many topics before, it is also likely X will follow Y again for a new topic. Generally speaking, the closer in time X followed Y, the more influence Y has on X based on this direct observation [2].

2.2 Indirect Observation

The direct observation helps to recover the influence network based on past topic adoption cascades, but it requires sufficient number of relevant cascades to learn the network. This limitation motivates the consideration of other factors that indirectly implies the relationship, hence the influence one author may have on another to adopt a new topic.

[1] Topic adoption is a complex process. Nevertheless, to some extent, we believe that publishing papers with topic terms are reflective of the nature of the adoption.

Social Connections. The basic idea is that authors are more likely to receive influence from their colleagues, co-authors and other socially connected peers, than unknown persons. Sun *et al.* [6] [7] suggested that the social connections could be the main reason of people establishing a new relation. Sun *et al.* [6] use meta-path to define the social connections. A meta-path is defined on the network schema, where nodes are object types and edges are relations between object types. For example, two authors are coauthored in a paper is defined as: Author-Paper-Author. To define the social relationships among authors, eight common social connections are defined: CI: cite peer's paper; CA: coauthor; CV: publish in the same venue; CSA: are co-authors of same authors; CT: write about the same topic; CICI: cites papers cite peer's; CIS: cite the same papers; SCI: cited by the same papers.

Topic Popularity. Besides the social connections affecting the adoption probability, the research topic itself is also a factor. Not only the match between the topic and the author's own interest, but also the popularity of the topic can affect how fast and easily it is being adopted. The topic popularity is defined as the average number of papers adopting the topic since the first year it appeared. Fig. 1 shows the popularity of four selected machine learning topics extracted from 20 conferences from 1989 to 2010.

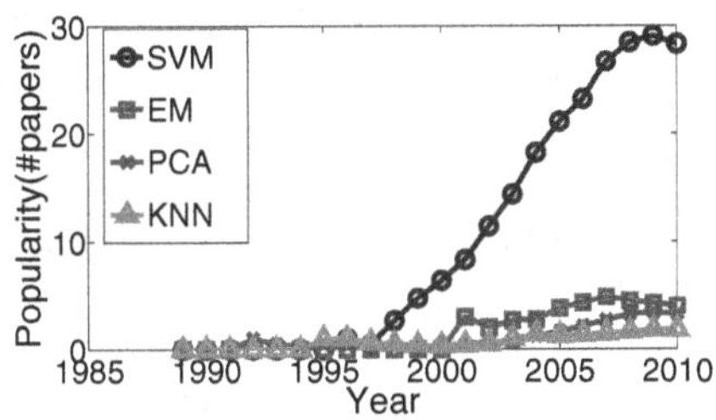

Fig. 1. Individual Popularity

As the figure shows, different terms have different popularity curves. Given the same social connections and the same initial authors who adopted the topic earliest, different topics could result in different adoption cascades. There also exists the case where an author adopts the topic without receiving any influence from her social peers. Based on these observations, this work hypothesizes that topic popularity can be a factor that affects the adoption process. To include the topic popularity in the model, a virtual author is added where the connection between any author to this virtual author represents the influence comes from the topic popularity.

3 Model Formulation

3.1 NetInf*

Traditional network inference algorithm such as NetInf [2] uses the direct observation *i.e.*, adoption time difference to estimate the influence probability, where the infection probability for each link (ij) on cascade c is calculated as:

$$Pr_{ij}^{c} = \exp(-\frac{\Delta_{ij}}{\alpha}) \tag{1}$$

where Δ_{ij} is the adoption time difference between i and j, and α is set to 1 according to [2]. Based on that, NetInf selects links which contribute mostly on infection probability over all cascades with a greedy algorithm. However, the purpose of NetInf is only to infer the most likely influence network based on past cascades. To do the prediction on future topic, NetInf* is developed as an extension based on NetInf that it estimates the influence probability for all links as:

$$Pr_{ij} = 1 - \prod_{c \in C}(1 - Pr_{ij}^{c}) \tag{2}$$

where Pr_{ij} is the probability i will adopt after j for any topic, and the transition probability matrix can be set accordingly.

3.2 HetNetInf

The second algorithm HetNetInf uses the indirect observation, *i.e.*, social connections and topic popularity to generate the influence probability. Every author can be infected (adopt a topic) by her infected neighbor with a probability. This neighbor could be the actual peer with social connections, or a virtual author which represents the topic popularity. So the problem becomes how to use the social connection features and topic popularity to represent the infection probability for each author-pair, and to reflect the time order they adopt the topic. Given authors i and j, author i is infected at time t_i and j is infected at time t_j ($t_i > t_j$). Let X_{ij} be the feature vector between author i and j, and $X_{ij} = [X_{ij,f1}X_{ij,f2}...X_{ij,fN}X_{pt}]$. $X_{ij,fk}$ is the kth social connection feature and X_{pt} is the popularity of topic t.

$$\text{if } j \text{ is the virtual author:} X_{ij} = [0...0X_{pt}]$$
$$\text{if } j \text{ is a normal author:} X_{ij} = [X_{ij,f1}X_{ij,f2}...X_{ij,fN}0]$$

The probability author j infects author i in topic cascade c is defined as:

$$Pr_{ij}^{c}(\beta) = \exp(-\frac{1}{X_{ij} * \beta}) \tag{3}$$

where X_{ij} is the feature vector between authors i and j; and $\beta > 0$ is the weight which transfers the features to the influence probability.

For every infected neighbor of i, they all have chances to infect author i. Then the likelihood of infecting author i in cascade c is:

$$L_i^c(\beta) = [1 - \prod_{j \in K(i)} (1 - Pr_{ij}^{c}(\beta))]^{I(t_i<T)} [\prod_{j \in K(i)} (1 - Pr_{ij}^{c}(\beta))]^{I(t_i \geq T)} \tag{4}$$

where $K(i)$ is the neighbor set of author i, I is the indicate function, and T is the observation period. If $t_i \geq T$, this means author i is not infected in the observation window and the L_i^c is interpreted as the probability i is not infected by any neighbor; otherwise, L_i^c stands for the probability i is infected.

Two concerns remain. Firstly, it is unclear whether one should use the same parameter β for all authors, which assumes the same adoption behavior, *i.e.*, weight distribution for all authors. So, each author has its own β_i, and HetNetInf runs the optimization for each author independently. Equation (3) is updated as:

$$Pr_{ij}^c(\beta_i) = \exp(-\frac{1}{X_{ij} * \beta_i}) \tag{5}$$

For cascade set C, the likelihood of author i being infected for all cascades $c \in C$ is:

$$L_i(\beta_i) = \prod_{c \in C} L_i^c(\beta_i) \tag{6}$$

Substituting (4) in the (6), the log likelihood is:

$$LL_i(\beta_i) = \sum_{c \in C} \log[1 - \prod_{j \in K(i)} (1 - Pr_{ij}^c(\beta_i))]^{I(t_i<T)} + \log[\prod_{j \in K(i)} (1 - Pr_{ij}^c(\beta_i))]^{I(t_i \geq T)} \tag{7}$$

Secondly, because the influence network is sparse, β need to be penalized to have fewer links which is same to have some links with very small link weight. Take it into consideration, the optimization is to minimize following function:

$$\min_{\beta_i} -LL_i(\beta_i) + ||\beta_i||_2 \quad \text{subject to } \beta_i > 0 \tag{8}$$

The ridge regularization term of $||\beta_i||_2$ is the penalty for the influence network sparsity. For each author i, ten optimizations with random initial points will be run to find the local optimals, and the best one is selected.

4 Experiments

4.1 The Dataset and Experimental Setting

This paper selects DBLP to evaluate how to infer the influence network and use that to predict who will adopt a new topic once it emerges. The experiment examines publications on top 20 computer science conferences in four areas,[2] and in two periods, $T_0 = [1991, 2000]$ and $T_1 = [2001, 2010]$. All topic terms are extracted from the paper title or the abstract, and filtered by the machine learning related keyword list in Microsoft Academic website [3]. A total of 111 machine-learning related topics that were first introduced in period T_0 are selected for training, and additional 57 topics from T_1 are selected for testing; all terms appear in more than 10 papers. For each topic, the time an author first adopted the topic is recorded. The author will be treated as not adopting the topic if either the author did not adopt the topic at all or adopted the topic beyond the monitored period. As mentioned in Section 2.2, eight social connection features are extracted from period T_0. Including the topic popularity (TP), the feature vector includes nine features in total: CI,CA,CV,CSA,CT,CICI,CIS,SCI,TP.

[2] Data mining: KDD, PKDD, ICDM, SDM, PAKDD; Database: SIGMOD Conference, VLDB, ICDE, PODS, EDBT; Information Retrieval: SIGIR, ECIR, ACL, WWW, CIKM; and Machine Learning: NIPS, ICML, ECML, AAAI, IJCAI.

[3] http://academic.research.microsoft.com

The experiment selects 196 authors who published at least 1 paper in both periods in these conferences as low productive authors (Low Pub Set), and an another set of 47 authors who published at least 3 papers in both periods as high productive authors (High Pub Set). The Low Pub Set is a superset of High Pub Set. The reason to select these authors is to have authors who are active in both periods, such that the relationships among them are relatively stable.

4.2 Prediction Study

Each algorithm uses the 111 topic cascades appeared in T_0 as training set to infer the influence network; and uses 57 terms in T_1 for testing. For each term, the authors who adopted the topic earliest are labeled as the initial authors. To predict who will follow the topic, both algorithms return the probabilities of adopting the topic for all authors. Then performance is measured by comparing all authors' adoption probabilities (excluding the initial adopted authors) to the actual adopted authors in Mean Average Precision (MAP) and Area under ROC Curve (AUC).

Table 1 shows the average MAP and AUC achieved by NetInf* and HetNetInf for the topics being tested. A high MAP value means that the author who adopted the topic is also predicted with a high probability. Looking at "High Pub Set" column in MAP, "% Authors Adopt = 0.0621" means that there are roughly only 1 out of 16 authors follow the topic. NetInf* has the chance of 0.1649 to have a correct prediction, while HetNetInf increases the chance to 0.2121. These numbers are significantly higher than 0.0621, and HetNetInf outperforms NetInf*. These numbers, however are not close to 1.0, which would be ideal for perfect prediction. This is because each past cascade only covers less than 10% of authors. The MAP results are worse for "Low Pub Set", because the same number of cascades are used to explain more authors.

In terms of AUC, HetNetInf also exhibits better performance than NetInf* for both sets. Interestingly, both algorithms achieve better AUC for "Low Pub Set" than for "High Pub Set". This is because, as the network becomes larger, though the percentage of authors adopt the topic decreases, the absolute number of authors followed the topic increases. It also increases the number of positive samples that true positive rate increases more smoothly, and thus, improves the AUC.

Table 1. Topic Adoption Prediction Performance

	MAP		AUC	
	High Pub Set	Low Pub Set	High Pub Set	Low Pub Set
NetInf*	0.1649	0.0970	0.5624	0.6333
HetNetInf	**0.2121**	**0.1044**	**0.6188**	**0.6376**
% Authors Adopt	0.0621	0.0305		

Fig. 2 and Fig. 3 show the MAP for each topic being tested on "High Pub Set" and "Low Pub Set", respectively. As the figures show, the result varies from topic to topic. The topic "gene expression data" shows an example of how social connections help predict the adoption of novel topic. In this case, authors S and M published about the topic in 2003 and 2007, respectively. In T_0, M only followed S once among 111 topics. Compared to other authors who followed S many times, the influence from S to M is relative weak in NetInf*. However, there is very strong social connection between them. As a result, this helps increase the probability for M to follow S using HetNetInf. In this case, HetNetInf has MAP $= 1$; while NetInf* has MAP $= 0.026$.

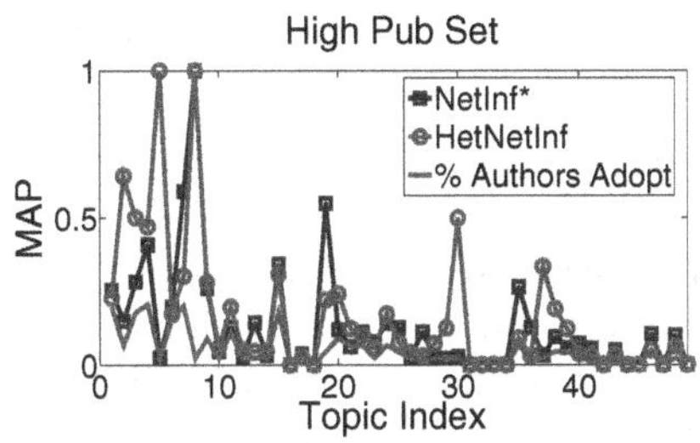

Fig. 2. Prediction Performance

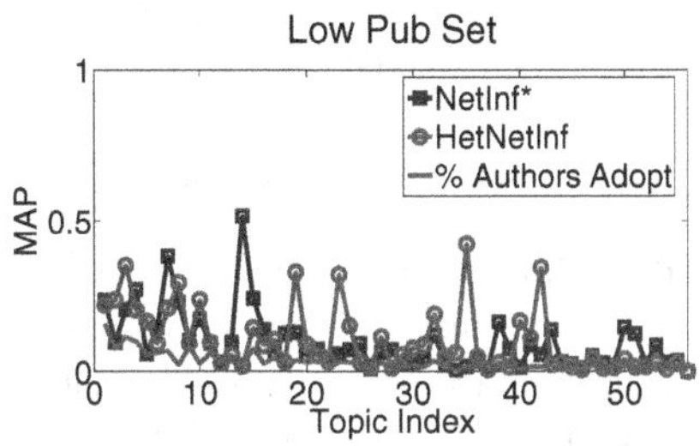

Fig. 3. Prediction Performance

HetNetInf also provides information about each author's preference (β_i) in which factors would decide the adoption process. For example, independent researchers may be used to discover topics by themselves, while other researchers step into a new topic under the influence of their research community. Table 2 shows two authors' normalized features weights. These two authors have very different following behaviors: author X's action is mostly affected by other authors' work in the same area; and author Y is mostly affected by the topic popularity.

Table 2. Individual Features Weight

Author	CI	CA	CV	CSA	CT	CICI	CIS	SCI	TP
X	**0.4471**	**0.6248**	0.0103	**0.6396**	0.0104	0.0103	0.0102	0.0102	0.0103
Y	0.2686	0.2514	0.2548	0.2654	0.3155	0.2540	0.2897	0.2517	**0.6465**

Fig. 4 and 5 further show the normalized feature weight distributions. There is no one feature that dominates the adoption process for all authors, and the weight distribution varies significantly from author to author. This also validates the approach of HetNetInf that uses different β_i for different authors.

5 Conclusion

This paper investigated the problem of predicting which author will adopt the novel topics, and tested two algorithms using the DBLP dataset. To solve this problem, the basis is to find the inherent influence network from past observations. The first approach NetInf* is based on the direct observations of past following behavior on topic adoption. It has the shortcoming that the prediction performance is much dependent on the amount of collected past cascades and whether the novel topic cascade is relevant to these past

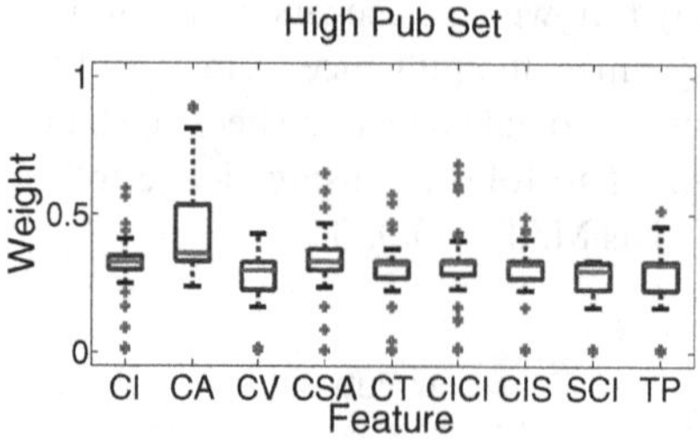

Fig. 4. Feature Weight Distribution

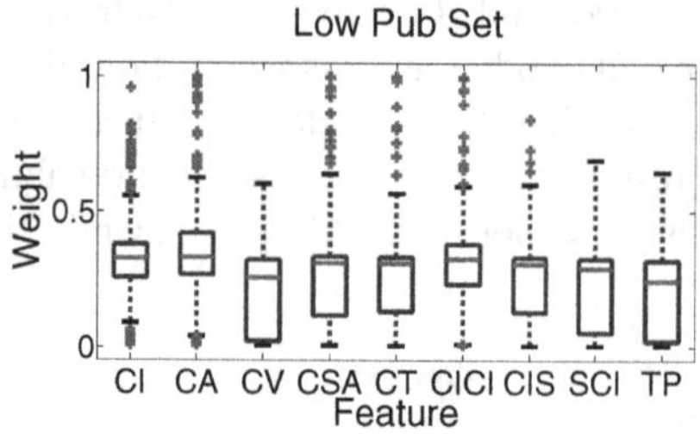

Fig. 5. Feature Weight Distribution

cascades. Though all topic terms selected are in the same area (machine learning), all these terms are unique. Each topic can have its own unique cascade which could be totally different from past ones. Alternatively, HetNetInf defines the author's following behavior based on the factors affecting the adoption process, such as social connections and topic popularity. Given the adoption information of the author's social connection peers and the topic popularity, it can estimate the probability the author adopting the topic, even the topic is totally new and irrelevant to past ones. The experiment using DBLP shows that HetNetInf outperforms NetInf* in predicting the novel topic adoption. Furthermore, HetNetInf provides information on individual author's preference on the influence factors.

References

1. Gomez-Rodriguez, M., Balduzzi, D., Schölkopf, B.: Uncovering the temporal dynamics of diffusion networks. In: International Conference on Machine Learning, pp. 561–568 (2011)
2. Gomez-Rodriguez, M., Leskovec, J., Krause, A.: Inferring networks of diffusion and influence. ACM Trans. Knowl. Discov. Data 5(4), 21:1–21:37 (2012)
3. Ji, M., Han, J., Danilevsky, M.: Ranking-based classification of heterogeneous information networks. In: Proceedings of the 17th ACM SIGKDD International Conference on Knowledge Discovery and Data Mining, KDD 2011, pp. 1298–1306. ACM, New York (2011)
4. Kempe, D., Kleinberg, J., Tardos, E.: Maximizing the spread of influence through a social network. In: Proceedings of the Ninth ACM SIGKDD International Conference on Knowledge Discovery and Data Mining, KDD 2003, pp. 137–146. ACM, New York (2003)
5. Myers, S.A., Leskovec, J.: On the convexity of latent social network inference. In: Advances in Neural Information Processing Systems, pp. 1741–1749 (2010)
6. Sun, Y., Barber, R., Gupta, M., Aggarwal, C.C., Han, J.: Co-author relationship prediction in heterogeneous bibliographic networks. In: Proceedings of the 2011 International Conference on Advances in Social Networks Analysis and Mining, ASONAM 2011, pp. 121–128. IEEE Computer Society, Washington, DC (2011)
7. Sun, Y., Han, J., Aggarwal, C.C., Chawla, N.V.: When will it happen?: relationship prediction in heterogeneous information networks. In: Proceedings of the Fifth ACM International Conference on Web Search and Data Mining, WSDM 2012, pp. 663–672. ACM, New York (2012)
8. Sun, Y., Han, J., Yan, X., Yu, P.S., Wu, T.: Pathsim: Meta path-based top-k similarity search in heterogeneous information networks. PVLDB 4(11), 992–1003 (2011)

Identifying Emergent Thought Leaders

Andrew Duchon[1] and Emily S. Patterson[2]

[1] Aptima, Inc., Woburn, MA
aduchon@aptima.com
[2] School of Health and Rehabilitation Sciences, College of Medicine,
Ohio State University, Columbus, OH
patterson.150@osu.edu

Abstract. Determining the emergent thought leader of an ad hoc organization would have enormous benefit in a variety of domains including emergency response. To determine if such leaders could be identified automatically, we compared an observer-based assessment to an automated approach for detecting leaders of 3-person ad hoc teams performing a logistics planning task. The automated coding used a combination of indicator phrases indicative of reasoning and uncertainty. The member of the team with the most reasoning and least uncertainty matched the observer-based leader in two thirds of the teams. This determination could be combined with other analyses of the topics of discussion to determine emergent thought leaders in different domains. As an example, a real-time user interface providing this information is shown which highlights communications by others that are relevant to the automatically detected, topic-specific, emergent thought leader.

Keywords: communications, leader, ad hoc teams, dialogue act analysis.

1 Introduction

An emergent leader is defined as an individual in an ad hoc group who is not assigned to a leadership position, but emerges over time as dominant in power, decision-making, and communications and is accepted by other members when acting in the capacity of how an assigned leader would be expected to act [1]. We define here an *emergent thought leader* (ETL) as one who emerges as a leader with respect to a particular topic or for a particular analytic task.

Detection of ETLs is critical, especially in situations without a clear command and control hierarchy or even knowledge of who is actually contributing. Being able to efficiently identify and communicate directly with such leaders is critical during time-pressured, high-consequence responses to emergencies to prioritize activities, allocate resources, and communicate with other entities. For example, following a hurricane, governmental and non-governmental organizations need to coordinate to distribute resources such as food and water, shelter, and medical services. The ETL with regards to medicine may be different than one that which emerges for shelter. The person most aware of these resources may change dynamically over a period of hours, and

W.G. Kennedy, N. Agarwal, and S.J. Yang (Eds.): SBP 2014, LNCS 8393, pp. 51–58, 2014.

may not be the officially designated person. For example, an ETL for medicine could be a physician on voluntary leave from their hospital in another state to provide emergency medical support and becomes the central figure working with the military branches and the Red Cross to provide medical resources to displaced civilians within a particular region. In turn, the ETL should also be aware of information and communication that may impact their decision-making.

We compare the observed leader to that assigned using an automated approach applied to transcripts of conversation from 12 three-person ad hoc teams doing a simulated logistics task. The automated approach used a form of dialogue act analysis previously shown to be useful with regards to military chat [2] and recognizes two-thirds of the ETLs correctly. We combined this process with statistical topic extraction [3] to show how a user interface could be powered by these analyses to highlight information appropriate for the ETL of each topic.

2 Related Work

There is a vast literature on the detection of leaders from text-based communications, both manually and automatically. For manual analysis, a leading approach [4], based solely on written text communications, is to predict the leader based upon activity level and written linguistic quality as a surrogate for communication/expression. Other indicators that can be manually detected are: lengthier messages, more complex and rich language, more confidence, and more emotional affect [5].

For automated analysis, a number of leadership indicators have been employed, the most common of which is dominance (the most utterances or longest turns) [5], [6]. Including variation in the tone of voice and energy improved accuracy in detecting emergent leaders over solely using dominance measures [7]. Indicators of initiative are also common, including being the first to speak [8], introducing new topics [9], and making activity-related utterances [6]. Finally, evidence of behaviors that are used by those higher in a power hierarchy when interacting with subordinates, such as giving praise, indicates leadership status (see [10], [11]).

Dialogue act analysis is a method for automatically capturing the domain-independent processes of interactions, regardless of the domain-related content. Taxonomies of dialogue acts (DAs) range from 10-20 [12] up to hundreds [13]. DAs, in their various conceptual frameworks, have been related to a range of group performance metrics (see [14]–[16]). Since DAs are useful for assessing performance, a number of groups have developed methods for automatically classifying them. These systems generally take one of two forms. The feature- or rule-based approach uses lists of text features (cue phrases) as indicators of a certain category [17]. Machine learning techniques predict the DA of each utterance typically using a wider range of features, but these technique require large amounts of labeled training data.

The rule-based system we use was first developed for extracting dialogue acts from chat data in an Air Force Dynamic Targeting Cell [2] in which the development of hand-tagged corpora was prohibitive in terms of access, time and cost, so a feature-based system was created. In military domains in which there is difficulty obtaining data (especially annotated data), this approach using specific cue phrases to classify dialogue acts has yielded state-of-the-art results on domain-specific corpora [18].

3 Team Communications Data and Observer-Based Leaders

The verbal transcripts were generated in a prior research study [19] in which 12 three-person ad hoc teams produced a solution to a military logistics task. The teams were randomly assigned to a face-to-face or a virtual (audio Skype) condition. The participants were undergraduate students specializing in homeland security and intelligence analysis and thus accustomed to analytical tasks. Each team had 90 minutes to create a written plan to transport military troops and cargo to a specified location, choosing the best combination of routes and vehicles, while minimizing cost and maximizing security. A hidden profile task was employed in which each participant had unique information which defined their roles as Fuel (e.g., available routes and fuel), Capability (e.g., carrying capacity of vehicles), or Security (e.g., access to intelligence reports). The sessions were audio- and video-taped and transcripts were generated by a research assistant from the audiotapes, reviewing the videotapes as necessary.

Two observers [EP and the research assistant] independently identified team leaders for all twelve teams by reviewing the transcript data using the manual indicators cited above (first to speak, dominance, praise, etc.). In cases in which the leader changed over time, the individual identified as the leader at the end of the session was used. For every session, one leader was uniquely identified, even when there was suggestive evidence of power-sharing between two team members, in order to facilitate a direct comparison with the automated approach. The two investigators initially agreed on the leader of 11 of 12 teams. Resolution on the leader of the remaining team was easily achieved via discussion.

4 Automated Coding of Team Leaders

This study builds upon prior research [2] investigating dialogue acts related to teamwork effectiveness, e.g., whether team members were making a correction to previous information, anticipating one another's information needs, confirming if information was correct, or indicating their uncertainty about some information. In that study, a multiple regression on these DAs accurately predicted observer-based measures of team performance. The categories and features were then updated to be applied to email from a large Army Command and Control exercise [26]. The current 29 categories represent more standard dialogue acts (question, acknowledgment, positive reply, negative reply), as well as categories related to command and control, such as intentions (command, alert, proposal), internal state or attitude (certainty, uncertainty), and process (reasoning, factual statement, clarification). The goal of these categories is to usefully classify all the linguistic features of communications that do not refer to the actual tasks being conducted. To capture that content—what is being talked about—a statistical topic analysis [3] is applied, which typically removes textual features using a stop word list. The dialogue acts defined here can then be used to assess process—how topics are being discussed—making use of many of the stop words excluded from topic analysis. To do this in practice with new data sets, such as the one here, all the unique utterances from the experiment are tagged with the current gazetteer, any untagged words or phrases that are not directly related to the task or content are then manually binned into one of the categories or left as a stop word. This process has been conducted by AD on over 80,000 messages from six data sets [2], [19]–[23].

Each message was analyzed with the subsequent gazetteers for 29 DAs. Each message was given a score for each category which was the proportion of all the message's DAs that were in that category. A participant's score for each category was determined by dividing the sum of their message scores in that category by the total sum of message scores in that category from the entire team. In addition, the proportion of the team's statements spoken by the participant was also calculated as an indicator of conversation dominance (i.e., speaking the most). From the DAs, it was predicted that Certainty, Command, and Proposal would serve as indicators of social dominance (i.e., a higher position in a social hierarchy). Uncertainty was hypothesized to have an inverse relationship (i.e., a higher position in a command hierarchy is associated with less expressed uncertainty). Finally, Reasoning was expected to be an indicator of cognitive dominance (i.e., having the most influence over how decisions are made and on what basis) and expected to be most related to thought leadership. Table 1 provides examples of the most frequent features in these categories present in the transcripts.

Table 1. Dialogue Acts Examined

Category	Description	Most common features
Certainty	Confident or extreme statements of absolutes	*all, only, need, we are, the most*
Command	Requests for others to do something	*just, have to, should, can be, we have to*
Proposal	Suggested actions	*we can, going to, could, we want, let's*
Reasoning	Determining or recognizing causal structure; relating pieces of information	*so, but, would, if, because*
Uncertainty	Indications of uncertainty, lack of confidence or vagueness about information	*think, or, which, some, don't know*

We hypothesized that these six independent variables from all 36 participants would be related to the binary value of being the leader in a logistic regression. The logistic regression analysis found that the overall model was significant ($\chi2(6, N=36) = 17.944, p<0.01$), with only Reasoning being individually significant ($B = 33.475, p < 0.04$). In a stepwise regression starting with Reasoning, only Uncertainty added significantly to the model ($\chi2 (1, N=36) = 4.6002, p<0.04$). With these two factors, the final model was significant ($\chi2 (2, N=36) = 12.688, p<0.002$) with estimates for

Reasoning = 30.906, and Uncertainty = -13.519. Note that the proportion of messages was not as significant as the proportion of Reasoning and Uncertainty features spoken. Using these estimates, each individual could be assigned a probability of being the ETL, p(ETL), of their team. The highest scoring team member was the same as the one assigned by observers in 8 of the 12 teams, and was in the top two in 11 of the 12 teams. If p(ETL) can discriminate between leaders and non-leaders, then large differences in the value should clearly determine the leader. To assess this idea, for each team we compared the max p(ETL) score to the second highest. We see in Fig. 1 that large differences (above 0.38) always correctly identify the leader.

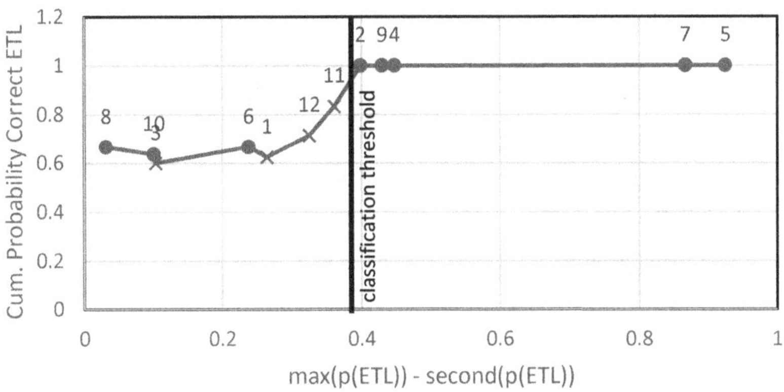

Fig. 1. Cumulative probability of determining the correct ETL based on the difference between highest p(ETL) on a team and the second highest

In three of the four discrepant cases (Teams 1, 3, 11, 12) marked with ×, the teams were single gender: two all-female and one all-male. Upon re-examination of the transcripts by EP, two of the three all-female teams (11 and 12) appeared to have more of a power-sharing arrangement between two members as compared to the other teams.

5 Example Interface Using the Results

While the results from the automated system were not as consistent as the observer-based results, they are far above chance (66% vs. 33%) which suggests this mechanism could be used to identify ETLs in ad hoc teams. To demonstrate how such information could be used, we combined this analysis with a six-topic analysis [3] of the features extracted that were not a DA or stop word. Each message was assigned a proportion of each topic present, the amount of ETL spoken, and these two values were multiplied and accumulated per participant, so the team member indicating the most leadership on each topic could be identified.

A prototype was created which read the transcripts as if they were running live. In Figure 2, the Fuel expert is logged into the chat system. His current expertise is displayed to the right of his name (*Troops-Cargo*, *Armor-Security*). The gold star next to Security's name indicates that this person is the current overall (non-topic weighted) leader of the conversation to that point. In the Chat tab, new messages are displayed on the top along with their sender and the topics of the message. In order to draw attention to important information, messages are highlighted in blue which contain a topic for which Fuel is currently considered an expert. Note that as chat messages stream through the viewer, the expert may change as a participant shows more or less expertise on a particular topic relative to the other team members.

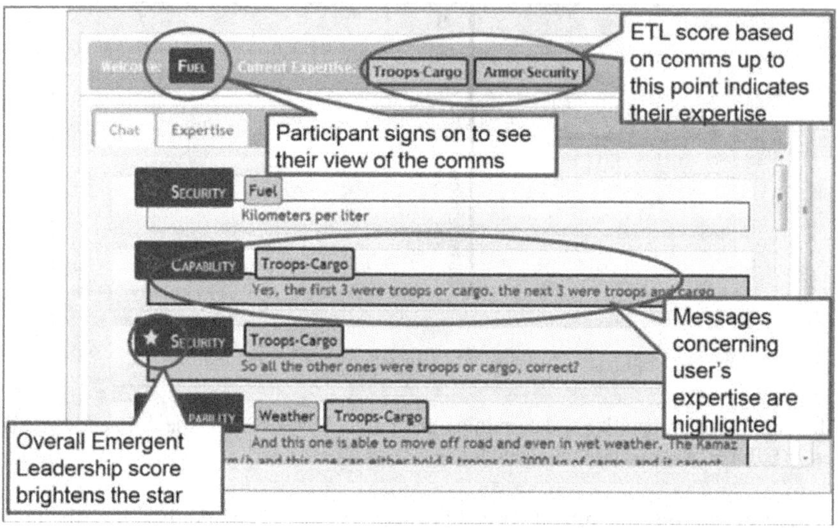

Fig. 2. Prototype interface highlights messages on topics for which the user is the ETL

6 Conclusion

The examination of the study's transcripts suggests that ETLs can be detected with domain-independent phrases representing acts of Reasoning and Uncertainty. Combined with a statistical, domain-dependent topic model, information and communications could be routed to or highlighted for those who can use it most—improving situational awareness and decision-making, and routed away from or de-emphasized for those who can use it least—reducing information overload.

Due to its domain-independence and simplicity, and the ability to get real-time access to chat data in a number of military command and control settings, this approach could be used with current technology for real-time support in actual work settings. For example, following a hurricane, the algorithm could identify a physician volunteer who is an ETL for providing medical services. Others looking for a point of contact to provide medical expertise could thus quickly find such a person efficiently by avoiding a search through a word-of-mouth network. In practice, this might be done only when large differences in p(ETL), and thus a clear leader, among the team members is present. We are currently applying this analysis to new datasets of ad hoc teams, both academic [21] and real-world [24] in order to test the generalizability to new domains and larger team sizes.

Acknowledgments. We thank Jeff Morrison, Ranjeev Mittu, Fernando Bernal, Robert Stephens, Marcela Borge, Sean Goggins, and Carolyn Rosé for insightful discussions about this research. Gabriel Ganberg and Michael Therrien helped develop the software. Funding was provided by the Office of Naval Research's Command Decision Making program (N00014-11-1-0222). The views are those of the authors and do not necessarily represent the views of the Navy.

References

[1] Guastello, S.J.: Managing emergent phenomena: Nonlinear dynamics in work organizations. Lawrence Erlbaum Associates Publishers, Mahwah (2002)
[2] Duchon, A., Jackson, C.: Chat analysis for after-action review, presented at the Interservice/Industry Training, Simulation, and Education Conference (I/ITSEC) (2010)
[3] Blei, D.M., Ng, A.Y., Jordan, M.I.: Latent Dirichlet allocation. J. Mach. Learn. Res. 3, 993–1022 (2003)
[4] Balthazard, P.A., Waldman, D.A., Atwater, L.E.: The mediating effects of leadership and interaction style in face-to-face and virtual teams. In: Weisband, S. (ed.) Leadership at a Distance, pp. 127–150 (2008)
[5] Huffaker, D.: Dimensions of leadership and social influence in online communities. Hum. Commun. Res. 36(4), 593–617 (2010)
[6] Simoff, S.J., Sudweeks, F., Amant, K.S.: The Language of Leaders: Identifying Emergent Leaders in Global Virtual Teams. In: Linguistic and Cultural Online Communication Issues in the Global Age, pp. 93–111. IGI Global (2007)
[7] Sanchez-Cortes, D., Aran, O., Mast, M.S., Gatica-Perez, D.: Identifying emergent leadership in small groups using nonverbal communicative cues. In: International Conference on Multimodal Interfaces and the Workshop on Machine Learning for Multimodal Interaction, pp. 39–43 (2010)
[8] Heckman, R., Crowston, K., Misiolek, N.: A structurational perspective on leadership in virtual teams. In: Virtuality and Virtualization, pp. 151–168. Springer (2007)
[9] Strzalkowski, T., Broadwell, G.A., Stromer-Galley, J., Taylor, S.: Modeling Socio-Cultural Phenomena in Online Multi-Party Discourse. In: Workshops at the Twenty-Fifth AAAI Conference on Artificial Intelligence (2011)
[10] Mayfield, E., Rosé, C.P.: Recognizing authority in dialogue with an integer linear programming constrained model. In: Proceedings of the Association for Computational Linguistics (2011)
[11] Bunt, H.: Multifunctionality in dialogue. Comput. Speech Lang. 25(2), 222–245 (2011)
[12] Stolcke, A., Ries, K., Coccaro, N., Shriberg, E., Bates, R., Jurafsky, D., Taylor, P., Martin, R., Ess-Dykema, C.V., Meteer, M.: Dialogue Act Modeling for Automatic Tagging and Recognition of Conversational Speech. Comput. Linguist. 26(3), 339–373 (2000)

[13] Bunt, H.: The DIT++ taxonomy for functional dialogue markup. In: AAMAS 2009 Workshop, Towards a Standard Markup Language for Embodied Dialogue Acts, pp. 13–24 (2009)
[14] Convertino, G., Mentis, H.M., Rosson, M.B., Carroll, J.M., Slavkovic, A., Ganoe, C.H.: Articulating common ground in cooperative work: content and process. In: Proceeding of the Twenty-Sixth Annual SIGCHI Conference on Human Factors in Computing Systems, pp. 1637–1646 (2008)
[15] Soller, A.: Supporting social interaction in an intelligent collaborative learning system. Int. J. Artif. Intell. Educ. IJAIED 12, 40–62 (2001)
[16] Fischer, U., McDonnell, L., Orasanu, J.: Linguistic correlates of team per-formance: toward a tool for monitoring team functioning during space missions. Aviat. Space Environ. Med. 78(suppl. 5), B86–B95 (2007)
[17] Pennebaker, J.W., Francis, M.E., Booth, R.J.: Linguistic inquiry and word count: LIWC. Lawrence Erlbaum Associates, Mahwah (2001)
[18] Budlong, E.R., Walter, S.M., Yilmazel, O.: Recognizing connotative meaning in military chat communications. In: Evolutionary and Bio-Inspired Computation: Theory and Applications III, Orlando, FL, USA, vol. 7347, pp. 73470–8 (2009)
[19] Patterson, E.S., Bernal, F., Stephens, R.: Differences in Macrocognition Strategies With Face to Face and Distributed Teams. In: Proceedings of the Human Factors and Ergonomics Society Annual Meeting, vol. 56, pp. 282–286 (2012)
[20] Duchon, A., McCormack, R., Riordan, B., Shabarekh, C., Weil, S., Yohai, I.: Analysis of C2 and 'C2-Lite' Micro-message Communications, San Francisco, CA. presented at the AAAI 2011 Workshop on Analyzing Microtext (2011)
[21] Borge, M., Ganoe, C.H., Shih, S.-I., Carroll, J.M.: Patterns of team proc-esses and breakdowns in information analysis tasks. In: Proceedings of the ACM 2012 Conference on Computer Supported Cooperative Work, pp. 1105–1114 (2012)
[22] Biron, H.C., Burkman, L.M., Warner, N.: A Re-analysis of the Collaborative Knowledge Transcripts from a Noncombatant Evacuation Operation Scenario: The Next Phase in the Evolution of a Team Collaboration Model. Naval Air Warfare Center Aircraft Division, Technical Report NAWCADPAX/TR-2008/43 (2008)
[23] Kiekel, P.A., Gorman, J.C., Cooke, N.J.: Measuring speech flow of co-located and distributed command and control teams during a communication channel glitch. In: Proceedings of the Human Factors and Ergonomics Society 48th Annual Meeting, pp. 683–687 (2003)
[24] Goggins, S.P., Mascaro, C., Mascaro, S.: Relief Work after the 2010 Haiti Earthquake: Leadership in an Online Resource Coordination Network (2012)

System-Subsystem Dependency Network for Integrating Multicomponent Data and Application to Health Sciences

Edward H. Ip[1], Shyh-Huei Chen[1], and Jack Rejeski[2]

[1] Department of Biostatistical Sciences, Wake Forest School of Medicine, Winston-Salem, U.S.A.
{eip,schen}@wakehealth.edu

[2] Department of Health and Exercise Sciences, Wake Forest University, Winston-Salem, U.S.A.
rejeski@wfu.edu

Abstract. Two features are commonly observed in large and complex systems. First, a system is made up of multiple subsystems. Second there exists fragmented data. A methodological challenge is to reconcile the potential parametric inconsistency across individually calibrated subsystems. This study aims to explore a novel approach, called system-subsystem dependency network, which is capable of integrating subsystems that have been individually calibrated using separate data sets. In this paper we compare several techniques for solving the methodological challenge. Additionally, we use data from a large-scale epidemiologic study as well as a large clinical trial to illustrate the solution to inconsistency of overlapping subsystems and the integration of data sets.

Keywords: Generalized dependency network, Gibbs sampler, System-subsystem modeling.

1 Introduction

Diametrically opposed to reductionist thinking, system thinking seeks both local and global models that incorporate evidence about the delayed and distal effects of human intervention, policies, and situations [1]. Accordingly, system science requires a new set of tools to aid the understanding of complexity, feedback loop, stock-and-flow, and interactions between system components. Yet, for empirical researchers in system sciences, one important barrier is the lack of tools for integrating models built for different components, or subsystems–meaningful and scientifically self-contained smaller systems–of the entire system. Virtual world modelers often assemble submodels, which are independently calibrated using ad hoc and diverse data, into a system model that is amenable to simulation experiments. The Foresight Tackling Obesities: Future Choices project [2] provides an illustrative example of a system-subsystem. The Tackling Obesities project is a U.K. government initiative to find a policy response to the obesity epidemic over the next several decades. A centerpiece of the project is a

W.G. Kennedy, N. Agarwal, and S.J. Yang (Eds.): SBP 2014, LNCS 8393, pp. 59–66, 2014.

collection of system maps, compiled by over 300 epidemiologists, nutritionists, geneticists, biologists, and social scientists for schematically capturing the different drivers of obesity. By not treating obesity only as a medical condition, the system approach redefines the nations health as a societal and economic narrative. Foresight researchers also created an overlay of subsystem contours. For example, physiology, individual physical activity, physical activity environment, individual psychology, and societal influences all exist as subsystems. Partly for interpretability and simplicity, the Foresight subsystem contours were drawn not to overlap with each other. However, in reality, most subsystems are neither closed nor independent of one another. More realistically, subsystems often share common drivers and (sub)components. Indeed, the interface between subsystems could be more interesting than the components; if a system can be compartmentalized into completely unrelated and independent subsystems, then the system approach would have almost become a moot exercise. From a model building perspective, it would be easier, perhaps as part of a divide-and-conquer strategy, to first develop models for subsystems and then "glue" the established subsystem models together to form a system model.

This paper attempts to explore methods for integrating subsystems, which may overlap with each other, into a coherent system through a generalized version of the dependency network (GDN) [3], a modeling tool developed for handling complex and reciprocal relationships. The subsystem method is primarily motivated by two applications - the study of childhood obesity, which we briefly described above, and that of the aging and disablement process in older adults. Because of space limitation, in this paper we only focus on the latter. A contemporary view of the human disablement that emerged from the literature is that it is a complex dynamic system that consists of multiple subsystems interacting with each other [4]. Two particular challenges when one applied a system science approach to the aging and disablement model are: (1) inference in the presence of a large number of variables that can be approximately categorized into different subsystems but the subsystems also have overlapping variables, and (2) integration of data sets collected from different studies, which possibly have different emphases on the types of data they respectively collect. In this paper we compare several techniques for solving the two issues within the system-subsystem dependency network framework. We respectively use data from a large-scale epidemiologic study to illustrate the solution to the former issue, and additional data from large-clinical trial to illustrate the latter.

2 Background for Generalized Dependency Network and the Techniques for Learning the Network

As an alternative modeling tool to Bayesian network or graphical model based on directed acyclic graphs (DAG)[5], dependency network is a probabilistic model that is represented by a product of conditional probabilities $p(x_i|pa(x_i))$, where x_i denotes a variable in the network, and $pa(x_i)$ denotes the parents of x_i. Consistent dependency networks are cyclic graphical models in which all the conditional probabilities can be derived from a joint probability distribution. The consistent dependency network has not attracted much attention because of the restriction that the set of conditional distributions has to be consistent with a joint probability distribution. Developed by a team of scientists at Microsoft Research [3], the GDN removes the consistency requirement. A (pseudo) Gibbs sampler for inference was also proposed to reconcile inconsistency between conditional distributions. The GDN is especially useful for encoding predictive relationships and provides better predictive accuracy because of its collective inference approach as opposed to inference based on individual conditional distributions.

To illustrate the idea, consider the simple (cyclic) GDN in Fig. 1 in which X, Y, and Z are variables. The parents for X, Y, and Z are, respectively, (Y, Z), X, and Y. Consider the following three regression models that are individually specified according to the GDN in Fig. 1: $x = \alpha_0 + \alpha_1 y + \alpha_2 z + \epsilon_1$, $y = \beta_0 + \beta_1 x + \epsilon$, and $z = \gamma_0 + \gamma_1 y + \epsilon_3$, where ϵ_1, ϵ_2, and ϵ_3 represent Gaussian noise, and α, β, γ denote regression coefficients. When empirical data from a data set are used to independently estimate the three submodels, the result may not lead to a proper joint distribution. It is argued that if a GDN is estimated from the same data set, which is assumed to be generated from a proper joint distribution, then the set of conditional distributions would be consistent when the sample size is large; in that case a generic Gibbs sampler would also accurately recover the joint distribution. By consistent (or compatible), we refer to the property that there exists a proper probability distribution that can be used to derive each individual conditional distribution. The Gibbs sampler is a simulation based estimation algorithm that iteratively draws sample x_i from the conditional distributions $p(x_i|pa(x_i))$. It can be proved that this simple scheme, regardless of where one starts, will converge and lead to samples drawn from the underlying joint distribution [6]. In Fig. 1, for example, the Gibbs sample would take the form of iteratively sampling from the following conditional distributions: $p(x|y, z)$, $p(z|y)$, and $p(y|x)$. It is worth pointing out that when creating a GDN, there is no restriction on the predictive tool - e.g., decision tree or neural network - can be used instead of linear regression.

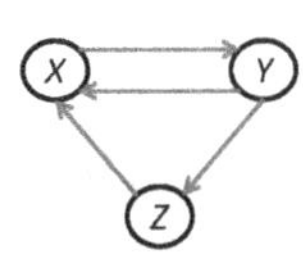

Fig. 1. A simple general dependency network

The system-subsystem dependency network we propose in this paper is even more general; we extend "variable" in GDN to "subsystem". Without resorting to formality, we illustrate the approach using a simple two-subsystem example from an aging study. Assume that there are two subsystems, S1 of self efficacy about functions, and S2 of physical performance as measured objectively by several variables, and that there is a single measure, indicated by variable X_0, that is common to both subsystems (Fig. 2). Furthermore, assume that there exist two distinct data sets D_1 and D_2 that are respectively associated with the two subsystems. Following the GDN approach for variables, the system-subsystem GDN proceeds as follows: (1) learn a GDN for subsystem S_1 using data set D_1 ; likewise, learn (independently) a GDN for subsystem S_2 using data set D_2 ; (2) apply a system-level pseudo-Gibbs sampler $p(S_1|S_2)$, $p(S_2|S_1)$, defined by the following implementation, which we call the ordered pseudo-Gibbs sampler: (a) for $p(S_1|S_2)$, draw a sample from the three conditional distributions $p(X_1|X_0, X_2)$, $p(X_2|X_0, X_1)$, and then $p(X_0|X_1, X_2)$; and (b) for $p(S_2|S_1)$, draw samples from $p(X_3|X_0, X_4, X_5)$, $p(X_4|X_0, X_3, X_5)$, $p(X_5|X_0, X_3, X_4)$, and then $p(X_0|X_3, X_4, X_5)$; (3) iterate; (4) when the ordered pseudo-Gibbs sampler reaches convergence, collect the sample $X^\star = (X^\star_{i0}, X^\star_{i1}, X^\star_{i2}, X^\star_{i3}, X^\star_{i4}, X^\star_{i5})$ for inference for the entire system.

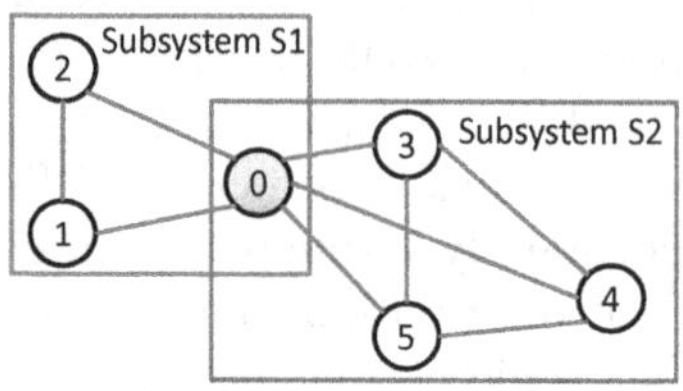

Fig. 2. Two subsystems with a common variable

The implementation of the ordered pseudo-Gibbs sampler for system-subsystem described above is an extension of the fixed-scan procedure in GDN for potentially incompatible distributions. Fixed-scan Gibbs samplers follow a prefixed order of sampling - e.g., for 3 variables, $X_1 \rightarrow X_2 \rightarrow X_3 \rightarrow X_1$ and so on [6]. However, fixed-scan pseudo-Gibbs sampler, which is commonly used in Bayesian estimation, is not optimal for potentially incompatible conditional distributions in the sense that the conditional distributions derived from the inferred joint distribution may have large error variances [7,8]. It has been shown that by exploiting the full range of possible scan orders, either through a technique called the Gibbs ensemble, or through randomizing the scan order during sampling [9] in the pseudo-Gibbs procedure, one can substantially reduce the level of incompatibility measured in error variances.

3 Analysis 1: Comparison of Methods

Here we expand the idea of randomizing scan order to include the subsystem-level (e.g., for a 3-subsystem system, instead of following a fixed order cycling through the subsystem $S_1 \rightarrow S_2 \rightarrow S_3 \rightarrow S_1$, use a random order such as $S_1 \rightarrow S_3 \rightarrow S_1 \rightarrow S_2$ and so on). This is in addition to implementing random scan at the variable level within a subsystem. Four different methods were compared: (1) Random scan at the subsystem level and fixed scan at the variable level (RF); (2) Random scan at the subsystem level and random scan at the variable level (RR); (3) Random scan at the subsystem level but with the restriction that each subsystem must appear once and only once in a cycle (e.g., $S_1 \rightarrow S_3 \rightarrow S_2$ for a cycle at the subsystem level), and fixed scan at the variable level (RF*); and (4)

Random scan at the subsystem level in which each subsystem must appear once and only once in a cycle, and random scan at the variable level (RR*). Because using fixed scan algorithm at the both the subsystem and the variable level often failed to converge or converged to apparently highly biased solution, we did not include it into our comparison study.

We used a data set from a multi-center epidemiological study - the Health Aging and Body Composition Study (Health ABC) - to anchor the comparisons. The Health ABC is a cohort study focused on understanding the relationship between body composition, weight-related health conditions, and incident functional limitation. This rich data set now contains more than 10 years of follow-up since its inception in 1997/98, when baseline data were collected. The study sample consisted of 3,075 well-functioning black and white men and women aged 68 to 80 years, recruited from two field centers in the Eastern part of the U.S. For the purpose of the current study, selected variables were used to represent four subsystems. Fig. 3 shows the four subsystems and the set of common variables across the subsystems - Disease status (S1), Mental status (S2), Self-efficacy in function (S3), and Physical performance (S4), whereas the common set of variables include demographic information (gender, age, and race) and anthropometric measure (BMI). The L-divergence and the L_1 discrepancy (e.g., see [7]) were used to measure error, which is defined as a function of the difference between the derived conditional distribution and the actual conditional distribution within each subsystem S1-S4. Errors were normalized by averaging the total cumulative errors across the number of cells used in each subsystem so that values across subsystems could be fairly compared. For each method, a total of 300 million Gibbs samples were used to estimate the cell probabilities.

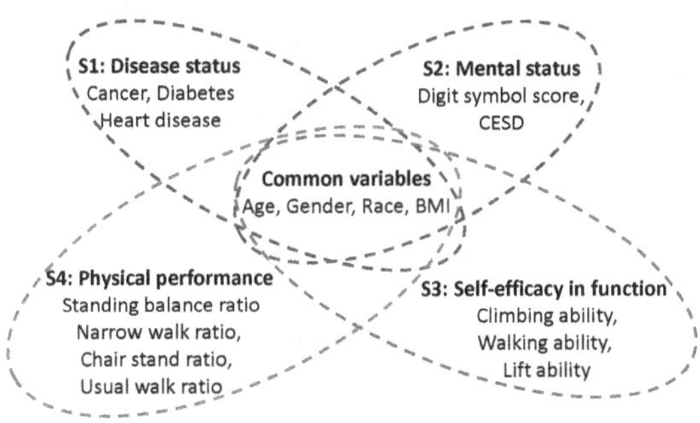

Fig. 3. Four subsystems with a set of common variables in HABC

3.1 Results for Analysis 1

Table 1 shows the errors (smaller is better) across the 4 methods. In general, the random scan methods are superior to their respective fixed scan counterparts. For example, RR, the method that uses random scans at both the subsystem and variable level has, on average, a reduction of 38% on L-divergence and 20% on L_1 divergence. We were surprised by the improvement in RR* over RR, which appears to be consistent across the subsystems, even though the magnitude is not as large as the improvement seen in comparing RR and RF, or RR* and RF*. Simply enforcing the requirement that each cycle walks through all subsystems appear to help overall performance, if not the rate of convergence.

Table 1. Normalized error rates ($\times 10^{-3}$) of 4 pseudo-Gibbs-sampler methods

	L-divergence				L_1			
Method	S1	S2	S3	S4	S1	S2	S3	S4
RF	5.25	13.6	7.56	6.69	53.0	79.3	53.7	58.6
RR	3.25	7.16	4.75	4.78	42.3	58.6	43.0	50.1
RF*	5.21	13.6	7.50	6.67	52.8	79.3	53.3	58.5
RR*	3.04	6.47	4.45	4.59	40.9	55.9	41.7	49.1

4 Analysis 2: Learning from Two Data Sets

The second example illustrates how the system-subsystem GDN approach could be used to learning networks from different data sources. In this example, besides some common variables (age, gender, BMI), we selected a few variables (lift index and narrow walking ratio) from the HABC study as well as a few variables (the physical component summary (PCS) and mental component summary (MCS)) from the Action For Health in Diabetes (LookAHEAD) study, a multi-site randomized clinical trial. We used the entire sample of HABC as in Analysis 1 and a sample N=770 from the LookAHEAD trial in our analysis. Unlike the HABC sample, which contains 18% of individuals with diabetes, the LookAHEAD sample only contains participants with diabetes. The LookAHEAD sample also represents a younger cohort than the HABC sample (50% below 71 years old in LookAHEAD versus 29% in HABC) as well as a more obese cohort (BMI $>$ 30 is 78% in LookAHEAD versus 25% in HABC). Such differences in the samples arose from the design of the respective studies - e.g., only individuals with diabetes and were overweight were eligible for LookAHEAD. As a result, the distributions in many of the variables considered in this analysis were not similar. We ran the several methods described above using both data sets and then estimated the strengths of associations between various factors included in the system-subsystem GDN. Because the variables were all discretized, we followed a commonly used approach in bioinfomatics and ordered their p-values. Then we used the magnitude of the p-values to indicate the strength of association.

4.1 Results from Analysis 2

An inspection of the error rates across the four methods showed that there were less of a difference between the error rates across the methods than in Analysis 1, although the overall trend still favored random scan methods. Because of limited space and given the evidence regarding the strength of the RR* method in the first analysis, here we only report results from the RR* method. Fig. 4 shows the relative strengths of the associations (bidirectional) between the variables within the two subsystems (indicated by boxes in Fig. 4). Diabetes status was included as a "common" variable because it was needed for both data sets for indicating the status of a participant. The strongest association appear to be between the variable pairs BMI and diabetes, and lift index and gender ($p < 10^{-30}$). Moderate association exist between PCS and MCS in subsystem S2. However, both PCS and MCS do not appear to have either strong or moderate associations with the other variables. We shall further discuss this in the Discussion section.

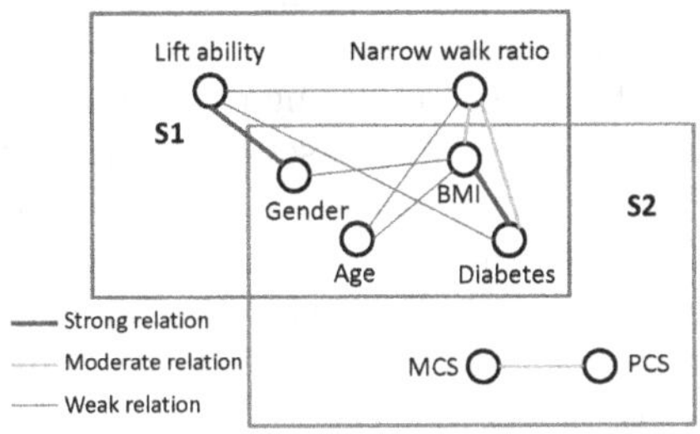

Fig. 4. Structure of two subsystems using data from two different sources

5 Discussion

The GDN has the promise to become an important tool set for system science because it is highly flexible, convenient to implement, and has the potential to integrate heterogeneous data sources. One contribution of this paper is to extend the concept of GDN from variable to subsystem. Currently the standard method for learning the joint distribution and making subsequent inference is the pseudo Gibbs sampler, which was also adopted in the current work to a system of componential networks, or subsystems. One surprising finding both from this work and previous work is that the usual fixed scan Gibbs sampling method does not work well. Analogous to the local optima problem in optimization, the fixed scan Gibbs sampler for potentially incompatible conditional distributions may be trapped in suboptimal paths, and as a result either does not converge or converges to a biased solution or solutions of high variance. The random scan method offers a relatively simple and quick fix. In this paper we offer further evidence that the pseudo Gibbs sampler can be enhanced when random scan is applied to the subsystem level. A third contribution of the paper is the illustration of using the GDN for integrating different data sources, as demonstrated in the joint analysis of the HABC and the LookAHEAD data sets.

The intuition behind the system-subsystem GDN is that it is often easier to study components of a system and subsequently integrate the components. The process of integration is typically "messy" so approximate method needs to be used, and the pseudo Gibbs sampler emerges as a powerful tool for this purpose. The idea of using the Gibbs sampler in a subsystem analysis is to allow information from one subsystem to "diffuse" to another subsystem through the repeated draws of samples from the common, or the overlapping part of the subsystems such that potentially incongruent subsystems will "reconcile" with each other. There are however many challenges that remain to be studied and solved.

First, by restricting the relationship between variables to be bidirectional and thus associative, we have not exploited the full modeling strength of the GDN. The GDN in general allows directed arcs in the corresponding graph. While it is true that typically only for a few well studied pathways of which causality could be well understood, in many cases only correlation could be readily established, resulting in a "picture" that contains a mixture of directed causal pathways and undirected relationships. The examples we used did not address possible causal relationships or mixture of relationships. We also have not used variable selection methods for simplifying potentially complex structures. In the pseudo Gibbs sampler, we used the full set of conditional variables. However, this limitation, while affecting interpretation of the structure, is not likely to have any meaningful impact on the evaluation of the performance of the methods.

The second challenge has to do more with the inherent heterogeneity that typically exist across different data sets. In Analysis 2, we observed that the data from the LookAHEAD study in the second subsystem (S2) did not seem to affect the parameters in the variables in the first subsystem (S1). The information exchange between subsystems might be more meaningful if for example, both diabetic and non-diabetic individuals are present in both studies. Heterogeneity of data from different sources is a significant issue that is beyond the scope of this paper and will require further research.

Acknowledgment. The study is supported by NIH grants 1R21AG042761-01 and 1U01HL101066-01 (PI: Ip).

References

1. Sterman, J.D.: Learning from evidence in a complex world. Am. J. Public Health 96, 505–514 (2006)
2. Vandenbroeck, P., Goossens, J., Clemens, M.: Foresight Tacking Obesity: Future Choices Building the Obesity System Map. Government Office for Science, UK (2013), `http://www.foresight.gov.uk` (last retrieved November 13, 2013)
3. Heckerman, D., Chickering, D.M., Meek, C., Rounthwaite, R., Kadie, C.: Dependency networks for inference, collaborative filtering, and data visualization. Mach. Learn. Res. 1, 49–75 (2000)
4. Lawrence, R.H., Jette, A.M.: Disentangling the disablement process. J. Gerontol. B-Psychol. 51, 173–182 (1996)
5. Lauritzen, S.L.: Graphical models. Oxford Press (1996)
6. Casella, G., George, E.I.: Explaining the Gibbs sampler. Am. Stat. 46, 167–174 (1992)
7. Chen, S.H., Ip, E.H., Wang, Y.: Gibbs ensembles for nearly compatible and incompatible conditional models. COMPUT. Stat. Data An. 55, 1760–1769 (2010)
8. Chen, S.H., Ip, E.H., Wang, Y.: Gibbs ensembles for incompatible dependency networks. WIREs Comp. Stat. 5, 475–485 (2013)
9. Levine, R.A., Casella, G.: Optimizing random scan Gibbs samplers. J. Multivariate Ana. 97, 2071–2100 (2006)

How Individuals Weigh Their Previous Estimates to Make a New Estimate in the Presence or Absence of Social Influence

Mohammad S. Jalali

Grado Department of Industrial and Systems Engineering, Virginia Tech
Northern Virginia Center, Falls Church, VA, 22043, USA
mj@vt.edu

Abstract. Individuals make decisions every day. How they come up with estimates to guide their decisions could be a result of a combination of different information sources such as individual beliefs and previous knowledge, random guesses, and social cues. This study aims to sort out individual estimate assessments over multiple times with the main focus on how individuals weigh their own beliefs vs. those of others in forming their future estimates. Using dynamics modeling, we build on data from an experiment conducted by Lorenz et al. [1] where 144 subjects made five estimates for six factual questions in an isolated manner (no interaction allowed between subjects). We model the dynamic mechanisms of changing estimates for two different scenarios: 1) when individuals are not exposed to any information and 2) when they are under social influence.

Keywords: Estimate aggregation, collective judgment, and social influence.

1 Introduction

The present study examines individuals' mechanisms for revising their estimates in the presence and absence of social influence. How do people weigh their previous estimates while forming new estimates? How do they account for the judgment of others in their next estimate? We base our modeling and estimation work on the data of an experiment by Lorenz et al. [1] where each individual, in 12 groups of 12 people, makes five estimates for six factual questions. Lorenz et al. [1] study different scenarios when individuals do not receive any feedback about others' estimates and when they are given feedback to some degree. Before reviewing their experiment (in Section 2) and presenting our modeling work (in Section 3), we review the key findings of the literature regarding this topic and identify the research questions to which we contribute.

1.1 Aggregation of Individual Estimates

One of the main research areas relevant to our study is the impact of aggregation of individual estimates. In general, individuals aggregate their opinion by averaging

W.G. Kennedy, N. Agarwal, and S.J. Yang (Eds.): SBP 2014, LNCS 8393, pp. 67–74, 2014.

[2, 3]. Although nineteenth century scientists did not trust averaging [3], recent studies have shown that the average of multiple estimates from different individuals is more accurate than the average of multiple estimates from one individual [4-11]. Surowiecki [12] demonstrates that the results of aggregating individual estimates are superior to even those provided by experts.

Averaging increases accuracy, because different individuals' estimates often bracket the true value and thus averaging yields a smaller error than randomly choosing one estimate. Only if significant bias is present across all individuals, and thus the estimates do not bracket the truth, the average would be as accurate as a random estimate [3, 5, 8, 9]. Research shows that averaging ensures that the result has lower variability, lower randomization error, lower systemic error, and converges towards the true forecast [see: 5, 13, 14]. Additionally, averaging not only increases the accuracy, but also some form of averaging is almost nearly optimal. Yaniv [6] notes the "independency of individual" as a central condition for obtaining optimal accuracy, and Hogarth [15] presents that groups containing between 8 and 12 individuals have predictive ability to the optimum. This simple mathematical fact of averaging individual estimates, the so-called "wisdom of crowds", can be easily missed or even if it is seen, it can be hard to accept [12].

1.2 Weighing Process in Aggregation

Research shows that people make decisions by weighing their own opinions with advice from other sources [16]. In the process of giving and receiving advice, individuals discount advice and weigh their own opinions more because they are usually egocentric in revising their opinions and have less access to reasons behind the advisor's view [6, 17, 18]. Harvey and Harries [19] observe a similar behavior in their experiment, where individuals put more weight on forecasts that are their own rather than on equivalent forecasts that are not theirs. In short, differential access to reasons (e.g. advisor's reasons) and egocentric beliefs are the two common causes of overweighing one's own opinions; however, Soll and Mannes [3] show that neither of these two can fully account for the tendency to overweigh one's own reasons.

Yaniv and Milyavsky [20] note that individuals with less information change their opinions based on advice more than more knowledgeable individuals. In their experiment, individuals were less likely to change their initial opinion if they had a strong and formed opinion than others who had not. Additionally, Mannes [21] believes that when individuals recognize the wisdom of crowds and place more weight on their opinions, their revised belief becomes more valid.

In sum, studying the effects of social influence on individual decision making is important to evaluate the reliability of their specific predictions. In fact, the internal mechanisms that drive individuals to update their estimates are not fully specified in the literature, especially when they are under social influence. To be more clear, social influence occurs when an individual changes her attitude because of the attitudes of others; that is, when an individual's beliefs, feelings, and behaviors are affected by other people [22].

2 The Experiment

Lorenz et al. [1] ask 144 students to answer six quantitative questions on geographical facts and crime statistics in Switzerland, including (1) population density, (2) border length, (3) the number of new immigrants, (4) the number of murders, (5) the number or rapes, and (6) the number of assaults. The participants were split randomly into 12 experimental sessions, each consisting of 12 participants. Each question was repeatedly answered in five rounds (time periods). The questions were designed in such a way that individuals were not likely to know the exact answer, but could still have some clue. Participants did not interact with each other and the only information they received about the others' estimates was provided by experimental manipulation (no information, aggregated information, and full information) through the software interface. Individuals were also asked about their confidence levels in their first and fifth estimates for each question on a six-point Likert scale (1, very uncertain; 6, very certain). The confidence level values were not provided to the others.

Three different scenarios were tested regarding information exposure, including "no information", "aggregated information", and "full information". In the no information scenario, individuals were not aware of others' estimates and, therefore, the five consecutive estimates were made with no additional information. In the aggregated information scenario, each subject was provided with the arithmetic average of others' estimates in the previous round. Finally, in the full information scenario, individuals received a figure of the trajectories of all others' estimates along with their numerical values from the previous round. For each group, two questions were posed in each information scenario. The order of questions in each information condition was randomized.

Subjects were encouraged to answer questions precisely by offering them financial rewards. Individuals received increasing monetary payments if their estimates fell into the 40%, 20%, or 10% intervals around the truth; otherwise, they received no reward. The correct answer and rewards were disclosed at the end of the experiment to avoid giving away a priori knowledge about the right answer.

3 Modeling

In this study we design two scenarios, no information (No Info) and aggregated information (Aggregated Info), to reproduce the dynamic mechanism of changing estimates, over five estimation making rounds. No Info and Aggregated Info models are discussed in Sections 3.1 and 3.2 respectively. Parameter estimation is then presented in section 3.3.

3.1 No Info Model

We present two alternative models for the No Info scenario. In the first model, estimates are generated anchoring on the initial estimate. In the second one, estimates are generated anchoring on all previous estimates (a weighted average of those).

No Info Model I

The key hypothesis here is that people change their estimates depending on the previous estimates and some random variations, where the higher their initial confidence level is the less variation will be observed in the estimates. Estimate of individual i at round r, ($\hat{E}_i^{(r)}$), $r \in \{1, \ldots, 5\}$, is generated based on the initial estimate (E_{0_i}) and a change rate ($C_i^{(r)}$):

$$\hat{E}_i^{(r)} = E_{0_i} + C_i^{(r-1)}, \quad r \in \{1, \ldots, 5\} \tag{1}$$

$C_i^{(r)}$ is the rate at which an estimate becomes the next desired estimate (note that $C_i^{(0)} = 0$, so $\hat{E}_i^{(1)} = E_{0_i}$). $C_i^{(r)}$ is calculated as:

$$C_i^{(r)} = DE_i^{(r)} - \hat{E}_i^{(r)} \tag{2}$$

Desired estimate ($DE_i^{(r)}$) is estimated based on E_{0_i} and some variations ($V_i^{(r)}$):

$$DE_i^{(r)} = E_{0_i} V_i^{(r)} \tag{3}$$

$V_i^{(r)}$ is then generated using a lognormal random variable to change the next desired estimate at each round, $V_i^{(r)} = exp(\xi)$ where $\xi \sim \mathcal{N}(0, \hat{\sigma}_i^{(r)})$. We hypothesize that $\hat{\sigma}_i^{(r)}$ is a function of individual's initial confidence level (CL_i is between one and six; it is normalized as $CL_i^* = \frac{CL_i - 1}{5}$, so $CL_i^* \in \{0, 0.2, 0.4, 0.6, 0.8, 1\}$). It can be simply assumed that $\hat{\sigma}_i^{(r)}$ is linearly changing in the range of [0,1] based on CL_i^* but the experimental data does not evidence such linearity. We use equation 4 to estimate the functional structure of $\hat{\sigma}_i^{(r)}$,

$$\hat{\sigma}_i^{(r)}(CL_i^*, \alpha, \beta, \gamma) = \gamma[\beta + (1 - CL_i^{*\alpha})(1 - \beta)], \quad \{\alpha, \beta, \gamma\} \in \mathbb{R} \tag{4}$$

α, β, and γ are control parameters (to be estimated) where: $\alpha > 0$ controls the convexity shape of the function (linear function if $\alpha = 1$); $0 \le \beta < 1$ controls the intersect of $\hat{\sigma}_i^{(r)}$ at $CL_i^* = 1$, e.g. $\left\{\hat{\sigma}_i^{(r)}(CL_i^* = 1) = 0 \middle| \beta = 0\right\}$; and $\gamma > 0$ controls the intersect of $\hat{\sigma}_i^{(r)}$ at $CL_i^* = 0$, e.g. $\left\{\hat{\sigma}_i^{(r)}(CL_i^* = 0) = 1 \middle| \gamma = 1\right\}$.

No Info Model II

In the second model, individuals take into account all previous estimates for making estimates rather than only their initial estimate as a pivot point. $\hat{E}_i^{(r)}$, estimate of individual i at round r, is a linear combination of weighted previous estimates ($E_i^{(j)}$), j^{th} previous estimate made by individual i, ($j < r$ and $j \in \{1, \ldots, 4\}$). Also, weights ($\widehat{W}^{(j)}$) are assigned to estimates of previous rounds, j^{th} previous estimate. $\hat{E}_i^{(r)}$ is calculated as shown in equation 5.

$$\hat{E}_i^{(r)} = \frac{\sum_2^r E_i^{(r-1)} \widehat{W}^{(r-1)}}{\sum_2^r \widehat{W}^{(r-1)}}, \quad r \in \{2, \ldots, 5\} \tag{5}$$

3.2 Aggregate Info Model

In this model, individuals are affected by social influence as they are given the group average ($GA^{(r)} = \frac{1}{12}\sum_{i=1}^{12} E_i^{(r)}$) of the previous round. The main idea is that individuals make estimates based on a weighted average of their own estimates and the group average. Individuals also consider the group average based on two variables: degree of compliance DC_i (how much they rely on the opinion of others) and threshold T_i (e.g. they do not take into account the group average when it is T_i times bigger/smaller than their own estimate). The degree of compliance ($0 \leq DC_i \leq 1$), defines whether the individual is willing to follow the group estimate and is based on CL_i^*:

$$DC_i = \vartheta[\zeta + (1 - CL_i^{*\eta})(1 - \zeta)]$$
$$\text{where, } \{\eta, \zeta, \vartheta\} \in \mathbb{R} \text{ and } \eta > 0,\ 0 \leq \zeta < 1,\ 0 < \vartheta \leq 1 \tag{6}$$

Defining $E_i^{(j)}$ and $\widehat{W}^{(j)}$ as same as in the No Info model II, $\hat{E}_i^{(r)}$ is calculated as:

$$\hat{E}_i^{(r)} = DC_i * GA^{(r-1)} + (1 - DC_i) * \frac{\sum_2^r E_i^{(r-1)} \widehat{W}^{(r-1)}}{\sum_2^r \widehat{W}^{(r-1)}} \tag{7}$$

We design five hypotheses to study the structural form of DC_i and T_i on estimates as: H_1: fixed DC_i without T_i, H_2: fixed DC_i and fixed T_i, H_3: fixed DC_i and variable T_i ($T_i = \rho + \omega CL_i^*, \{\rho, \omega\} \in \mathbb{R}$), H_4: variable DC_i and fixed T_i, H_5: variable DC_i and variable T_i. The results of F-tests show that H_5 has less error and variance. Note that in equation (7), $DC_i = 0$ if the group average is T_i times bigger/smaller than individual's estimate.

3.3 Parameter Estimation

No Information Model

The maximum likelihood estimation is used to estimate parameters so that the model optimally fits the experimental data. The results of the parameters' estimation are shown in Table 1. Comparison of the two models indicates that although they do not significantly differ in variability, model II provides a better fit with the experimental data.

Table 1. Parameters estimated in No Info models

No Info, model I		**No Info, model II**	
Parameter	Estimation (95% CI)	Parameter	Estimation (95% CI)
α	2.07 (2.05-2.09)	$W^{(2)*}$	0.33 (0.27-0.40)
β	0.80 (0.75-0.91)	$W^{(3)}$	0
γ	1.06 (1.05-1.07)	$W^{(4)}$	0

* $W^{(j)}$ is weight of jth previous estimate. $W^{(1)} = 1$.

Aggregated Information Model
Parameters are estimated using the maximum likelihood estimation. Given estimated ρ and ω as 5 and 3 respectively, individuals' threshold T_i to follow the group average varies between 5 and 8 with respect to their confidence level. Table 2 shows parameters estimated in the Aggregated Info model.

Table 2. Parameters estimated in Aggregated Info model

Aggregated Info			
$\widehat{W}^{(j)}$		DC_i parameters	
Parameter	Estimation (95% CI)	Parameter	Estimation (95% CI)
$W^{(2)}$*	0.21 (0.14-0.22)	η	1.25 (0.59-1.90)
$W^{(3)}$	0	ϑ	0.54 (0.51-0.60)
$W^{(4)}$	0	ζ	0

* $W^{(j)}$ is weight of jth previous estimate. $W^{(1)} = 1$

4 Discussion

In this study we aim to understand how individuals weigh previous estimates of their own in presence and absence of social influence. We first modeled the No Info scenario where individuals were not aware of others' estimates. Two models were tested for this scenario—in model I individuals rely only on their initial estimate while in model II they take into account all previous estimates rather than only their initial estimate. Our analysis shows that Model II provides a better fit with the experimental data. The main difference between the No Info model II and the Aggregated Info model is that in the Aggregated Info model each subject was provided with the arithmetic average of others' estimates in the previous round, in other words, individuals were affected by social influence.

We show that in both scenarios, the initial estimate has no significant influence on the final estimate, which is not consistent with Mannes's [21] findings. One reason for this could be the best guess efforts are the most updated ones for individuals and not their initial ones. Another reason could be that individuals do not know the true value and as they also do not receive any feedback about others' estimates, they change their ideas based on their more recent thoughts.

Our results also show that when individuals are not affected by social influence they use one and two previous estimates to generate new estimates, where the two previous estimate has lower weight, 0.33 (0.27-0.40). When they are affected by social influence, they still use only one and two previous estimates and the two previous estimate has lower weight 0.21 (0.14-0.22). Comparison between those situations reveals that the weight of the two previous estimate is lower when individuals are given the group average—they tend to make a combination of their own estimates and of the others. To do so individuals consider two variables, threshold and degree of compliance as functions of their confidence level, to include estimates of others in their next estimate. Our analysis reveals that individuals have a threshold of about 6.5

(they do not take into account the group average when it is 6.5 times bigger or smaller than their own estimate), to consider the group average as significant. This result is in close accordance with Yaniv's study [20].

One of our major limitations in this study is the small sample size of the experimental data. The precision and accuracy of our estimates can be potentially improved as larger data are analyzed. Future research can also apply our models in different settings to test the robustness of the findings. Analysis of inconsistencies of research findings such as the influence of initial estimate on the final estimate could be also another research work.

References

1. Lorenz, J., et al.: How social influence can undermine the wisdom of crowd effect. Proceedings of the National Academy of Sciences of the United States of America 108(22), 9020–9025 (2011)
2. Budescu, D.V., Rantilla, A.K.: Confidence in aggregation of expert opinions. Acta Psychologica 104(3), 371–398 (2000)
3. Soll, J.B., Mannes, A.E.: Judgmental aggregation strategies depend on whether the self is involved. International Journal of Forecasting 27(1), 81–102 (2011)
4. Rauhut, H., Lorenz, J.: The wisdom of crowds in one mind: How individuals can simulate the knowledge of diverse societies to reach better decisions. Journal of Mathematical Psychology 55(2), 191–197 (2011)
5. Herzog, S.M., Hertwig, R.: The Wisdom of Many in One Mind: Improving Individual Judgments With Dialectical Bootstrapping. Psychological Science 20(2), 231–237 (2009)
6. Yaniv, I.: The benefit of additional opinions. Current Directions in Psychological Science 13(2), 75–78 (2004)
7. Yaniv, I.: Receiving other people's advice: Influence and benefit. Organizational Behavior and Human Decision Processes 93(1), 1–13 (2004)
8. Larrick, R.P., Soll, J.B.: Intuitions about combining opinions: Misappreciation of the averaging principle. Management Science 52(1), 111–127 (2006)
9. Soll, J.B., Larrick, R.P.: Strategies for Revising Judgment: How (and How Well) People Use Others' Opinions. Journal of Experimental Psychology-Learning Memory and Cognition 35(3), 780–805 (2009)
10. Wright, G., Rowe, G.: Group-based judgmental forecasting: An integration of extant knowledge and the development of priorities for a new research agenda. International Journal of Forecasting 27(1), 1–13 (2011)
11. Lee, M.D., Zhang, S., Shi, J.: The wisdom of the crowd playing The Price Is Right. Memory & Cognition 39(5), 914–923 (2011)
12. Surowiecki, J.: The wisdom of crowds: why the many are smarter than the few and how collective wisdom shapes business, economies, societies, and nations, 1st edn., vol. xxi, p. 296. Doubleday, New York (2004)
13. Bonaccio, S., Dalal, R.S.: Advice taking and decision-making: An integrative literature review, and implications for the organizational sciences. Organizational Behavior and Human Decision Processes 101(2), 127–151 (2006)

14. Hourihan, K.L., Benjamin, A.S.: Smaller Is Better (When Sampling From the Crowd Within): Low Memory-Span Individuals Benefit More From Multiple Opportunities for Estimation. Journal of Experimental Psychology-Learning Memory and Cognition 36(4), 1068–1074 (2010)
15. Hogarth, R.M.: Note on Aggregating Opinions. Organizational Behavior and Human Performance 21(1), 40–46 (1978)
16. Gino, F., Moore, D.A.: Effects of task difficulty on use of advice. Journal of Behavioral Decision Making 20(1), 21–35 (2007)
17. Krueger, J.I.: Return of the ego–Self-referent information as a filter for social prediction: Comment on Karniol (2003)
18. Yaniv, I., Kleinberger, E.: Advice taking in decision making: Egocentric discounting and reputation formation. Organizational Behavior and Human Decision Processes 83(2), 260–281 (2000)
19. Harvey, N., Harries, C.: Effects of judges' forecasting on their later combination of forecasts for the same outcomes. International Journal of Forecasting 20(3), 391–409 (2004)
20. Yaniv, I., Milyavsky, M.: Using advice from multiple sources to revise and improve judgments. Organizational Behavior and Human Decision Processes 103(1), 104–120 (2007)
21. Mannes, A.E.: Are We Wise About the Wisdom of Crowds? The Use of Group Judgments in Belief Revision. Management Science 55(8), 1267–1279 (2009)
22. Reed-Tsochas, F., Onnela, J.P.: Spontaneous emergence of social influence in online systems. Proceedings of the National Academy of Sciences of the United States of America 107(43), 18375–18380 (2010)

Two 1%s Don't Make a Whole: Comparing Simultaneous Samples from Twitter's Streaming API

Kenneth Joseph, Peter M. Landwehr, and Kathleen M. Carley

Carnegie Mellon University, Pittsburgh, PA, USA
{kjoseph,plandweh,kathleen.carley}@cs.cmu.edu

Abstract. We compare samples of tweets from the Twitter Streaming API constructed from different connections that tracked the same popular keywords at the same time. We find that on average, over 96% of the tweets seen in one sample are seen in all others. Those tweets found only in a subset of samples do not significantly differ from tweets found in all samples in terms of user popularity or tweet structure. We conclude they are likely the result of a technical artifact rather than any systematic bias.

Practically, our results show that an infinite number of Streaming API samples are necessary to collect "most" of the tweets containing a popular keyword, and that findings from one sample from the Streaming API are likely to hold for all samples that could have been taken. Methodologically, our approach is extendible to other types of social media data beyond Twitter.

1 Introduction

A common method for collecting data from Twitter is to provide a set of keywords representative of a current event or trend to the "Streaming API"[1]. The Streaming API provides only a portion of the tweets matching the proscribed keywords[2], but delivers these messages in near-real time.

The Streaming API provides "enough" data for most analyses. However, situations do arise where this is not the case. For example, this limit is generally assumed to be too low when the research is aimed at testing data-hungry algorithms for pattern identification [1]. Additionally, in disaster situations, a given sample from the Streaming API may only contain peripheral chatter from the greater Twitter-sphere, therefore missing the relatively small number of tweets sent by victims. Recent work also suggests that even where a researcher does not want more data, her sample from the Streaming API may be a biased representation of the full data [2]. Consequently, datasets that have been pulled from the

[1] `https://dev.twitter.com/docs/streaming-apis`

[2] We find it provides on the order of 50 tweets per second, though the more common assumption is that it provides no more than 1% of the entire volume of *all* tweets sent within a particular interval.

W.G. Kennedy, N. Agarwal, and S.J. Yang (Eds.): SBP 2014, LNCS 8393, pp. 75–83, 2014.

Streaming API and analyzed as representative of the entire collection of relevant tweets may lead to inaccurate conclusions.

The simplest solution to these issues is to get the full set of tweets pertaining to a given keyword set via the "Twitter Firehose". However, Firehose access is often prohibitively expensive. Consequently, the most popular way to access Twitter data is to use the Streaming API and ignore or design around these limitations [3]. As boyd and Crawford [4, pg. 669] note, however, "[i]t is not clear what tweets are included in... different data streams... Without knowing, it is difficult for researchers to make claims about the quality of the data that they are analyzing". Our work addresses three important open questions in this area:

- *RQ1:* How different are Streaming API samples from others taken at the same time tracking the same keywords?
- *RQ2:* Can one obtain more tweets by employing multiple Streaming API samples?
- *RQ3:* If Streaming API samples are different, do the features of tweets shared across samples differ from those that are not?

To address these issues, we use a pool of five connections to the Streaming API to track the same popular keywords at the same time. We repeat this process several times with different terms to test the robustness of our findings. With respect to *RQ1*, on average over 96% of the tweets captured by any sample are captured by all samples. This differs significantly from what is expected under uniform sampling of the full set of relevant tweets, a quantity we derive. Thus, it appears that Twitter provides *nearly* the same sample to all Streaming API connections tracking the same term at the same time.

Given the magnitude of the overlap across samples, it is not surprising that with respect to *RQ2*, a practitioner would need nearly an infinite number of Streaming API samples to capture the full set of tweets on a popular topic. However, this does not rule out the possibility that the small percentage of tweets unique to each sample are somehow different from those seen by all. To this end, and with respect to *RQ3*, we compare tweets found by different subsets of our five connections. Across metrics covering user popularity and tweet structure, we find no practically interesting differences in tweets seen in different numbers of samples. Rather, the only difference observed relates to the time tweets are delivered within sub-intervals of the sampling period (described below). We take this as evidence that Streaming API samples are slightly different only because of a technical artifact in how Twitter constructs samples on the fly.

Our results have two important practical implications. First, we show that research conducted on a single sample of the Streaming API is likely to generalize to any other Streaming API sample taken at the same time tracking the same terms. Second, we show that if one desires more data than the Streaming API provides and is not willing to pay for it, trying to obtain more tweets using more Streaming API connections on the same keyword is not a feasible solution. Beyond these practical implications, the methodology used here is applicable to similar questions regarding the quality and quantity of data obtained via other social media APIs.

2 Related Work

Geo-spatial filters, network-based sampling algorithms [5] and user-based sampling [6], amongst several other approaches, have all been used to capture data from Twitter. Interest has also increased recently in *how* to perform sampling effectively under the constraints imposed by Twitter [7,3], leading to innovative solutions for capturing more complete data. However, it still appears that the most common method for extracting data from Twitter is simply to specify a set of manually-defined keywords to the Streaming API and capture the resulting messages. Thus, we focus on this sampling methodology here.

There are only two articles we are aware of that explicitly consider bias across samples on Twitter. In [8], the authors compare a sample from the Streaming API to one collected from the Search API[3], stating "[t]he alternative to comparing samples to the full stream of information is to compare the two available API specifications: streaming and search". This characterization of the options available for study is not, in our opinion, complete. In understanding biases that may exist across APIs, one must first understand potential biases of each API individually, as is done here. More recently, [2] found that the set of tweets returned from a Streaming API sample provided aggregate network, topic and hashtag based metrics that did not comply with those computed on the full set of tweets matching the proscribed parameters that were sent during the sampled same time period. Our work complements their efforts by showing that findings would almost surely have extended to any sample from the Streaming API that might have been taken at that time.

3 Methodology

Answering our three research questions required that we draw multiple, simultaneous samples of tweets from the Streaming API using identical keywords as search terms. Because drawing each sample from the API required a unique Twitter user, we obtained access to five accounts for the purposes of this experiment. We used the Twitter API as of November 15, 2013 and carried out all connections to it using Hosebird[4], Twitter's open-source library for accessing the Streaming API.

We ran all five Streaming API connections simultaneously for two hours for each of the configurations listed in Table 1. Each configuration was run twice for a total of fourteen independent sampling periods. The configurations tested include adding a non-sensical term ("thisisanonsenseterm") to each connection, adding different terms (specifically, the name of the Twitter account for the given connection) to each connection, using multiple keywords as opposed to just one and staggering the starting time of the different connections by .15 seconds to alter sampling intervals. While one would never be interested in the keyword sets shown in Table 1, ongoing work on disasters suggests samples taken here

[3] `https://dev.twitter.com/docs/api/1.1`

[4] `https://github.com/twitter/hbc`

Table 1. The configurations tested in our experiment. Note each is run twice. All runs are in a separate two-hour period with all five connections.

Keywords	Staggered	Keywords	Staggered	Keywords	Staggered
the	no	the	yes	i	no
the, thisisanonsenseterm	no	be	no	the, i, be	no
the, {user name}	no				

are of comparable size to collections of disaster-related tweets collected in the few hours after a natural disaster.

After generating our samples and before analysis, we checked to ensure that there were no disconnections from the API during sampling (occasionally, Twitter will disconnect users from the API and make them re-connect). This information, along with all code and public data for the present work, can be found at https://github.com/kennyjoseph/sbp_14.

4 Results

Table 2 shows the variables we consider throughout this section- note that each variable is defined for each individual run configuration. Across thirteen of the fourteen configurations, n was almost exactly 367K for all $c \in C$. Indeed, averaged across all configurations, $\sigma_n \approx 7$ tweets. For the theoretical derivations that follow, we thus use the simplifying assumption that n is constant across all c. The one configuration in which n was not close to 367K was one of the two configurations using the term "be", where we captured only around 335K tweets. This presumably occurred because the sample was taken early in the morning (EST), when the volume of English tweets was low.

Table 2. Variables used in this section

T	The set of all tweets in the interval *matching any keyword being tracked*
C	The set of samples of T obtained via our Streaming API connections
$L_{t,c}$	Random variable denoting the likelihood that a tweet $t \in T$ is in any $c \in C$
X_t	Random variable denoting the number of samples in C in which we see tweet t
n	The size of each sample in C
σ_n	The standard deviation of n across C

Values of $\frac{\bar{n}}{|T|}$, the mean percentage of all relevant tweets captured in a given configuration, ranged from .06-.30 across all runs except for this outlier, where the value raises to .94. The size of T is determined by making use of the "limit notice"[5] from the Streaming API, a property not available in previous studies [2,8,3] and one that consequently warrants mention. Approximately once per second, the Streaming API sends a limit notice (instead of a tweet) detailing how many tweets matching the given keywords have been skipped because of rate limiting since the connection began.

[5] `https://dev.twitter.com/docs/streaming-apis/messages#Limit_notices_limit`

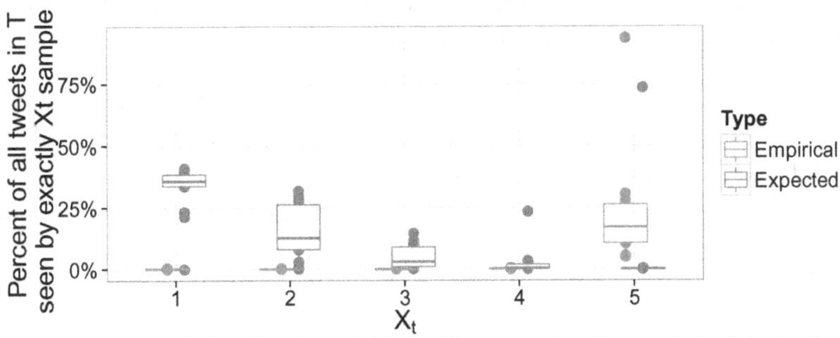

Fig. 1. The empirical distribution of X_t (red) versus the theoretical distribution (cyan) presented as box plots of values for the different configurations

4.1 RQ1: How Different Are Streaming API Samples from Others Taken at the Same Time?

One way to understand how different samples are from each other is to compare to a theoretical baseline of how different we would expect them to be if all c provided an independent, uniform sample of T. Under this assumption, it follows that $\forall t, c$, $L_{t,c} \sim Bernoulli(\frac{n}{|T|})$. That is, the likelihood we get any tweet in T in a given c is equal to the number of tweets in the sample divided by the number of all tweets matching our keywords. Since $X_t = \sum_{c \in C} L_{t,c}$, we can say that $\forall t$, $X_t \sim Binomial(|C|, \frac{n}{|T|})$. For all of the values of $\frac{n}{|T|}$ except for our one outlier, we would thus expect the probability of a tweet appearing in exactly X_t samples to decrease rapidly as X_t increases.

Figure 1 shows the distribution governing X_t as two sets of box plots[6]. Cyan-colored box plots depict the range of possible theoretical values taken on across run configurations, while red denotes the same information for the empirical data. Data falling outside the inter-quartile range depicted by the box plots are shown as points. The theoretical values for each configuration are determined by via the binomial distribution, where the second parameter is set to $\frac{|C|}{5|T|}$, the mean size of the five samples. Because the estimated shape of the distribution was different for each configuration, box plots are used for the theoretical distribution as opposed to showing a single probability density function.

As is clear, the empirical data is not binomially distributed. Over 96% of the tweets found by any c were found by all c, something that would occur on average less than .000001% of the time if sampling were to occur randomly. Even in the most extreme outlier case described above, where $n \approx |T|$ and we would thus expect "most" tweets to be seen by all samples, odds of this were still much higher in the real data than would be theoretically expected under uniform sampling. This shows the value in comparing to a theoretically derived result and perhaps best of all indicates that one should expect samples taken at the same time on the same terms to be nearly identical, regardless of the size of T.

[6] $X_t = 0$, the likelihood that no c observes t, is not shown.

4.2 RQ2: Can One Obtain More Tweets Simply by Employing Multiple Streaming API Samples?

Given that Twitter makes money selling their data, it is not particularly surprising that Streaming API samples are almost identical. Again, a theoretical comparison is of use to prove this point. Equation 1 below derives $\mathbb{E}[\frac{\sum_{t \in T} I_{\neq 0}(X_t)}{|T|}]$, the expected proportion of tweets in T that we will get with $|C|$ samples, under the assumption of random sampling. Note that in this quantity, $I_{\neq 0}(X_t)$ is the indicator function that is 1 if and only if $X_t > 0$ and is 0 otherwise. Line 1 of Equation 1 uses the law of iterated expectations. Line 2 uses the fact that $\mathbb{E}[I_{\neq 0}(X_i)] = P(X_i \neq 0)$. Line 3 substitutes in the value for $P(X_t = 0)$, which equals the odds that a single c does not see a given tweet raised to the $|C|^{th}$ power.

$$\begin{aligned}
\mathbb{E}[\frac{\sum_{t \in T} I_{\neq 0}(X_t)}{|T|}] &= \frac{\sum_{t \in T} \mathbb{E}[I_{\neq 0}(X_t)]}{|T|} \\
&= \frac{\sum_{t \in T} P(X_t \neq 0)}{|T|} = \frac{\sum_{t \in T} 1 - P(X_t = 0)}{|T|} \\
&= \frac{\sum_{t \in T}(1 - (1 - \frac{n}{|T|})^{|C|})}{|T|} = 1 - (1 - \frac{n}{|T|})^{|C|} \qquad (1)
\end{aligned}$$

Using the result of Equation 1, we can compute that under random sampling we would need only 12 connections to capture more than 95% of the full stream when $\frac{n}{|T|} \approx .23$, the average across all configurations. Alternatively, we can use empirical data to find $n + |T| * \frac{\mathbb{E}[P(X_t=1)]}{|C|}$, the expected number of unique tweets we will get when $|C| > 1$. Using this empirical estimate, one million connections to the Streaming API would provide us with only approximately 25% of the full data. Thus, the answer to *RQ2* for all practical situations is simply "no".

4.3 RQ3: Do the Features of Tweets Shared Across Samples Differ from Those That Are Not?

RQ3 asked how tweets unique to a subset of C differ from those observed in all samples in C. While there are only a limited number of these tweets, systematic differences between them and tweets seen in all samples could still bias analyses. Figure 2a shows summary statistics with 95% confidence intervals for the number of hashtags, URLs and mentions per tweet and the logarithm of follower and followee counts of users for tweets having different values of X_t (across all runs). While differences are statistically significant, they are so small in magnitude in each metric that they are not of practical interest. In addition, values show no obvious pattern across X_t that would indicate a systematic bias in which tweets are sent to which number of Streaming API connections.

Further support for the lack of such a systematic bias comes from Figure 2b, which plots histograms of the "position" metric calculated for each tweet for each value of X_t. The position metric specifies the number of tweets after a limit

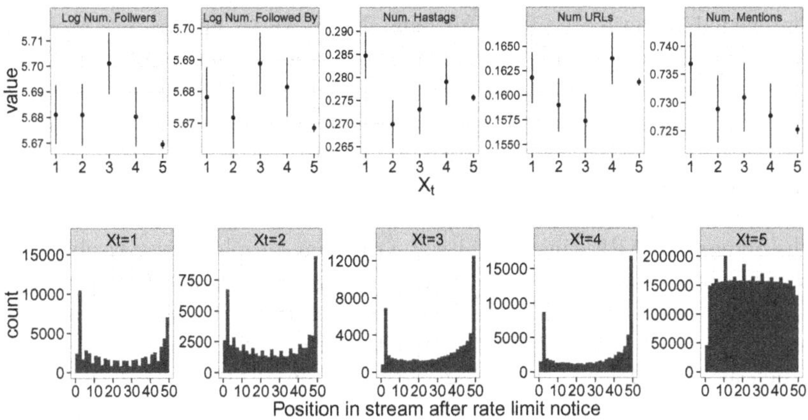

Fig. 2. a) Top row; Comparison of metrics for tweets seen by a different number of connections (each metric is a different plot). 95% confidence intervals are given using the standard error and assuming normality b) Bottom row; histogram for the position metric for all tweets with a given value of X_t. Each value of X_t is a different plot.

notice before each tweet is seen. Thus, for example, tweets with a position metric of 1 were the first tweet to be seen after a limit notice in a particular connection. Figure 2b shows the position metric for the range 0-50, which encompasses 96% of the tweets received. As we can see, tweets seen by all samples are almost equally likely to be seen any time after a limit notice. In contrast, tweets seen by a subset of C are disproportionately likely to be seen right before or after a limit notice.

This observation holds when ignoring tweets sent in the first and last few minutes of the overall sampling period, showing that the observation is unrelated to start-up or shutdown time differences across streams. While outside of the range of 0-50, distributions for the metric are much more similar across values of X_t, differences within this range suggest that tweets seen by a particular subset of C may simply be a technical artifact in how Twitter constructs samples for the Streaming API between rate limit notices.

5 Conclusion

The present work gives evidence that Streaming API samples are not a uniform sample of all relevant tweets- rather, Twitter's technological infrastructure includes the capacity to send all connections tracking the same keywords approximately the same result. Because of this, using a larger number of Streaming API connections to track a particular keyword will not significantly increase the number of tweets collected. Other sampling methodologies, like user-based approaches [9,6], may therefore be vital if one is to capture an increased number of tweets without resorting to purchasing data. Unfortunately, future work is needed to better understand what biases these approaches themselves introduce, and how much more data they really allow one to obtain. We also show that the few tweets unique to particular samples from the Streaming API are similar at a practical level to those seen in all samples. This holds across metrics associated with user popularity and several measures of tweet structure.

Differences between Streaming API samples consequently appear to be both slight and uninteresting. However, our work should not be taken as an indication that the Streaming API is in and of itself a random sample- it is entirely possible that Twitter holds out tweets from *all* Streaming API samples. Our findings are also restricted in that we run all samples on a single IP in a single location, use a very particular set of keywords and do not test other features of the Streaming API, such as searching via bounding boxes. While Twitter has stated that IP restrictions are not used and we expect our work to extend to other approaches to using the Streaming API, future work is still needed.

Regardless, efforts here and those we have built on lead the way to interesting future work on providing error bars for data from Twitter and sites with similar rate limiting techniques, for example on network metrics and the number of "needles in a haystack" we are likely to miss in disaster scenarios. As Twitter is more likely to further restrict their data than to provide more of it for free, such research is critical in our understanding of findings resulting from this increasingly popular social media site [1].

References

1. National Research Council: Frontiers in Massive Data Analysis. The National Academies Press (2013)
2. Morstatter, F., Pfeffer, J., Liu, H., Carley, K.M.: Is the sample good enough? comparing data from twitter's streaming API with twitter's firehose. In: The 7th International Conference on Weblogs and Social Media (ICWSM 2013), Boston, MA (2013)
3. Li, R., Wang, S., Chen-Chuan, K.: Towards social data platform: Automatic topic-focused monitor for twitter stream. Proceedings of the VLDB Endowment 6(14) (2013)
4. Boyd, D., Crawford, K.: Critical questions for big data: Provocations for a cultural, technological, and scholarly phenomenon. Information, Communication & Society 15(5), 662–679 (2012)
5. Wu, S., Hofman, J.M., Mason, W.A., Watts, D.J.: Who says what to whom on twitter. In: Proceedings of the 20th International Conference on World Wide Web, WWW 2011, pp. 705–714. ACM, New York (2011)
6. Vieweg, S., Hughes, A.L., Starbird, K., Palen, L.: Microblogging during two natural hazards events: what twitter contribute to situational awareness. In: Proceedings of the 28th International Conference on Human Factors in Computing Systems, CHI 2010, pp. 1079–1088. ACM, New York (2010)

7. Ghosh, S., Zafar, M.B., Bhattacharya, P., Sharma, N., Ganguly, N., Gummadi, K.P.: On sampling the wisdom of crowds: Random vs. expert sampling of the twitter stream. In: CIKM (2013)
8. González-Bailón, S., Wang, N., Rivero, A., Borge-Holthoefer, J., Moreno, Y.: Assessing the bias in communication networks sampled from twitter. Available at SSRN (2012)
9. Bakshy, E., Hofman, J.M., Mason, W.A., Watts, D.J.: Everyone's an influencer: quantifying influence on twitter. In: Proceedings of the Fourth ACM International Conference on Web Search and Data Mining, WSDM 2011, pp. 65–74. ACM, New York (2011)

Estimating Social Network Structure and Propagation Dynamics for an Infectious Disease

Louis Kim[1,2], Mark Abramson[1], Kimon Drakopoulos[2], Stephan Kolitz[1], and Asu Ozdaglar[2]

[1] Draper Laboratory, Cambridge, MA 02139, USA
{Lkim,kolitz,mabramson}@draper.com
[2] Massachusetts Institute of Technology, Cambridge, MA 02139, USA
{Kimondr,asuman}@mit.edu

Abstract. The ability to learn network structure characteristics and disease dynamic parameters improves the predictive power of epidemic models, the understanding of disease propagation processes and the development of efficient curing and vaccination policies. This paper presents a parameter estimation method that learns network characteristics and disease dynamics from our estimated infection curve. We apply the method to data collected during the 2009 H1N1 epidemic and show that the best-fit model, among a family of graphs, admits a scale-free network. This finding implies that random vaccination alone will not efficiently halt the spread of influenza, and instead vaccination and contact-reduction programs should exploit the special network structure.

Keywords: network topology, disease dynamics, parameter estimation.

1 Introduction

Many diseases spread through human populations via contact between infective individuals (those carrying the disease) and susceptible individuals (those who do not have the disease yet, but can catch it) [1]. These contacts form a social network, along which disease is transmitted. Therefore, it has long been recognized that the structure of social networks plays an important role in understanding and analyzing the dynamics of disease propagation [2]. In this paper, we present an algorithm to estimate the structure of the underlying social network and the dynamics of an infectious disease. Better understanding the social network and transmission parameters will help public officials devise better strategies to prevent the spread of disease.

Many previous studies of disease propagation assume that populations are "fully mixed," meaning that an infective individual is equally likely to spread the disease to any other susceptible member of the population to which he belongs [3]. In the same line of work, Larson et al. enriched the aforementioned models by incorporating different types of agents [4]. In these works, the assumption of "full mixing" allows one to generate nonlinear differential equations to approximate the number of infective individuals as a function of time, from which the behavior of the epidemic can be studied. However, this assumption is clearly unrealistic, since most individuals have contact with only a small fraction of the total population in the real world.

W.G. Kennedy, N. Agarwal, and S.J. Yang (Eds.): SBP 2014, LNCS 8393, pp. 85–93, 2014.

Building on this insight, a number of authors have pursued theoretical work considering network implications. These models replace the "fully mixed" assumption of differential equation-based models with a population in which contacts are defined by a social network [2, 5-9]. Nonetheless, to the best of our knowledge, there hasn't been work on inferring network structure characteristics from epidemics data.

Another strand of work employed large-scale experiments to map real networks by using various sources of data such as email address books, censuses, surveys, and commercial databases. However, this often requires extensive amount of time and resources collecting, manipulating, and combining multiple data sources to capture large size networks and estimate connections within those networks [10-12, 24, 25]. In this work, we use much lower dimensional data, temporal infection data, to infer the network characteristics assuming the network follows scale-free or small-world model.

The contribution of our paper is twofold. Firstly, we develop a method to extract the network structure from the observed infection data. Specifically, our approach assumes a parameterized network model and disease process parameters to simulate expected infection curve. Then, the algorithm greedily searches for the parameter values that will generate an expected infection curve that best fits the estimated real infection curve. We demonstrate that our suggested algorithm, assuming a scale-free network, closely estimates the network characteristics and disease dynamic parameters for the 2009 H1N1 influenza pandemic. Our results confirm that the network-based model performs better in estimating the propagation dynamics for an infectious disease compared to the differential equation-based models with the "fully mixed" assumption.

Secondly, given this finding we shed light on designing efficient control policies: For example, due to high asymmetry in degree distribution for scale-free graphs, degree vaccination will be superior to random vaccination in stopping the spread of disease.

The outline of the paper is as follows. In Section 2 we introduce the disease spread model and the proposed estimation algorithm. In Section 3 we evaluate the performance of the algorithm and suggest efficient control policies to mitigate the disease spread. In Section 4 we provide our conclusions.

2 Methods

In this section, we describe a discrete-time stochastic multi-agent SIR model, and propose a corresponding inference algorithm to fit the disease dynamics generated by the model to real H1N1 infection data. The inference algorithm learns the social network structure and key disease spread parameters, such as the rate of infection and the rate of recovery, for a given infectious disease. This enables us to make useful predictions about the contact network structure and disease propagation for similar types of diseases and allows us to devise appropriate control strategies.

2.1 Data

We obtained data from state health departments, including the weekly percentage of all hospitalizations and outpatient visits resulting from influenza-like illness (%ILI) over the 2009-2010 flu seasons [14, 15]. Each point on the %ILI temporal curve represents the percentage of the total number of hospitalizations and outpatient visits that are specific to H1N1. We also obtained total estimated cases of H1N1 and total estimated H1N1 related hospitalizations from the Center for Disease Control [15]. Using these data, we estimated the number of H1N1 infections for each week as follows:

Assuming that the flu wave first grows then declines after the peak of the infection while the number of non-H1N1 hospitalizations remains relatively stable, we estimate the number of non-H1N1 hospitalizations. Finally, the above allow us to estimate the number of H1N1 related hospitalizations during each period.

Given the number of H1N1 related hospitalizations during each period and the total estimated cases of H1N1, we can estimate the number of H1N1 infections at each period, assuming that the number of H1N1 infections are proportional to the number of H1N1 related hospitalizations [4]. (Refer to the Epidemic curve estimation section of [4] for more details.)

We used the estimation method described above to estimate the infection curve for the state of Massachusetts, in which the estimated true infection curve includes the effects of vaccines as administered. Figure 1 shows the estimated temporal infection curve and the temporal curve of vaccines as administered [17].

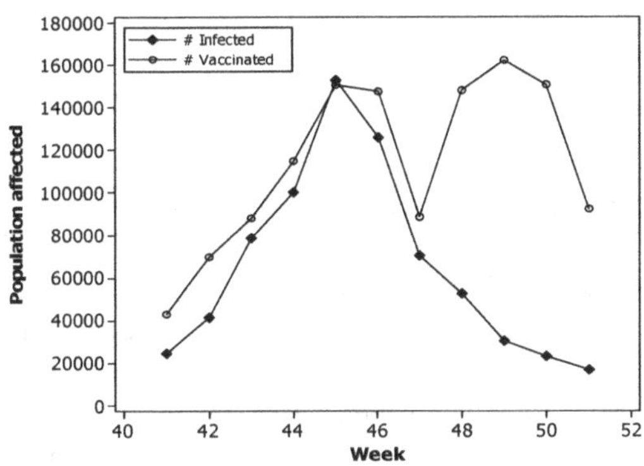

Fig. 1. The infection curve estimated from %ILI data and the number of vaccines administered during the observation period (October, 2009 – December, 2009) for Massachusetts

2.2 Disease Process

We employ a variation of susceptible-infective-removed (SIR) model first proposed by Kermack and McKendrick [13]. Individuals in the network, represented by nodes, are assigned one of the three states: the susceptible state (S) in which individuals are not infected but could become infected, the infective state (I) in which individuals are currently carrying the disease and can spread the disease to susceptible individuals upon contact, and the recovered state (R) in which individuals have either recovered from the disease and have immunity or have died. Edges connecting nodes in the network indicate contacts between individuals – contacts may occur through conversation between friends, co-workers, family members, etc. Alternatively, contacts can also occur between two strangers passing by chance. When a contact occurs between an infective individual and a susceptible individual, the susceptible will become infective with probability β. Each infective individual recovers from the disease and becomes immune with probability δ after a period of time, a week in our case.

There exists a population of individuals,V connected by a graph $G = (V, E)$. We define $X_i(t) \in \{S, I, R\}$ to be the state of individual $i \in V$ at time t. And let $\eta_i(t) = \sum_{(j\setminus i)\in E} \mathbb{I}\, X_j(t) = I$, where $\mathbb{I}$ is the indicator function, denote the number of infected neighbors the individual i has at time t. Then given that the individual i is susceptible at time t, he will become infected at time t+1 with the following probability:

$$\mathbb{P}(X_i(t+1) = I | X_i(t) = S) = 1 - (1-\beta)^{\eta_i(t)} \quad (1)$$

since with probability $(1-\beta)^{\eta_i(t)}$ all infection attempts fail. Given that the individual i is infected at time t, he will recover at time t+1 with the probability:

$$\mathbb{P}(X_i(t+1) = R | X_i(t) = I) = \delta. \quad (2)$$

We assumed independence in infection attempts between neighbors. We also assume that if a susceptible individual is vaccinated, then he or she will immediately become immune to the disease and the individual's state will change to recovered state. Given a network and set of initial infections, the disease propagation process can be simulated according to the described probabilities.

2.3 Estimation Algorithm

The estimation algorithm uses the disease process described above to simulate infections. The simulated results are compared to the real H1N1 infection data, and we optimize over the network and disease spread parameters to obtain a best-fit simulated curve. The purpose of the algorithm is to find network characteristics, such as degree distribution, and disease spread parameters, β and δ that will help us make useful predictions about the network and how the disease spreads within the community.

Many real-world social networks such as citation networks, internet and router topologies, sexual contact networks are expected to exhibit small-world or scale-free properties [9-12, 18-20]. We tested both small-world and scale-free networks in our algorithm for the contact network.

Input

The inputs to the algorithm include:

- A parameterized disease spread network structure, where nodes represent people and undirected edges represent contacts between people. In our simulations, the network structure is assumed to be either scale-free or small-world, though the algorithm could be applied to other network structures.
- Initial values of the network parameters, the average degree, k_0 and, for the small-world network structure, p_0, the probability of a long-range contact.
- Initial values of the disease process parameters β_0 and δ_0.
- Real temporal infection data to fit the model generated infection dynamics.
- Data on vaccines administered (if administered).

Output

- The algorithm outputs a simulated expected infection curve, which fits the real data as closely as possible, and the network and infection parameters used to generate the expected infection curve.

Procedure

Begin with the given initial values of the social network and disease spread parameters: k_0, p_0, β_0 and δ_0. Let Δk, Δp, $\Delta\beta$ and $\Delta\delta$ be the amounts by which k, p, β and δ are changed at each step in the optimization. Let $\hat{e}_i$ and e_i each denote the number of infections for the real infection curve and the estimated expected infection curve at time i, respectively. We define the error, E, between the simulated expected infection curve and the true infection curve as:

$$E = \sum_{i=1}^{max.period} |\hat{e}_i - e_i| \quad (3)$$

Repeat the following steps until the error can no longer be reduced by changes to the parameters (we define the optimal output parameters as k^*, p^*, β^* and δ^*):

1. Given k_0, p_0, β_0 and δ_0, search in all possible directions to find a direction that improves E. That is, evaluate E at all possible combinations of k, p, β and δ, where $k \in \{k_0, k_0 + \Delta k, k_0 - \Delta k\}$, $p \in \{p_0, p_0 + \Delta p, p_0 - \Delta p\}$, $\beta \in \{\beta_0, \beta_0 + \Delta\beta, \beta_0 - \Delta\beta\}$ and $\delta \in \{\delta_0, \delta_0 + \Delta\delta, \delta_0 - \Delta\delta\}$. Evaluate E by doing the following for each set of parameters:
   (a) Generate the network according to the given network type (small-world or scale-free) and network characteristics (k, the average degree for a scale-free network; k, the average degree and, p, the short-cut probability for a small-world network).
   (b) Simulate R realizations of the disease process. For each realization, initialize the disease simulation infection by assigning N_I nodes to the infected states, where N_I is the number of people infected at the beginning of the observation period in the data. We assume that the initial infected nodes are selected uniformly at random from among all the nodes.[1] Update the disease states for each time period, according to the disease process parameters and the vaccine administration data (We assume that those who receive vaccines at each time period are chosen uniformly at random).
   (c) Generate an expected infection curve by averaging the number of infected individuals at each time period over the R realizations of the disease process.
   (d) Calculate E.
2. Determine which search direction resulted in the minimum error. Update k_0, p_0, β_0 and δ_0 to the values of k, p, β and δ that achieved the lowest sum of residuals. The algorithm is summarized as a flow chart in Figure 2.

[1] Commonly used assumption in epidemic simulation. [25]

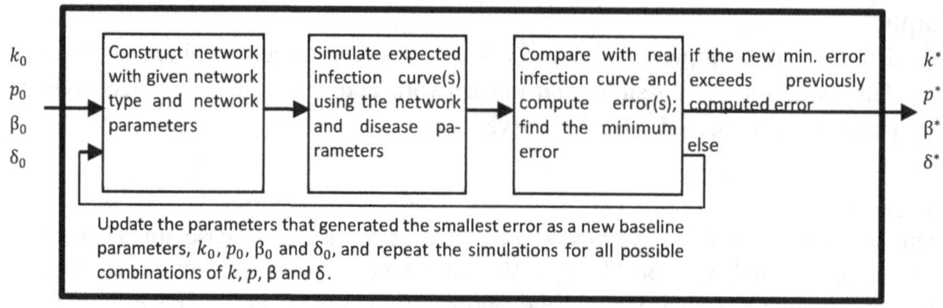

Fig. 2. Flow diagram showing the estimation algorithm details

3 Results

We have applied our algorithm to data from the 2009 H1N1 outbreak to demonstrate how our algorithm finds realistic network and infection parameters that can approximate the dynamics of an infectious disease. We scaled down the population size by a factor of 10,000 uniformly at random in order to reduce the computation time for infection simulation. In our simulations, we set R, the number of realizations per set of parameters, equal to 1,000. This effects how well the simulated curve approximate the expected curve. We began our search with relatively large values of Δk, Δp, $\Delta\beta$ and $\Delta\delta$ (changes in average degree, long-range connection probability, infection probability and recovery probability, respectively) and then manually decreased them as the sums of residuals began to converge. Specifically, initially $\Delta k = 10$, Δp, $\Delta\beta$, $\Delta\delta = 0.1$. We narrowed the search by reducing Δk to 1 and Δp, $\Delta\beta$ and $\Delta\delta$ to 0.01.

3.1 Estimation Algorithm Results

Figure 3 shows the resulting infection curves generated by the algorithm, compared to the estimated infection curve from data and from using differential equations with "fully-mixed" assumption. In addition to the error measure described above, we used the difference in total expected number of infections,

$$\left|\sum_{i=1}^{max.\ period} \hat{e}_i - \sum_{i=1}^{max.\ period} e_i\right| \quad (4)$$

and the difference in peak number of infections to compare the curves:

$$|\hat{e}_w - e_w|, \text{ where } w \text{ represents the time period of infection peak} \quad (5)$$

For the small-world network, the estimated parameter values were 8 for average degree, 0 for short-cut probability, .2 for infection probability (β), and .35 for recovery probability (δ). The error measured was close to 25 infections, which is 35% of total number of infections. Compared with the data-generated infection curve, the simulated infection curve for the small-world network had an 8% lower expected total number of infections and a 43% lower expected peak infections. Overall, the small-world network model did not provide a good fit to the estimated infection curve from data.

On the other hand, the best-fit infection curve generated using a scale-free network fits the data well. The estimated parameter values were 2 for average degree (k), .43 for infection probability (β), and .62 for recovery probability (δ). Measured error was about 9 infections, constituting 12% of the total infections. Compared to the estimated infection curve from data, we measured a 1.3% difference in the expected total number of infections and a 2.1% difference in the expected peak infections. This result is a significant improvement over the estimation under the "fully-mixed" assumption, which had a measured error of 26 infections (36% of total infections), a 25.7% difference in the expected total number of infections, and a 10% difference in the expected peak infections.

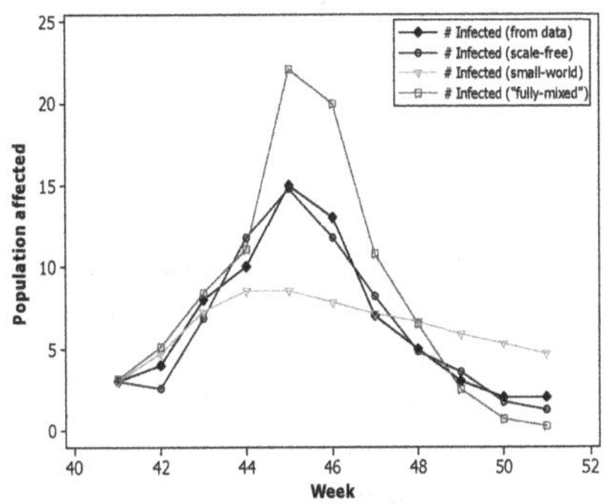

Fig. 3. Best fit curves generated by the algorithm using small-world network (green) and scale-free network (blue) compared to the estimated infection curve from data (black) and from using differential equations with "fully-mixed" assumption (red).

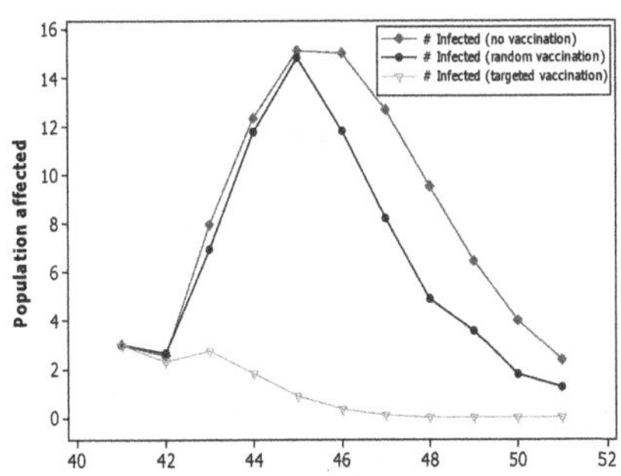

Fig. 4. A simulated curve that closely reflects the estimated infection curve from data with random vaccination scheme (blue) is compared to simulated curves with no vaccination (red) and degree-based targeted vaccination (green).

3.2 Control

Understanding the network structure and disease dynamics facilitates the adoption of efficient control measures to contain or stop the propagation of the disease on the network. Many studies have shown that, since scale-free networks have some nodes with a very large number of connections compared to the average degree, targeting those high-degree nodes to vaccinate will effectively reduce the propagation of the disease [21-23].

Now that our algorithm has successfully learned contact network and disease dynamics parameters, we can use the model to study the effect of different control methods. Figure 4 compares the disease processes with no vaccination, random vaccination, and targeted vaccination (where we selectively vaccinate those individuals with high degree). These results validate the claim that vaccinating high-degree nodes with very large connections is effective in stopping the disease propagation. Randomly vaccinating individuals reduced the expected number of infections by about 22%, whereas targeting highly connected nodes for vaccination reduced the expected number of infections by around 88%.

4 Conclusions

Understanding the network structure and the disease dynamics on the network has important implications both for refining epidemic models and for devising necessary control measures in order to effectively utilize resources to prevent the spread of disease. The spread of infection is often complex to analyze due to the lack of information about the contact network on which it occurs. This paper showcases a methodology to learn network and disease propagation parameters of an infectious disease, H1N1 influenza. The findings for H1N1 give us useful insight into the infection dynamics of similar diseases and assist in analyzing effect of different vaccination policies. We hope that this study will benefit future efforts in infection prevention.

References

1. Newman, M.E.J.: The spread of epidemic disease on networks. Phys. Rev. E 66, 16128 (2002)
2. Moore, C., Newman, M.E.J.: Epidemics and percolation in small-world networks. Phys. Rev. E 61, 5678–5682 (2000)
3. Anderson, R.M., May, R.M.: Infectious diseases of humans. Oxford University Press, Oxford (1991)
4. Larson, R.C., Teytelman, A.: Modeling the effects of H1N1 influenza vaccine distribution in the United States. Value in Health 15, 158–166 (2012)
5. Pastor-Satorras, R., Vespignani, A.: Epidemic spreading in scale-free networks. Phys. Rev. Lett. 86, 3200–3203 (2001)
6. Kuperman, M., Abramson, G.: Small world effect in an epidemiological model. Phys. Rev. Lett. 86, 2909–2912 (2001)
7. Keeling, M.J., Eames, K.T.D.: Networks and epidemic models. J. R. Soc. Interface 2, 295–307 (2005)
8. Eubank, S.: Network based models of infectious disease spread. Jpn. J. Infect. Dis. 58, S9–S13 (2005)
9. Liljeros, F., Edling, C.R., Åmaral, L.A.N., Stanley, H.E., Aberg, Y.: The web of human sexual contacts. Nature 411, 907–908 (2001)

10. Salathé, M., Kazandjieva, M., Lee, J.W., Levis, P., Feldman, M.W., Jones, J.H.: A high-resolution human contact network for infectious disease transmission. Proc. Natl Acad. Sci. USA 107, 22020–22025 (2010)
11. Dong, W., Heller, K., Pentland, A.(S.): Modeling Infection with Multi-agent Dynamics. In: Yang, S.J., Greenberg, A.M., Endsley, M. (eds.) SBP 2012. LNCS, vol. 7227, pp. 172–179. Springer, Heidelberg (2012)
12. Newman, M.E.J., Forrest, S., Balthrop, J.: Email networks and the spread of computer viruses. Phys. Rev. E 66, 035101 (2002)
13. Kermack, W., McKendrick, A.: A contribution to the mathematical theory of epidemics. Proc. R. Soc. A 115, 700–721 (1927)
14. Seasonal influenza (flu). Center for Disease Control and Prevention (2010), http://www.cdc.gov/flu/weekly/
15. CDC estimates of 2009 H1N1 influenza cases, hospitalizations and deaths in the United States, April 2009 – March 13, 2010. Center for Disease Control and Prevention (2010), http://www.cdc.gov/h1n1flu/estimates/April_March_13.htm
16. Weekly influenza update, May 27, 2010. Massachusetts Department of Public Health (2010), http://ma-publichealth.typepad.com/files/weekly_report_05_27_10.pdf
17. Table and graph of 2009 H1N1 influenza vaccine doses allocated, ordered, and shipped by project area. Center for Disease Control and Prevention (2010), http://www.cdc.gov/h1n1flu/vaccination/vaccinesupply.htm
18. Faloutsos, M., Faloutsos, P., Faloutsos, C.: On power-law relationships of the internet topology. Comput. Commun. Rev. 29, 251–262 (1999)
19. de Solla Price, D.J.: Networks of scientific papers. Science 149, 510–515 (1965)
20. Newman, M.E.J.: Networks: an introduction. Oxford University Press (2010)
21. Albert, R., Jeong, H., Barabasi, A.-L.: Error and attach tolerance of complex network. Nature 406, 378–382 (2000)
22. Pastor-Satorras, R., Vespignani, A.: Immunization of complex networks. Phys. Rev. E 65, 036104 (2002)
23. Dezsõ, Z., Barabási, A.-L.: Halting viruses in scale-free networks. Phys. Rev. E 65, 055103(R) (2002)
24. Barrett, C.L., Beckman, R.J., Khan, M., Kumar, V.A., Marathe, M.V., Stretz, P.E., Dutta, T., Lewis, B.: Generation and analysis of large synthetic social contact networks. In: Rossetti, M.D., Hill, R.R., Johansson, B., Dunkin, A., Ingalls, R.G. (eds.) Proceedings of the 2009 Winter Simulation Conference. IEEE Press, New York (2009)
25. Bisset, K., Chen, J., Feng, X., Kumar, V.A., Marathe, M.: EpiFast: a fast algorithm for large scale realistic epidemic simulations on distributed memory systems. In: Proceedings of the 23rd International Conference on Supercomputing (ICS), pp. 430–439 (2009)

Predicting Social Ties in Massively Multiplayer Online Games

Jina Lee and Kiran Lakkaraju

Sandia National Laboratories*
P.O. Box 969, MS 9406, Livermore, CA 97551, USA
P.O. Box 5800, MS 1327, Albuquerque, NM 87185, USA
{jlee3,klakkar}@sandia.gov

Abstract. Social media has allowed researchers to induce large social networks from easily accessible online data. However, relationships inferred from social media data may not always reflect the true underlying relationship. The main question of this work is: How does the public social network reflect the private social network? We begin to address this question by studying interactions between players in a Massively Multiplayer Online Game. We trained a number of classifiers to predict the social ties between players using data on public forum posts, private messages exchanged between players, and their relationship information. Results show that using public interaction knowledge significantly improves the prediction of social ties between two players and including a richer set of information on their relationship further improves this prediction.

1 Introduction

Social media (such as Facebook, Twitter) has allowed researchers to induce large social networks from easily accessible online data and study the dynamic social patterns among people. However, relationships inferred from social media data may not always reflect the private social tie among people. First, individuals who interact using social media understand (to a certain extent) that it is a public forum for communication, and hence may limit what they say. Secondly, the relations expressed may not represent the full set of relations individuals have. One may have friends who do not use Facebook, and thus that relationship would be missing. Thirdly, extraneous relations may be present in social media that do not occur in the real world. For instance, I may "follow" a twitter account of a celebrity, but that is not a real indicator of a relationship. We can view the social network inferred from social media as reflecting some of the relationships, but what we really want is the private, hidden social network that identifies the true bond.

* Sandia National Laboratories is a multi-program laboratory managed and operated by Sandia Corporation, a wholly owned subsidiary of Lockheed Martin Corporation, for the U.S. Department of Energy's National Nuclear Security Administration under contract DE-AC04-94AL85000.

W.G. Kennedy, N. Agarwal, and S.J. Yang (Eds.): SBP 2014, LNCS 8393, pp. 95–102, 2014.

The main question of this work is: "How does the public social network reflect the private social network?" In this paper we are interested in uncovering the covert social ties among people from their public interaction behaviors. Previous research has shown that one can predict tie strengths between two people using their communication data from social media [1]. Here we investigate whether we can find similar patterns in a Massively Multiplayer Onling Game (MMOG) by studying game players' public and private interaction data. Our data set contains public posts by players in a Usenet-like forum within the game, private messages exchanged between players, and their relationship data collected from an online game *Game X* . We hypothesize that if players communicate with each other publicly (e.g. posting on the same forum topic, referencing another player in a post, etc.), they are more likely to have a social tie such as being in the same nation or interest group (i.e. guild).

We begin by reviewing related work, followed by an overview of the online game used for this work. Section 4 outlines our data set and models trained to predict social ties among game players. Section 5 discusses the results and suggests future work.

2 Related Work

Massively multiplayer online games (MMOG) attract millions of players to a shared, virtual world. MMOGs have several advantages as a method of gathering data on human behavior and social patterns. For instance, we can gather data on a large number of people with a diverse background, and the data gathered from MMOGs are high in experimental realism [2] with rich social dynamics. In addition, MMOGs provide a unique way to observe communication and behavior of players; in-game forums and messaging data can be gathered along with player action data, which allows us to study the correlation between communication and behavior.

The main criticism against MMOG data is that in-game player behaviors and social patterns are not the same as those of real world. This is an ongoing endeavor, however, there has been a number of studies showing that in-game patterns do reflect real world patterns. For instance, Yee et. al [3] found high correlations between players' in-game behaviors and their personality traits. Castronova et. al [4] analyzed economic data from the virtual world in EverQuest2 and found that in-game economy follows real-world patterns.

As mentioned above, we are interested in predicting the covert social ties from game users' public interaction patterns. This problem is characterized as the *link prediction problem* in social network analysis, which is concerned with predicting links among nodes in a social network that may form in the future. There is a family of sub-problems within this area, including predicting the sign of the links [5], recency of links [6], and existence of links based on node proximities [7]. Most of the previous work focuses on predicting links within the same social network. Here, we investigate using links from one network (public interaction) to predict those of another network (social tie).

3 Description of Game X

Game X is a browser-based exploration game which has players acting as adventurers owning a vehicle and traveling a fictional game world. There is no winning in Game X, rather players freely explore the game world and can mine resources, trade, and conduct war. There is a concept of money within Game X, which we refer to as marks. To buy vehicles and travel in the game world players must gather marks. There is also a vibrant market-based economy within Game X.

Players can communicate with each other through in-game personal messages, public forum posts and chat rooms. Players can also denote other players as friends or as hostiles. They can take different actions, such as: move vehicle, mine resources, buy or sell resources, build vehicles, products, and factory outlets. Players can use resources to build factory outlets and create products that can be sold to other players.

Unlike other MMOG's like World of Warcraft and Everquest, Game X applies a "turn system." Every day each player gets an allotment of "turns." Every action (except communication) requires some number of turns to execute. For instance, if a player wants to move their vehicle by two tiles, this would cost, say, 10 turns. Turns can be considered a form of "energy" that players have.

Players can also engage in combats with non player characters, other players, and even market centers and factory outlets. They can modify their vehicles to include new weaponry and defensive elements. Players have "skills" that can impact their ability to attack/defend.

The world in Game X is 2D and does not wrap around. Players move from tile to tile in their vehicles. Tiles can contain resources, factory outlets, and market centers. Only one factory outlet or market center may exist on a tile.

3.1 Groups in Game X

There are four types of groups a player may belong to.

Nations. There are three nations a player may join, which are fixed and defined by the game creators. We label them A, B, and C. Nation membership is *open*; players may join or leave any nation they wish at any time. A player may also choose not to join a nation. Joining a nation provides several benefits such as: 1) access to restricted, "nation controlled" areas, 2) access to special quests, and 3) access to special vehicles and add-ons. Nations have different strengths; one nation may be better suited for weaponry, and thus have more weaponry related add-ons. Another may be suited for trading. Completing quests for a nation increases a player *stature* towards the nation, which leads to access to special vehicles and add-ons. Wars can occur between nations.

Agency. An agency can be thought of as a social category. There are two agencies, X and Y. A player can only be a part of one agency at any time. To join an agency, certain requirements need to be met, but if those are met anyone can join the agency. Certain vehicles are open to particular agencies.

Race. Player may choose their race when they create a character. Different races have strengths in certain areas, implemented as different initial levels of skill. Once chosen, race is fixed and cannot be changed. Race also determines the starting location.

Guild. Game X also allows the creation of player-led guilds. Guilds allow members to cooperate to gain physical and economic control of the game world. Guilds are comprised of a leader and a board who form policy and make decisions that impact the entire guild members. Guilds can be created by any player once they have met the experience and financial requirements. Apart from officers, there are "privileged guild members," a special set of players who are considered important, and the rest are regular members. Guilds are *closed* – players must submit an application and can be denied of membership. Once accpeted, members have access to private communication channels. Guilds have a "guild account" which can store marks from players (taken in the form of tax). These marks can be redistributed at the will of the CFO.

3.2 Communication in Game X

Game X includes 3 methods by which players can communicate with each other:

Personal Messages: An email-like system for communicating with other players, or in some cases groups of players.
Public Forum: A Usenet-like system where players can post topics and replies.
Chat: An IM-like system for players to chat with others in their guild.

Each forum post includes the name of the player who posted, an image of their avatar in the game, and their guild affiliation.

4 Experiments

In this section we investigate whether players' public interaction knowledge can help us discover the private social ties among players. To do this, we choose to train a number classifiers to predict whether two players are from the same *guild*. Guilds are mainly created to allow members to cooperate and gain physical and economic control in the game. Unlike nation membership, guild membership is closed; one has to be approved by other members in the guild to join. In the real world, guilds are analogous to special interest groups such as economic alliances. For these reasons, we think guild membership would be a good criteria for uncovering the private social ties among the game players.

4.1 Methods

We collected data from 50 days in the game (days 500-550) on pairs of players who have interacted either publicly or privately. This period has a relatively stable rate of posts per day, new players per day, and active players per day. For this work, we further defined three types of public interaction between two players as the following:

Co-posters: Co-posters of player p are all players who have posted in a topic that player p has also posted in.
Co-quoters: Co-quoters of player p are all players who have posted in a topic and quoted any of player p's posts.
Co-referencers: Co-referencers of player p are all players who have posted in a topic and referenced player p in the post.

For each pair of players who have publicly or privately interacted, we collected the following information, which constructs our dataset:

Membership Information (yes/no): whether two players are from the *same guild, same nation, same race*, or the *same agency.*
Relationship Information (yes/no): whether two players are *friends, foes*, or of the *same sex.*
Public Interaction (numerical): how many times two players have *co-posted, co-quoted*, or *co-referenced.*
Private Interaction (numerical): how many times two players have exchanged *private messages*, or got involved in *tradings* and *combats.*
Player Proximity (numerical): *a* measure of *geographical distance* between two players.

To study whether public interaction information improves prediction of the private network, we constructed different feature sets to train the classifiers. Feature Set 1 serves as the baseline, consisting of features that are easily available to all other players in the game. Feature Set 2 adds public interaction information in addition to Feature Set 1. Feature Set 3 includes a richer set of information about the players. Our hypothesis is that models trained on Feature Set 2 will predict same guild membership significantly better than models trained on Feature Set 1 and models trained on Feature Set 3 will further improve the prediction.

Feature Set 1 (Baseline): same-nation, same-race.
Feature Set 2: same-nation, same-race, co-posts, co-references, co-quotes.
Feature Set 3: same-nation, same-race, co-posts, co-references, co-quotes, same-sex, same-agency, are-foes, are-friends, num-private-msg, num-trades, num-combats, distance.

We trained three types of classifiers for each feature set using models available in R: generalized linear models (LM), boosted decision trees (BT), and support vector machines (SVM). The caret package [8] for R was used for pre-processing the data and tuning/training the models. For boosted trees, the tuning parameter 'shrinkage' was set to 0.1 and for SVMs, sigma was set to 0.3069, 0.2621, and 0.1711 for each feature set.

Because the original dataset is highly unbalanced, mostly consisting of data points where two players are not in the same guild (i.e. negative samples), a classifier trained on this data will achieve high performance by predicting that the sample belongs to the negative class. To avoid this, we balanced the data by

Table 1. Paired-samples t-tests for resampling results

Feature Set	Models	Accuracy	Precision	Recall	F-Score
Feature Set 1	Linear Classifier	.694	.673	.753	.711
	Boosted Tree	.694	.673	.753	.711
	SVM	.694	.673	.753	.711
Feature Set 2	Linear Classifier	.706	.693	.740	.716
	Boosted Tree	.709	.698	.736	.717
	SVM	.724	.685	.828	.750
Feature Set 3	Linear Classifier	.739	.697	.842	.763
	Boosted Tree	.767	.721	.873	.790
	SVM	.745	.678	.934	.786

randomly sampling the same number of negative samples and positive samples, which resulted in 65076 samples in total. 75% of the samples were used for training and 25% for testing. Training was done using a 5-fold cross validation process repeated 5 times.

4.2 Results

We use accuracy, precision, recall, and F-score to discuss the performance of the trained models. Table 1 shows the results. Comparing the results of Feature Set 1 (baseline) to Feature Set 2, we see that including public interaction information does improve predicting whether two people are from the same guild. Adding a richer set of information about the players (Feature Set 3) further improves the prediction. SVM had the highest F-Score for Feature Set 2, however decision tree outperformed SVM for Feature Set 3. Linear models generally had the poorest performance in all three feature sets. Further investigation on which features had large contributions to the training process revealed that for Feature Set 1, all three models only used *same-nation*. For Feature Set 2, *same-nation, same-race, co-reference, co-quote* were the major contributors for linear classifier and SVM, whereas *same-nation* and *co-posts* were the main contributors for Boosted Tree. For Feature Set 3, the contributing features vary among models, however *num-trades* was the most contributing feature for all three models.

To better compare the models, we look at their resampling distributions with the area under ROC curves as a metric. Since we used a 5-fold cross validation method repeated 5 times, we have 25 resampling measurements for each model. Figure 1 plots the resampling results. A paired two-sample t-tests were also conducted to see whether these differences are statistically significant, which is shown in Table 2. The upper right hand side of the table shows the mean differences between a pair of models and the lower left hand side shows the p-values. The bottom three rows show that for each model type, the models trained on different feature sets are statistically different from each other (e.g. SVM1 vs. SVM2 vs. SVM3). The results shown in Table 1 and Table 2 let us conclude that using public interaction information significantly improves the prediction

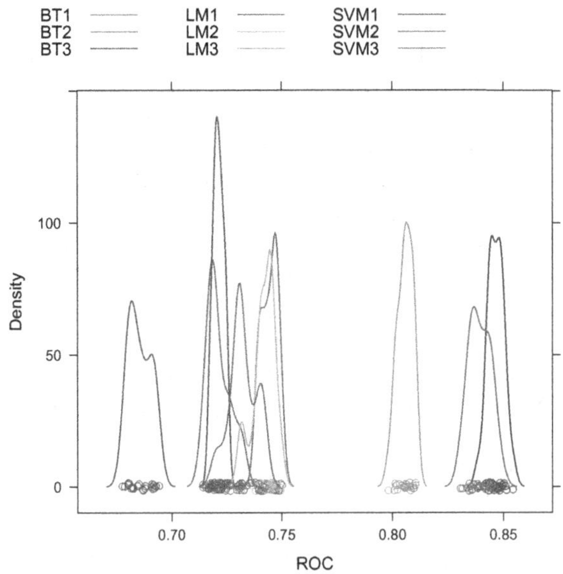

Fig. 1. Density distribution plot of model resampling

Table 2. T-tests for the resampling results (LM: Linear Model, BT: Boosted Tree, SVM: Support Vector Machine)

	LM1	BT1	SVM1	LM2	BT2	SVM2	LM3	BT3	SVM3
LM1		0.000	0.036	-0.021	-0.023	-0.011	-0.084	-0.125	-0.085
BT1	1.000		0.036	-0.021	-0.023	-0.011	-0.084	-0.125	-0.085
SVM1	<2.2E-16	<2.2E-16		-0.056	-0.059	-0.047	-0.120	-0.161	-0.121
LM2	0.000	0.000	<2.2E-16		-0.003	0.010	-0.064	-0.105	-0.064
BT2	<2.2E-16	0.000	<2.2E-16	1.000		0.012	-0.061	-0.102	-0.062
SVM2	0.000	0.000	<2.2E-16	0.000	0.000		-0.073	-0.114	-0.074
LM3	<2.2E-16	<2.2E-16	<2.2E-16	<2.2E-16	<2.2E-16	<2.2E-16		-0.041	0.000
BT3	<2.2E-16	<2.2E-16	<2.2E-16	<2.2E-16	<2.2E-16	<2.2E-16	<2.2E-16		0.041
SVM3	<2.2E-16	<2.2E-16	<2.2E-16	<2.2E-16	<2.2E-16	<2.2E-16	<2.2E-16	1.000	

of whether two people are from the same guild, thus supporting our hypothesis. Results also show that using a richer set of information about the players further improves the prediction.

5 Conclusions and Future Work

The accessibility and abundance of data generated from various social media and online games have fostered active research studies for understanding social dynamics. However, oftentimes these data are limited to constructing public networks among users, which may or may not reflect the private hidden networks that represent the true relationship.

In this paper we address the question of whether we can use public social interaction data to uncover the private social ties using data taken from a Massively Multiplayer Online Game. To answer this, we trained models that classify whether two players are from the same guild using their public and private interaction data, as well as their relationship data. Results show that by using public interaction information, we can significantly improve the prediction of guild co-membership compared to only using easily available, extrinsic information such as nation or race association. Using the relationship data improves the prediction even more.

This work is the first step for uncovering private networks from public interaction data, showing that we can predict whether two people belong to the same social group. Next we would like to investigate whether we can predict the dynamics of the private interaction, such as how often two people exchange private messages or participate in trades/combats. Another extension would be to analyze the context of players' posts instead of merely looking at the frequency to better understand the relationship among players. Finally, we may analyze player characteristics to study whether there are certain interaction patterns (e.g. leaders of nation/guild may get more incoming messages) depending on these characteristics.

References

1. Gilbert, E., Karahalios, K.: Predicting tie strength with social media. In: Proceedings of the SIGCHI Conference on Human Factors in Computing Systems, CHI 2009, pp. 211–220. ACM, New York (2009)
2. Reis, H.T., Judd, C.M.: Handbook of research methods in social and personality psychology. Cambridge University Press (2000)
3. Yee, N., Ducheneaut, N., Nelson, L., LIkarish, P.: Introverted elves & conscientious gnomes: The expression of personality in world of warcraft. In: Proceedings of CHI 2011 (2011)
4. Castronova, E., Williams, D., Shen, C., Ratan, R., Xiong, L., Huang, Y., Keegan, B.: As real as real? macroeconomic behavior in a large-scale virtual world. New Media and Society 11(5), 685–707 (2009)
5. Leskovec, J., Huttenlocher, D., Kleinberg, J.: Predicting positive and negative links in online social networks. In: Proceedings of the 19th International Conference on World Wide Web, WWW 2010, pp. 641–650. ACM, New York (2010)
6. Chen, H.-H., Gou, L., Zhang, X.(L.), Giles, C.L.: Predicting recent links in FOAF networks. In: Yang, S.J., Greenberg, A.M., Endsley, M. (eds.) SBP 2012. LNCS, vol. 7227, pp. 156–163. Springer, Heidelberg (2012)
7. Liben-Nowell, D., Kleinberg, J.: The link-prediction problem for social networks. Journal of American Society for Information Science and Technology 58(7), 1019–1031 (2007)
8. Kuhn, M.: Building predictive models in R using the caret package. Journal of Statistical Software 28(5), 1–26 (2008)

Winning by Following the Winners: Mining the Behaviour of Stock Market Experts in Social Media

Wenhui Liao[1], Sameena Shah[1], and Masoud Makrehchi[2]

[1] Thomson Reuters, USA
{wenhui.liao,sameena.shah}@thomsonreuters.com
[2] University of Ontario Institute of Technology, Canada
masoud.makrehchi@uoit.ca

Abstract. We propose a novel yet simple method for creating a stock market trading strategy by following successful stock market expert in social media. The problem of "how and where to invest" is translated into "who to follow in my investment". In other words, looking for stock market investment strategy is converted into stock market expert search. Fortunately, many stock market experts are active in social media and openly express their opinions about market. By analyzing their behavior, and mining their opinions and suggested actions in Twitter, and simulating their recommendations, we are able to score each expert based on his/her performance. Using this scoring system, experts with most successful trading are recommended. The main objective in this research is to identify traders that outperform market historically, and aggregate the opinions from such traders to recommend trades.

1 Introduction

Social media and other "non-professionally" curated content are much larger (in terms of volume, type, sources, etc) and are growing at a much higher rate than "professionally" curated content such as those in traditional media. Even companies that historically relied only on highly curated content are now seeking to enhance their value propositions by extracting valuable information from Twitter and other social media. What makes this data even more interesting is the fact that a good portion of it is created by professionals who are increasingly relying on social media to disseminate, consume, and contextualize information. In this paper, we propose a technique that allows us to recognize experts amongst others, so that one can extract signal from these expert's recommendations.

The core idea in this paper is that we assume those stock market experts who are active in social media as recommenders. They recommend certain actions (for example "buy" or "sell") on some targeted stocks. The problem is to answer which expert recommends better action. In other words how to choose the best expert who leads us to a maximum gain. In order to select the best recommender, we exercise the market based on the experts' recommendations and compare our

W.G. Kennedy, N. Agarwal, and S.J. Yang (Eds.): SBP 2014, LNCS 8393, pp. 103–110, 2014.

performance with the market performance. The more gain, the more promising is the expert recommendation. By scoring the expert based on their performance, we are able to find the best recommender helping us to gain most.

The paper consists of five sections. Following the introduction, related literature is briefly reviewed in Section 2. The proposed algorithm, mining winning traders, in Section 3. Experimental results and conclusions are presented in Sections 4 and 5 respectively.

2 Related Work

Researchers have proposed extracting various types of signals from Twitter. With regards to financial signals, Oh et al. [10] forecast future stock price movement with microblog postings from StockTwits and Yahoo Finance. Bandari et al. [1] propose a method to predict the popularity of news items on the social web. For each news article, features are generated based on the source of the news story, news category, subjectivity of the news, and named entities mentioned in the news. In contrast to [1], who predict future popularity of news stories, Hong et al. [4] predict popularity of recent messages on twitter. They model the problem as a classification problem, and construct features based on message content, temporal information, metadata of messages and users, as well as the userss social graph.

Gilbert et al. [3] estimate anxiety, worry and fear from over 20 million posts on LiveJournal, and find that an increase in these negative expressions predict downward pressure on the S&P 500 index. Zhang et al. [14] describe a simple approach to predict stock market indicators such as Dow Jones, NASDAQ and S&P 500 by analyzing twitter posts. Bollen et al. [2] also uses twitter mood to predict the stock market. The sentiment analysis is based on OpinionFinder [13] and POMS [9]. A time series of mood is constructed using collective tweet sentiment per day. Their analysis shows strong correlation between 'calm' mood and DJIA data while others including OpinionFinder sentiment shows weak correlation. They also showed that a combination of certain moods was correlated with the stock market. Ruiz et al. [11] use tweets about specific stocks and represent tweets through graphs that capture different aspects of the conversation about those stocks. Two groups of features are then defined based on these graphs: activity-based and graph-based features. In [8], the authors propose a novel approach to finding event based sentiment and using that to create stock trading strategies that outperform baseline models.

There is a growing body of literature that emphasizes on mining experts or higher quality information from twitter. The value of such a task is clear, if experts are better recommenders, then using only their opinion, or giving higher weights to their opinion should generate better models. In [12], design a *who-is-who* service for inferring attributes that characterize individual Twitter users. The authors present an interesting approach of using crowdsourced 'Lists' that contain metadata about a user to infer topics related to the user's expertise. This allows them to remove mundane topics from profiles to improve the accuracy of the system. Hsieh et. al [5] manually identify a group of experts who consistently recommend interesting news stories. They report that the expert group did not perform any better than the wisdom of the crowd. However, they propose eliminating overly active users and augmenting expert opinion in certain conditions to enhance the overall accuracy. In [7], the authors aim to recognize news story curators from the general Twitter crowd. The main focus of the paper is to study different features and propose a model that can successfully classify news curators from the general crowd. Kumar et. al [6] address a practical problem of identifying a niche group who should be followed in times of crises to gain more relevant information. They use a corpus of tweets relevant to the Arab Spring and show how information can be better obtained if users are first recognized as "experts" along the two dimensions of user's location and user's affinity towards topics of discussion.

A question that comes to mind is why would experts share valuable information about future outcomes. Our reasoning is that outcomes those that are reached to by public opinion or consensus (like price of a stock), are not necessarily negated if people predict so in advance. For example, if you have already bought a stock, and you tell others who believe you to buy that stock as well, then prices might be driven up further, and so will your profits. Hence a lot of traders might be promoting their ideas on social media to reinforce beliefs and hence reap returns. This kind of herding or positive reinforcement or momentum is a favourable outcome in many other domains as well like opinions about products. The motivations of analysts, reviewers, etc are different. They analyze and make a sincere recommendation, and would like more people to know about it. This also helps them build their personal brand if more people follow them, popularize them by re-tweeting etc.

3 Mining Winning Traders

There is a lot of information generated in social media. For example, at present, Twitter generates approximately 500 million tweets per day. A lot of people who are interested in stock markets share their opinions on stocks in twitter. These people can be real traders, or amateurs. Real traders or financial professionals have a strong incentive to express their true opinions on stocks. One big incentive is to build a good reputation. If a person's past predictions, as expressed via tweets, turn out to be true, then that person's opinions will be valued more by the people in his/her network. The person can then either use his/her reputation to gain more business, such as, managing people's money; or sell subscription services of their recommendations, and may even affect stock price by attracting people's attention to the stocks that they buy or short. This is why platform like stocktwits.com are increasingly becoming an almost must go-to place for traders to share their trading ideas. Our goal is to identify traders that outperform market historically, and aggregate their opinions to recommend trades. In our context, a trader is defined as a twitter user who tweets about stock market and their trading activities. They don't need to be a professional trader in real life.

3.1 Identifying Target Tweets

We first need to identify trading-related tweets since there is a lot of noise in tweets in general. Each tweet has the following fields: {twitter_id, created_time, from_user, to_user, twitter_text, hashtags, url}.

Given a great amount of tweets every day, it is critical to have a method that is: 1) simple; 2) fast; and 3) high precision. It is ok to miss tweets that talk about trading since there are an almost unlimited number of trading related tweets. We believe some simple rules serve this purpose. Specifically, the rules are:

1. If a tweet ends with "$$" sign, it is a trader-related tweet
2. If a tweet contains one of the predefined recommendation word and at least one stock ticker, it is a trading-related tweet

The predefined recommendation word list is created as following. We count the frequency of verbs that occur within several word windows of a stock symbol. Words with high frequency are put in to a candidate list. Also, we collect tweets from a set of well-known traders in real world, and identify the frequently used verbs in their tweets as well. We manually check the list to make sure the words are commonly used trading words, and cover both selling and buying.

3.2 Identifying Traders

Once we have relevant tweets, we can use them to classify traders. We can look at how often a user tweets, how many stocks he/she covers, how often does this person claim to have traded versus just expressing opinion, etc. For each user, we store the following information: the overall number of tweets, the overall number of trading-related tweets, the frequency of trading related tweets per week, per day, and per month. Only users that pass a threshold are considered to be a trader.

The next step is to identify trades from each trader's tweet. Let's look at some example tweets: Traders typically talk about trades in certain patterns. And we extract templates from the tweets of known traders who are very active. Such templates are then used to extract trades indicated by a tweet. The trades for the tweets could be "buy $SPY", "buy $GLD", "short $AAPL","short $JPM", and "buy $T".

One main reason that we use template-based method is because it is simple, fast, and has a high precision. One further step we could do is, not only identify a specific trade, but also associate each trade with a confidence level. For example, words such as, "strong", "ok", "maybe", "whats happening", "?", can be used to extract how confident/committed the user is to a trade. The confidence level could be decided by two factors: the sentiment level of a tweet, and how many times the tweet is retweeted by others. To be clear, our experiments do not use confidence level.

Evaluating Traders: Next, we evaluate a trader's performance. Since the historical data of stock markets is available, it can be easily used to check how profitable a trader's recommended trade is. We evaluate traders as follows:

1. Evaluating each single trade: For example, if a trade is "BUY $AAPL" at 11:00am 06/12/2012, one could evaluate the performance of this specific trade using different holding periods (for example, holding for 1 day, 3 days, a week, month, quarter, year, etc). For example, for a 1 day holding period, we calculated returns of the trader as percentage change of AAPL closing price compared to the price at the time the recommendation was made, that is, 11am on 06/12/2012 in this case. If a trade is "short $JPM" at 09:30am 05/10/2012 , we define it as a short position and evaluate it accordingly. Therefore, if a trader made 10 trades in the past, for each trade, we have a score vector that measures the trades based on historical stock market data. Each dimension of the vector corresponds to different time period (day/week/month, etc.), associated with the corresponding stocks.
2. Integrating multiple trades: Once we have the score vector for each trade for each trader, we integrate the vectors to several scores.
   - Scores based on different trading timeline: For example, if the trading timeline is daily, the overall score is a combination of each individual daily trading score, weighted by recency of trade. The past trades will be weighted lesser than recent ones. We update scores each time that a new trade is made.
   - Scores based on different types of stocks: The type can be large/small/mid cap, growth/value, industries the stocks are in (tech/financial/consumer/etc.), countries (developed country, emerging, etc).

   The idea is to rate the traders based on their trading styles. For example, some traders may have a high score if evaluated based on long-term stock performance, some traders may have a high score only in tech domain. We want to pick the right winning traders for different trading purposes.

3.3 Ranking Traders and Recommending Trades

Now we are ready to rank traders. Given a trading time period and stock type, a score of a trader is the sum of the weighted score from the previous steps. The winning traders are the ones with top scores. The scores are updated for each trader when the trader makes new trade, and thus the list of top traders change overtime.

Once we have top traders, we can recommend trades based on these traders' trades, as expressed in their tweets. The candidate trades are the recent trades from the top traders (See Table 1). Based on Table 1, we recommend stock actions that have the highest scores. The score of each stock action is the sum of the traderScore from traders who recommend this stock action. The percentage of money allocated to each stock action is its score dividend by the overall score of all recommended actions.

Table 1. Example recommended trades

stockTicker	action	time	traderId	traderScore	tradingStyle
AAPL	BUY	2012/03/05 12:13pm	1234567	0.12	daily,tech
GOOG	SELL	2012/03/05 13:04pm	1234568	0.34	daily, tech

4 Experimental Results

We collected tweets for about 60 stocks (30 of them are DJI30) from 27th March to 20th June 2012 using Twitters Search API. The Search API returns all recent tweets that include the search query. We manually curated a query list in which we included all the terms that are used to refer to a company on Twitter. For example, for Apple Inc. our search query was Apple Inc OR AAPL OR #AAPL OR $AAPL OR AAPL. Since the Search API only returns a recent history, we ran a cron job every 15 minutes that iteratively searched and stored all unique tweets in a database. Tweets returned by the Search API have unique identifiers and hence are easily distinguishable. This led to a total collection of about 30 million tweets.

Since one tweet can mention multiple stocks, overall, there are around 1700 different stocks mentioned. The distribution of tweets across stocks is not uniform with some stocks being far more popular than others. For example, $AAPL is far more popular (mentioned 31508 times) than $IBM (mentioned 4644 times). Even though we didn't search for $SPY (S&P 500) specifically, it got mentioned 48397 times. All together we identified 6512 twitter users as "traders". It doesn't matter whether such users are real traders or if they executed their trades in real life or not. If a user who never trades, but always talk about trading, and somebody can make money just following his trades, then this user is a valuable trader.

To test the profitability of following winning traders, we assume we start with one million dollar. Each day, we execute suggested trades (buy or short stocks based on recommendations obtained previously) when market is open, and clear all the positions when market closes. In other words, no stocks are kept or shorted more than one day. We compare our proposed method (Winners) with 5 different strategies as explained below:

- SPY: A buy-and-hold strategy, holding SPY through the whole period.
- AllTrades: Follow all trades recommended by identified traders. In this strategy, we treat each trader equally, regardless of her historical performance inferred from her tweets.
- TopTrades: Follow the top common trades. In this strategy, only trades that are agreed by most of the traders are executed.
- UniqueTrades: Follow unique trades (opposite of TopTrades).
- Majority: In this strategy, we compare the number of buy/sell trades regardless of specific stocks. If buying is dominant among traders's trades, buy SPY; if selling is dominant, short SPY.
- Winners: This is the proposed strategy. We rank traders who made good trades in the past, and follow trades from such "winning traders".

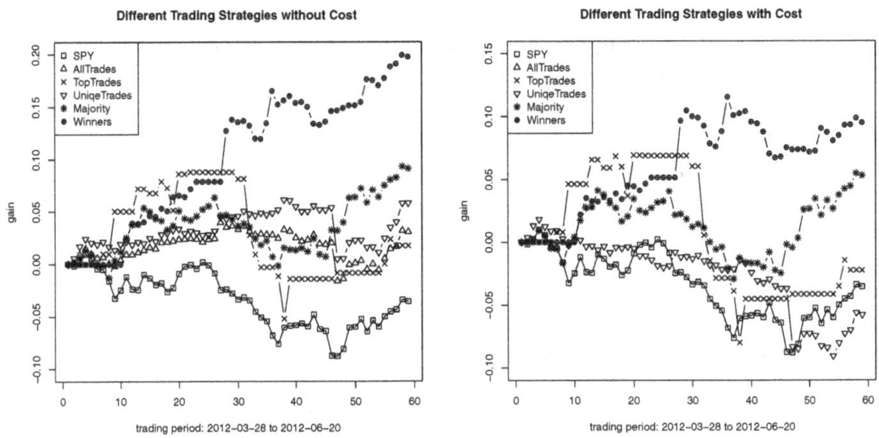

Fig. 1. Trading Performance with different trading strategies: (a) Without cost. The total gains in the 3 month period for different trading strategies are: Winners 19.76%, Majority 9.14%, UniqueTrades 5.80%, TopTrades 1.74%, AllTrades 3.05%, and SPY -3.55%. The winners strategy significantly outperforms other strategies. (b) With cost (0.2% of each trading amount per trade). The total gains over around 3 months of period for different trading strategies are: Winners 9.48%, Majority 5.29%, UniqueTrades -5.79%, TopTrades -2.25%, AllTrades -8.24%, and SPY -3.55%. The winners strategy significantly outperforms other strategies.

Figure 1 compares the performance of the different trading strategies. In Figure 1, we ignore trading cost. In Figure 2, we subtract the trading cost from the returns. We assume the cost is 0.2% of trading amount. For example, if you buy \$50k worth of stocks, the cost is \$1k. With and without costs, the winners strategy significantly outperforms other strategies with much less volatility.

As shown in Figure 1, over the 3 month period, the total return of the winners strategy is 19.76%, vs a buy-and-hold strategy of -3.55%, which makes the winners strategy returns 23.31% more. The second best strategy is the majority strategy, whose returns are 9.14%. This tells us that following the "sentiment" of various traders does have an advantage. And surprisingly, following the minority (UniqueTrades) actually performs better than following the majority (TopTrades).

Even with cost being considered, the winners strategy still outperforms all other strategies, and outperforms buy-and-hold by 13.03%. Our cost considering is quite aggressive, because we assume we clear all the stock positions everyday, even though in reality it won't be the case. For example, in our setup, even if the next day's recommended trade is to buy the same stock as the day before, we subtract the cost of selling the stock at the end of the previous day, and buying the same stock back at the beginning of the next day. As shown in Figure 2, with cost, UniqueTrades, TopTrades, and AllTrade actually lost money, because a significant amount of trading is involved.

5 Conclusion

In our current work, we didn't factor in the relationships between traders. In twitter, a trader can follow other traders, or followed by others. And their tweets can be retweeted and commented. We can build a graph to model the relationships between each trader. In the graph, each node is a trader, and a link between two nodes indicate one trade interacts (follow, or retweet, or comment) with another. Such network has potential to help more to identify winning traders.

References

1. Bandari, R., Asur, S., Huberman, B.A.: The pulse of news in social media: Forecasting popularity. CoRR (2012)
2. Bollena, J., Maoa, H., Zengb, X.: Twitter mood predicts the stock market. Journal of Computational Science (2011)
3. Gilbert, E., Karahalios, K.: Widespread worry and the stock market. In: Int. AAAI Conf. on Weblogs and Social Media (2010)
4. Hong, L., Dan, O., Davison, B.D.: Predicting popular messages in twitter. In: Int. Conf. Companion on WWW (2011)
5. Hsieh, C.-C., Moghbel, C., Fang, J., Cho, J.: Experts vs the crowd: Examining popular news prediction performance on twitter. In: Int. Conf. on WWW (2013)
6. Kumar, S., Morstatter, F., Zafarani, R., Liu, H.: Whom should I follow? Identifying relevant users during crisis. In: ACM Conf. on Hypertext and Social Media (2013)
7. Lehmann, J., Castillo, C., Lalmas, M., Zuckerman, E.: Finding news curators in twitter. In: Int. Conf. on WWW Companion, pp. 863–870 (2013)
8. Makrehchi, M., Shah, S., Liao, W.: Stock prediction using event-based sentiment analysis. In: IEEE/WIC/ACM Int. Conf. on Web Intelligence (2013)
9. McNair, D., Heuchert, J.P., Shilony, E.: Profile of mood states. Bibliography, 1964–2002 (2003)
10. Oh, C., Sheng, O.: Investigating predictive power of stock micro blog sentiment in forecasting future stock price directional movement. In: ICIS, pp. 57–58 (2011)
11. Ruiz, E.J., Hristidis, V., Castillo, C., Gionis, A., Jaimes, A.: Correlating financial time series with micro-blogging activity. In: Int. Conf. on Web Search and Data Mining, WSDM 2012, pp. 513–522 (2012)
12. Sharma, N.K., Ghosh, S., Benevenuto, F., Ganguly, N., Gummadi, K.: Inferring who-is-who in the twitter social network. SIGCOMM Comput. Commun. Rev. 42(4), 533–538 (2012)
13. Wilson, T., Hoffmann, P., Somasundaran, S., Kessler, J., Wiebe, J., Choi, Y., Cardie, C., Riloff, E., Patwardhan, S.: Opinionfinder: a system for subjectivity analysis. In: HLT/EMNLP on Interactive Demonstrations, HLT-Demo 2005, pp. 34–35 (2005)
14. Zhang, X., Fuehres, H., Gloor, P.A.: Predicting stock market indicators through twitter "I hope it is not as bad as I fear". In: Innovation Networks Conference-COINs 2010 (2010)

Quantifying Political Legitimacy from Twitter*

Haibin Liu and Dongwon Lee

College of Information Sciences and Technology
The Pennsylvania State University, University Park, PA
{haibin,dongwon}@psu.edu

Abstract. We present a method to quantify the *political legitimacy* of a populace using public Twitter data. First, we represent the notion of legitimacy with respect to k-dimensional probabilistic topics, automatically culled from the politically oriented corpus. The short tweets are then converted to a feature vector in k-dimensional topic space. Leveraging sentiment analysis, we also consider the polarity of each tweet. Finally, we aggregate a large number of tweets into a final legitimacy score (i.e., L-score) for a populace. To validate our proposal, we conduct an empirical analysis on eight sample countries using related public tweets, and find that some of our proposed methods yield L-scores strongly correlated with those reported by political scientists.

1 Introduction and Related Work

The term *political legitimacy* in political science refers to the acceptance of authority by a law, government, or civil system, and has been the subject of extensive study in the discipline. The concept is often viewed as "central to virtually all of political science because it pertains to how power may be used in ways that citizens consciously accept"[1]. As such, in political science, many proposals have been made to quantify the legitimacy of a populace. Some recent works such as [1, 2] have been well received in the community. While useful, however, such existing works are largely based on hand-picked small-size data from governments or UN based on an ad hoc formula. Therefore, it is still challenging to renew or expand the results from [1, 2] to other regions if there exist no reliable base data. To address this limitation, in this research, we ask a research question "*if it is possible to quantify political legitimacy of a populace from social media data*", especially using Twitter data. As a wealth of large-scale public tweets are available for virtually all populaces, if such a quantification is plausible, the application can be limitless. For instance, in the stochastic simulation environment such as NOEM [3], a quantified legitimacy score forms one of important input parameters. While there is currently no good way to synthetically generate a

* Part of the work was done while Dongwon Lee visited the Air Force Research Lab (AFRL) at Rome, NY, in 2013, as a summer faculty fellow. Authors thank John Salerno at AFRL for the thoughful feedback on the idea and draft. This research was also in part supported by NSF awards of DUE-0817376, DUE-0937891, and SBIR-1214331.

W.G. Kennedy, N. Agarwal, and S.J. Yang (Eds.): SBP 2014, LNCS 8393, pp. 111–118, 2014.

legitimacy score of a populace, one may be able to estimate it from the tweets generated from or closely related to the populace.

In recent years the exploitation of social media such as Twitter and Facebook to predict latent patterns, trends, or parameters has been extensively investigated. For instance, [4] computationally tried to classify tweets into a set of generic classes such as news, events, or private messages. In addition, [5–7] attempted to track and analyze the status of public health via social media data. Some even tried to predict stock market from public mood states collected from Twitter [8]. Studies have also been carried out about the correlation between tweets' political sentiment and parties and politicians' political positions [9, 10]. The case study about 2009 German federal election [9] reported a valid correspondence between tweets' sentiment and voters' political preference. Such studies also verify that the content of tweets plausibly reflects the political landscape of a state or region. Another paper [11] also aggregates text sentiment from tweets to measure public opinions.

While closely related, our method focuses on quantifying the political legitimacy, that is related to not only politics and elections, but also other concepts such as governments, laws, human rights, democracy, civil rights, justice systems, etc. To our best knowledge, this is the first attempt to computationally quantify the political legitimacy of a populace from a large amount of big social media data and conduct a correlation analysis against the results in political science.

2 The Proposed Method

Our goal is to build and validate a model to accurately quantify the political legitimacy score of a populace using tweet messages. The underlying assumption is that some fraction of populace would occasionally express their opinions on the status of political legitimacy. Two such examples are shown in Figure 1.

Fig. 1. Tweets related to legitimacy

Let us use the term **L-score** to refer to the political legitimacy score of a populace, scaled to a range of [0,10]. Then, our overall method consists of three steps: (1) identify and convert relevant tweets into computable feature space, (2) compute L-score of each tweet, and (3) aggregate L-scores to form a time series and compute final L-score of a populace. This overall workflow is illustrated in Figure 2.

2.1 Step 1: Vectorizing Tweets

Each tweet can be up to 140 characters but often very terse. The challenge of this step is to be able to accurately capture and extract critical features from

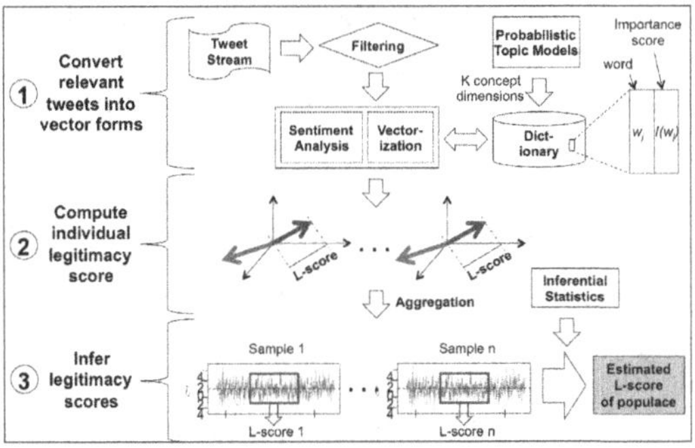

Fig. 2. Overview of the proposed method

Concept: War

states,war,international,military,soviet,state,foreign,crisis,force,security,domestic,united,nuclear,power,strategy,threat,conflict,attack,first,world,leaders,likely,decision,policy,defense,arms,two,relations,deterrence,american,strategic,press,costs,university,balance,bargaining,peace,behavior,relative,forces,threats,interests,challenger,system,ing,crises,national,politics,con,country,tion,three,science,important,union,second,capabilities,political,gains

Concept: Election

vote,election,party,presidential,candidates,elections,candidate,voters,campaign,political,voting,incumbent,american,republican,democratic,electoral,science,model,congressional,senate,partisan,house,effects,incumbents,effect,table,results,data,state,variables,identification,review,two,national,spending,variable,partisanship,change,time,democrats,president,analysis,challenger,new,year,voter,tion,support,approval,challengers,personal,reagan

Fig. 3. Two prominent topics found from political science journal articles

short tweets that can indicate the opinion of a writer toward the status of legitimacy. Since there is no widely-accepted "computable" definition of legitimacy, we assume that the notion of political legitimacy is related to k-dimensional topics such as justice system, human rights, democracy, government, etc. While treating k as a tunable parameter in experiments, then, we simply attempt to represent each tweet as a k-dimensional vector, where the score in each dimension indicates the relevance of the tweet to the corresponding topic. Further, we use a dictionary of k dimension where each dimension (i.e., topic) contains a set of keywords belonging to the topic. Finally, we run a probabilistic topic modeling technique such as Latent Dirichlet Allocation (LDA) [12] over politically oriented corpus[1] and build such a k-dimensional dictionary.

Figure 3 illustrates two example topics found by LDA and prominent keywords within each topic (the labels such as "war" and "election" are manually assigned). Note that, although found automatically, such topics represent the main themes of the corpus reasonably well and can be viewed as related to the legitimacy. In addition, prominent keywords within each topic also make sense. Therefore, if a tweet mentions many keywords found in either topic, then the tweet is used to quantify the legitimacy. Suppose k topics are first manually selected and corresponding keywords in each topic are found using LDA. Imagine a k-dimensional dictionary such that a membership of a keyword can be

[1] http://topics.cs.princeton.edu/polisci-review/

quickly checked. For instance, one can check if the keyword "military" exists in the "war" dimension of the dictionary. Furthermore, suppose each keyword, w, in the dictionary is assigned an importance score, $I(w)$. In practice, a frequency-based score or LDA-computed probability score can be used to measure the importance of keywords. For instance, an importance of a word can be computed using the following frequency-based formula: $I(w) = \frac{freq(w)}{\sqrt{1+(freq(w))^2}}$. Using this data structure of the k-dimensional dictionary, we can convert tweets into vectors and then compute the L-score.

With such a topic dictionary, we can convert each tweet into a k-dimensional vector by checking membership of words in each dimension. Assume that a tweet, t, is pre-processed using conventional natural language processing (NLP) techniques such as stemming and represented as a bag-of-words, w, with n words: $t \Rightarrow w = \{w_1, w_2, \cdots, w_n\}$. Then, the k-dimensional vector representation of a tweet, v_t, is:

$$v_t \in R^k = [\alpha_1 \sum\nolimits_{\forall m_1 \in |w \cap D_1|} I(m_1), \cdots, \alpha_k \sum\nolimits_{\forall m_k \in |w \cap D_k|} I(m_k)]$$

such that $\sum_1^k \alpha_i = 1$, D_i refers to the i-th dimension of the dictionary, and α_i is the weighting parameter for the relative importance of the i-th dimension.

2.2 Step 2: Computing L-Scores of Tweets

The intuition to compute L-score of a tweet is that when a tweet either positively or negatively mentions keywords related to k-dimensions of the legitimacy, their "strength" can be interpreted as the legitimacy score. The L-score of the tweet, $L-score(t)$, is then defined as the magnitude (i.e., L2-norm) of v_t, with the sign guided by the sentiment of the tweet t–Δ_{sent}. Suppose $v_t = (x_1, ..., x_k)$. Then,

$$L - score(v_t) = \Delta_{sent}\|v_t\| = \Delta_{sent}\sqrt{x_1^2 + \cdots + x_k^2}$$

where Δ_{sent} indicates a $[-1, 1]$ range of sentiment polarity score of the tweet. Note that an alternative to this single Δ_{sent} per tweet is to allow for different sentiment polarity per dimension, Δ_i , in each tweet. However, in our preliminary study, as typical tweets are rather short and there are usually simply not enough information to determine different polarity score per dimension, we maintain a single sentiment score per tweet.

2.3 Step 3: Aggregating L-Scores of Tweets

Once the L-score has been computed for all tweets, we next need to aggregate all the L-scores per some "group" and determine the representative L-score of the group. One example grouping constraint can be a region (e.g., country such as Egypt or city such as Detroit). Suppose we want to aggregate all L-scores of the day d. Assuming the distribution of the daily L-scores follow the Gaussian Distribution, then, we compute the mean L-score of the day and apply the interval-based Z-score normalization, similar to [13], to the L-score.

Table 1. L-scores published in [2]

Country	Score	Country	Score	Country	Score	Country	Score
Norway	7.97	Japan	6.13	India	5.21	Bulgaria	3.21
Canada	7.26	Thailand	5.89	France	5.03	Peru	3.44
Vietnam	7.07	United States	5.83	Brazil	4.68	Iran	2.04
New Zealand	6.78	South Africa	5.45	Slovenia	4.33		
Spain	6.64	China	5.36	Turkey	3.96		

Collecting such normalized L-scores over a time interval, finally, we derive a time series and employ standard time series analysis techniques to either compute the overall representative score of the entire time series, or predict future L-scores. For instance, in the current implementation, we used both moving average (MA) and auto-regressive MA (ARMA) models.

3 Empirical Validation

Since there is no ground truth to L-scores of populaces, as an alternative, we aim to see "if our method yields L-scores of populaces similar to those reported in [2]." For instance, Table 1 shows example L-scores reported in [2]. This, computed from UN and WHO data, is widely accepted in political science community. We chose eight countries with varying L-scores in [2]–i.e., Brazil, Iran, China, Japan, Norway, Spain, Turkey, and USA. We prepared two sets of data: (1) *Geo* dataset contains tweets generated within the bounding box of the geo-coordinates of each country of interest, and (2) *Keyword* dataset contains tweets that mention terms related to each country (e.g., a hash tag of "#USA"), regardless of their geo-coordiates. From 9/28/2013 to 11/6/2013, we collected a total of 300,450 tweets using Twitter streaming API that are written in English, and relatively meaningful (e.g., terse tweets with less than 4 words or location-based tweets having the form of "I'm at location" are removed). Figure 4, for instance, shows the geo-coordinates of tweets in the Geo dataset for USA and China. Table 2 summarizes tweets that we used in the experiments. We first present the aggregated mean L-score of crawled tweets during the monitored period.

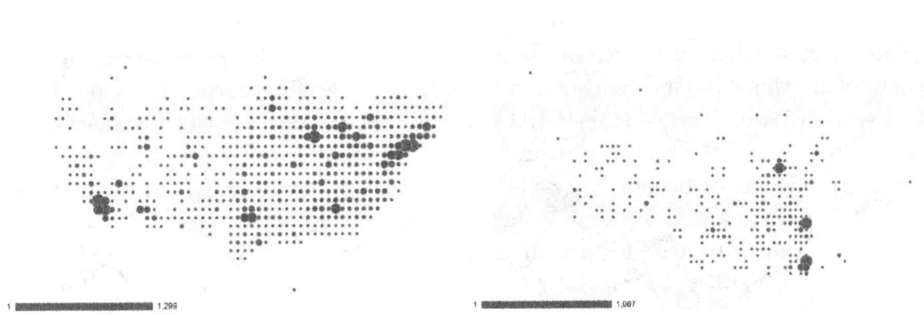

(a) USA (b) China

Fig. 4. Geo-cordinates of tweets in Geo datasets

Table 2. Summary of crawled tweets

	# Keyword tweets	# Geo tweets	# Filtered Geo Tweets
Brazil	10,924	18,788	14,715
China	17,848	8,060	7,569
Iran	51,743	9,600	6,594
Japan	13,112	9,948	9,427
Norway	6,561	5,633	5,554
Spain	15,845	13,094	12,477
Turkey	13,281	38,187	14,634
USA	28,801	39,025	38,662

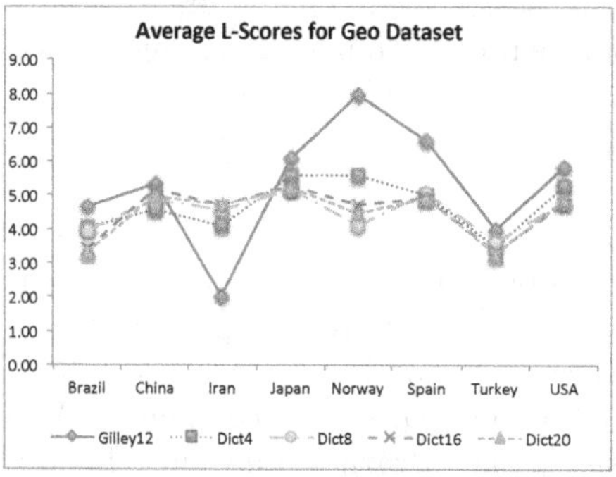

(a) L-scores of Geo dataset

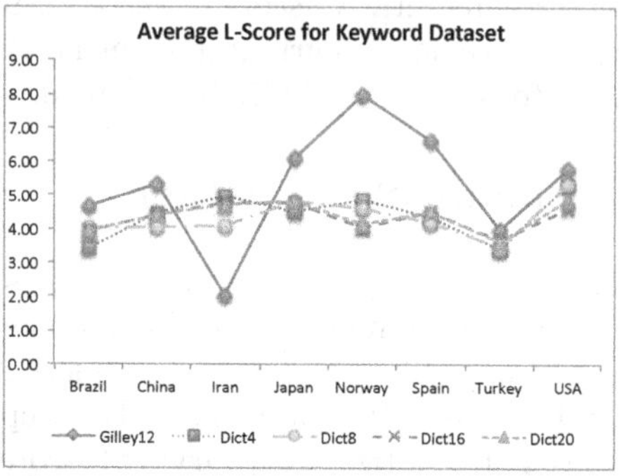

(b) L-scores of Keyword dataset

Fig. 5. Aggregated mean L-scores

Several factors are studied that may affect the final L-score. First, the number of topics obtained from LDA may play an important role in quantifying tweets' score. We tried different number of topics from 4 to 20, and the results are shown in Figure 5, where *Dict*4 means result from dictionary with 4 topics. Note that the range of the L-scores are rescaled to $[0, 10]$ to be compliant with the results of [2]. We can see that different number of topics lead to slightly different L-scores on both Geo and Keyword datasets. Studies are also carried out to see the impact of granularity of sentiment analysis in calculating L-scores. While previous results are calculated using sentiment polarity scaled in range $[-1, 1]$, we also tested with only extreme sentiment values of $\{-1, 1\}$. However, the L-scores using this extreme sentiment values show little difference. Figure 6 shows time-series of 4 countries using 4 LDA topics on Geo and Keyword datasets. In most cases, L-scores estimated from Geo tweets match better than those estimated from keyword tweets. Note that compared to L-score of [2], our estimation of L-score matches well for some countries but poor for others (e.g., Norway).

To see the overall correlation with [2], we computed the *Pearson correlation coefficient* (PCC) [14] between the L-scores of all of our methods (using different number of topics or sentiment values) and [2]. As shown in Table 3, the best performer is the *Dict*4 over Geo dataset. With the coefficient value of 0.7997887 (P-value = 0.01717), we can claim a significant correlation between L-score computed

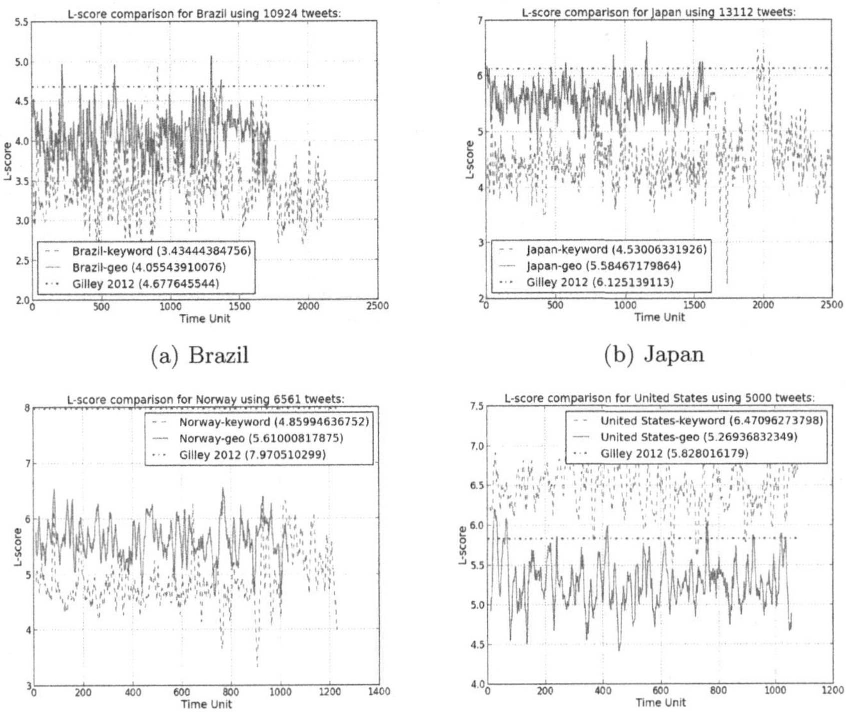

(a) Brazil (b) Japan

(c) Norway (d) USA

Fig. 6. L-score time series of 4 countries with 4 LDA topics on Geo dataset

Table 3. PCC values between L-scores of our proposed methods and [2]

	Keyword	Geo	Keyword-Extreme	Geo-Extreme
Dict4	0.203461755	**0.799788652**	0.214170058	-0.452748888
Dict8	0.472864444	0.233350916	0.401502605	-0.538886828
Dict16	-0.063411723	0.375603634	0.27090464	-0.594296533
Dict20	0.070540651	0.307136136	0.188031801	-0.631019398

using *Dict*4 and Geo dataset and that reported in [2]. This discovery also indicates that tweets directly generated from the territory of a region (i.e., Geo dataset) is a better source to quantify L-score than those conceptually related to a region (i.e., Keyword dataset).

4 Conclusion

We study the problem of quantifying political legitimacy of a populace based on public Twitter data. We propose a solution that converts short tweet text messages into a number of topic dimensions using probabilistic topic modeling. We leverage sentiment analysis to evaluate polarity of each tweet, and aggregate a large number of tweets into the final legitimacy score of a populace. Our experiments over real tweets collected about eight countries reveal that some configuration of our proposal shows a strong correlation to results reported in political science community. Despite the promising result, there are a set of *limitations* to our study: (1) To derive a more definite conclusion on the validity of our proposed method in quantifying the legitimacy, a more comprehensive experiment is needed–e.g., more number of countries, larger tweet datasets, or topics derived from different corpus; (2) While [2] is a reasonable "beta" ground truth for our study, there is no formal analysis why or how accurate it is. As such, more correlation analysis of our proposal using different methods to compute the legitimacy is needed; (3) In addition to social media such as Twitter, other large-scale data can be used as a source of legitimacy. For instance, a dataset such as GDELT[2] contains a large-scale rich data on world-wide conflicts and can be used to infer the legitimacy status expressed by a populace.

References

1. Gilley, B.: The meaning and measure of state legitimacy: Results for 72 countries. European J. of Political Research 45(3), 499–525 (2006)
2. Gilley, B.: State legitimacy: An updated dataset for 52 countries. European J. of Political Research 51(5), 693–699 (2012)
3. Salerno, J.J., Romano, B., Geiler, W.: The national operational environment model (noem). In: SPIE Modeling and Simulation for Defense Systems & App. (2011)
4. Sriram, B., Fuhry, D., Demir, E., Ferhatosmanoglu, H., Demirbas, M.: Short text classification in twitter to improve information filtering. In: ACM SIGIR, pp. 841–842 (2010)
5. Paul, M.J., Dredze, M.: You are what you tweet: Analyzing twitter for public health. In: ICWSM (2011)
6. Lamb, A., Paul, M.J., Dredze, M.: Investigating twitter as a source for studying behavioral responses to epidemics. In: AAAI Fall Symp. (2012)
7. Evans, J., Fast, S., Markuzon, N.: Modeling the social response to a disease outbreak. In: Greenberg, A.M., Kennedy, W.G., Bos, N.D. (eds.) SBP 2013. LNCS, vol. 7812, pp. 154–163. Springer, Heidelberg (2013)
8. Bollen, J., Mao, H., Zeng, X.: Twitter mood predicts the stock market. J. of Computational Science 2(1), 1–8 (2011)
9. Tumasjan, A., Sprenger, T.O., Sandner, P.G., Welpe, I.M.: Predicting elections with twitter: What 140 characters reveal about political sentiment. In: ICWSM, pp. 178–185 (2010)
10. Tumasjan, A., Sprenger, T.O., Sandner, P.G., Welpe, I.M.: Election forecasts with twitter how 140 characters reflect the political landscape. Social Science Computer Review 29(4), 402–418 (2011)
11. O'Connor, B., Balasubramanyan, R., Routledge, B.R., Smith, N.A.: From tweets to polls: Linking text sentiment to public opinion time series. In: ICWSM (2010)
12. Blei, D.M., Ng, A.Y., Jordan, M.I.: Latent dirichlet allocation. The Journal of Machine Learning Research 3, 993–1022 (2003)
13. Bollen, J., Mao, H., Pepe, A.: Modeling public mood and emotion: Twitter sentiment and socio-economic phenomena. In: ICWSM (2011)
14. Rodgers, J., Nicewander, A.: Thirteen ways to look at the correlation coefficient. The American Statistician 42(1), 59–66 (1988)

[2] `http://gdelt.utdallas.edu/`

Path Following in Social Web Search

Jared Lorince[1,*], Debora Donato[2], and Peter M. Todd[1]

[1] Cognitive Science Program
Department of Psychological & Brain Sciences
Indiana University, Bloomington, Indiana 47405, USA
{jlorince,pmtodd}@indiana.edu
[2] StumbleUpon, Inc.
San Francisco CA 94107, USA
debora@stumbleupon.com

Abstract. Many organisms, human and otherwise, engage in path following in physical environments across a wide variety of contexts. Inspired by evidence that spatial search and information search share cognitive underpinnings, we explored whether path information could also be useful in a Web search context. We developed a prototype interface for presenting a user with the "search path" (sequence of clicks and queries) of another user, and ran a user study in which participants performed a series of search tasks while having access to search path information. Results suggest that path information can be a useful search aid, but that better path representations are needed. This application highlights the benefits of a cognitive science-based search perspective for the design of Web search systems and the need for further work on aggregating and presenting search trajectories in a Web search context.

Keywords: Search Paths, User Study, Path Following, Social Search.

1 Introduction

Path following is ubiquitous among social species in natural environments, be it mediated by stigmergic pheromone trails of ants and termites [1], emerging from crowd dynamics [2], or evidenced by the the reinforcement of worn paths through grass and snow on college campuses [3]. On the Web, too, we follow paths — albeit implicitly — when the search results we encounter, videos we watch, and products suggested to us all depend on the interaction patterns of the users who went before us. In this paper we take inspiration from work on path following in physical environments to explore whether sharing explicit search paths between Web searchers can be a useful search aid.

Research in cognitive science suggests that goal-directed cognition is an evolutionary descendant of spatial foraging capacities [4], and an increasing number of studies show that the way humans search in information spaces is deeply

* The first author was an intern at, and the second employed by, Yahoo! Research during study development. Data collection/analyses were done at Indiana University.

W.G. Kennedy, N. Agarwal, and S.J. Yang (Eds.): SBP 2014, LNCS 8393, pp. 119–127, 2014.

linked to the way we search in spatial environments [5, 6]. This conclusion is also bolstered by the embodied view of cognition, which highlights the connections between information processing and bodily movement in space [7].

With this in mind, we developed and tested a prototype interface for applying the notion of path following to a Web search environment. A path, by definition, carries a special kind of information typically lacking from Web-based recommendations and other search tools: It provides not only a destination, but a route from one's current location to that destination, thus delineating what lies between. This may seem a simple point, but the vast majority of tools for guiding information search on the Web — from product recommendations on Amazon to "Also try:" suggestions on Yahoo! search — are pointillistic: You should issue *this* query, or buy *this* product. This is not to say that such recommendation systems are not utilizing path-like data "under the hood" to generate suggestions, but to our knowledge there exist no systems in regular use that explicitly share paths — sequences of activity extended over time — between users.

In a Web context, sharing path information creates opportunities for serendipitous discoveries by exposing the user to content that would be missed by "teleporting" directly to a recommended resource. When these paths are relevant to the current search context, they can provide windows on how to approach a search task that the user might not consider otherwise, and that likely could not be readily communicated via pointillistic recommendations. This of course holds little value in cases where a user's query has a clear, discrete answer ("What is the capital of North Dakota?"). Many search tasks we engage in, however, are simultaneously more complex and less explicitly defined ("What car should I buy?", "What is fun to do in North Dakota?"). In these cases, paths can capitalize on modern Web users' interest in shared social content and propensity for social copying. We hypothesized that the incorporation of path information into the search interface would lead to increased levels of (1) user engagement and (2) satisfaction with solutions to assigned search tasks. To explore our hypotheses, we developed a custom search engine interface that incorporated path information. Study participants were assigned a series of search tasks, and presented with the paths taken by previous users performing the same task.

A full understanding of search path use requires work at three levels: The cognitive-behavioral (what is the theoretical case for using path information in search and how do people respond to it), the algorithmic (how can search paths be generated and coherently aggregated across multiple users), and design-centric (how should such paths be presented to users). Here we address the first level, as a preliminary attempt to explore how path-like information can be translated to a Web search context. While some of our positive results are suggestive of the power of this approach, our other negative results also indicate that it will be crucial to determine better ways of presenting path information if it is to be helpful to users. Thus a principal goal of this paper is to encourage future work that explores methods for creating and presenting useful path information to individuals searching the Web an other information spaces.

2 Related Work

Cognitive Science: Though following a path in a physical environment bears little surface similarity to a Web searcher's "movement" on the Internet, the activity of Web searchers create valuable signals that can facilitate future users' search efforts, much as animals create physical trails. Web path signals are utilized by many modern search systems, both when they are left behind explicitly (as in collaborative tagging or when people share links on a social network) and, more commonly, when they are implicit. These implicit signals, formed as users issue queries and click on results, are integral to intelligent query suggestions and to the ranking of results on modern Web search engines.

In the mid-1990s, Pirolli applied optimal foraging theory — a theory of how organisms search for resources in a physical environment — to Web search with considerable success [8, 9]. More recent work [5] found that participants could be primed by a spatial search task to behave in predictably different ways on a subsequent mental search task. The authors hypothesized the existence of generalized cognitive search processes, and molecular and behavioral evidence [4, 6, 10] supports the hypothesis that evolved capacities for spatial search deeply influence the way we search in other domains. This suggests the usefulness of spatially-inspired data representations, like paths, for information search.

Path-Based Web Search: Recently a few works [11–14] have studied algorithms inspired by physical spatial search to improve web search engine performance, modifying page ranking by enriching link data with collective intelligence information. For each page, the information about Web trails taken by other users (often called Web pheromones) is accumulated and used to modify the global rank of the page. This differs from our approach of showing the paths used by others, but leaving page ranks unchanged. In terms of methodology, only one other study [14] conducted a controlled experiment on real users as we did, but again, participants were not directly presented with search paths.

Search Tool Evaluation: Social search tools can be evaluated via two main criteria: effectiveness (and hence user satisfaction) and elicited engagement [15–18]. Often, shorter time to completion (i.e. the time spent on a search task) is used to assess effectiveness. In a social setting, however, time to completion is not always a good metric: Social interactions can lead to increased engagement, which can in turn increase time to completion, such as through distortions in the subjective perception of time [15, 19]. Since evaluations of social search tools depend on subjective measures, they are typically tested with user studies [15–17], which are limited in number of participants and constrained by the need for extended experience with a new tool [20]. Despite these problems, there is typically no viable alternative for testing users' subjective responses to search tools.

3 Methodology

Participants completed a sequence of search tasks either with social search information (BestSearcher paths condition) or without (baseline condition). We ran the baseline condition first, and used data from those participants to generate the search path information for the experimental BestSearcher condition. All participants completed the same set of search tasks (in randomized order) for one condition or the other. The study was administered in a modified[1] web browser that allowed for display of path information, presentation of search tasks, collection of task responses, survey administration, and clickstream logging. Baseline condition participants used a standard version of the Yahoo! SERP (Search Engine Results Page), while participants in the experimental condition also saw a sidebar with social search path data (Figure 1). Paths were socially generated sequences of clicks and queries from the baseline condition, and participants could click path elements to visit a URL or issue a query from the path, respectively.

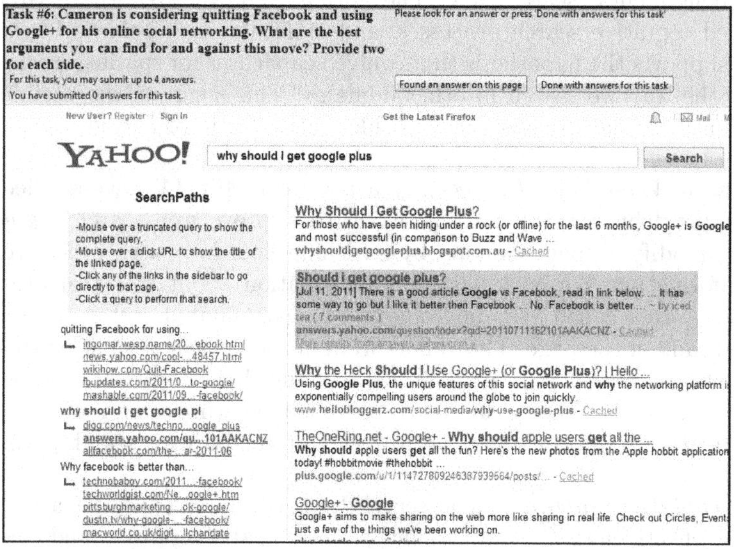

Fig. 1. Screenshot of experiment interface

Participants were given search tasks that we deliberately selected to be both complex and minimally specific; that is, none of them had a particular set of "correct" answers. The goal was to use questions that would enable participants to utilize social information to aid their searches, without the social information leading to a single best answer for any question. Thus all tasks incorporated a level of subjectivity ("find the best...", etc.) and required multiple answers. Table 1 shows several examples of the tasks used (participants completed eight in total).

[1] Via the HCIbrowser extension [21] and a variety of CSS and Javascript tools that allowed for visual modification of the SERP.

Table 1. Examples of search tasks

"austria": You're on a backpacking tour of Europe, and will be stoping in Innsbruck, Austria, but unfortunately you'll only have a few hours to spend there. Find the two most interesting activities that could both be done in the 4 hours you'll have.
"disney": Tammy is planning a two-day trip to Disneyland with her three-year-old daughter (who loves princesses) and is looking for the must-see attractions. She's already been to disneyland.com, and had little luck, so find three appropriate pages to help her in her trip planning.
"facebook": Cameron is considering quitting Facebook and using Google+ for his online social networking. What are the best arguments you can find for and against this move? Provide two for each side.
"metal": A friend wants to take up metal detecting as a hobby. Find the three best resources (books, online tutorials, videos, etc.) you can to get her started.

3.1 Generating Search Paths

To generate the social search path data displayed in the sidebar, we had to extract meaningful paths from users' search activity in the baseline condition. After comparing several options, we settled on "BestSearcher" paths for this study, which show the complete path (sequence of queries and clicks) of the "best" participant from the baseline condition for each task — requiring a measure of query success.[2] Our ranking metric used the total number of queries (because the tasks require multiple answers, issuing more queries should increase the probability of finding more unique pages), the total number of long dwell-time (i.e. time spent on page) clicks per query, and the inverse of the time required to reach the first long dwell-time result. The path followed by the baseline condition participant with the greatest score on this metric for each task was then used as the BestSearcher path for that task, such that all participants in the experimental condition saw that same path for any given task (but the source "best" searcher for paths varied from task to task).

3.2 Participants and Procedure

Participants were Indiana University undergraduates compensated with course credit. 26 female and 42 male students (12 female and 12 male in the baseline condition, 14 female and 30 male in the BestSearcher condition) participated. All were between 18 and 24 years old. Participants in the baseline condition were informed that they would be presented with a series of questions for which they should search the web for answers. Those in the experimental condition were given the same instructions, but were also told that they would have "access to information about how previous IU students have completed the search tasks."

[2] The technical details of this metric, along with expanded discussion of our methodology and results, appear in an extended version of this paper available online at `http://mypage.iu.edu/ jlorince/papers/ lorince.donato.todd.2014.sbp.extended.pdf`

Participants first completed a practice trial to get familiar with the interface, then eight experimental search tasks (in randomized order). They then rated task difficulty, satisfaction with results, engagement with the task, and, for the BestSearcher condition, the usefulness of the search path information.

4 Results and Discussion

We focus here on determining if participants found the social path data engaging and/or helpful. Analyses discussed below reflect only participants who utilized the social path information (by clicking a query or URL) at least once (33 of 44).

Subjective Ratings: Unsurprisingly, we found a general pattern of anticorrelation between task difficulty and satisfaction (baseline: ($r(209) = -.59, p < .0001$), experimental: ($r(317) = -.60, p < .0001$), as well as weak but significant correlation between engagement and search satisfaction (baseline: $r(209) = .35, p < .0001$, experimental: $r(317) = .26, p < .0001$) across both conditions. As subjective difficulty went up, engagement went down in the Baseline condition ($r(209) = -.29, p < .0001$), but not in the BestSearcher condition. This suggests that social facilitation did ameliorate the negative effect of task difficulty on engagement. In contrast to our initial predictions, we found no significant difference in mean satisfaction or engagement between conditions. Problematically, participants did not report the experimental tasks to be of strong personal relevance, rating them on average below the midpoint of a Likert scale (i.e. disagreeing with the statement "This is a realistic search task for you in particular.").

Behavioral Measures: The key metrics for each task (Figure 2) were mean time to completion, mean dwell time (i.e. the average time spent on each clicked page), proportion of trials successfully completed, and total number of search events (i.e sum of queries and clicked URLs for each task). Again, there was little difference between conditions. The data suggest a trend towards faster completion times and fewer total search events when path information was available, with the notable exception of the "indiana" task, which required significantly more time and search events in the BestSearcher condition compared to baseline. This may stem from the difficulty of the task (highest subjective difficulty rating), along with the possibility that the information in the sidebar was not particularly useful, but participants still explored the social data in detail in an effort to solve the difficult task. This task also had the greatest proportion of activity originating from the sidebar across participants. There is, in fact, a weak but significant ($r(209) = -.29, p < .0001$) correlation between the proportion of activity originating from the sidebar and the perceived difficulty of tasks, indicating that participants relied more on socially available data when search tasks were more challenging.

Evaluation of Search Paths: Participants in the experimental condition also rated the usefulness of seeing search paths, whether it made the task more interesting/engaging, whether they actually used paths, and whether path information allowed them to complete the task more quickly than they would have

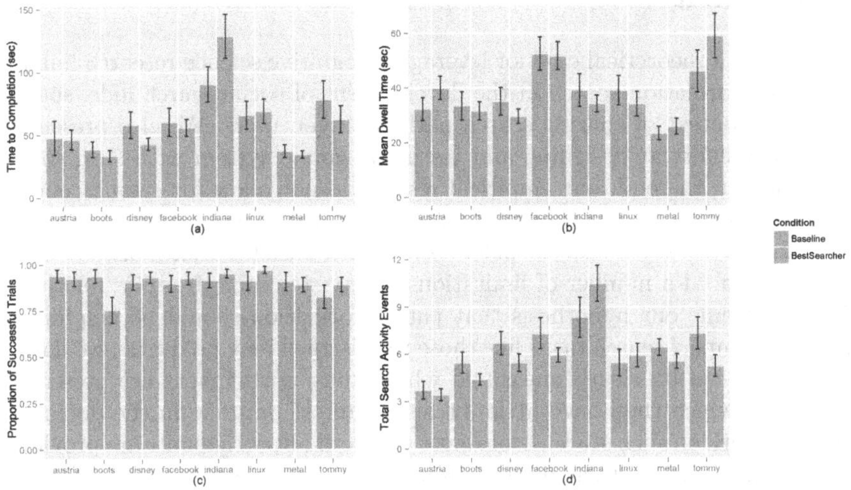

Fig. 2. Summary of behavioral measures by condition and task. (a): Mean time to complete task (seconds). (b): Mean dwell time (seconds). (c): Mean proportion of successfully completed trials. (d): Total search activity (number of clicks + number of queries). (a),(b), and (d) normalized by the number of responses required for each task. Error bars show +/- 1 standard error.

otherwise. Responses hovered around the middle of the response scale on average, indicating that participants found the search paths moderately helpful overall. There appears to be a general pattern of paths being more positively evaluated on the more difficult tasks, though the only measure here that correlates (weakly) with difficulty in a statistically reliable way is participants' reported usage levels ($r(258) = .23, p < .001$). These responses were not particularly strongly aligned with the respective behavioral measures we collected, though; ratings of how much participants actually used the paths, for instance, had only a weak correlation ($r(258) = .22, p < .001$) with their total sidebar activity (i.e. sum of clicked queries and URLs from the sidebar).

Notable here is that all responses to the search paths evaluation questions were moderately to highly inter-correlated ($r > .6, p < .0001$ in all pairwise correlations). So, even though their subjective responses may not correlate well with their behavioral patterns, these results indicate (consistent with our hypotheses) that a useful search tool is one that enhances both engagement and the speed with which a user can achieve his or her search goal. Unexpected here is the weak correlation between how much participants reported using the sidebar by actually clicking queries and URLs and the true use of the sidebar we logged experimentally. The unexpected low correlation between perceived and actual use could have come about because participants had an inflated sense of how much they used the sidebar when they found the sidebar path information to be useful. Nevertheless, these data suggest that the SearchPaths tool may have been of help to participants in ways not apparent from our collected behavioral measures.

5 Conclusions

We have made a theoretical case for leveraging cognitive science research linking spatial and information search in the development of social search aids, specifically in the context of sharing search paths between users. We also presented a preliminary effort at designing and testing a simple system with such social functionality. In the end, our empirical results do not allow for strong conclusions to be drawn from our user study, but our methods will likely be useful in future comparative work that considers other path-based search tools.

Our study faced a number of limitations, many stemming from its relatively small scale. While our hypothesis that path information should be helpful for moderately complex search tasks like those we assigned may hold true, we doubt such an effect can be clearly measured when study participants are presented with tasks in which they have little intrinsic interest or stake in the outcome. Subsequent work on such search tools must ensure that participants are provided with tasks that capture their interest in an ecologically valid manner. Further, larger-scale work is also required to determine how to aggregate path information from many searchers and how to effectively present that information to users.

Our study does nonetheless suggest that path information may be useful to Web searchers. Research in cognitive science has revealed that human search mechanisms in non-physical environments remain deeply connected to evolved foraging and spatial search processes, and work of this nature thus can inform both our understanding of how individuals interact with information search systems, and the design of tools to facilitate search in such environments. Our study focused on one particular application, namely applying notions of spatial path following to a Web search environment. Our hope is that this work can serve as inspiration for further exploration of how path information can be leveraged in Web search, and for applications of cognitive science research about search behavior to the improvement of online information search systems more generally.

References

1. Theraulaz, G., Bonabeau, E.: A brief history of stigmergy. Artificial Life 5(2), 97–116 (1999)
2. Helbing, D., Molnar, P., Farkas, I.J., Bolay, K.: Self-organizing pedestrian movement. Environment and Planning B: Planning and Design 28(3), 361–384 (2001)
3. Goldstone, R.L., Roberts, M.E.: Self-organized trail systems in groups of humans. Complexity 11(6), 43 (2006)
4. Hills, T.T.: Animal foraging and the evolution of goal-directed cognition. Cognitive Science 30(1), 3–41 (2006)
5. Hills, T.T., Todd, P.M., Goldstone, R.L.: Search in external and internal spaces: Evidence for generalized cognitive search processes. Psychological Science 19(8), 802–808 (2008)
6. Todd, P.M., Hills, T.T., Robbins, T.W. (eds.): Cognitive Search: Evolution, Algorithms, and the Brain. MIT Press (2012)
7. Glenberg, A.M.: Embodiment as a unifying perspective for psychology. Wiley Interdisciplinary Reviews: Cognitive Science 1, 586–596 (2010)

8. Pirolli, P., Card, S.: Information foraging in information access environments. In: Proc. of the SIGCHI Conf. on Human Factors in Computing Systems, pp. 51–58 (1995)
9. Pirolli, P.: Information foraging theory: Adaptive interaction with information. Oxford University Press, USA (2007)
10. Hills, T.T., Todd, P.M., Goldstone, R.L.: The central executive as a search process: priming exploration and exploitation across domains. Journal of Experimental Psychology: General 139(4), 590 (2010)
11. Wu, J., Aberer, K.: Swarm intelligent surfing in the gweb. In: Cueva Lovelle, J.M., González Rodríguez, B.M., Gayo, J.E.L., del Pueto Paule Ruiz, M., Aguilar, L.J. (eds.) ICWE 2003. LNCS, vol. 2722, pp. 431–440. Springer, Heidelberg (2003)
12. Kantor, P.B., Boros, E., Melamed, B., Menkov, V.: The information quest: A dynamic model of user's information needs. In: Proc. of the American Society For Information Science Annual Meeting, vol. 36, pp. 536–545 (1999)
13. Furmanski, C., Payton, D., Daily, M.: Quantitative evaluation methodology for dynamic, web-based collaboration tools. In: Proceedings of the 37th Annual Hawaii International Conference on System Sciences, p. 10. IEEE (2004)
14. Gayo-Avello, D., Brenes, D.J.: Making the road by searching - a search engine based on swarm information foraging. CoRR abs/0911.3979 (2009)
15. Attfield, S., Kazai, G., Lalmas, M., Piwowarski, B.: Towards a science of user engagement. In: WSDM Workshop on User Modelling for Web Applications (2011)
16. McCay-Peet, L., Lalmas, M., Navalpakkam, V.: On saliency, affect and focused attention. In: Proceedings of the ACM Conference on Human Factors in Computing Systems, pp. 541–550 (2012)
17. O'Brien, H.: Exploring user engagement in online news interactions. Proceedings of the American Society for Information Science and Technology 48(1), 1–10 (2011)
18. O'Brien, H.L., Toms, E.G.: Is there a universal instrument for measuring interactive information retrieval?: The case of the user engagement scale. In: Proceedings of the Third Symposium on Information Interaction in Context, pp. 335–340 (2010)
19. O'Brien, H., Toms, E.: The development and evaluation of a survey to measure user engagement. Journal of the American Society for Information Science and Technology 61(1), 50–69 (2009)
20. Hoeber, O., Yang, X.D.: A comparative user study of web search interfaces: HotMap, Concept Highlighter, and Google. In: IEEE/WIC/ACM International Conference on Web Intelligence, pp. 866–874 (2006)
21. Capra, R.: HCI Browser: A Tool for Administration and Data Collection for Studies of Web Search Behaviors. In: Marcus, A. (ed.) HCII 2011 and DUXU 2011, Part II. LNCS, vol. 6770, pp. 259–268. Springer, Heidelberg (2011)

Multi-objective Optimization for Multi-level Networks

Brandon Oselio, Alex Kulesza, and Alfred Hero*

Department of Electrical Engineering and Computer Science,
University of Michigan, Ann Arbor, MI, USA
{boselio,kulesza,hero}@umich.edu

Abstract. Social network analysis is a rich field with many practical applications like community formation and hub detection. Traditionally, we assume that edges in the network have homogeneous semantics, for instance, indicating friend relationships. However, we increasingly deal with networks for which we can define multiple heterogeneous types of connections between users; we refer to these distinct groups of edges as layers. Naïvely, we could perform standard network analyses on each layer independently, but this approach may fail to identify interesting signals that are apparent only when viewing all of the layers at once. Instead, we propose to analyze a multi-layered network as a single entity, potentially yielding a richer set of results that better reflect the underlying data. We apply the framework of multi-objective optimization and specifically the concept of Pareto optimality, which has been used in many contexts in engineering and science to deliver solutions that offer tradeoffs between various objective functions. We show that this approach can be well-suited to multi-layer network analysis, as we will encounter situations in which we wish to optimize contrasting quantities. As a case study, we utilize the Pareto framework to show how to bisect the network into equal parts in a way that attempts to minimize the cut-size on each layer. This type of procedure might be useful in determining differences in structure between layers, and in cases where there is an underlying true bisection over multiple layers, this procedure could give a more accurate cut.

1 Introduction

Network analysis has been an important research field for many communities in the past few decades. In a sociology context, social network analysis has been useful in examining complex relationships in communities. There is also a strong tradition of network analysis in the complex systems community, whose motivation lies in the application to physics. Increasingly, researchers find themselves able to frame interesting problems using a network structure.

When operating in a network structure, we often assume that the relations we define between nodes (e.g., users) are homogeneous. However, we are increasingly faced with networks that present the challenge of heterogeneous links between

* This work was partially supported by ARO grant number W911NF-12-1-0443.

W.G. Kennedy, N. Agarwal, and S.J. Yang (Eds.): SBP 2014, LNCS 8393, pp. 129–136, 2014.

nodes. The naïve method of simply ignoring the different types of links in a network would potentially result in a substantial loss of information about the network itself. An example on social networks is easily imagined. First, we could have multiple social networks on overlapping sets of users that we wish to analyze simultaneously, a situation that is considered in [1]. Second, even links derived from a single social network may not be homogenous. Direct relationships, for example, might be derived from explicit user actions that establish a link, from observed contact within a time period, or from other measures that capture the *extrinsic* connections between agents. We could also obtain *intrinsic* connections based on ancillary data; these types of connections attempt to measure how similarly two agents behave. This separation of extrinsic and intrinsic relationships has also been explored in the past [2].

Our proposed formalism for networks with multiple types of links is to assign each type of link to its own *layer*. In each layer we are able to once again impose the homogeneous assumption that is used in single-layer network analysis. In addition to the term multi-layer, multi-relational and multiplex networks have been used to describe this heterogeneous edge structure. When using the term multi-layer, we do not mean separate entities of a larger structure (for instance, when we say layers of TCP/IP protocol), but rather networks of homogeneous edges on the same nodes. While this multi-layer structure is still ill-suited to the current single-layer network tools, it allows us to extend those tools in several ways.

One technique for extending these single-layer network tools into a multi-layer network paradigm comes from multi-objective optimization (MOO). Methods that can be distilled into optimization problems can be solved in the multi-layer setting through a MOO framework that trades off between different objective functions from each layer, finding solutions that are appropriately optimal. One characterization of optimality in this problem is defined by the Pareto front; by exploring this front we can traverse the solutions that are optimal across all possible trade-offs between layers.

In the following section, we will briefly discuss some of the literature on multi-layer networks, multi-objective optimization, and some applications of this framework to networks from prior work. We lay out a mathematical formulation for multi-layered networks, introduce some key concepts of multi-objective optimization. We then describe a well known algorithm for graph bisection, and show how they can be modified for multi-layer networks, including some empirical results for synthetic and real datasets. Finally, we draw some conclusions about the applications of these type of methods for multi-level networks in the future.

2 Related Work

While single-layer network analysis has a large body of work behind it, multi-level network analysis is a field that is just beginning to come into its own as a separate research field. The work in [2,3,1] applies some multi-layer concepts to social networks. Results in [4] propose a multi-level framework based on tensors to represent the new levels of information, and discuss different metrics on this structure.

Some results have been shown in the multi-level setting, including spectral properties [5] and clustering coefficients [6]. [7,8] have explored propagation through a multi-level network. [1] studied specifically how to extend shortest path distances to the multi-level framework, and also used the concept of Pareto optimality to analyze the multilevel structure; their main focus of application was social networks.

Multi-objective optimization has a large body of research to support it. Most current work in this field is in solution attainment. By far the most popular method of solving MOO problems is through some form of evolutionary algorithm (for a popular example, see [9]). Other less ubiquitous measures are summarized in [10]. In the case study that we present, we are actually only interested in a sorting algorithm used for domination points. The method used in this paper is part of the evolutionary algorithm described in [9]. Some interesting application work has been done using MOO [11], including supervised and unsupervised learning.

3 Multi-level Networks

A multi-layer network $G = (\mathcal{V}, \mathcal{E})$ comprises vertices $\mathcal{V} = \{v_1, \ldots, v_p\}$, common to all layers, and edges $\mathcal{E} = (\mathcal{E}_1, \ldots, \mathcal{E}_M)$ in M layers, where $\mathcal{E}_k$ is the edge set for layer k, and $\mathcal{E}_k = \{e^k_{v_i v_j}; \quad v_i, v_j, \in V\}$. A visual depiction of the layers is shown in Figure 1. We allow for an edge $e^k_{v_i v_j} \in \mathcal{E}_k$ to be a real number in $[0, 1]$. We will consider only undirected edges, though extensions to the directed case are not difficult. The degree of a node i is $d^i \in \mathbb{R}^M$, with each entry $[d^i]_k$ being the degree of node i on the single layer k.

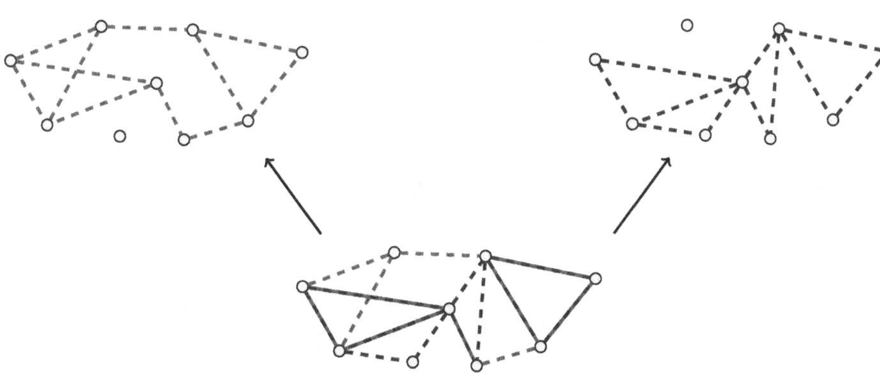

Fig. 1. A visual depiction of a multi-layer network with two types of links. The types of links are separated into separate layers, but the nodes remain the same.

We can now define the adjacency matrix and degree matrix for each layer:

$$[[A^k]]_{ij} = e^k_{v_i v_j} \qquad D^k = \text{diag}([d^1]_k, [d^2]_k, \ldots, [d^p]_k) \tag{1}$$

Note that D^k is simply a $p \times p$ diagonal matrix with the layer-specific node degrees on the diagonal. The purpose of defining both the adjacency and degree matrices will become clear when we attempt the graph bisection across multiple layers; we will use the Laplacian matrix of each layer.

4 Multi-objective Optimization

Multi-objective optimization is a general framework for solving optimization problems when there is more than one objective function to be minimized. Often, these objective functions can contradict each other, so that their individual minimizations lead to solutions that are far away. Thus, the first step in this type of optimization problem is to define what an optimal solution is; we do this with the concept of non-dominated solutions.

Formally, we define the following multi-objective optimization problem:

$$\hat{x} = \underset{x}{\operatorname{argmin}}[f_1(x), f_2(x), \ldots, f_n(x)] . \tag{2}$$

We are interested in solutions that are called *non-dominated* solutions. A solution y^* is dominated by the solution x^* if for all i between 1 and n, $f_i(x^*) \leq f_i(y^*)$, and for at least one j between 1 and n, $f_j(x^*) < f_j(y^*)$. We call the set of feasible solutions that are not dominated by any other solutions the first Pareto front. The Pareto front contains solutions that are not dominated by any other feasible solution; it is not possible to do better than a solution on the Pareto front in all objective functions. In our MOO framework, we say that solutions in the first Pareto front are optimal. A visualization of a Pareto front for $n = 2$ is shown in Figure 2.

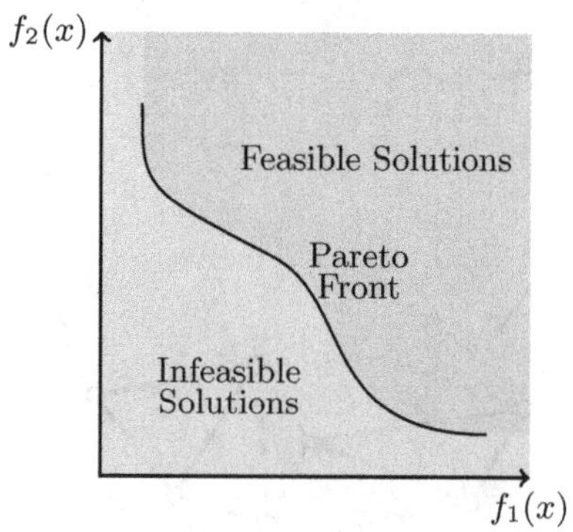

Fig. 2. An example of a Pareto front for two objective functions. An important aspect of this example is that the Pareto front is non-convex; therefore, a weighted linearization search strategy will not explore the entire front.

There are a variety of approaches to obtaining solutions for Equation 2. Perhaps the most basic is to linearize the problem and solve the corresponding scalar optimization

$$\hat{x} = \underset{x}{\operatorname{argmin}} \sum_{i=1}^{n} \alpha_i f_i(x) \tag{3}$$

for some set of weights $\{\alpha_i\}$. This approach is advantageous because it distills the problem down to a single optimization problem for which there are many standard methods. There are two main disadvantages, however. First, it is up to the user to choose the weights α_i in advance, or through trial and error, which can be difficult in practice. The second and perhaps more pervasive problem is that this procedure will only recover the Pareto front if the solution space and all objective functions are convex [12]. When convexity does not hold, this procedure can only find a subset of the feasible Pareto solutions. Two approaches that avoid this problem are ϵ methods and goal attainment, although both are very sensitive to parameter settings. The most popular methods for finding an approximate Pareto front are evolutionary algorithms. These algorithms use heuristic concepts from biology, along with some parameters and randomly selected seed cases to attempt to find solutions on the Pareto front by propagation. More details can be found in [9,10] and references therein.

Another strategy is to avoid the heavy computational and analytical burden of computing an exact Pareto front. If it is possible to obtain a sample of solutions that are likely to be on or near the front, we can sort these points for non-domination. In this way, we can filter a large set of solutions to find the optimal ones that are worth further consideration. In the next section we show how, given two solutions that are assumed to be approximately Pareto optimal, a greedy, recursive algorithm can be used to find more approximately non-dominated points.

5 Approximate Network Bisection of Social Networks

We first start with a brief introduction to spectral graph bisection, including the approximate solution given by spectral clustering. A more detailed overview is given in [13]. For now, let us assume our network has only one layer. We wish to find the vector $C^* \in \{1,2\}^p$ that forms the minimum ratio cut [13], i.e., that minimizes

$$f(C) = \frac{1}{2}\sum_{k=1}^{2} \frac{\text{cut}(C)}{|\{i : C(i) = k\}|} \qquad \text{cut}(C) = \sum_{C(i)=1, C(j)=2} [A]_{ij} \tag{4}$$

It can be shown that an approximately optimal solution $\hat{C}^*$ to this problem is found by using the second to last eigenvector u_{n-1} of the Laplacian matrix $L = D - A$ as an indicator vector, so that

$$\hat{C}^*(i) = \begin{cases} 1 : u_{n-1}(i) \leq 0 \\ 2 : u_{n-1}(i) > 0 \end{cases} . \tag{5}$$

Since finding an optimal partition on even one layer is already computationally demanding, using any type of comprehensive search technique across the solution space to find a Pareto front would be infeasible for multiple layers. However, we can use the approximately optimal solution above to our advantage. For this algorithm we assume there are two layers, and we have the spectral clustering solution from both. If we assume that these solutions are exactly optimal for the individual layers, then they must also be weakly Pareto optimal on the multi-level network since for at least one layer they cannot be improved.

The goal now is to leverage these Pareto optimal solutions to discover an approximate Pareto front. Beginning from one of the spectral solutions, we consider a greedy algorithm that attempts to improve in the other layer as much as possible by swapping two nodes from the different partitions with each other. We only swap nodes that differ in the two spectral solutions. To determine which nodes to swap, we calculate the cost of each node with respect to the non-optimized layer, where the cost function is defined as

$$\mathrm{Cost}(v_i) = \sum_{j:C(j)=C(i)} e_{ij}^2 - \sum_{j:C(j)\neq C(i)} e_{ij}^2 \tag{6}$$

Note that this procedure, while giving us the optimal one-step swaps in the network, is in no way guaranteed to give us Pareto optimal solutions. It will, however, create a solution path between the two spectral solutions, and a non-dominated sorting algorithm like that described in [9] can be used. We then have multiple approximately optimal solutions to choose from.

6 Results

We tested this greedy algorithm on synthetic multi-level networks. For our experiments, we used an unweighted network of 500 nodes, whose average degree was 50. The first layer was constructed using an Erdös-Rényi model with each node having an average degree of 50. We then changed a variable percentage of the edges in that layer to create the second layer, and ran the greedy algorithm to construct an approximate Pareto front. Figure 3 shows some example results for differing levels of variation between layers.

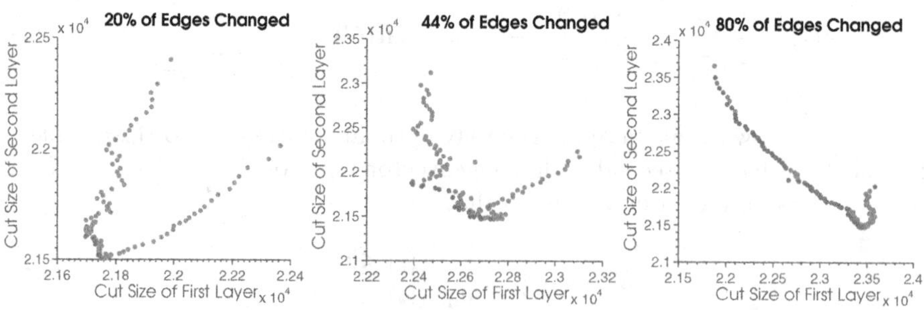

Fig. 3. Pareto fronts for different levels of similarity. The greedy path between the spectral solutions is shown in blue; those points that are weakly non-dominated, and thus make up the approximate Pareto front, are shown in red.

Changing the variation between layers changes the nature of the solution path that is tested, as well as the resulting non-dominated set. Layers that were more similar actually were able to do better than their initializations; the Pareto front in these cases does not include the points that we assumed to be approximately optimal. As the layers become dissimilar, we are not able to improve as much on the starting points; at 80% dissimilarity almost every solution explored was part of the non-dominating set. This implies that with almost every swap, the tradeoffs to be had could almost never do better in both cut-sizes. Moreover, as the layer become more dissimilar, the overall cut-size increases.

Figure 4 displays the results of running the same algorithm on the ENRON email dataset. This dataset is a collection of emails that were publicly released as a result of an SEC investigation; it consists of approximately half a million messages sent to or from a set of 150 employees. It covers a span of approximately 4 years from 1998 to 2002, though the density of emails is varied over that time. We split the emails into week-by-week periods and built a multi-layer graph for each period. The first (extrinsic) layer was created by placing an edge of weight of 1 between individuals that had correspondence over the course of that week. The second (intrinsic) layer was created by measuring semantic correlation in the email body using the TF-IDF score. These values are then thresholded to form edges, with the threshold dependent on the desired sparsity level.

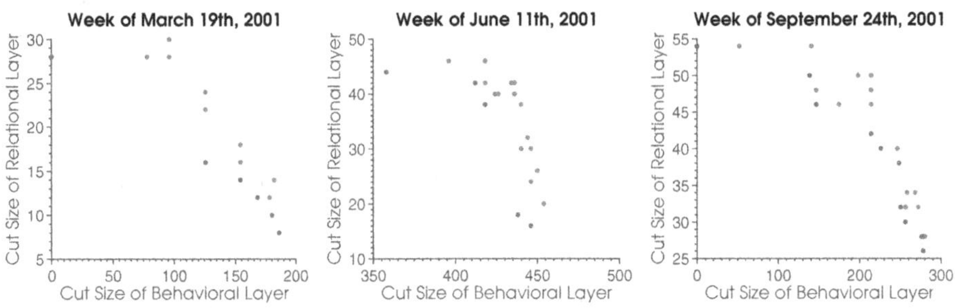

Fig. 4. Pareto fronts from ENRON Email Dataset. These Pareto fronts are derived from the cut sizes of extrinsic and intrinsic layers.

Note that the Pareto fronts in Figure 4 do not appear to be convex. This is interesting because it implies that simply minimizing a weighted combination of objective functions would not generate the full space of potentially interesting solutions. By exploring the Pareto front we get a more nuanced view of the data. We also see a large variation in cut-sizes; in two of the cases, we see that the cut-size in one layer reaches 0, while the other cut size is still much larger. This implies that that the layers are sparse enough to be bisected almost exactly. The difference in optimal bisections between the two layers implies that that the layers have distinct properties. We also notice that the cut-size on the behavioral layer tends to be much larger than that of the relational layer; this is because the behavioral layer is less sparse.

7 Conclusion

Multi-level network analysis is of growing interest as we are faced with increasingly complex data. The framework we describe offers one possible methodology, and can also be applied to allow extensions of single-layer algorithms in the multi-layer setting. We introduced a simple algorithm for finding an approximate Pareto front in two-layer graph bisection; this algorithm in some cases used layer similarity to improve what could be done using the layers individually. In ongoing work we are extending this framework to more complicated models and algorithms.

References

1. Magnani, M., Rossi, L.: Pareto distance for multi-layer network analysis. In: Greenberg, A.M., Kennedy, W.G., Bos, N.D. (eds.) SBP 2013. LNCS, vol. 7812, pp. 249–256. Springer, Heidelberg (2013)
2. Jung, J.J., Euzenat, J.: Towards semantic social networks. In: Franconi, E., Kifer, M., May, W. (eds.) ESWC 2007. LNCS, vol. 4519, pp. 267–280. Springer, Heidelberg (2007)
3. Mika, P.: Ontologies are us: A unified model of social networks and semantics. Web Semantics: Science, Services and Agents on the World Wide Web 5(1) (2007)
4. De Domenico, M., Solè-Ribalta, A., Cozzo, E., Kivelä, M., Moreno, Y., Porter, M.A., Gòmez, S., Arenas, A.: Mathematical Formulation of Multi-Layer Networks (September 2013)
5. Radicchi, F., Arenas, A.: Abrupt transition in the structural formation of interconnected networks. Nature Physics (September 2013)
6. Cozzo, E., Kivelä, M., Domenico, M.D., Solé, A., Arenas, A., Gómez, S., Porter, M.A., Moreno, Y.: Clustering coefficients in multiplex networks. CoRR abs/1307.6780 (2013)
7. Granell, C., Gómez, S., Arenas, A.: Dynamical interplay between awareness and epidemic spreading in multiplex networks. Phys. Rev. Lett. 111, 128701 (2013)
8. Yağan, O., Gligor, V.: Analysis of complex contagions in random multiplex networks. Phys. Rev. E 86, 036103 (2012)
9. Deb, K., Pratap, A., Agarwal, S., Meyarivan, T.: A fast and elitist multiobjective genetic algorithm: Nsga-ii. IEEE Transactions on Evolutionary Computation 6(2), 182–197 (2002)
10. Caramia, M., Dell'Olmo, P.: Multi-objective optimization. In: Multi-objective Management in Freight Logistics, pp. 11–36. Springer, London (2008)
11. Jin, Y., Sendhoff, B.: Pareto-Based Multiobjective Machine Learning: An Overview and Case Studies. IEEE Transactions on Systems, Man, and Cybernetics, Part C: Applications and Reviews 38(3), 397–415 (2008)
12. Geoffrion, A.M.: Proper efficiency and the theory of vector maximization. Journal of Mathematical Analysis and Applications 22(3), 618–630 (1968)
13. Luxburg, U.: A tutorial on spectral clustering. Statistics and Computing 17(4), 395–416 (2007)

Cover Your Cough! Quantifying the Benefits of a Localized Healthy Behavior Intervention on Flu Epidemics in Washington DC

Nidhi Parikh, Mina Youssef, Samarth Swarup, Stephen Eubank, and Youngyun Chungbaek

Network Dynamics and Simulation Science Laboratory,
Virginia Bioinformatics Institute,
Virginia Tech,
1880 Pratt Drive, Blacksburg, VA 24060, USA
{nidhip,myoussef,swarup,seubank,ychungba}@vbi.vt.edu

Abstract. We use a synthetic population model of Washington DC, including residents and transients such as tourists and business travelers, to simulate epidemics of influenza-like illnesses. Assuming that the population is vaccinated at the compliance levels reported by the CDC, we show that additionally implementing a policy that encourages healthy behaviors (such as covering your cough and using hand sanitizers) at four major museum locations around the National Mall can lead to very significant reductions in the epidemic. These locations are chosen because there is a high level of mixing between residents and transients. We show that this localized healthy behavior intervention is approximately equivalent to a 46.14% increase in vaccination compliance levels.

Keywords: disease dynamics, intervention strategies, synthetic social network, transient population.

1 Introduction

Influenza transmission is a big concern for society. There are about 24.7 million influenza cases annually in the US and its economic burden is estimated to be about $87.1 billion [1]. The CDC-recommended policy is for everyone over 6 months of age to be vaccinated (with rare exceptions). However, the effectiveness of such pharmaceutical interventions depend upon public adherence. The compliance level varies quite a bit by age-group, and is below 50% for ages 13-64 across the entire United States (see Table 1).

For big cities, such as Washington DC, apart from the contact patterns among residents, the spread of such infectious diseases is also affected by the presence of transients (tourists and business travelers), because they provide a constant pool of susceptible people who can get infected and pass on the disease easily as they visit crowded areas and come into contact with each other and with residents [2]. In the present work, we evaluate the effects of augmenting the vaccination policy with a relatively mild, location specific, healthy behavior intervention

W.G. Kennedy, N. Agarwal, and S.J. Yang (Eds.): SBP 2014, LNCS 8393, pp. 137–144, 2014.

Table 1. Compliance levels for 2012-13 influenza season, reported by CDC

Age group	Compliance for District of Columbia	United States
6 months to 4 years	85.9	69.8
5 to 12 years	71.4	58.6
13 to 17 years	56.7	42.5
18 to 49 years	34.5	31.1
50 to 64 years	50.2	45.1
65 years and over	63.6	66.2

(like the use of hand sanitizers, covering coughs, minimizing contact with potential fomites) at major tourist locations. We use a synthetic population-based model for resident and transient populations in Washington DC.

Some previous studies have suggested that transients play an important role in epidemics [3–5]. However, they are limited either in assuming homogeneous mixing among subpopulations or assuming limited mixing among residents and transients (at hotels). In reality, big cities have many tourist destinations which are quite crowded and visited by many transients. In our previous work, we have modeled tourists and business travelers in detail for Washington DC using a synthetic population model [2], where they visit various places to perform activities like tourism, eating, night life, and work. We also show that closing museums (a social distancing intervention) does not help in reducing the epidemic but promoting healthy behavior at major tourism locations could significantly reduce the epidemic.

In this work, we attempt to quantify the effect of healthy behavior in terms of vaccination compliance for the Washington DC metro area. The results show that if, in addition to the current level of vaccination compliance, people engage in healthy behavior at four major museums, it can reduce peak and cumulative resident infections by 51.2% and 34.58%. It can also delay the epidemic by 3 weeks. We also show that this location specific intervention is approximately equivalent to having 70-75% compliance rate for vaccination across the entire population, which is equivalent to a 46.14% increase in the number of people vaccinated.

2 The Synthetic Population

A synthetic population represents individuals along with demographics (i.e., age, income, family structure) and mobility related (i.e., type of activity, time, location) information. Hence it models the population dynamics (including interactions among individuals) of a region. It is constructed using data from multiple sources (Table 2) and has been used to study epidemic outbreaks [6].

We use a previous developed synthetic resident population for Washington DC metro area (about 4.1 million) and synthetic transient population (about 50000). The process of generating resident and transient populations is briefly outlined below.

Table 2. Datasets used for population generation

Synthetic Population	Data source	Information provided
Base US population	American Community Survey	Demographic and sample household data
	National Center for Education Stat.	Data about schools and children
	National Household Travel Survey	Activity/travel survey
	Navteq	Street data
	Dun & Bradstreet	Business location data
Transient population	Destination DC	Demographic data
(additional)	Smithsonian Institution	Visit counts at Smithsonian locations

2.1 Base Population (Residents)

Generating Synthetic Population: This step generates synthetic individuals (along with demographics) grouped into households. This is done by combining marginal demographic distributions and sample household data from the American Community Survey (ACS) using iterative proportional fitting algorithm [7]. The resulting population matches marginal demographic distributions from ACS at census block group level and preserves the anonymity of individuals.

Locating Households: Each synthetic household is then assigned a home location using street data from Navteq and housing structure data from the ACS.

Assigning Activities: In this step, each synthetic individual is assigned a schedule of activities (i.e., home, work, school, shop) to perform during the course of a day. Activity templates are created using the National Household and Travel Survey (NHTS) and data from the National Center for Education Statistics (NCES). Each synthetic household is mapped to a survey household based in its demographic and synthetic individuals are assigned corresponding activities.

Locating Activities: An appropriate activity location (building) is assigned for each activity of an individual. This is done using the gravity model and SIC (Standard Industrial Classification) codes from Dun and Bradstreet data.

Sublocation Modeling and Constructing Synthetic Social Contact Network: Each location is subdivided into sublocations (similar to rooms within a building). Each sublocation has a capacity which depends upon the type of activity performed there. The number of sublocations for a location is obtained by taking into account the maximum number of people present at that location at any time and the capacity of sublocations. Each person is assumed to be in contact with all individuals presents at the same sublocation at the same time, which induces a synthetic social contact network.

2.2 Transient Population

We use the same method for creating the synthetic transient population as for creating the synthetic resident population, but with different data sources. According to Destination DC[1], there are about 50000 visitors (55% of which are

[1] `http://washington.org`

tourists and the rest are business travelers) visiting Washington DC on any given day. It also provides marginal demographic distributions about transients which are used in conjunction with some rules about the party structure using sampling without replacement technique to generate transients along with their demographics. Tourists are grouped into parties (i.e., groups of people traveling together) and business travelers are assumed to be by themselves. All individuals in a party are assumed to stay at the same hotel, preferentially near the down town. Hotel locations have been identified using DNB data.

As all individuals in a party are assumed to travel together, they are assigned the same set of activities to perform during a day like staying in a hotel, visiting museums and other tourist destinations (or work activities, for business travelers), going to restaurants, and night life activities. Activity locations have been identified using SIC codes from DNB data and activity assignment is calibrated to match the visit counts at Smithsonian Institution locations. Sublocation modeling is done in a similar manner as for the base population, except for four major tourist locations (the National Air and Space Museum (NASM), the National Museum of Natural History (NMNH), the National of American History (NMAH), and the National Gallery of Art (NGA)), where we looked up floor plans to decide the number of sublocations. People are assumed to move from one sublocation (similar to exhibition) to other at 5 to 15 minutes intervals, at these four locations. Details can be found in [2].

3 Experiments

We use EpiSimdemics [8], an interaction-based high performance computing software for simulating flu-like disease in Washington DC. The disease model is as shown in Figure 1. Each node represents a disease state and edges represent transition from one state to other. Edges are labeled with the probability of transition and the nodes are labeled with the number of days for which a person remains in this state and the probability of him infecting others. The histogram in the upper right corner shows the probability of being in the given state versus the number of days for symptom1, symptom2, symptom3, and asymptomatic states. The value of transimissibility (4×10^{-5}) is calibrated to obtain about 30% of the residents to be infected over the period of 120 days, in the absence of transients and interventions.

We assume efficacy of vaccine to be 60%[2], i.e., when a person takes vaccine, with 60% probability, he becomes immune (goes to removed state). Healthy behavior is a mild, location specific intervention, implemented at four major museums: NASM, NMNH, NMAH, and NGA. We choose these locations as they are locations of high mixing, having about 40000 to 80000 visits per day. We assume that whenever a person engages in healthy behavior (e.g., covering cough, using hand sanitizers), his susceptibility and infectivity reduce to 60% of their original values. This is only effective for the time a person is within these museums. We assume 50% of the people visiting these museums engage in healthy behavior.

[2] `http://www.cdc.gov/flu/about/qa/vaccineeffect.htm`

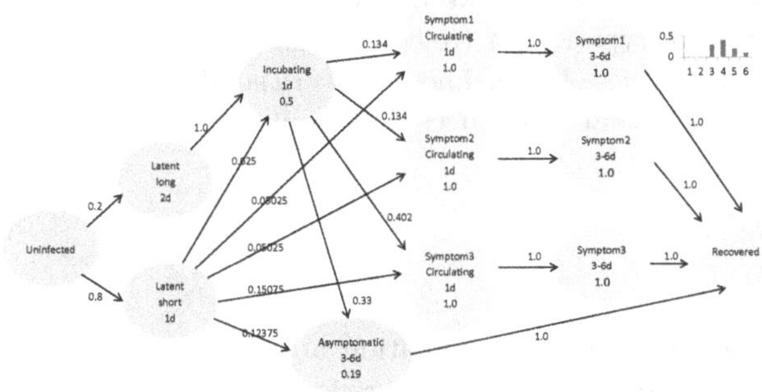

Fig. 1. The 12-state Probabilistic Timed Transition System (PTTS) disease model

Our goal is to quantify the benefit of healthy behavior intervention in addition to current vaccination compliance levels. We present results for the following scenarios: no intervention, vaccination at current compliance levels (Table 1), vaccination at current compliance levels plus healthy behavior at four major museums, and vaccination at 70% and 75% compliance levels. Only residents are initially infected. So transients do not bring disease to the city. However during their stay in the city, they may get infected and infect others. This represents the best case scenario, epidemic outcomes are worse if transients are allowed to be infected when they arrive. As per Destination DC, the average length of stay for transients is 5 days, hence we assume that every day approximately 20% of the transients leave the city and they are replaced by new, incoming transients with exactly the same demographics and activity patterns.

Table 3. The number of simulations when there was an outbreak

Scenario	Total number of simulations	Number of simulations with epidemic outbreak
No intervention	100	99
Vaccination at CDC compliance levels	100	69
Vaccination at CDC levels + healthy behavior	100	49
Vaccination with 70% compliance	100	40
Vaccination with 75% compliance	100	31

We run 100 simulations for each scenario. In some cases, initial infections do not result into an outbreak. Table 3 shows the number of cases with epidemic outbreak. For statistical tests, we only consider cases when there was an outbreak and also remove outliers (Outliers are detected by SPSS as points which fall outside the whiskers of box-plots. They shift the mean of the group which is used in t-test for comparison.). Pairwise comparison using t-tests require the data to be normally distributed. As we have enough number of samples (even after removing outliers), we can assume data to be normally distributed using central limit theorem. We perform Levene's test (for equality of variance) for determining which t-test to perform. If two groups have equal variances then we do a two sample t-test, otherwise we do Welch's t-test. Doing multiple t-tests can potentially lead to type I error (rejecting null hypothesis when it is actually true). However, all p-values obtained here are very small ($\leq 4.085 \times 10^{-6}$). Hence, we are fairly confident that differences are significant whenever the t-test test suggests so.

3.1 Results

First, we evaluate the effect of vaccination at current compliance levels. We use 2012-13 compliance rates by age, as reported by the CDC (Table 1). We use compliance rates for DC and US for residents and transients, respectively. 69 cases resulted in an outbreak with current vaccination compliance levels as opposed to 99 cases in absence of it. Figure 2 shows the fraction of residents infected at peak, cumulatively over 120 days, and the day of peak as box-plots. Results (Table 4) show that vaccination at current compliance rate reduces the fraction of residents infected at peak and cumulatively over the period of 120 days by 65.45% and 67.82% (statistically significant), respectively. It also delays the peak by about 2 weeks.

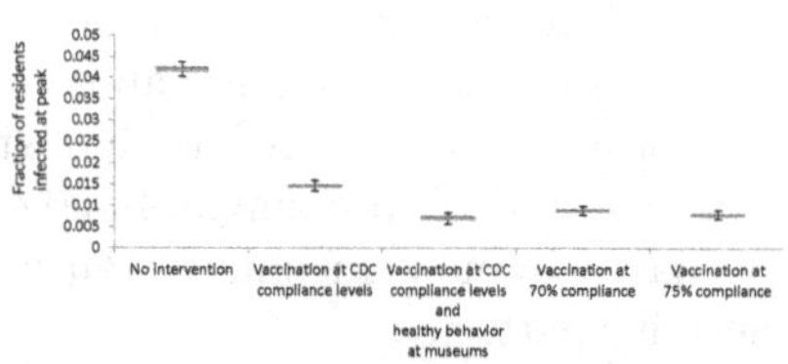

(a) Fraction of residents infected at peak.

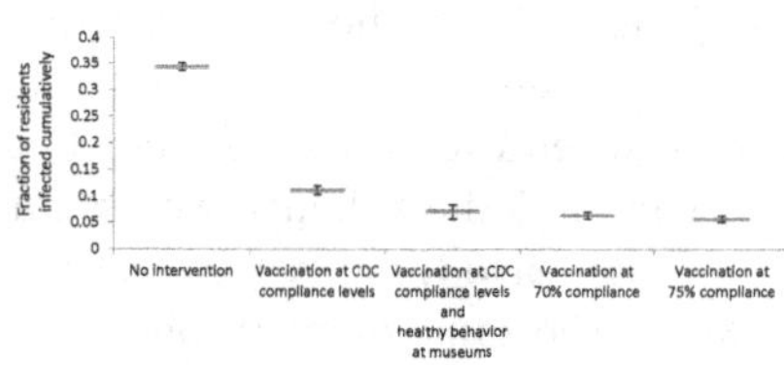

(b) Fraction of residents infected cumulatively over 120 days.

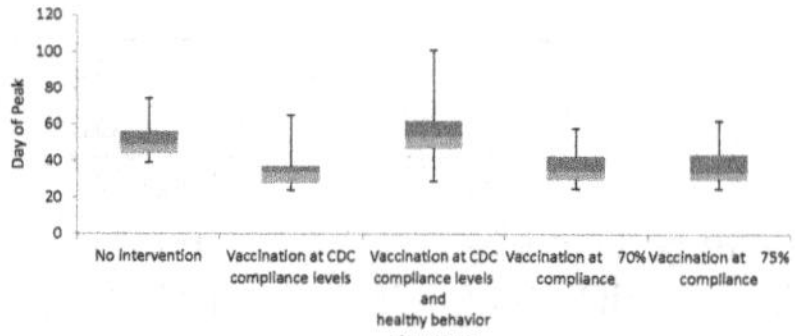

(c) Day of peak.

Fig. 2. Comparison of various scenarios (in case of an outbreak and without outliers) in terms of box-plots

Next, we evaluate how much benefit it would yield if people engage in healthy behavior at four major museum locations in addition to current vaccination compliance levels. As opposed to 69, 49 cases resulted in an outbreak when healthy behavior is encouraged at museums. Distributions of the fraction of residents infections at peak, cumulatively over 120 days, and the day of peak are as shown in Figure 2. Results (Table 4) show that it can significantly reduce the peak and total resident infections by 51.2% and 34.58%, respectively. It also delays epidemic by 3 weeks.

Table 4. Statistical tests for comparing various scenarios. Scenarios are numbered as follows: 1 - No intervention, 2 - Vaccination at current compliance levels, 3 - Vaccination at current compliance levels and healthy behavior at major museums, 4 - Vaccination at 70% compliance, and 5- Vaccination at 75% compliance.

Scenarios	Variable	Levene test		t-test						
		statistic	p-val	t	df	p-value	groups means		95% CI	
							group-1	group-2	Lower	Upper
1 and 2	peak infections	3.951	0.049	251.956	146.296	<2.2e-16	0.042	0.015	0.027	0.028
	total infections	0.79	0.376	373.089	156	<2.2e-16	0.344	0.111	0.232	0.234
	day of peak	0.226	0.635	10.282	156	<2.2e-16	50.804	35.443	12.410	18.313
2 and 3	peak infections	1.204	0.275	59.440	105	<2.2e-16	0.015	0.007	0.0071	0.0076
	total infections	6.064	0.015	35.486	73.494	<2.2e-16	0.111	0.072	0.036	0.040
	day of peak	8.795	0.004	-8.444	72.098	2.248e-12	35.442	57.978	-27.855	-17.216
3 and 4	peak infections	5.248	0.025	-13.475	80.673	<2.2e-16	0.007	0.0087	-0.0019	-0.0014
	total infections	10.052	0.002	8.082	70.415	1.241e-11	0.072	0.064	0.007	0.011
	day of peak	10.531	0.002	8.172	69.451	9.265e-12	57.978	36.132	16.514	27.179
3 and 5	peak infections	2.78	0.099	-4.977	74	4.085e-06	0.0071	0.0078	-0.001	-0.0004
	total infections	9.822	0.0024	13.984	69.966	<2.2e-16	0.072	0.057	0.013	0.018
	day of peak	5.87	0.018	7.524	73.218	1.083e-10	57.978	36.367	15.887	27.336

Last, we want to see what level of compliance would be required to get results similar (or comparable) to healthy behavior intervention at targeted tourism locations (on top of current vaccination compliance). If compliance for vaccination is increased to 70% and 75% (for both residents and transients; for all age groups except the groups that already have higher that 70% or 75% compliance), there are still 24.38% and 10.40% (statistically significant; Table 4) more residents infected at peak, respectively, and peak is about 3 weeks earlier as compared to the use healthy behavior in addition of current vaccination compliance levels. It reduces the total number of resident infections over 120 days by 12.05% and 21.21% (statistically significant), respectively, and only 41 and 31 cases resulted in an epidemic outbreak as opposed to 49 cases.

The tradeoff between increased vaccination and healthy behavior is that there are more number of residents infected and more chances of an outbreak, in the latter case, while the number of infections at peak is reduced (which is very important from healthcare planning perspective as it decides the number of resources (i.e., antivirals, beds, etc.) required at a time) and the epidemic is delayed significantly (which gives more time for preparation). The healthy behavior intervention would presumably also be much cheaper to implement.

4 Conclusions

While the CDC recommends flu vaccinations for everyone over 6 months of age, compliance levels remain relatively low. In this work we show that instead of focusing effort on increasing compliance, it may be easier to implement a localized healthy behavior intervention in the Washington DC metro area.

We quantify the effects of the healthy behavior interventions by conducting simulations to find a comparable level of reduction in the epidemic through vaccinations alone. Results suggest that if, in addition to current vaccination compliance levels, people engage in healthy behavior at major tourism locations, it can yield significant benefits (in terms of peak, cumulative infections and the day of the peak). We show that this is comparable to increasing vaccination compliance rate to 70% to 75%.

Limitations: Though the model constructed is as detailed and high-fidelity as possible, it has some limitations. In real world, apart from museums, some other locations (i.e., public transport) also have high mixing and have similar influence on epidemics. However, we expect that our results would hold qualitatively if we include them. We do not differentiate between different mechanisms of flu transmission (i.e., direct, droplet). We assume that engaging in healthy behavior reduces infectivity and susceptibility to 60% of the original values. But in reality, the reduction in infectivity and susceptibility depend upon the type of healthy behavior (i.e., covering cough, using hand sanitizer, etc.). Here also, we expect that this would not change the results qualitatively.

Acknowledgements. We thank our external collaborators and members of the Network Dynamics and Simulation Science Laboratory (NDSSL) for their suggestions and comments. This work is supported in part by DTRA CNIMS Contract HDTRA1-11-D-0016-0001, NIH MIDAS Grant 2U01GM070694-09, and NSF HSD Grant SES-0729441.

References

1. Molinari, N.A.M., Ortega-Sanchez, I.R., Messonnier, M.L., Thompson, W.W., Wortley, P.M., Weintraub, E., Bridges, C.B.: The annual impact of seasonal influenza in the US: Measuring disease burden and costs. Vaccine 25, 5086–5096 (2007)
2. Parikh, N., Youssef, M., Swarup, S., Eubank, S.: Modeling the effect of transient populations on epidemics in Washington DC. Sci. Rep. 3, article 3152 (2013)
3. Colizza, V., Barrat, A., Barthélemy, M., Vespignani, A.: The role of the airline transportation network in the prediction and predictability of global epidemics. PNAS 103 (2006)
4. Colizza, V., Barrat, A., Barthélemy, M., Valleron, A.J., Vespignani, A.: Modeling the worldwide spread of pandemic inuenza: baseline case and containment interventions. PLoS Med. 4 (2007)
5. Ferguson, N.M., Cummings, D.A.T., Fraser, C., Cajka, J.C., Cooley, P.C., Burke, D.S.: Strategies for mitigating an influenza pandemic. Nature 442(7101), 448–452 (2006)
6. Eubank, S.G., Guclu, H., Kumar, V.S.A., Marathe, M.V., Srinivasan, A., Toroczkai, Z., Wang, N.: Modelling disease outbreaks in realistic urban social networks. Nature 429(6988), 180–184 (2004)
7. Beckman, R.J., Baggerly, K.A., McKay, M.D.: Creating synthetic baseline populations. Transportation Research Part A: Policy and Practice 30(6), 415–429 (1996)
8. Bisset, K.R., Feng, X., Marathe, M., Yardi, S.: Modeling interaction between individuals, social networks and public policy to support public health epidemiology. In: Proc. Winter Simulation Conference, pp. 2020–2031 (2009)

Integrating Epidemiological Modeling and Surveillance Data Feeds: A Kalman Filter Based Approach

Weicheng Qian, Nathaniel D. Osgood, and Kevin G. Stanley

Department of Computer Science, University of Saskatchewan
first.last@usask.ca

Abstract. Infectious disease spread is difficult to accurately measure and model. Even for well-studied pathogens, uncertainties remain regarding dynamics of mixing behavior and how to balance simulation-generated estimates with empirical data. While Markov Chain Monte Carlo approaches sample posteriors given empirical data, health applications of such methods have not considered dynamics associated with model error. We present here an Extended Kalman Filter (EKF) approach for recurrent simulation regrounding as empirical data arrives throughout outbreaks. The approach simultaneously considers empirical data accuracy, growing simulation error between measurements, and supports estimation of changing model parameters. We evaluate our approach using a two-level system, with "ground truth" generated by an agent-based model simulating epidemics over empirical microcontact networks, and noisy measurements fed into an EKF corrected aggregate model. We find that the EKF solution improves outbreak peak estimation and can compensate for inaccuracies in model structure and parameter estimates.

Keywords: Kalman Filter, simulation, epidemiology.

1 Introduction

Infectious diseases are notoriously difficult to manage, because they can exhibit great instability, with periods of quiescence interspersed by sudden outbreaks. Anticipating the future behavior of the outbreak and how interventions will affect the disease spread is important for policy makers who must marshal prophylactic and treatment campaigns. However, in the case of emerging pathogens such as SARS or H1N1, the disease dynamics and appropriate treatment regime are unknown [1].

Dynamic models can project possible epidemic evolutions, and aid assessment of tradeoffs between interventions. Models are traditionally parameterized and calibrated when they are constructed, but frequently the underlying parameters are dependent on hard-to-predict dynamic factors such as human contact patterns, diagnosing practices, or even the weather [2]. In particular, dynamic human contact patterns have been shown to play a significant role in the spread of disease [3, 4], and are poorly captured by even the best open-loop models.

Statistical filtering and estimation methods for dynamic models, such as sequential Monte Carlo (SMC) method and Markov chain Monte Carlo (MCMC) methods, can

W.G. Kennedy, N. Agarwal, and S.J. Yang (Eds.): SBP 2014, LNCS 8393, pp. 145–152, 2014.

be used to estimate model parameters as information becomes available [5]. The typical formulation identifies posterior distributions for parameters or outputs conditional on the model, and do not explicitly recognize growing model projection errors. In this paper we demonstrate that the Extended Kalman Filter can be used to adapt a model to better approximate an epidemic outbreak even when the parameters in question are not empirically or logically observable. In particular, we provide the first demonstration of an EKF-enhanced dynamic disease model evaluated against empirically observed and evolving interpersonal contact data. We demonstrate that the EKF enhanced system provides significantly better estimates of the number of infected individuals and the peak timing of the disease than calibrated models, particularly when the model contact rate is allowed to vary with incoming data.

2 Related Work

Dynamic epidemiology models have been used to study a variety of conditions, and have made contributions to all areas of epidemiology [6, 7, 8]. Recently, research has highlighted the importance of population heterogeneity and network structure in shaping outbreak emergence and progression [4]. While this work has elevated the recognized value of agent-based models (ABMs), such models exhibit diverse and textured tradeoffs with aggregate models [9], including the use of ABMs as synthetic ground truth whose dynamics the aggregate model is seeking to adequately characterize [10,11].

Some practitioners have sought to address aspects of parameter uncertainty in dynamic models using Monte Carlo statistical methods [6, 8]. While such approaches can offer great insight into parameter-related uncertainty, using them to understand model-related uncertainty is more involved [9]. Similarly, while SMC methods are applied in the context of stochastic processes [12], the Kalman Filter has been employed to address shortcomings in disease model or calibration including temporal-spatial integration [13], epidemic evolution [14] and time variation of covariates [15].

Recently, researchers have detailed empirical data on dynamic human contact patterns [16, 4, 17, 18] and have indicated that there is a strong heterogeneity in contact patterns [18, 4, 17] and that contact dynamics impact the spread of disease [5, 4]. However, methods for incorporating such statistics tend to work best with ABM and not aggregate dynamic models.

3 Model Description

For simplicity, and to demonstrate that even stylized models benefit from the closed loop design provided by Kalman Filter, we employed classic Susceptible, Infected, Recovered (SIR) aggregate dynamic models. To test these aggregate models, we used agent-based models that are more textured, but still adhere to a similar SIR characterization of health status.

3.1 Agent Model

We used an ABM to provide synthetic ground truth data against which we could compare aggregate model estimates. The ABM employed classic dynamic-network-based SIR formulation whose details as given in [18]. In this model, 36 agents were connected by a dynamic network whose connectivity mirrored that recorded between the 36 participants in [18] over time. Each infectious agent generated exposure events according to a Poisson process with λ=0.003; for each such event, a connected susceptible agent experienced a 25% likelihood of infection.

3.2 Population Models

The SIR model is the most widely known and well-studied model in mathematical epidemiology. In this model, all infected individuals are assumed to mix randomly and recover from infection in a memoryless fashion. The model contains 3 state variables: the number of Susceptible (S), Infectious (I), and Removed (R) individuals, and the parameter τ, the mean time to recovery. The SIR model represents the baseline open loop system for comparison. The system of state equations is:

$$\begin{aligned} \dot{S} &= -\frac{c\cdot\beta\cdot I\cdot S}{S+I+R} \\ \dot{I} &= \frac{c\cdot\beta\cdot I\cdot S}{S+I+R} - \frac{I}{\tau} \\ \dot{R} &= \frac{I}{\tau} \end{aligned} \tag{1}$$

The SIR model assumes a constant mixing parameter, c, which implies emergent behavior unlikely to obtain in many empirical dynamic contact networks. Reflecting the observation that the mean contact rate can change substantially over the course of an outbreak due to both behavioral changes [19] and network heterogeneity [3, 4], we generalize to an SIRc model in which c is changed from a fixed parameter to a state variable of the model, allowing it to change over time. Rather than seeking to impose a flaw-prone behavioral model, the model initially estimates no change in c, forcing updates to c to be entirely data-driven via the EKF.

$$\begin{aligned} \dot{S} &= -\frac{c\cdot\beta\cdot I\cdot S}{S+I+R} \\ \dot{I} &= \frac{c\cdot\beta\cdot I\cdot S}{S+I+R} - \frac{I}{\tau} \\ \dot{R} &= \frac{I}{\tau} \\ \dot{c} &= 0 \end{aligned} \tag{2}$$

3.3 Extended Kalman Filter Model

The EKF provides estimates combining information from both model and measurements. Both SIR and SIRc models employing EKF estimates represent closed loop models. To evaluate EKF effectiveness, we used the ABM to generate noise-corrupted measurements of the count of infected and removed agents at integral times k. For each measurement time k, the EKF were used to derive the new state estimates for k based on a weighted combination of the model output at time k (as shaped by previous EKF updates), and the incoming measurement at time k.

Because the system dynamics are non-linear, we employed an EKF variant known as the continuous-discrete Extended Kalman Filter [20], referred to herein simply as the EKF. The EKF consists of two processes for updating state variables and process covariance estimates (which specify the degree of accuracy of model estimates): continuous time updates – which govern the behavior of the aggregate SIR model and process covariance estimates between measurement points – and discrete-time measurement updates – which modify system state and process covariance estimates with a consensus estimate derived from both measurement and model estimates. Interested readers are referred to [20] for details. A standard EKF formulation posits a state vector $x(t)$ governed by a set of state equations with a non-linear right hand side, $f(x(k),k)$ disturbed by Gaussian process noise, and a measurement-state mapping function which maps states to expected measurements $h(x(k),k)$. Typically, f and h are non-linear, so the EKF approximates each with a first-order Taylor series approximation.

Initial values for most covariant matrices were based on random values about a fixed point, based on the assumed confidence in the state estimate in the dynamic model. The process noise covariance matrix was initialized to expect contagious contacts with between 0 and 3 individuals, based on empirical data [18] and observation.

In the model employed in these simulations, we measure I and R using artificially corrupted measurements of the ABM ground truth at time k, and estimate I, R, and c as model state variables. Starting from the initial conditions or the "consensus" results of the previous measurement update (whichever is later), the ODE for both the process state equations and covariance matrix P are numerically integrated from the previous state until the next measurement update. Reflecting ongoing process noise, the process covariance generally grows during such times. At the next measurement, the gain matrix K is calculated, and the EKF equations are used to generate new consensus state estimate based on the measurements, model state estimates and covariance matrices. In the event of physically impossible situations – such as if S, I, and R exceed their upper and lower bounds, or move in a direction that is impossible under the model (for example the number of recovered decreasing) – we reset the Kalman filter to entirely weight towards the measurements.

4 Experimental Setup

For each trial we assume an initial contact rate c of 1. β and τ are assumed to hold the same value in the aggregate and ABM model (β=0.25, τ=7). In the open loop SIR model and between measurements using the EKF, we used MATLAB's function **ode15s** to numerically integrate I and R over time. Since we have a closed population, for time t, $S(t) = N-I(t)-R(t)$.

4.1 Simulation Setup

The SIR model is setup with β=0.25, τ=7 and N=36, providing it with perfect information about the system. The set-by-step solution to the SIR equations at time k are solved using MATLABs **ode15s** solver. These values are directly output for the SIR solution. A variation based on the evolution of the system is used as the input for the EKF. We used similar parameters to [18,3] for key parameters of the ABM: $\beta = 0.25$, $\tau = 7$ and $\alpha = 2$. All possible cases with either one or two initially infected agents were the initial conditions.

4.2 Kalman Filter Configuration

Within the EKF, we used the aggregate model to generate the next state estimate, and assumed that β, τ, α and the noise distribution were well estimated. Three rounds of experiments were performed, first between a traditional SIR model with constant c with and without EKF compensation investigate the filter's performance compensating for measurement noise. The second experiment compared SIR models with estimated c (SIRc models) with ground truth to determine the ability of the EKF enhanced system to compensate for a varying, but unmeasured parameter. For each trial, the initial contact rate c of 1 was used. This was initialized based the output of the ABM to provide a highly plausible starting point for the open loop system.

5 Results

We compared the performance of our EKF-enhanced system against simulated ground truth and the open loop population model. It is important to note that in these comparisons, the only difference is the addition of the EKF mediated feedback. We chose to focus our evaluation on the ability of each algorithm to correctly estimate the number of infected people at each time point. Fig. 1 a and b show two example infection trajectories.

Fig. 1a is an example of a relatively normal trajectory tracking, while Fig 1b in an example of an exceptionally good tracking performance. As is apparent from the figure, the EKF estimation does a substantially better job of tracking the infection than the open loop variant, which diverges from the ground truth rapidly. The measurement provides a better estimate than either the EKF or open loop models, due to the conservative error model we employed; however, the measurement is not predictive, and cannot be used to forecast into the future.

Fig. 1 c and d show two-dimensional histograms of the difference between the synthetic ground truth and the open loop model and the synthetic ground truth and the EKF estimate. The EKF and open loop systems have similar (but opposite in sign) errors at the onset of the outbreak, but the EKF simulations converge more quickly to the desired state, reaching near zero error after 10 days rather than 20 days. Because the SIR system is run open loop, it makes exactly the same prediction regardless of the initial point of infection for the outbreak in the ABM. Variability in the open loop SIR case is entirely due to the variation of the outbreak dynamics with different seed agents and different disease dynamics. The EKF on the other hand attempts to reground the simulation at every measurement point, and the error is determined by the capacity of the EKF to track the epidemic given the model and data. That the variance of the two cases is similar indicates that any additional variance introduced by the EKF estimates is similar to normal variation due to disease dynamics.

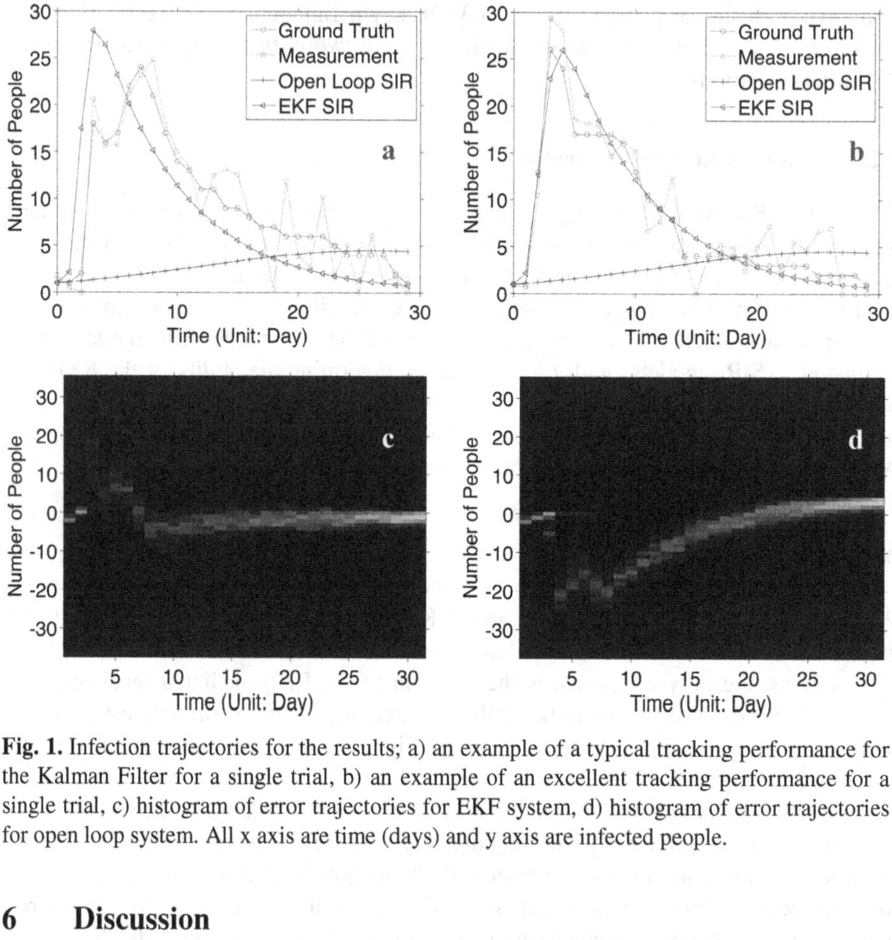

Fig. 1. Infection trajectories for the results; a) an example of a typical tracking performance for the Kalman Filter for a single trial, b) an example of an excellent tracking performance for a single trial, c) histogram of error trajectories for EKF system, d) histogram of error trajectories for open loop system. All x axis are time (days) and y axis are infected people.

6 Discussion

We have proposed a method for integrating epidemiological surveillance data with population based mathematical epidemiology models. This approach creates consensus estimates of underlying epidemiological situation from measured data and model estimates in a way that reflects the confidence modelers place in each. Compared to MCMC implementations, the EKF is computationally parsimonious, allowing for ready incorporation in sensitivity analyses and large-scale scenario exploration.

However, the current approach is accompanied by limitations. The EKF's requirement for the process noise and measurement covariance matrices may be difficult to obtain in practice, but bounding distributions can be readily derived. The assumptions that the system is corrupted by white Gaussian noise could render the model temporarily obsolete if presented with non-Gaussian disturbances. The EKF will attempt to correct for the deviations due to model inaccuracies, distorting the result. However, these risks are no worse than in existing open loop systems.

We believe that there is the potential for such systems to broaden the contributions of models to practical decision-making. The capacity to keep models up to date raises the potential for much greater trust being placed in them. More fundamentally, closed loop models lower the divisions between epidemiological data collection and modeling, encouraging decision makers to consider an integrated process. Finally, modelers have traditionally sought model precision through additional model complexity [9] or ABMs [21]. Closed loop models offer a third way for enhancing prediction reliability.

7 Summary

We have presented a technique for creating closed loop epidemiological models, compensating for dynamic human contact patterns. The technique described here is based on well-established stochastic optimization techniques, and permits flexibly incorporating ongoing data streams data directly into the model. We have established the efficacy using cross-validation across different model types in light of empirical data on observed human contact patterns. The approach is easily integrated with population level models; provides better than open loop estimates with noisy measurements, and is more computationally efficient that MCMC approaches. This work has the potential to serve as a valuable tool for policy makers and researchers alike, when attempting to compensate for changing parameters and flawed models during epidemic outbreaks.

References

1. Tuite, A., Greer, A., Whelan, M., Winter, A., Lee, B., Yan, P., Wu, J., Moghadas, S., Buckeridge, D., Pourbohloul, B., et al.: Estimated epidemiologic parameters and morbidity associated with pandemic H1N1 influenza. Canadian Medical Association Journal 182(2), 131–136 (2010)
2. Keeling, M.: The implications of network structure for epidemic dynamics. Theoretical Population Biology 67(1), 1–8 (2005)
3. Hashemian, M., Qian, W., Stanley, K.G., Osgood, N.D.: Temporal aggregation impacts on epidemio- logical simulations employing microcontact data. BMC Medical Informatics and Decision Making 12(1), 132 (2012)
4. Machens, A., Gesualdo, F., Rizzo, C., Tozzi, A.E., Barrat, A., Cattuto, C.: An infectious disease model on empirical networks of human contact: bridging the gap between dynamic network data and contact matrices. BMC Infectious Diseases 13(1), 185 (2013)
5. Mbalawata, I.S., Särkkä, S., Haario, H.: Parameter estimation in stochastic differential equations with markov chain monte carlo and non-linear kalman filtering. Computational Statistics, 1–29 (2012)

6. Dorigatti, I., Cauchemez, S., Pugliese, A., Ferguson, N.M.: A new approach to characterising infectious disease transmission dynamics from sentinel surveillance: Application to the italian 2009–2010 a H1N1 influenza pandemic. Epidemics 4(1), 9–21 (2012)
7. Coelho, F.C., Codeço, C.T., Gomes, M.G.M.: A bayesian framework for parameter estimation in dynam- ical models. PloS One 6(5), e19616 (2011)
8. Osgood, N., Liu, J.: Bayesian parameter estimation of system dynamics models using markov chain monte carlo methods: An informal introduction. In: The 30th International Conference of the System Dynamics Society, p. 19. Curran Associates, Inc, New York (2013)
9. Tian, Y., Osgood, N.: Comparison between individual-based and aggregate models in the context of tuberculosis transmission. In: The 29th International Conference of the System Dynamics Society, Washington, D.C, p. 29 (2011)
10. Wang, F.Y.: Toward a revolution in transportation operations: Ai for complex systems. IEEE Intelligent Systems 23(6), 8–13 (2008)
11. Rahmandad, H., Sterman, J.: Heterogeneity and network structure in the dynamics of diffusion: Comparing agent-based and differential equation models. Management Science 54(5), 998–1014 (2008)
12. Obeidat, M.: Bayesian estimation of time series of counts. Presentation at the 41st Annual Meeting of the Statistical Society of Canada, Edmonton, May 26-29 (2013)
13. Chiogna, M., Gaetan, C.: Hierarchical space-time modelling of epidemic dynamics: an application to measles outbreaks. Statistical Methods and Applications 13(1), 55–71 (2004)
14. Cazelles, B., Chau, N.: Using the kalman filter and dynamic models to assess the changing hiv/aids epidemic. Mathematical Biosciences 140(2), 131–154 (1997)
15. Chiogna, M., Gaetan, C.: Dynamic generalized linear models with application to environmental epidemiology. Journal of the Royal Statistical Society: Series C (Applied Statistics) 51(4), 453–468 (2002)
16. Eagle, N., Pentland, A.S., Lazer, D.: Inferring friendship network structure by using mobile phone data. Proceedings of the National Academy of Sciences 106(36), 15274–15278 (2009)
17. Salathé, M., Kazandjieva, M., Lee, J.W., Levis, P., Feldman, M.W., Jones, J.H.: A high-resolution human contact network for infectious disease transmission. Proceedings of the National Academy of Sciences 107(51), 22020–22025 (2010)
18. Hashemian, M., Stanley, K., Osgood, N.: Leveraging H1N1 infection transmission modeling with proximity sensor microdata. BMC Medical Informatics and Decision Making 12(1), 35 (2012)
19. Funk, S., Salathé, M., Jansen, V.A.: Modelling the influence of human behaviour on the spread of infectious diseases: a review. Journal of The Royal Society Interface 7(50), 1247–1256 (2010)
20. Gelb, A.: Applied optimal estimation. MIT Press (1974)
21. Osgood, N.: Using traditional and agent based toolset for system dynamics: Present tradeoffs and future evolution. System Dynamics (2007)

Identifying Users with Opposing Opinions in Twitter Debates

Ashwin Rajadesingan and Huan Liu

Arizona State University
{arajades,huan.liu}@asu.edu

Abstract. In recent times, social media sites such as Twitter have been extensively used for debating politics and public policies. These debates span millions of tweets and numerous topics of public importance. Thus, it is imperative that this vast trove of data is tapped in order to gain insights into public opinion especially on hotly contested issues such as abortion, gun reforms etc. Thus, in our work, we aim to gauge users' stance on such topics in Twitter. We propose ReLP, a semi-supervised framework using a retweet-based label propagation algorithm coupled with a supervised classifier to identify users with differing opinions. In particular, our framework is designed such that it can be easily adopted to different domains with little human supervision while still producing excellent accuracy.

Keywords: label propagation, semi-supervised, opinion mining, polarity detection.

1 Introduction

With the advent of online social networks, almost every topic of importance is being constantly discussed and debated by millions of people online. Because of their immense reach and their ability to quickly disseminate information, social networking sites, such as Twitter, have emerged as perfect platforms for discussions and debates. With more than 800 million registered users[1] of whom 232 million are active users[2], Twitter has become the platform of choice for most discussions and debates of public interest. For example, during the first U.S presidential debate between Barack Obama and Mitt Romney on October 3, 2012, 10.3 million tweets were generated in only 90 minutes[3].

This huge, high-velocity data is absolutely invaluable as it provides a deep insight into public opinion without the need for explicit surveys and polls. As inferences from social network data are made through passive observations in which users voluntarily express their opinions, this resource, in spite of the inherent selection bias involved, may provide a useful insight into public opinion

[1] http://twopcharts.com/twitteractivitymonitor

[2] http://www.sec.gov/Archives/edgar/data/1418091/000119312513424260/d564001ds1a.htm

[3] https://blog.twitter.com/2012/dispatch-from-the-denver-debate

W.G. Kennedy, N. Agarwal, and S.J. Yang (Eds.): SBP 2014, LNCS 8393, pp. 153–160, 2014.

as seen in existing work[1][2]. Thus, it is imperative that algorithms and systems are built to analyze discussions and opinions expressed in Twitter.

In this paper, we focus on developing a framework to identify users' position on a specific topic in Twitter. The dynamic nature of content in Twitter is very different from content in conventional media such as news articles, blogs, etc., and poses new problems in identifying user opinion. Any framework catering to Twitter needs to overcome the following challenges: (i) The sheer volume and velocity of tweets generated by millions of users, (ii) the usage an ever-changing set of slang words, abbreviations, memes etc., which are not used in common parlance, and (iii) the considerably fewer word cues to identify opinion as a result of the 140 character limit on the size of the tweet. However, social networks do provide other important information such as retweets, mentions etc., which may be used in identifying users' opinion. Furthermore, the primary limiting factor in applying most supervised learning algorithms to identify opinions in Twitter is simply the lack of sufficient, reliable training data. Currently, researchers manually label or use services such as Amazon Mechanical Turk[4] to build training data for their supervised classifiers. However, such an approach is not practically feasible considering the diversity in topics and the huge volumes of tweets we are typically interested in. To overcome this obstacle, we design our framework such that very little manual effort is required to produce sufficient training data by exploiting patterns in the users' retweeting behavior.

Our framework consists of two parts: semi-supervised label propagation and supervised classification. The core idea is to use a very small set of labeled seed users to produce a set of annotations using the proposed label propagation algorithm. The resulting set of annotations will then be used for training a supervised classifier to label the other users. The label propagation algorithm ensures that sufficient data with high quality labels is produced with little manual effort. It must be noted that any supervised classifier may be used in the second step. However, as proof of concept, we use a Multinomial Naïve Bayes classifier trained on unigrams, bigrams and trigrams to showcase the efficiency of our framework. The main contributions of this paper are summarized below:

- Present **ReLP**, a semi-supervised **Re**tweet-based **L**abel **P**ropagation framework, to identify users' opinion on a topic in Twitter;
- Drastically reduce the manual effort involved in constructing reliable training data by using users' retweet behavior;
- Comprehensively evaluate our framework on both visibly opinionated users such as politicians, activists etc., and moderately opinionated common users who make up the majority producing excellent results.

The rest of the paper is organized as follows. In Section 2, we discuss about related research. In Section 3, we discuss our framework in detail. In Section 4, we evaluate our framework on a Twitter dataset. In Section 5, we conclude and explore further research avenues.

[4] `https://www.mturk.com/mturk/`

2 Related Work

In recent years, many studies on detecting opinions in Twitter have been performed using lexical, statistical and label propagation based techniques with reasonably successful results. Closely related to our work, Speriosu et al.[3] devised a semi-supervised label propagation algorithm using the follower graph, n-grams etc., with seeds from a variety of sources including OpinionFinder[4], a maximum entropy classifier etc., to identify the users' polarity on a topic. Tan et al.[5] exploit the theory of homophily in determining the sentiment of users on a particular topic by using a semi-supervised framework involving the twitter's follower and mention network. Somewhat related, Wong et al.[6] detected the political opinion of users by formulating it as an ill-posed linear inverse problem using retweet behavior. Techniques[7][8] using emoticons and hashtags to train supervised classifiers for sentiment and opinion analysis have also been studied. Researchers have also used lexicon-based techniques using SentiWordNet[9], OpinionFinder[4] etc., to identify user opinions.

Our semi-supervised framework differs from the previous approaches as it requires very little manual effort, utilizing the users' retweet behavior to obtain annotations. Unlike supervised techniques, very little seed data is required and hence, the framework may be ported to different domains with little effort. Also, our framework uses the retweet network for label propagation unlike some semi-supervised techniques which require constructing the follower graph which may not be feasible in real-time because of the volume and velocity of tweets.

3 Methodology

The proposed ReLP framework consists of two parts: semi-supervised label propagation and supervised learning. The seeds inputted by the user is used to produce an expanded training set which is then used to train a supervised classifier. A sketch of the framework described is shown in Fig. 1.

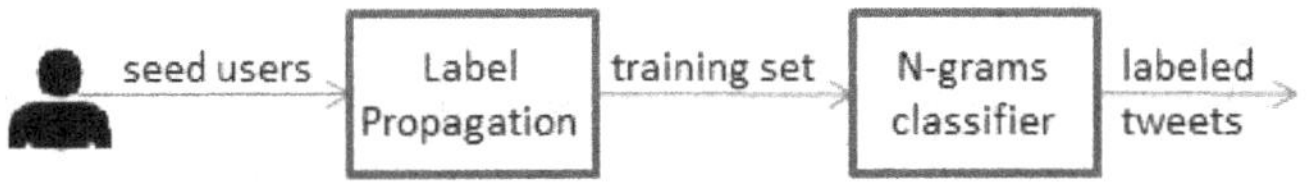

Fig. 1. Schematic sketch of the ReLP framework

We base our approach on our observation that if many users retweet a particular pair of tweets within a reasonably short period of time, then it is highly likely that the two tweets are similar in some aspect. In the context of Twitter debates, naturally, it is highly likely that this pair of tweets share the same opinion. For example, we are more likely to find a large number of users retweeting a pair of tweets that support gun reforms than retweeting a pair, one of which is for gun reforms while the other is against gun reforms. Therefore, we model this observation by constructing a column normalized retweet co-occurrence matrix

M. Let t_1, $t_2 ... t_n$ represent the tweets in the dataset. Then, each element $M_{i,j}$ of matrix M represents the fraction of users who retweeted both t_i and t_j.

$$M_{i,j} = \frac{number\ of\ users\ retweeting\ t_i\ and\ t_j}{total\ number\ of\ users\ retweeting\ t_j}$$

The steps involved in label propagation are described below and the pseudocode is given in algorithm 1 and algorithm 2.

1. Initially, we label the seed users' tweets using the label of the seeds themselves as it is extremely unlikely that seed users (who are major voices) would actually contradict their viewpoint in their own tweets. Each tweet label has values for two fields, *for* and *against*, which are updated during label propagation. If the tweet is in support of the topic, the *for* value is higher than the *against* value and vice-versa. The value of the fields may vary from 0 to 1. The seed users' tweets are initialized with the field values based on the input label of the seed users. All other tweets have both field values initialized to 0.
2. In each iteration, we select the tweets whose labels are to be propagated using an increasing hash function h. We define the hash function $h(v) = \lfloor v * n \rfloor$ where $v \in [0, 1]$. Here, we use $n = 10$, a greater precision may be obtained if n is set lower, however, this may result in extra iterations taking more time. The value of n can be set based on scalability requirements.
3. Using each selected tweet, we update the label of its co-occurring tweets weighted based on values of M.
4. Steps 2 and 3 are repeated until no tweets are returned by the hash function.

Finally, the tweets are labeled as "for" or "against" depending on their *for* and *against* field values, whichever has higher value. While this method does not label all the tweets, it provides a very accurate classification of the tweets that it labels (as shown in Section 4). Therefore, the tweets labeled through this process are then used to train a supervised classifier to classify the remaining tweets. We use a Multinomial Naïve Bayes classifier with unigrams, bigrams and trigrams as features. As any out-of-the-box supervised classifier may be used, we do not go into details of the classifier's workings. We used Scikit-learn (a machine learning library in Python)'s[10] implementation of the classifier.

To elucidate the workings of the framework, we provide an example scenario: a user would like to view tweets from users in both sides of a particular debate, say gun reforms, to obtain a balanced and informed viewpoint. The user has some idea of who the main actors are, for example, Barack Obama and Piers Morgan who are for gun reforms and the National Rifle Association (NRA) and Rand Paul who are against gun reforms. The framework aims to take as input the aforementioned users and churn out tweets from other users having similarly opposing views.

Algorithm 1. Label propagation algorithm

```
procedure LabelPropagation(M)
    labels ← labeled tweets of seed users
    final_labels ← None
    tweets ← SeedSelection(labels)
    while tweets ≠ None do
        for each t_i in tweets do
            for each t_j in M[t_i] do
                labels[t_j]['for'] ← labels[t_j]['for'] + labels[t_i]['for'] * M[t_i][t_j]
                labels[t_j]['against'] ← labels[t_j]['against']+
                                        labels[t_i]['against'] * M[t_i][t_j]
            end for
            final_labels[t_i] ← labels[t_i]
            delete labels[t_i]
        end for
        tweets ← SeedSelection(labels)
    end while
end procedure
```

Algorithm 2. Selecting seeds for label propagation

```
procedure SeedSelection(labels)
    for each t_i in labels do
        max_value ← max(labels[t_i]['for'], labels[t_j]['against'])
        store t_i using h(max_value)
    end for
    return tweets with the highest hash value
end procedure
```

4 Evaluation

In order to evaluate our framework, we use a real-world twitter dataset. We collected over 900,000 tweets[5] over a period of five days during the hotly contested gun reforms debate from April 15th, 2013 to April 18th, 2013. The dataset was collected using Twitter's Streaming API[6] using the keywords "gun" and "#guncontrol". We deliberately used non-partisan keywords so as not to create any bias in the data collected. We filtered out the non-english tweets and ignored users who posted only one tweet during the entire data collection period as we found that those tweets to be very noisy (which maybe due to the keywords used in data collection). The filtered and processed dataset consists of 543,404 tweets from 116,033 users and contains 246,454 retweets.

We evaluate and compare the performance of the ReLP framework on the collected dataset using three competitive baselines. The baselines were designed so

[5] The dataset may be obtained by contacting the first author.

[6] https://dev.twitter.com/docs/streaming-apis

as to require the same/similar amount of manual effort needed by our framework for opinion identification. Furthermore, it is important to note that the baselines (except B3) function using the same supervised classifier used in our framework. The only difference between the methods is in the way in which they are trained. Therefore, any difference in the performance of our framework with respect to the baselines can be directly attributed to the efficiency (or inefficiency) of our label propagation algorithm. We use the following three baselines:

- **Baseline 1 (B1):** Multinomial Naïve Bayes classifier with unigrams, bigrams and trigrams as features trained using the seed users' tweets.
- **Baseline 2 (B2):** Multinomial Naïve Bayes classifier with the same features as above, trained using partisan hashtags (#Protect2A, #NewtonBetrayed).
- **Baseline 3 (B3):** K-means clustering algorithm with the same features as above. The initial centroids were chosen from seed users' tweets. The final labels are assigned to the clusters after sampling instances from both clusters.

As it is practically impossible to obtain the ground truth labels for all the users, we evaluate our framework over smaller subsets of users. In order to obtain a representative sample of the users involved, we divide and sample users from two groups (inspired by Cohen et al.[11]): visibly opinionated users and moderately opinionated users.The rationale behind such a division is to evaluate the framework on not only obviously opinionated users such as politicians, organizations etc., but also on the relatively inactive common users who, in fact, constitute the majority. Therefore, a good classifier must be capable of predicting the opinion of users from both brackets. Details on how the subsets of users were curated are given below.

- **Visibly Opinionated Users:** This group consists of the most vocal and opinionated users such as senators, activists, activist organizations etc. We collected a subset of these users using Twitter lists. In Twitter, any user can create lists and add other users to these lists. We manually identified lists such as "Protect 2nd Amendment", "Guns Save Lives" etc., whose users were clearly against gun reforms and other lists such as "Prevent Gun Violence", "Gun Safety" etc., whose users were clearly for gun reforms. We collected the list members and filtered out users whose tweets were not present in the dataset. In total, we obtained 363 users with 88 users for gun reforms and 255 users against gun reforms.
- **Moderately Opinionated Users:** We define this group as users who posted between 2-4 tweets during the collection cycle, do not belong to any relevant lists and do not label themselves as for/against gun reforms in their Twitter profile page. We manually annotated a randomly selected sample of 500 users based on the tweets that they posted. Out of 500 users, 276 users were for gun reforms, 120 users were against gun reforms and the rest shared tweets information (such as gun reforms related news) but did not voice out their personal opinions and were hence, ignored.

The difference in the number of users on both sides between the two subsets can be attributed to the well studied observation of the silent majority and vocal minority[12]. Using Barack Obama, White House and Gabrielle Giffords as seed users in support of gun reforms and the National Rifle Association (NRA), Ted Cruz and Pat Dollard as seed users against gun reforms, we obtain the results shown in Table 1. We chose these users as seeds based on the large number of retweets generated by their tweets while making sure that the total number of retweets in support of and against gun reforms are fairly balanced. We use precision, recall and f-measure to compare the various methods used.

Table 1. Performance evaluation and comparison with baselines

Methods	**Moderately opinionated users**			**Visibly opinionated users**		
	precision	**recall**	**F-measure**	**precision**	**recall**	**F-measure**
ReLP	96.73	93.36	95.01	94.18	97.59	95.85
B1	81.09	80.79	80.94	75.75	88.23	81.51
B2	81.70	86.04	83.81	56.10	97.64	71.25
B3	66.26	78.98	72.06	38.73	50.58	43.86

From Table 1, we observe that ReLP framework clearly outperforms the baselines in classifying both visibly opinionated and moderately opinionated users in almost every measure. Importantly, the f-measure (harmonic mean of the precision and recall values) of our framework is consistent across the subsets of users unlike other approaches whose f-measures fluctuate. This shows the versatility of the framework in efficiently classifying users across the spectrum.

5 Conclusions and Future Work

The semi-supervised ReLP framework showcases the effectiveness of combining a simple label propagation algorithm with existing supervised classifiers. The framework greatly reduces need for manually labeled training sets which are the main obstacles preventing extensive, large-scale use of supervised methods in the ever-evolving Twittersphere. This ensures that, in order to port the framework to another domain, the user only needs to supply a list of major players in that domain. Interestingly, by choosing only the major players as seeds, the user need not even understand the language of the tweets supplied as training data.

In the future, we aim to perform experiments to determine the optimum number of seed users required by our framework to provide good classification results and examine its sensitiveness to seed selection. Furthermore, we wish to incorporate other activities in Twitter such as mentioning, replying, into our framework to improve performance. We also wish to build an automated system which will be capable of identifying and showcasing in real-time, conversations/debates between users with opposing views on a particular topic.

Acknowledgements. This research is supported by ONR (N000141110527) and (N000141410095).

References

1. O'Connor, B., Balasubramanyan, R., Routledge, B.R., Smith, N.A.: From tweets to polls: Linking text sentiment to public opinion time series. In: ICWSM, vol. 11, pp. 122–129 (2010)
2. Tumasjan, A., Sprenger, T.O., Sandner, P.G., Welpe, I.M.: Predicting elections with twitter: What 140 characters reveal about political sentiment. In: ICWSM, vol. 10, pp. 178–185 (2010)
3. Speriosu, M., Sudan, N., Upadhyay, S., Baldridge, J.: Twitter polarity classification with label propagation over lexical links and the follower graph. In: Proceedings of the First workshop on Unsupervised Learning in NLP, pp. 53–63. Association for Computational Linguistics (2011)
4. Wilson, T., Hoffmann, P., Somasundaran, S., Kessler, J., Wiebe, J., Choi, Y., Cardie, C., Riloff, E., Patwardhan, S.: Opinionfinder: A system for subjectivity analysis. In: Proceedings of HLT/EMNLP on Interactive Demonstrations, pp. 34–35. Association for Computational Linguistics (2005)
5. Tan, C., Lee, L., Tang, J., Jiang, L., Zhou, M., Li, P.: User-level sentiment analysis incorporating social networks. In: Proceedings of the 17th ACM SIGKDD International Conference on Knowledge Discovery and Data Mining, pp. 1397–1405. ACM (2011)
6. Wong, F.M.F., Tan, C.W., Sen, S., Chiang, M.: Quantifying political leaning from tweets and retweets (2013)
7. Go, A., Bhayani, R., Huang, L.: Twitter sentiment classification using distant supervision. CS224N Project Report, Stanford, pp. 1–12 (2009)
8. Davidov, D., Tsur, O., Rappoport, A.: Enhanced sentiment learning using twitter hashtags and smileys. In: Proceedings of the 23rd International Conference on Computational Linguistics: Posters, pp. 241–249. Association for Computational Linguistics (2010)
9. Esuli, A., Sebastiani, F.: Sentiwordnet: A publicly available lexical resource for opinion mining. In: Proceedings of LREC, vol. 6, pp. 417–422 (2006)
10. Pedregosa, F., Varoquaux, G., Gramfort, A., Michel, V., Thirion, B., Grisel, O., Blondel, M., Prettenhofer, P., Weiss, R., Dubourg, V., Vanderplas, J., Passos, A., Cournapeau, D., Brucher, M., Perrot, M., Duchesnay, E.: Scikit-learn: Machine learning in Python. Journal of Machine Learning Research 12, 2825–2830 (2011)
11. Cohen, R., Ruths, D.: Classifying political orientation on twitter: It's not easy! In: Seventh International AAAI Conference on Weblogs and Social Media (2013)
12. Mustafaraj, E., Finn, S., Whitlock, C., Metaxas, P.T.: Vocal minority versus silent majority: Discovering the opionions of the long tail. In: Privacy, Security, Risk and Trust (passat), IEEE Third International Conference on Social Computing (SocialCom), pp. 103–110. IEEE (2011)

A New Approach for Item Ranking Based on Review Scores Reflecting Temporal Trust Factor

Kazumi Saito[1], Masahiro Kimura[2], Kouzou Ohara[3], and Hiroshi Motoda[4,5]

[1] School of Administration and Informatics, University of Shizuoka
k-saito@u-shizuoka-ken.ac.jp
[2] Department of Electronics and Informatics, Ryukoku University
kimura@rins.ryukoku.ac.jp
[3] Department of Integrated Information Technology, Aoyama Gakuin University
ohara@it.aoyama.ac.jp
[4] Institute of Scientific and Industrial Research, Osaka University
motoda@ar.sanken.osaka-u.ac.jp
[5] School of Computing and Information Systems, University of Tasmania

Abstract. We propose a new item-ranking method that is reliable and can efficiently identify high-quality items from among a set of items in a given category using their review-scores which were rated and posted by users. Typical ranking methods rely only on either the number of reviews or the average review score. Some of them discount outdated ratings by using a temporal-decay function to make a fair comparison between old and new items. The proposed method reflects trust levels by incorporating a trust discount factor into a temporal-decay function. We first define the *MTDF (Multinomial with Trust Discount Factor) model* for the review-score distribution of each item built from the observed review data. We then bring in the notion of z-score to accommodate the trust variance that comes from the number of reviews available, and propose a z-score version of MTDF model. Finally we demonstrate the effectiveness of the proposed method using the MovieLens dataset, showing that the proposed ranking method can derive more reasonable and trustable rankings, compared to two naive ranking methods and the pure z-score based ranking method.

Keywords: Item-ranking, Trust discount factor, Z-socre.

1 Introduction

The emergence of Social Media has provided us with the opportunity to collect a large number of user reviews for various items, e.g., products and movies. Many social media sites offer the review score for an individual item which a user rated and posted. Reliable ranking is needed in order to efficiently identify high-quality items from among a set of items in a given category. Typical item-ranking methods are very naive, i.e., ranking items according to either the number of reviews or the average review score. It is often the case that these methods discount outdated ratings by using some temporal-decay function (see, e.g., [1,9]) in order to fairly compare old and new items. In fact, the idea of temporal-decay has already been exploited in several different contexts

W.G. Kennedy, N. Agarwal, and S.J. Yang (Eds.): SBP 2014, LNCS 8393, pp. 161–168, 2014.

of social media mining, aiming at better performance. For example, Koren [6] proposed several time-drifting user-preference models using temporal-decay functions in the context of recommendation system (collaborative filtering). Temporal-decay effects of information diffusion processes have often been treated by introducing propagation probabilities through links [3,4,10] in the context of modeling information diffusion in social networks. We have enhanced the original voter model [11,2] and proposed the temporal-decay voter model [5] by incorporating a temporal-decay function in the context of modeling opinion formations in social networks on the ground that people may decide their opinions by taking account not only of their neighbors' latest opinions, but also of their neighbors' past opinions with some discount.

We believe that bringing in the notion of trust factors in temporal-decay is important in the context of item-ranking. For example, consider predicting the future review-score distribution of an item from its past observed trend. An item that has maintained a stable review-score distribution over time will remain so in the near future and the past information is trustable (high trust level) for making prediction. On the other hand, an item that had gone through unstable fluctuating review-score distribution over time is not easily predictable and the past information is not trustable (low trust level). In this paper, we propose a novel item-ranking approach reflecting such trust levels by incorporating a trust discount factor into a temporal-decay function, and evaluate its effectiveness using the MovieLens dataset. [1] Note that the trust we introduced here is on items and is different from the work in recommender systems such as [8,7] where the trust is on users.

We first define the *MTDF (Multinomial with Trust Discount Factor) model*, which is a variant of extended voter model assuming a multinomial probability distribution, and estimate the review-score distribution of each item from the observed review data with trust discount. We then bring in the notion of z-score to accommodate the trust variance that comes from the number of reviews available, and propose a z-score version of MTDF model, i.e. we cannot trust the score if the number of reviews is too small. We finally demonstrate the effectiveness of the proposed method using the MovieLens dataset and show that the proposed ranking method can derive more reasonable and trustable rankings by comparing our results with the results of two naive ranking methods and the pure z-score based ranking method.

The rest of the paper is organized as follows: We describe the proposed ranking method in section 2 and report the results of our experiments using the MovieLens dataset in section 3. In the experiments, we examine the exponential and the power-law decay functions as two typical temporal-decay functions, and demonstrate the effectiveness of the proposed method by comparing our results with the results of three other simpler methods. We conclude the paper by summarizing the main results and needed future work in section 4.

2 Ranking Method

Let $\mathcal{V}$ be a set of items evaluated by users with one of integer scores $\mathcal{K} = \{1, \cdots, K\}$ during time interval $\mathcal{T}$. Then, we can describe a set of review results by $\mathcal{D} = \{(v, k, t) \mid v \in$

[1] `http://www.grouplens.org/datasets/movielens/`

$\mathcal{V}, k \in \mathcal{K}, t \in \mathcal{T}\}$. For any $v \in \mathcal{V}$ and $t \in \mathcal{T}$, we consider the set $M(v, t)$ consisting of the time τ at which an item v was reviewed before time t, i.e., $M(v, t) = \{\tau \mid (v, k, \tau) \in \mathcal{D}, \tau < t\}$. Let $g(v, t) \in \mathcal{K}$ be the score of item v at time t. Then, for $k \in \mathcal{K}$, we also consider a subset of $M(v, t)$, $M_k(v, t) = \{\tau \in M(v, t) \mid g(v, \tau) = k\}$, where $M_k(v, t)$ is the set of item v's evaluated time instances before time t at which v was evaluated to be the score k. Now, we can define a multinomial model which takes all the past scores into consideration. Namely, we consider the following model for predicting the review-score distribution of item v at time t from its observed data.

$$P(g(v, t) = k) = \frac{1 + |M_k(v, t)|}{K + |M(v, t)|}, \quad (k = 1, \cdots, K), \tag{1}$$

where we employed a Bayesian prior known as the Laplace smoothing. Here we note that the Laplace smoothing of Eq. (1) corresponds to the assumption that each item is initially evaluated by one of the K scores with equal probability. Note also that the Laplace smoothing corresponds to a special case of Dirichlet distributions that are very often used as prior distributions in Bayesian statistics, and in fact the Dirichlet distribution is the conjugate prior of the categorical distribution and multinomial distribution. We refer to this model as the *base multinomial model*.

Thus far, we assumed that all the past reviews are equally weighted. However, it is naturally conceivable that some of the quite old reviews are almost out-of-date and their trust levels might be low. In order to reflect this kind of effects into the model, we consider introducing some trust discount factors. The simplest one is an exponential discount factor defined by $\rho(\Delta t; \lambda) = \exp(-\lambda \Delta t)$, where $\lambda \geq 0$ is a parameter and $\Delta t = t - \tau$ stands for the time difference between t and τ. Another natural one would be a power-law discount factor defined by $\rho(\Delta t; \lambda) = (\Delta t)^{-\lambda} = \exp(-\lambda \log \Delta t)$, where $\lambda \geq 0$ is a parameter. Now, we construct a more general discount factor as described in [5]. For a given positive integer J, we consider a J-dimensional vector consisting of linearly independent features, $\boldsymbol{F}_J(\Delta t) = (f_1(\Delta t), \cdots, f_J(\Delta t))^T$, and a parameter vector with non-negative elements for these features, $\boldsymbol{\lambda}_J = (\lambda_1, \cdots, \lambda_J)^T$. Then, we define a general discount factor by $\rho(\Delta t; \boldsymbol{\lambda}_J) = \exp\left(-\boldsymbol{\lambda}_J{}^T \boldsymbol{F}_J(\Delta t)\right)$. Using this general discount factor $\rho(\Delta t; \boldsymbol{\lambda}_J)$, we define the *MTDF (Multinomial with Trust Discount Factor) model* in the following way. In this model, Eq. (1) is replaced with

$$P(g(v, t) = k) = \frac{1 + \sum_{\tau \in M_k(v,t)} \rho(t - \tau; \boldsymbol{\lambda}_J)}{K + \sum_{\tau \in M(v,t)} \rho(t - \tau; \boldsymbol{\lambda}_J)}, \quad (k = 1, \cdots, K). \tag{2}$$

Note that Eq. (2) is reduced to Eq. (1) when $\boldsymbol{\lambda}_J$ is the J-dimensional zero-vector $\boldsymbol{0}_J$, that is, the MTDF model of $\boldsymbol{\lambda}_J = \boldsymbol{0}_J$ coincides with the base multinomial model. Here, we can estimate the trust discount parameter values $\hat{\boldsymbol{\lambda}}_J$ of the MTDF model by maximizing the likelihood function based on Eq. (2) for a given observed review results $\mathcal{D}$. Note that the MTDF model of $\hat{\boldsymbol{\lambda}}_J \approx \boldsymbol{0}_J$ for some item v means that this item does not need to introduce a trust discount factor, which maintains a high trust level.

Below, we propose a method of ranking items based on the MTDF model using the observed review results. Here note that the average score and its standard deviation during time interval $\mathcal{T}$ are calculated by $\mu = \sum_{k \in \mathcal{K}} k p(k)$ and $\sigma = \sqrt{\sum_{k \in \mathcal{K}} (k - \mu)^2 p(k)}$, respectively, where $p(k) = \sum_{v \in \mathcal{V}} |M_k(v, T)| / \sum_{v \in \mathcal{V}} |M(v, T)|$ and T is the final observation time defined by $T = \max\{t \in \mathcal{T}\}$. Note that under the assumption that each review score is independently given according to the distribution defined by $p(k)$, the expected deviation of the average score after Q reviews $S = \{k_1, \cdots, k_Q\}$ becomes

$$\sqrt{\sum_{k_1 \in \mathcal{K}} \cdots \sum_{k_Q \in \mathcal{K}} \left(\mu - \frac{1}{Q} \sum_{q=1}^{Q} k_q \right)^2 \prod_{q=1}^{Q} p(k_q)} = \frac{\sigma}{\sqrt{Q}}.$$

Then, we can consider the z-score $z(v, t)$ of the average score for item v at time t by

$$z(v,t) = \frac{\mu(v,t) - \mu}{\sigma / \sqrt{|M(v,t)|}}, \quad \mu(v,t) = \sum_{k \in \mathcal{K}} k \frac{|M_k(v,t)|}{|M(v,t)|}. \tag{3}$$

Clearly, we can regard an item with a larger z-score $z(v, t)$ of Eq. (3) as the one having significantly higher scores. We refer to this ranking method as *base ranking method*. Now, we extend the z-score of Eq. (3) by introducing the idea of the MTDF model as follows:

$$z_\rho(v,t) = \frac{\mu_\rho(v,t) - \mu}{\sigma / \sqrt{\sum_{\tau \in M(v,t)} \rho(t - \tau; \hat{\boldsymbol{\lambda}}_J)}}, \quad \mu_\rho(v,t) = \sum_{k \in \mathcal{K}} k \frac{\sum_{\tau \in M_k(v,t)} \rho(t - \tau; \hat{\boldsymbol{\lambda}}_J)}{\sum_{\tau \in M(v,t)} \rho(t - \tau; \hat{\boldsymbol{\lambda}}_J)}, \tag{4}$$

where recall that $\hat{\boldsymbol{\lambda}}_J$ stands for the trust discount parameter values estimated from a given observed review results $\mathcal{D}$. Evidently, we can regard an item with a larger z-score $z_\rho(v, t)$ of Eq. (4) as the one having significantly better review scores as well as having a high trust level. We refer to this ranking method as *proposed ranking method*.

3 Experiments

To evaluate the proposed ranking method, we applied it to the real world dataset, and show its usefulness by comparing the resulting rankings with those by the naive and base ranking methods.

3.1 Dataset

We employed the MovieLens 10M/100k dataset to experimentally evaluate our ranking methods. MovieLnes is one of the online movie recommender services, and the dataset consists of 10,000,054 ratings with time stamps that are made on a 5-star scale with half-star increments for 10,681 movies by 71,567 users[2]. We, henceforth, simply refer to this dataset as the MovieLens dataset. Figure 1 shows the frequency of each review score. Assuming that it is drawn from a multinomial distribution with $K = 10$, the average score over all movies is 3.51 and the standard deviation is 1.06. Interestingly, the user is more likely to evaluate movies without using a half-star. Next, we plot the movies in Fig. 2, using the number of their reviews and the average score over them. In this figure, we can observe that many of the movies having over 10,000 reviews get relatively high scores greater than the overall average 3.51. The top-5 titles in descending order of

[2] http://files.grouplens.org/datasets/movielens/ml-10m-README.html

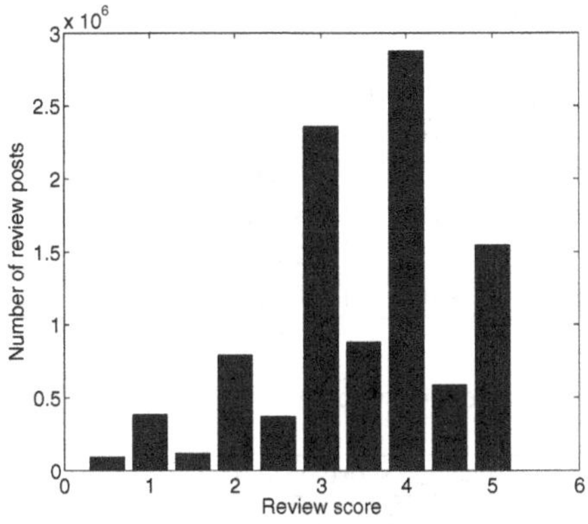

Fig. 1. Frequency of each review score

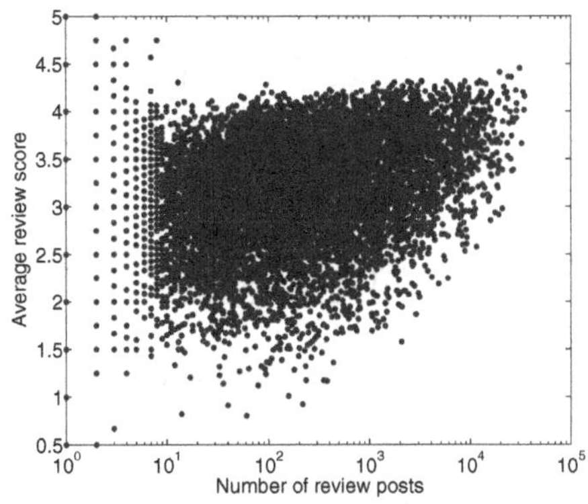

Fig. 2. Scatter plot of movies based on the number of reviews and the average score

the number of posts are summarized in Table 1, which is one of naive ranking results. These movies appear in the rightmost area in Fig. 2, and attract much attention from many users during the whole period, but seem to be a little bit old. This is because old movies stay in the system for a longer period of time than recent ones, and thus they are likely to get many reviews.

Another naive way of ranking movies is to use their average scores, and the top-5 movies in that ranking are shown in Table 2. They appear around the top-left corner of Fig. 2. Notice that their average scores are all 5.00, but all are derived only from at most 2 review scores. Since these average scores are meaningless from a statistical viewpoint, we should exclude such movies that have only a few reviews from rankings by specifying a certain threshold, say 10. Table 3 shows the revised top-5 titles after removing such movies that have less than 10 reviews. This threshold is reasonable in a sense that the distribution of the movies with less than 10 reviews is quite different from the rest in Fig. 2. In this case, one can see that the average scores of the 5 movies are derived from a sufficient amount of reviews, which makes them statistically reliable. Still these movies seem to be a little bit old, too, due to the same reason as the above.

3.2 Results

First, we applied the base ranking method to this dataset by setting the ranking evaluation time t in Eq. (3) at the latest available review time. The resulting top-5 titles are shown in Table 4. It is noted that the top-4 movies are identical to those in Table 3 except that the second and third titles are swapped. This is attributed to the fact that z-scores of such items that have only a few reviews become close to 0 as shown in Fig. 3, where the resulting z-scores for all the movies are plotted in two dimensional space in which the horizontal axis is the number of reviews and the vertical axis is z-score. On the other hand, we can observe in Fig. 3 that z-scores spread more widely as the number of reviews becomes larger. From these results, we can say that the z-score defined by Eq. (3) implicitly takes into account both the review frequencies and the average scores

Table 1. Top 5 movies in the number of review posts

	Title (year of release)	Average score	**# of posts**
1	Pulp Fiction (1994)	4.16	**34,864**
2	Forrest Gump (1994)	4.01	**34,457**
3	The Silence of the Lambs (1991)	4.20	**33,668**
4	Jurassic Park (1993)	3.66	**32,631**
5	The Shawshank Redemption (1994)	4.46	**31,126**

Table 2. Top 5 movies in the average review score

	Title (year of release)	**Average score**	# of posts
1	Satan's Tango (1994)	**5.00**	2
2	Shadows of Forgotten Ancestors (1964)	**5.00**	1
3	Fighting Elegy (1966)	**5.00**	1
4	Sun Alley (1999)	**5.00**	1
5	The Blue Light (1932)	**5.00**	1

Table 3. Top 5 movies in the average review score over at least 10 posts

	Title (year of release)	**Average score**	# of posts
1	The Shawshank Redemption (1994)	**4.46**	31,126
2	The Godfather (1972)	**4.42**	19,814
3	The Usual Suspects (1995)	**4.37**	24,037
4	Schindler's List (1993)	**4.36**	25,777
5	Sunset Blvd. (1950)	**4.32**	3,255

of items, and prevents rarely-reviewed items from being ranked high without the need of any threshold for the number of reviews.

Next, we applied the proposed ranking method to the MovieLens dataset. First, we tested the MTDF models with the exponential and power-low discount factors, and evaluated which model is better for this dataset. To do this, we computed the log-likelihood ratio statistic of each model against the basic multinomial model for each movie in the same way as described in [5]. Figure 4 illustrates the relationship between the statistics, where movies are plotted based on those statistics. From this figure, we can observe a positive correlation, but cannot see a big difference between them, meaning that both decays are equally good and acceptable. Thus, due to the page limitation, we present only the rankings derived from the MTDF model with the exponential discount factor in Table 5. Compared to Tables 3 and 4, it is remarkable that the relatively new movies rank in the top-5 thanks to the trust discount factor of the MTDF model that degrades the effects of old reviews, while keeping their average scores comparable with those in Tables 3. In deed, the ranking of the first-ranked movie is thought reasonable as it is such an acclaimed movie that it won the Academy Awards. On the other hand, the second-ranked movie in Table 5 is common to both Table 3 and 4 and it is relatively old. This implies that this movie maintains high ratings even in the recent period, and thus it has a high trust level. In summary, the proposed ranking method is useful and

Table 4. Top 5 movies in the z-score

Ranking	Title (year of release)	z-score	Average score	# of posts
1	The Shawshank Redemption (1994)	**157.19**	4.46	31,126
2	Schindler's List (1993)	**128.85**	4.36	25,777
3	The Usual Suspects (1995)	**124.96**	4.37	24,037
4	The Godfather (1972)	**119.82**	4.42	19,814
5	The Silence of the Lambs (1991)	**119.70**	4.20	33,668

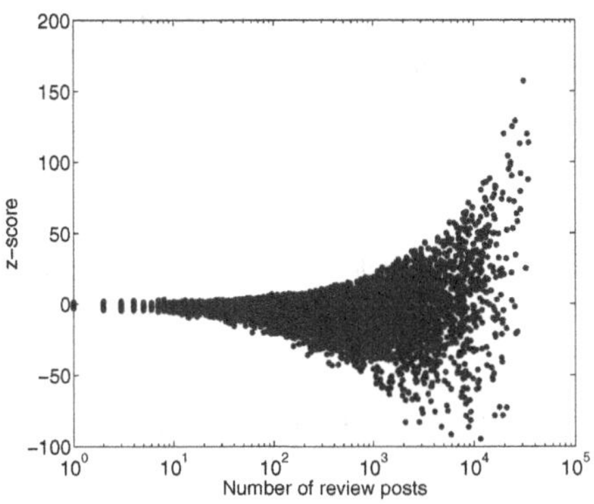

Fig. 3. Relationship between the number of posts and the z-score of movies

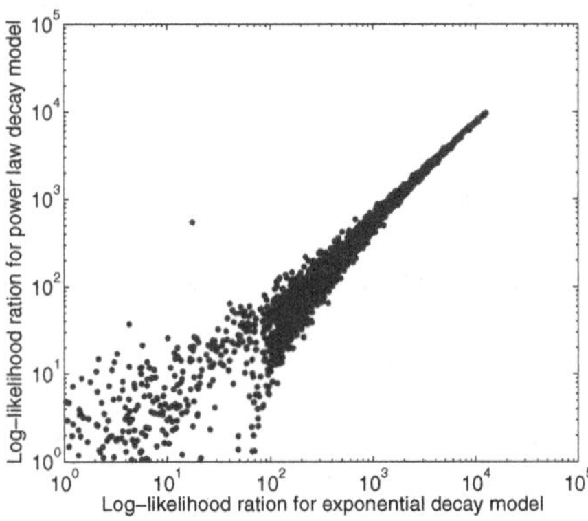

Fig. 4. Comparison between the exponential decay and power-low decay models in log-likelihood ratio of movies

Table 5. Top 5 movies in the z-score with exponential decay

Ranking	Title (year of release)	z-score	Average score	# of posts
1	The Departed (2006)	**35.48**	4.10	4,145
2	The Shawshank Redemption (1994)	**27.74**	4.46	31,126
3	The Lives of Others (2006)	**25.95**	4.30	1,230
4	The Prestige (2006)	**23.06**	4.00	2,808
5	Fight Club (1999)	**22.55**	4.19	16,334

derive more reasonable and trustable rankings by placing more weight on recent stable ratings. It is statistically reliable without a need of any threshold.

4 Conclusion

In this paper, we presented the *MTDF (Multinomial with Trust Discount Factor) model* for the review-score distribution of each item built from its observed data. We further introduced the z-score to account for the trust level variation due to the variation of the number of reviews. The final model combines these two that takes both the effect of 1) the time that a review was posted for an item (trust discount) and 2) the number of reviews users posted for that item (trust variation). In our experiments using the

MovieLens dataset, we demonstrated that the proposed ranking method can derive more reasonable and trustable rankings, compared to the two naive ranking methods and the pure z-score based ranking method. We believe that our MTDF model will play an important role not only for ranking tasks as shown in this paper, but also for other tasks such as predicting evolution of social networks. However, in order to support our claim, we need to evaluate our model on a wide variety of problems. To this end, we plan to elaborate functional forms of trust discount factor as our future study.

Acknowledgments. This work was partly supported by Asian Office of Aerospace Research and Development, Air Force Office of Scientific Research under Grant No. AOARD-13-4042, and JSPS Grant-in-Aid for Scientific Research (C) (No. 23500194).

References

1. Cormode, G., Shkapenyuk, V., Srivastava, D., Xu, B.: Forward decay: A practical time decay model for streaming systems. In: Proceedings of the 25th IEEE International Conference on Data Engineering (ICDE 2009), pp. 138–149 (2009)
2. Even-Dar, E., Shapira, A.: A note on maximizing the spread of influence in social networks. In: Deng, X., Graham, F.C. (eds.) WINE 2007. LNCS, vol. 4858, pp. 281–286. Springer, Heidelberg (2007)
3. Goldenberg, J., Libai, B., Muller, E.: Talk of the network: A complex systems look at the underlying process of word-of-mouth. Marketing Letters 12, 211–223 (2001)
4. Kempe, D., Kleinberg, J., Tardos, E.: Maximizing the spread of influence through a social network. In: Proceedings of the 9th ACM SIGKDD International Conference on Knowledge Discovery and Data Mining (KDD 2003), pp. 137–146 (2003)
5. Kimura, M., Saito, K., Ohara, K., Motoda, H.: Opinion formation by voter model with temporal decay dynamics. In: Flach, P.A., De Bie, T., Cristianini, N. (eds.) ECML PKDD 2012, Part II. LNCS, vol. 7524, pp. 565–580. Springer, Heidelberg (2012)
6. Koren, Y.: Collaborative filtering with temporal dynamics. In: Proceedings of the 15th ACM SIGKDD International Conference on Knowledge Discovery and Data Mining (KDD 2009), pp. 447–456 (2009)
7. Ma, H., Zhou, D., Liu, C., Lyu, M.R., King, I.: Recommender systems with social regularization. In: Proceedings of the Fourth ACM International Conference on Web Search and Data Mining (WSDM 2011), pp. 287–296. ACM, New York (2011)
8. O'Donovan, J., Smyth, B.: Trust in recommender systems. In: Proceedings of the 10th International Conference on Intelligent User Interfaces (IUI 2005), pp. 167–174. ACM, New York (2005)
9. Papadakis, G., Niederée, C., Nejdl, W.: Decay-based ranking for social application content. In: Proceedings of the 6th International Conference on Web Information Systems and Technologies (WEBIST 2010), pp. 276–281 (2010)
10. Saito, K., Kimura, M., Ohara, K., Motoda, H.: Learning asynchronous-time information diffusion models and its application to behavioral data analysis over social networks. Journal of Computer Engineering and Informatics 1, 30–57 (2013)
11. Sood, V., Redner, S.: Voter model on heterogeneous graphs. Physical Review Letters 94, 17801 (2005)

Segmenting Large-Scale Cyber Attacks for Online Behavior Model Generation

Steven Strapp and Shanchieh Jay Yang

Department of Computer Engineering
Rochester Institute of Technology, Rochester, New York 14623

Abstract. Large-scale cyber attack traffic can present challenges to identify which packets are relevant and what attack behaviors are present. Existing works on Host or Flow Clustering attempt to group similar behaviors to expedite analysis, often phrasing the problem as offline unsupervised machine learning. This work proposes online processing to simultaneously segment traffic observables and generate attack behavior models that are relevant to a target. The goal is not just to aggregate similar attack behaviors, but to provide situational awareness by grouping relevant traffic that exhibits one or more behaviors around each asset. The seemingly clustering problem is recast as a supervised learning problem: classifying received traffic to the most likely attack model, and iteratively introducing new models to explain received traffic. A graph-based prior is defined to extract the macroscopic attack structure, which complements security-based features for classification. Malicious traffic captures from CAIDA are used to demonstrate the capability of the proposed attack segmentation and model generation (ASMG) process.

1 Introduction

High-profile cyber assets often receive large-scale malicious traffic with diverse behaviors from many sources. Protecting these assets is contingent on analysts' ability to quickly isolate relevant traffic from the large traffic capture, determine the characteristics of the attack, and estimate the best defense strategy for the critical attack behaviors. The problem extends beyond understanding only the traffic incident on the critical asset; traffic emitted by the same attack sources to other destinations may be important to understanding the attack behavior, and so could traffic from other sources to other targets.

Unfortunately, characteristics of malicious IP traffic further intensify this problem. For example, IP address "spoofing" is a routine malicious behavior [10,12], where the source IP addresses of malicious packets are fabricated. From an analyst's perspective this common behavior greatly diminishes the reliability of source IP as a feature: a single malicious host may utilize many addresses, and multiple malicious activities may probabilistically, and inadvertently, spoof the same address. Given the high volumes of malicious traffic likely at a high-profile asset, either sophisticated or due to simple Malware, probabilistic intersection of different attack behaviors at a target is not unlikely.

W.G. Kennedy, N. Agarwal, and S.J. Yang (Eds.): SBP 2014, LNCS 8393, pp. 169–177, 2014.

Similar phenomena may exist in other problem domains where actor identity is not a reliable feature and independent or spoofed actors can exhibit similar behaviors. Such uncertain and probabilistic intersection calls for a departure from traditional clustering works that only aims at grouping similar behaviors; it requires identifying relevant observations for each target of interest and generating one or more models for the collective behaviors surrounding the target.

In the case of cyber attacks, host clustering works have utilized a broad spectrum of features and approaches to perform unsupervised learning [4,8,13,14,1,3]. While this set of work successfully groups similar attack behaviors, the clustered traffic may or may not be relevant and can dilute or distract the discovery of critical attacks. In addition, it is not uncommon to have sources perform related, sometimes complementary attacks. To group the potentially collaborative attacks, this work proposes to expand upon the use of "Attack Social Graph (ASG)" - a graphical representations of cyber attack traffic [5,1]. Graph-based measures are defined to capture the macroscopic attack structure between nodes, complementing security features to group relevant and potentially collaborative actions to produce attack models. Finally, this works aims at producing online processing to enhance situation awareness as malicious traffic is being observed.

2 Attack Segmentation and Model Generation

The Attack Segmentation and Model Generation (ASMG) processing aims at enhancing online situational awareness to security analysts: to assess the subset of traffic relevant to a target of interest as it is incident, and to estimate the attack behaviors exhibited by this subset of traffic. The resulting attack behavior models deviate significantly from those produced by existing Host Clustering works, in that, each attack model produced by ASMG is not just a description of source addresses that generate similar traffic or with similar characteristics; it models a single collective attack behavior produced by collaborating hosts or spoofed addresses of these malicious hosts. Similarly behaving source addresses may or may not be grouped into a single attack model, depending on how "close" the traffic is to the target of interest and how it contributes the collective behavior.

This objective drives the two principle design decisions of the ASMG: the processing must be performed online as new traffic is incident, and macroscopic information about the spatial or graphical structure must be captured to distinguish between likely collaborating addresses and addresses that simply behave similarly. Finally, these objectives are set in an attack environment where probabilistic intersection between sources or targets is routine. This environment drives a need to segment irrelevant behaviors that are probabilistically attached to the ASG of interest, and also drives the attack features away from incorporating IP address or its derivatives, like node degree.

Figure 1 presents a top-level diagram of the ASMG components. The ASMG is built from the principles of naïve Bayes classification, and maintains a series of empirically constructed non-parametric attack models, that are generated online from the observed traffic; as new traffic is incident on the current ASG of interest,

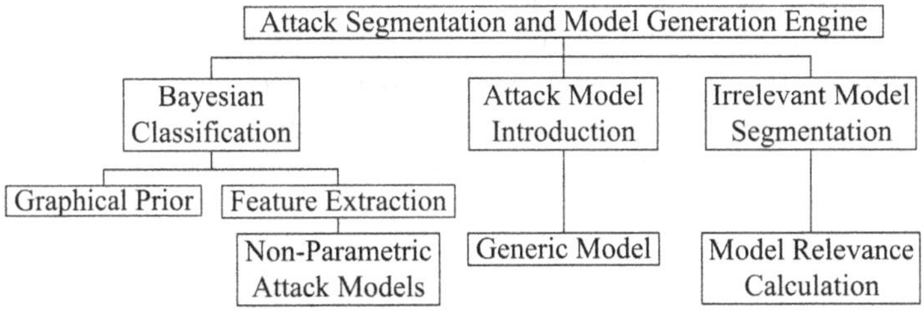

Fig. 1. ASMG Top-Level Diagram

it is classified to the attack model that maximizes the posterior probability at the time of observation.

2.1 Attack Model

In the context of ASMG, an **attack model** is defined as a collection of feature probability distributions. For large-scale cyber attacks, a set of features is defined based on packet header information, including:

Protocol: A discrete probabilistic feature illustrates how likely the collective behavior is due to TCP, UDP, or ICMP traffic.

ICMP Type: A discrete probabilistic feature models the likeliness of the type of ICMP packets non-parametrically when there exists ICMP packets.

SRC and DST Port: Malicious cyber traffic exhibits many different port selection behaviors, which may be categorized as deterministic and stochastic port selection behaviors. For deterministic selection, packets are sent from a certain port on the source machine to a particular port on the destination machine, likely indicating a particular service is being targeted. In this case, the port values can directly be compared to gauge the similarity of attack behaviors. A variety of stochastic behaviors [12,15,11] may be less obvious to treat: a malicious source may perform a "vertical" scan over many different destination ports; some attacks may randomly select source ports to obscure the behavior profile; "Backscatter" traffic generated from denial of service attacks may also give the appearance of randomly selecting source port. In the case of stochastic port selection, a comparison of the actual port values is unlikely to be meaningful.

Let $\mathcal{P}$ be the random variable on taking a specific port number, $P(D)$ the probability of making a deterministic port selection, and $P(S)$ the probability of making a stochastic port selection. $P(D)$ and $P(S)$ are estimated separately for the source and destination ports, and together fill the hypothesis space for entropy on the source and destination port selections. Equation (1) gives the probabilistic feature of port selection, by applying the intuition that the actual value of $\mathcal{P}$ is not important if the port is chosen stochastically.

$$P(\mathcal{P} = X) = P(S) + P(\mathcal{P} = X|D)P(D) \tag{1}$$

Graph-Position: The graph-position feature attempts at incorporating graphical structure into the ASMG framework. The intuition defining this feature is that the nodes in the attack social graph associated to an attack model should have similar "position" or "roles" in the graph. This feature extends from the concept of closeness centrality in Social Network Analysis [9], and the related concept of Graph Efficiency [6].

For a given node i, its "position" in a given subgraph $\mathcal{G}_M$ as defined by prior observations associated with a model M is determined by the inverse harmonic mean of distances, $d_{i,j}$, between node i to all other nodes $j \in \mathcal{G}_M$. Let $\mathcal{P}_{i,j}$ be the path from nodes i to j and $\mathcal{P}_{i,j}^k$ be the kth edge along the path. The distance $d_{i,j}$ is defined as $\sum_{k \in \mathcal{P}_{i,j}} \frac{1}{P(M|\mathcal{P}_{i,j}^k)}$ to reflect the probability the edges along the path between i and j belong to the model M. Furthermore, the harmonic mean of distances should not be unweighted, as the contribution of the path(s) between i and j depends on how likely the attack behavior shown on the terminal edges belong to the model M. As a result, the Graph Position for a node i with respect to a model M is shown in (2).

$$\text{GP}(i) = \sum_{j \in \mathcal{G}_M} P(M|\mathcal{P}_{i,j}^0) P(M|\mathcal{P}_{i,j}^{-1}) \frac{1}{d_{i,j}} \tag{2}$$

In summary, each attack model is a collection of empirically constructed probability density estimates for the features defined. Together these feature distributions provide the likelihood term for calculating the posterior in Bayes' Rule, as shown in (3). For each new packet incident on the attack social graph of interest the feature values described are extracted and represented as an observation vector $\mathbf{X}$. The goal of the Naïve Bayes classification is to determine the probability of each model after receiving this new evidence; based both on the likelihood of the feature values, and the prior probability of the attack model before observing the packet features.

$$\underbrace{P(M|\mathbf{X})}_{\text{posterior}} = \underbrace{P(\mathbf{X}|M)}_{\text{likelihood}} \underbrace{P(M)}_{\text{prior}} \tag{3}$$

2.2 Graph-Based Prior Probability

The graph-based prior probability, $P(M)$, is defined to signify a departure from the frequentist prior in routine Naïve Bayes classification. It utilizes macroscopic information about ASG to determine whether a collective set of evidences is spatially cohesive to infer a collaborating behavior at a target. ASMG supports this concept by extending the measure of Graph Efficiency defined by Latora and Marchiori [6]. The formulation is very similar to the concept of Closeness Centrality used to define the packet-level graph-position feature, but an extrapolation to measure the entire graph.

For each model M, the efficiency of the corresponding subgraph, $\mathcal{G}_M$, is determined by the inverse of the distances between all pairs of nodes in $\mathcal{G}_M$, and further weighted by the probability that both the start and terminal edges between the pairs belong to $\mathcal{G}_M$. Specifically, it is defined in (4) using the same $P(M|\mathcal{P}^0_{i,j})$, $P(M|\mathcal{P}^{-1}_{i,j})$, and $\frac{1}{d_{i,j}}$ as defined for the graph position feature.

$$E(\mathcal{G}_M) = \sum_{i \neq j \in \mathcal{G}_M} P(M|\mathcal{P}^0_{i,j})P(M|\mathcal{P}^{-1}_{i,j})\frac{1}{d_{i,j}} \tag{4}$$

The prior probability of each attack model, $P(M)$, is derived by normalizing over the set of all current attack models. Letting M_{all} be the current set of attack models, the prior probability of each attack model is given by (5).

$$P(M) = \frac{E(\mathcal{G}_M)}{\sum_{M_i \in M_{\text{all}}} E(\mathcal{G}_{M_i})} \tag{5}$$

2.3 Model Introduction Strategy

Sections 2.1 and 2.2 define a classification strategy for associating new observed packets with a set of attack models to maximize the posterior probability. This prompts questions about constructing the initial set of attack models, or alternatively, how to treat new behaviors that are not well explained by any of the existing models. This process is enabled by the adoption of a *Generic Attack Model*, a hypothesis to compete against the empirically constructed models during the classification phase. This new hypothesis intends to fit all behaviors with some modest probability, but not as well as a tailored empirical model. The feature distributions for the generic model are defined based on offline analysis of the traffic characteristics [7] and intuition about the features defined.

When new traffic is observed, the generic model is evaluated as a possible class. A higher posterior probability with the generic model suggests that none of the existing empirical models provide a probable explanation, and a new model should be introduced. This also applies to the initial start-up phase whereas the first packets observed must be classified to the generic model, and will be used to create a new empirical model. Certainly, if one of the empirical models maximizes the posterior, the new packets will be incorporated to update the corresponding empirical model. In effect, the result of unsupervised learning, creating clusters of behaviors based on observed data, has been reproduced with an online supervised approach, where the problem of determining the number of models has been mitigated through the generic model.

2.4 Segmentation of Irrelevant Behaviors

Modeling attack behaviors is necessary for segmenting irrelevant traffic that is probabilistically incident on the ASG of interest. An analyst may be concerned with traffic several hops away from the target of interest, but not if it is a dramatic departure in behavior from the activity present at the target. If traffic behavior changes abruptly as analysis moves more hops away from the target of interest, it is likely that this path exists only because of probabilistic intersection.

Packets should be segmented from the attack graph of interest if they do not belong to any relevant model with high probability. Some packets may have high probability with an attack model, but must pass through several edges with low relevance to be reached; these packets are also segmented. Let $\mathcal{P}_{toi,i}$ be the path between a node i and the target of interest, the irrelevance of an attack social graph component is given by (6), where $R(k) = \frac{1}{|E|}\sum_{e \in E} P(k|e)$, and E is the set of edges incident on the target. Node i is segmented if the irrelevance is greater than a fixed threshold.

$$I(i) = \sum_{j \in \mathcal{P}_{toi,i}} \min_{k \in M_{\text{all}}} \left\{ \frac{1}{P(k|j)R(k)} \right\} \tag{6}$$

3 Experimental Results

A prototype of ASMG was implemented and tested using the traffic capture produced by the Cooperative Association for Internet Data Analysis (CAIDA) UCSD Network Telescope over two days in November 2008 [2]. Network Telescope traffic [11,15] serves to analyze probabilistically connected behaviors, as it is primarily composed of Malware scanning or spoofed-address backscatter. Only the packet header information is available, which is in part responsible for the limited feature set used. Different target IP addresses were randomly selected, not preferentially, to provide a survey of performance across different attack scenarios. Traffic experienced by each target and the surrounding addresses was replayed to emulate an online flow of new traffic.

Figure 2 shows an example of ASG generated by ASMG processing for an example target. Three empirical models were generated, as indicated by different line groupings. A small set of segmented traffic is shown outside of the groupings for illustration purpose. In this case, Empirical0 and Empirical8 include hosts multiple hops away from the target of interest, which is situated in the center of the three groupings. Empirical8 includes a set of suspicious nodes that participate in the attacks against two additional targets in a similar fashion.

Figure 2 also shows a parallel coordinate plot exhibiting the features discussed earlier. Note that the attack behavior within a single model is not entirely deterministic, which is expected. For example, Empirical0 and Empirical1 show vertical scans across destination port, which can also be a result of backscatter. More interestingly, Empirical8 contains a mix of different destination port

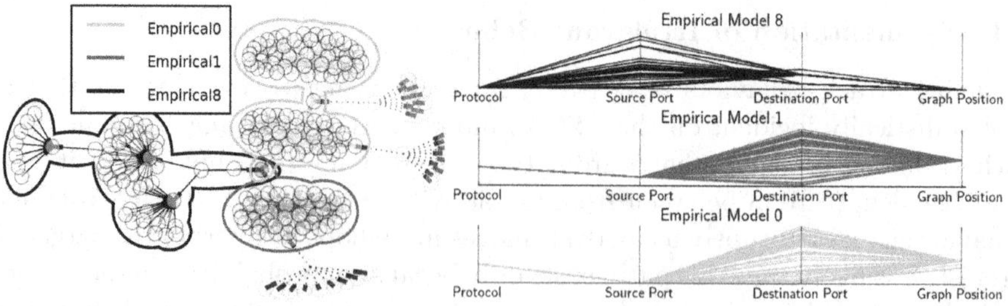

Fig. 2. An Attack Social Graph and Parallel Coordinate Plot produced by ASMG

behaviors. The attack as a whole appears to select source port stochastically, but there are also clearly three different destination port behaviors. One might argue that the outcome could be improved by introducing one or two additional attack models to cover each destination port selection. However, the graph structure of Empirical8 is very "efficient" and the prior probability is very high, so the differences in the feature space are weighed less heavily. This case exemplifies the difference between the objectives of clustering similar behaviors versus identifying nodes serving complementary purposes in a coordinated attack.

In addition to the specific example shown above, Figure 3 shows an overall performance assessment of ASMG over a large set of targets. The y-axis shows the expected value of packet likelihood; that is, the expected value of just the feature based term of the Bayesian calculation used to perform the modeling and segmentation. This expectation is taken over all the received packets. More colloquially, this measures on average how well the packets fit to the assigned model. Different, randomly selected, targets of interest are shown along the x-axis, sorted by the expected packet likelihood achieved by ASMG.

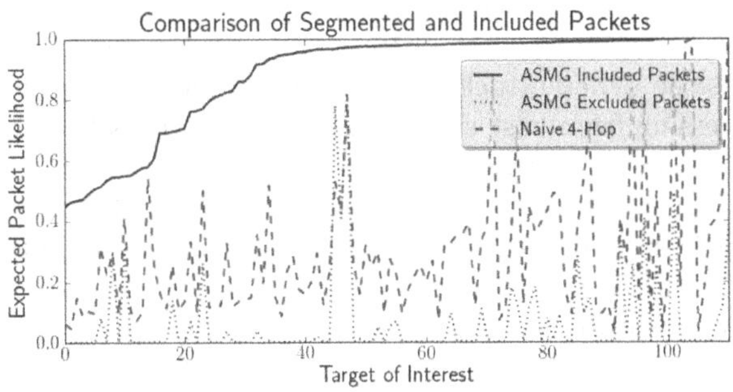

Fig. 3. Performance of ASMG Processing Over Many Targets of Interest

The plot shows three lines. The solid black line displays the expected likelihood of the packets deemed relevant and grouped into one of the empirical models by ASMG. The thick dashed line depicts the expected likelihood under a naïve 4-hop model that includes all traffic within a 4-hop perimeter from each target of interest. The idea behind the naïve 4-hop process is to reflect that analysts may choose to assess a subset of traffic without sophisticated algorithms that segment or cluster traffic. Four-hops were chosen to provide a fair comparison over the range of number of hops commonly produced by ASMG. The comparison between the solid and dashed lines shows that ASMG consistently outperforms the naïve 4-hop processing, by simultaneously determining the set of packets to be included for each target and the behavior group each packet belongs. The thin dotted line shows the expected packet likelihood for the packets that were segmented out by ASMG, but re-associated with one of the relevant attack models, whichever maximized the posterior probability. The consistently low expected likelihood of these ASMG excluded packets further demonstrates that the segmented packets were appropriate to be left out.

Note that, without regard for model complexity, an ideal result would produce an expected packet likelihood of one, and in fact ASMG nearly achieves this in the majority of cases. However, there are several less than ideal cases within the first third of the targets, which are primarily due to two reasons. First, the current ASMG processing is implemented in an online, greedy manner, and, therefore, there are cases where the empirical models would have been better if they were further segregated or merged when more traffic is observed. This is an inherent tradeoff between offline and online processing. Second, ASMG currently does not put emphasis on the graph-based prior or one feature versus the other. Further improvements can be achieved by investigating the tradeoff between the graph-based prior and the various behavior features.

4 Conclusion

Together, the re-characterization of a seemingly unsupervised problem as an online supervised problem, the novel introduction of a graph-based prior probability, and the inclusion of the generic attack model as a threshold for introducing new attack models accomplish the objective of simultaneously segmenting and modeling large-scale cyber attack traffic. The graph-based prior distinguishes the ASMG processing from host clustering; the focus shifts from grouping hosts with similar behaviors to diagnosing groups of likely collaborators into the same attack model. Targets of interest were randomly selected from Network Telescope traffic provided by CAIDA, and used to illustrate the quality and superior performance of ASMG. The proposed method is expected to be applicable to other domains where uncertain and probabilistic intersection of actor identities obfuscates the determination of potentially collaborative behaviors.

References

1. Du, H., Yang, S.J.: Discovering Collaborative Cyber Attack Patterns Using Social Network Analysis. In: Salerno, J., Yang, S.J., Nau, D., Chai, S.-K. (eds.) SBP 2011. LNCS, vol. 6589, pp. 129–136. Springer, Heidelberg (2011)
2. Aben, E., et al.: The CAIDA UCSD Network Telescope Two Days in November 2008 Dataset. Technical report (access Date: May 2012-August 2013)
3. Fukuda, K., Hirotsu, T., Akashi, O., Sugawara, T.: A PCA Analysis of Daily Unwanted Traffic. In: Proceedings of 24th IEEE International Conference on Advanced Information Networking and Applications, pp. 377–384 (April 2010)
4. Gu, G., Perdisci, R., Zhang, J., Lee, W.: BotMiner: Clustering Analysis of Network Traffic for Protocol- and Structure-Independent Botnet Detection. In: Proceedings of the 17th USENIX Conference on Security Symposium, Berkeley, CA, USA, pp. 139–154 (2008)
5. Karagiannis, T., Papagiannaki, K., Faloutsos, M.: BLINC: multilevel traffic classification in the dark. ACM Comp. Commun. Rev. 35(4), 229–240 (2005)
6. Latora, V., Marchiori, M.: Efficient behavior of small-world networks. Phys. Rev. Lett. 87, 198701 (2001)

7. Lee, D., Carpenter, B., Brownlee, N.: Observations of udp to tcp ratio and port numbers. In: Proceedings of Fifth International Conference on Internet Monitoring and Protection, pp. 99–104 (2010)
8. McGregor, A., Hall, M., Lorier, P., Brunskill, J.: Flow Clustering Using Machine Learning Techniques. In: Proceedings of Passive and Active Measurement Workshop, pp. 205–214 (2004)
9. Newman, M.E.J.: Scientific collaboration networks. ii. shortest paths, weighted networks, and centrality. Phys. Rev. E 64, 016132 (2001)
10. Ohta, M., Kanda, Y., Fukuda, K., Sugawara, T.: Analysis of Spoofed IP Traffic Using Time-to-Live and Identification Fields in IP Headers. In: Proceedings of IEEE International Conference on Advanced Information Networking and Applications, Washington, DC, USA, pp. 355–361 (2011)
11. Shannon, C., Moore, D.: Network Telescopes: Remote Monitoring of Internet Worms and Denial-of-Service Attacks Technical report (2013) (Technical Presentation-access Date: May 2012-August 2013)
12. Treurniet, J.: A Network Activity Classification Schema and Its Application to Scan Detection. IEEE/ACM Tran. on Networking 19(5), 1396–1404 (2011)
13. Wei, S., Mirkovic, J., Kissel, E.: Profiling and Clustering Internet Hosts. In: Proceedings of International Conference on Data Mining (DMIN) (June 2006)
14. Xu, K., Wang, F., Gu, L.: Network-aware behavior clustering of Internet end hosts. In: Proceedings of IEEE INFOCOM, pp. 2078–2086 (April 2011)
15. Zseby, T.: Comparable Metrics for IP Darkspace Analysis. In: Proceedings of 1st International Workshop on Darkspace and UnSolicited Traffic Analysis (May 2012)

Emergent Consequences: Unexpected Behaviors in a Simple Model to Support Innovation Adoption, Planning, and Evaluation

H. Van Dyke Parunak[1] and Jonathan A. Morell[2]

[1] Soar Technology, Inc.
3600 Green Court, Suite 600
Ann Arbor, MI 48105
van.parunak@soartech.com
[2] Fulcrum Corporation
4075 Wilson Blvd, Suite 550
Arlington, VA 22203
jmorell@fulcrum-corp.com,
jamorell@jamorell.com

Abstract. Many proven clinical interventions that have been tested in carefully controlled field settings have not been widely adopted. We study an agent-based model of innovation adoption. Traditional statistical models average out individual variation in a population. In contrast, agent-based models focus on individual behavior. Because of this difference in perspective, an agent based model can yield insight into emergent system behavior that would not otherwise be visible. We begin with a traditional logic of innovation, and cast it in an agent-based form. The model shows behavior that is relevant to successful implementation, but that is not predictable using the traditional perspective. In particular, users move continuously in a space defined by degree of adoption and confidence. High adopters bifurcate between high and low confidence in the innovation, and move between these groups over time without converging. Based on these observations, we suggest a research agenda to integrate this approach into traditional evaluation methods.

Keywords: Agent-based models, emergent behavior, innovation adoption, clinical evaluation.

1 Introduction

There is an extensive literature on factors that facilitate the adoption of innovations [8]. Two research traditions underlie this work. (1) Traditional statistical analysis applies familiar statistical methods to test hypotheses concerning the implementation and continued use of innovations in social and organizational settings. (2) Case studies perform in-depth observation and analysis on innovation cases.

Collectively, this research has successfully revealed which innovations will succeed. Despite this research, however, funders of research in health, mental health,

W.G. Kennedy, N. Agarwal, and S.J. Yang (Eds.): SBP 2014, LNCS 8393, pp. 179–186, 2014.

substance abuse, and social betterment face a dilemma. Even with decades of funding, many interventions that have proved to be successful in carefully controlled field settings have not been widely adopted. Why is this so, and what can be done about it? Traditional models of innovation adoption have not helped.

Our research applies systems thinking to innovation adoption in clinical settings, and is part of a trend in applying systems methodologies to study health and public health interventions [2]. In particular, we apply agent-based modeling to address two questions. First, does an agent-based approach reveal anything about innovation adoption that traditional research does not? Second, can agent-based models add value to traditional efforts to evaluate exercises in innovation adoption?

Agent-based models might add value because they can be based on knowledge from traditional research, but they look at phenomena in a way that is fundamentally different from statistical methods. [4,5]. Traditional statistics obviates the contribution of individual variation in a population. In contrast, the agent-based view is based on the interaction of individual behaviors. We seek to use this interaction to explore emergent system-level behavior not accessible to a mean-field model. (We do not claim to predict the detailed behavior of an individual human therapist.)

This paper presents a laboratory-based proof of concept that agent-based modeling can reveal unanticipated behavior that may affect adoption. Section 2 introduces a simple model of innovation adoption. Section 3 documents our agent-based implementation of this model, and Section 4 describes its behavior. Section 5 discusses the research implications of the observed behavior from two perspectives: the adequacy of our underlying logic model, and the implications for clinical management of innovation, including introduction, motivation, and evaluation.

2 A Logical Model of the Adoption of Innovations

Fig. 1 illustrates the type of program theory that might characterize evaluation of a program designed to facilitate the adoption of a new, proven best practice in a clinical setting. This model depicts a logic in which therapists' use of a new treatment is based directly on his or her confidence that the treatment will work, combined with organizational support for use. Confidence in turn is a function of the influence of a therapists' colleagues, quality of training in the new treatment, and the therapist's conclusion about success or failure in clinical outcome. Colors reflect the psychological (salmon), organizational (blue), and social psychological (yellow) factors that combine to facilitate use of the innovation. As with all evaluation "logic models," Fig. 1 is based on input from the stakeholders involved in planning the program, and on relevant research findings in health care and in

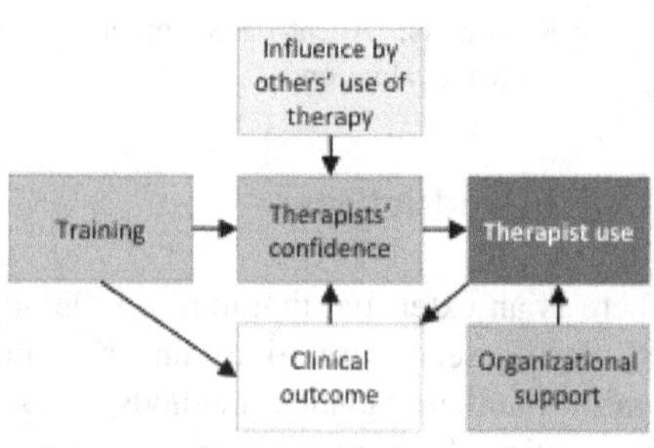

Fig. 1. Simple Model of Adoption of an Innovation

many other domains. That research makes it clear that successful innovation is a function of the characteristics of: 1) adopters, 2) the innovation, and 3) the setting in which adoption takes place [1,6]. Building models such as Fig. 1 is standard practice in the field of evaluation, a practice with a long history of success in leading to insightful understanding of program behavior and outcome. Such models help planners articulate their (often implicit) assumptions about what they are doing and why; provide a framework for drawing on the results of prior research; identifying the constructs that need to be operationalized and measured; and assuring that all parties involved have a common understanding of what the program will be and how it will be evaluated.[1]

This approach has limitations. The first is epistemological, i.e., innovation efforts based on theories such as these have provided disappointing results. Clearly, they do not provide the knowledge needed to achieve the desired level of implementation of best practices. Second, they are limited in their capacity to help stakeholders consider the likely behavior of the programs they are implementing. The work reported here probes the value of improving the application of traditional logic models by integrating an agent-based simulation into the development of those models.

3 An Agent-Based Implementation

We implement this model in NetLogo [7], a framework for multi-agent modeling that is widely used in the social sciences. Each therapist t_i is a software agent with six characteristics. We represent them as functions over the therapist index. Where there is no ambiguity about the therapist in question, we refer simply to the function name.

Three agent characteristics are assigned when the agent is initialized.

- **Group membership $G(t_i)$ = set of therapists:** Therapists exchange their experiences only within their group.
- **Adaptability $a(t_i) \in [0, 1]$:** Indicates how susceptible a therapist is to the opinions of her peers.
- **Training $t(t_i) \in [0, 1]$:** Models the quality of the training received by the therapists in how to apply the innovative best practice. Each therapist is assigned a level of training, which can modulate his initial confidence and the outcome of his application of the innovation. Training is a number in a specified range, and can be assigned uniformly randomly over all therapists, randomly chosen between the maximum and minimum level, or held constant within agent groups.

Three characteristics vary during the therapist's career:

- **Skill $s(t_i) \in [0, 1]$:** This variable is initialized to the therapist's training level, then increases as the therapist exercises the innovation and decays while she does not

[1] In a real evaluation this model would be more detailed. For instance "confidence" could be modulated by a random variable representing clinical judgment about the value of the new treatment with respect to a specific client. The level of detail presented here is more than adequate for our objective of showing how modeling can yield insights not available through the static model alone.

exercise it. The higher the therapist's skill, the more successful an exercise of the innovation is likely to be.

- **Confidence $c(t_i) \in [0, 1]$:** The central variable in Fig 1 is confidence, driven by the therapist's training, clinical experience, and input from colleagues in her group.
- **Adoption $d(t_i) \in [0, 1]$:** This variable records the percentage of opportunities on which the therapist decides to use the innovation.

In addition, 11 variables govern the entire model:

- the total number of therapists
- the number of groups
- **Organizational Support $o \in [0, 1]$:** the degree to which the therapists' employer encourages the innovation
- **Training Success $\tau \in [0, 10]$:** how training influences success
- **Maximum Success $m \in [0, 1]$:** the maximum success rate for this innovation
- **Success impact $i_s \in [0, 1]$:** how much a successful outcome increases c
- **Failure impact $i_f \in [0, 1]$:** how much an unsuccessful outcome decreases c
- **Skill increment $k_i \in [0, 1]$:** how much exercising an innovation increases s
- **Skill decrement $k_d \in [0, 1]$:** how much s decreases an innovation is not exercised
- **History length $h_l \in [1, 20]$:** how many rounds contribute to adoption level
- **History decay $h_d \in [0, 1]$:** the rate at which history decays

In the real world, adoption of innovation depends on more variables than these. However, these are adequate to capture our logic model (Fig. 1), and even this simplified model generates behavior that we might not have anticipated from the logic model.

At each cycle of the model, each therapist takes all four of the following actions. Then the next therapist takes them, and so forth. In incrementing or decrementing bounded variables, we modulate the impact by the amount of range left to move in the direction of the change, consistent with commonly observed ceiling and floor effects.

1. Update confidence level based on peer interaction. The therapist's confidence is moved toward that of her peers by a fraction of how far it is from theirs:

$$\gamma = \alpha * \left(\frac{\sum_{t_i \in G} c(t_i)}{|G|} - c\right)/2 \tag{1}$$

If $\gamma > 0$ (indicating that others in the group are, on average, more confident than ego), c is replaced by $c + \gamma(1 - c)$. If $\gamma < 0$, c is replaced by $c\ (1 + \gamma) < c$.

2. The therapist decides whether to adopt the innovation on her next engagement, by computing a probability of adoption p_d and then comparing it to a uniformly distributed random number in [0,1]. The model provides two ways to compute p_d: either as the average of c and o, or as a logistic function of their sum:

$$p_d = 1/\left(1 + e^{-s(c+o-1)}\right) \tag{2}$$

s is steepness. For $c + o = 1$, $p_d = 0.5$, and ranges from 0 to 1 as the sum of these values moves from 0 to 2. The logistic models widely-observed saturation effects.

3. If the therapist decides to adopt the innovation:
   (a) Her adoption level increases. We provide two ways to monitor adoption. An array remembers whether (1) or not (0) the therapist used the innovation on the last h_l trial, and we compute h as the average of this array. Alternatively, we increment d by $h_d(1-d)$, which realizes an exponential weighted adoption rate.
   (b) Her skill is incremented by $k_i(1-s)$.
   (c) The intervention succeeds with probability $x(1-e^{-\tau * s})$.
   (d) Optionally, her confidence is incremented or decremented depending on whether the intervention was successful or not. On success, confidence is incremented by $i_s(1-c)$. On failure, confidence is decremented by $i_f c$.
4. If the therapist decides not to adopt the innovation:
   (a) Her adoption level decreases, by recording a 0 in the history or decrementing d by $h_d d$.
   (b) Her skill is decremented by multiplying it by k_d.

Our dependent variables are adoption and confidence of the therapists, and evolve as the model executes. We do not track outcome, since it is not an emergent effect of therapist interactions, but is determined directly by individual skill and training level.

Our implementation is much more detailed and specific than the logical model of Fig 1, and another realization of that model might behave very differently than what we report in the next section. We have posted the full model online,[2] and encourage other researchers to explore variations of our implementation.

4 Model Behavior

At a gross level, confidence and adoption behave as one might expect. For example,

- Increased training raises the highest level of confidence attained;
- Increased organizational support raises both the upper and the lower limits of the adoption observed;
- High adaptability (susceptibility to colleagues' opinions) narrows the distribution of confidence across therapists.

But some other behaviors of the model are counterintuitive.

A therapist's confidence is initialized by training level, and then changes in only two ways: by attraction to the opinion of peers, and by experience with the innovation. Fig 2 shows how these effects interact. In these runs, 50 therapists are all in a single group, with training assigned uniformly randomly in [0, 1]. While these shapes reflect convergence of the model, individual therapists do not settle down to a single location in the state space of Adoption vs. Confidence, but continue to move.

In the bottom row, therapists' experience with the innovation does not change their confidence. At the right, with no peer influence (***a*** = 0 for all therapists), therapists are distributed continuously over both adoption and confidence, with a reasonable

---

[2] `http://www.abcresearch.org/abcresearch.org/models/InnovationAdoption.nlogo`

increase in adoption with confidence. Therapists change adoption over time, but remain at their initial confidence.

At the lower left, therapists still do not change confidence with experience, but adopt the confidence of their peers (to varying degrees). Eventually all therapists converge to the same confidence, within which they continue to change in adoption. Again, this result is not surprising.

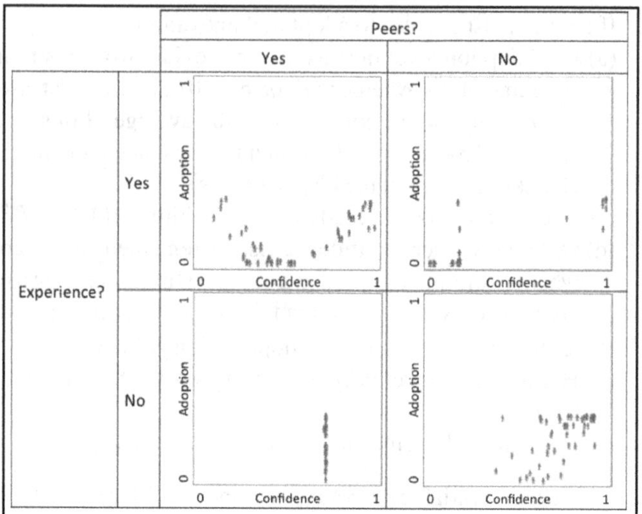

Fig. 2. Impact of peer and experience factors on distribution of therapist adoption and confidence

The top row is less intuitive. At the right, confidence changes by experience but not by peer interaction. Confidence and adoption are still correlated, but the distribution is no longer continuous. Instead, the therapists bifurcate in confidence, moving to extremes with a wide unpopulated area in between. While individual therapists continue to move in both dimensions, the bifurcation in confidence remains. Experimentation with the model shows that the bifurcation is the result of fairly high levels of i_f and i_s (0.5 or higher). When the impact of success or failure on confidence is high, a few successive successes or failures quickly drive the therapist to an extreme in confidence. This observation is relevant to training therapists in understanding the dynamics of best practices that demonstrate overall positive impact, but varying consequences for individual patients. We can understand the dynamic, but we would be unlikely to recognize the potential for this behavior from Fig 1 alone.

At the top left, both peers and experience modulate confidence. High levels of confidence show the same bifurcation as in the previous condition. Again, the bifurcation is driven by higher levels of i_f and i_s. However, low adopters tend to have middling levels of confidence. Because they seldom try the innovation, the main influence on their confidence is peer pressure, which pushes them toward intermediate confidence.

Interestingly, while confidence and adoption are positively correlated for therapists on the right-hand arm of the distribution, the relative shift in impact of outcome and peer confidence leads to a *negative* correlation on the left-hand arm. Detailed observation of the agents as the model executes shows that they tend to move counter-clockwise around the space. Consider an agent who starts on the right-hand arm, where adoption and confidence are correlated.

- The agent may move back and forth along this arm, as the peer effect pulls confidence lower, decreasing adoption, and as success pushes confidence higher.
- At some point, a failed adoption cuts the agent's confidence dramatically (because of the high value of i_f), moving the agent to the left-hand arm.
- Due to lower confidence, adoption drops. At the same time, peer pressure pulls confidence higher, moving the agent down the right-hand arm toward the center.
- At some point the agent's confidence grows enough that it resumes the therapy, moving higher. If successful, it climbs the right-hand arm; if not, it moves back to the left-hand arm.

This agent-based model shows that the logic model of Fig. 1 implies dynamics that would not be anticipated by simple examination, and suggests research hypotheses and practical innovations inaccessible from the static model alone.

5 Implications

Committing to action is a leap of faith that commits material resources and intellectual capital, incurs opportunity costs, and forces planners and evaluators to confront unintended consequences [3]. In theory, innovative programs can be changed if they do not work, and program evaluation provides data so that rational decisions can be made about program improvement. In practice, commitment makes change difficult. Therefore it is important to do the most rigorous possible planning, and to conduct evaluations whose results will be as potent as possible. We have shown that one way to increase rigor and potency is to reformulate the kind of model commonly used in planning and evaluation practice as an agent-based simulation.

This justification for modeling acknowledges that no matter how much research we draw upon, and no matter how well we conduct group processes with experts, we may miss important insights. Models are useful for catching some of what we have missed, but another possibility may be in play. We may have erred not because we failed to understand what we have, but because what we need to know is not yet known.

Consider two dimensions. First, neither previous research nor expert judgment has considered all the variables that matter, those that could account for unexplained variance in observed outcomes, or that could lead to consideration of variables that account for some of the unexplained variance. Our program theory may either be incorrect, or not strong enough to help us make a big enough practical difference. For instance, existing theory about "adoption" does not consider the built ecology in which therapists work, though traffic patterns affect interactions between therapists and their colleagues, their patients, and their patients' families. Second, traditional theorizing about social phenomena does not consider the importance of emergent dynamics. The *only* way to consider relevant dynamics is to employ an agent based model. All of these possibilities are illustrated in our findings about adoption and confidence.

Fig. 1 does not explicitly model peer interaction. Perhaps it should have been considered, given what we know about the importance of boundary spanners, mavens, leaders, and other such roles. But hindsight is seductive. Given the results of the simulation, *of course* planners should have considered role relationships. We do not know whether paying attention to these roles would affect the particular group behavior observed, but the results of the model might have led planners to consider peer dynamics more than they otherwise would have. In any case, who could have foreseen that the model would lead people to think of peer behavior? For all anyone knew, the model might have yielded surprising results with respect to clinical judgment of success, or critical thresholds of effectiveness between the old and new treatment, or any one many behaviors that the model might have unearthed. Agent-based models are valuable for planning and evaluation precisely because these dynamics are not easily predicted, despite the best expert judgment and extensive literature reviews.

Our results demonstrate that a simple logic model may have unexpected dynamics. The degree to which our results (and the underlying model) reflect the behavior of real therapists is an important question that lies beyond this paper, but should be confirmed by a broader survey of the empirical literature and clinical validation.

We have presented this model in a clinical context. However, discussions with practitioners in other areas (e.g., educational reform) suggest that the same basic pattern of influences will be widespread, and our research agenda includes studying the degree to which this model can in fact be generalized across disciplines.

References

[1] Berwick, D.M.: Disseminating Innovations in Health Care. Journal of the American Medical Assoiation 289(15), 1969–1975 (2003)

[2] Mabry, P.L., Milstein, B., et al.: Opening a Window on Systems Science Research in Health Promotion and Public Health. Health Education and Behavior 40(1S), 5S–8S (2013)

[3] Morell, J.A.: Evaluation in the Face of Uncertainty: Anticipating Surprise and Responding to the Inevitable. Guilford Press, New York (2010)

[4] Morell, J.A., Hilscher, R., et al.: Integrating Evaluation and Agent-Based Modeling: Rationale and an Example for Adopting Evidence-Based Practices. Journal of Multi Disciplinary Evaluation 6(14), 35–37 (2010)

[5] Van Dyke Parunak, H., Savit, R., Riolo, R.L.: Agent-Based Modeling vs. Equation-Based Modeling: A Case Study and Users' Guide. In: Sichman, J.S., Conte, R., Gilbert, N. (eds.) MABS 1998. LNCS (LNAI), vol. 1534, pp. 10–25. Springer, Heidelberg (1998)

[6] Rogers, E.M.: Diffusion of Innovations, 5th edn. Free Press, New York (2003)

[7] Wilensky, U.: NetLogo. Center for Connected Learning and Computer-Based Modeling, Northwestern University, Evanston, IL (1999), http://ccl.northwestern.edu/netlogo

[8] Wisdom, J.P., Chor, K.H.B., et al.: Innovation Adoption: A Review of Theories and Constructs. Administration and Policy in Mental Health and Mental Health Services Research (2013) (in press)

Deriving Population Assessment through Opinion Polls, Text Analytics, and Agent-Based Modeling

Ian Yohai, Bruce Skarin, Robert McCormack, and Jasmine Hsu

Aptima, Inc., 12 Gill Street, Suite 1400, Woburn, MA 01801
{iyohai,bskarin,rmccormack,jhsu}@aptima.com

Abstract. Surveys are an important tool for measuring public opinion, but they take time to field, and thus may quickly become out of date. Social and news media can provide a more real-time measure of opinions but may not be representative. We describe a method for combining the precision of surveys with the timeliness of media using an agent-based simulation model to improve real-time opinion tracking and forecasting. Events extracted through text analytics of Afghan media sources were used to perturb a simulation of representative agents that were initialized using a population survey taken in 2005. We examine opinions toward the U.S., the Afghan government, Hamid Karzai, and the Taliban, and evaluate the model's performance using a second survey conducted in 2006. The simulation results demonstrated significant improvement over relying on the 2005 survey alone, and performed well in capturing the actual changes in opinion found in the 2006 data.

Keywords: Population assessment, agent-based modeling, public opinion polls, text analytics, social identity theory.

1 Introduction

For over sixty years traditional surveys have been the gold standard in opinion research. These opinion polls represent a detailed description of how a population perceives the state of the world and thus provide insights into how the public may behave given certain situations. These polls, however, can only capture a static snapshot of a population living in a world that is changing ever faster. As soon as a survey is completed, the accuracy to which it represents the current state of the population quickly begins to decay. Events, such as competitor product releases, debates, protests, attacks, or even natural disasters, can drastically affect opinions. As such, surveys conducted just a few months past may often be an unreliable representation of the affected population. This is especially true in regions facing conflict or other situations of high volatility.

To help fill these gaps, a significant amount of attention is being given to the growing popularity of social media and twenty-four hour news coverage. For instance, one study was able to "use the chatter from Twitter.com to forecast box-office revenues for movies" and "show that a simple model built from the rate at which tweets are created about particular topics can outperform market-based predictors" [1]. Yet a

W.G. Kennedy, N. Agarwal, and S.J. Yang (Eds.): SBP 2014, LNCS 8393, pp. 187–194, 2014.

year-long study by the Pew Research Center [3] found that the "reaction on Twitter to major political events and policy decisions often differs a great deal from public opinion as measured by surveys." In the analysis of major news events such as presidential debates and the outcome of the 2012 presidential election, Pew found that Twitter posts were often more liberal or conservative than what was indicated in surveys.

Given the kind of incongruences seen in the Pew study, it is clear that social media analysis alone is not always a reliable means for estimating the current state of a population. This is especially true in populations where technology access is limited or other significant socioeconomic disparities exist. Even in traditional surveys conducted with scientific sampling, non-response can cause substantial bias in opinion estimates; the problem is magnified enormously in social media where only a limited subset of citizens post regularly, and "self-select" into the sample. As such, significant caution should be given to its use in situations requiring critical decision making.

2 Methodology

To help extend the shelf-life of opinion polls and mitigate the misrepresentations found in social and news media, this paper discusses a methodology for combining text analysis with simulations of an agent-based model. Our approach begins by initializing the agent-based model directly from polling data to create a synthetic population that is representative of all the respondent's social identities (e.g. gender, ethnicity, education, etc.) and initial opinions. This model is then simulated to allow the agents to interact with each other and to respond to significant events that have been extracted from news and social media. The intended result is then an updated assessment of population opinions. In the following sections we describe the development of the simulation model, the employment of text analysis techniques, and the results of an ongoing study based on surveys conducted in Afghanistan from 2005 to 2008.

2.1 Simulating Population Opinion Dynamics with an Agent-Based Model

The original motivation for developing a model of population scale opinion changes was to help decision makers in defense and security environments better understand how local populations affect the outcome of current and future missions. To model these changes, we first developed a system dynamics model of individual to individual interactions based on social identity theory [4]. Given the importance of geographic and social distributions however, an approach based solely on system dynamics modeling would have required immensely complex subscripts to keep track of the numerous combinations of population cohorts and physical locations. As such, after developing the individual opinion change behaviors in the system dynamics model, the resulting equations were adapted to provide discretized versions within an agent-based model. These versions are then evaluated when an agent receives an opinion from another agent or as one generated in response to a given event.

To create our agent population, we first produced an array of survey respondents that is statistically weighted by census data for the regions they live in. In this study, we use a representative survey conducted by the polling firm Charney Research in October 2005 to initialize the model; the simulation results are then compared with a second representative survey conducted by the same firm in November 2006. Both surveys had approximately 1,000 respondents and were conducted using face-to-face interviews. After analyzing the array, we determine the minimum number of agents required to ensure that each survey respondent is represented by at least one agent. We then initialize each agent using the identities of a respondent (gender, ethnicity, education) and by sampling a continuous range that corresponds to his or her response on a four point Likert scale for each opinion of interest (opinion toward the U.S., the Afghan government, Hamid Karzai, and the Taliban). The overall opinion scale ranges from negative one (strongly opposed) to positive one (strongly for). Each agent is also assigned a distribution of reactions to different event types (e.g. attacks, reconstruction, elections, etc.) that is based on their identity characteristics.

To define this reaction distribution, we fit a statistical model to derive the correlation between the opinion of interest and another item in a survey that taps attitudes related to the event type. For example, to estimate reactions to security incidents, we use an item that asks respondents' to rate the security situation in their area on a four point Likert scale. We then fit an ordered probit regression model with the opinion (e.g., the U.S.) as the dependent variable and the security item as one of the independent variables, along with the identity characteristics. Space requirements preclude details on the regression models, but the demographic and reaction variables explained a good deal of the variance in opinion. The stronger the coefficient on the security item, the stronger the estimated reaction. To allow reactions to vary among identity characteristics, we use interaction terms between all identities and the event item (e.g., the security question). Thus, for example, Pashtun uneducated males can have different reactions than Tajik educated females.

After initializing the agents, they are then distributed over geographic regions and connected to one another so that they may interact throughout the simulation. In more general terms, agents that share similar identity characteristics are more likely to interact with one another and update their opinions to match one another than agents with different identity characteristics. With the synthetic population initialized, we are now ready to simulate the population changes over a period of time, allowing them to interact at regular intervals and to respond to events that occur throughout the period.

As mentioned earlier, the process by which each agent updates its opinion in response to the signal sent by another agent or a given event was first developed in a system dynamics model (Fig. 1). This model was based on an extensive literature review of different social identity and social influence theories as described in [5]. The product of this study was an algorithmic interpretation of the theory of bounded confidence, where changes in an individual's opinion are moderated by his or her level of certainty. Depending on the disparity between two individuals and the level of certainty held by the receiver of a given opinion, a variety of outcomes are possible.

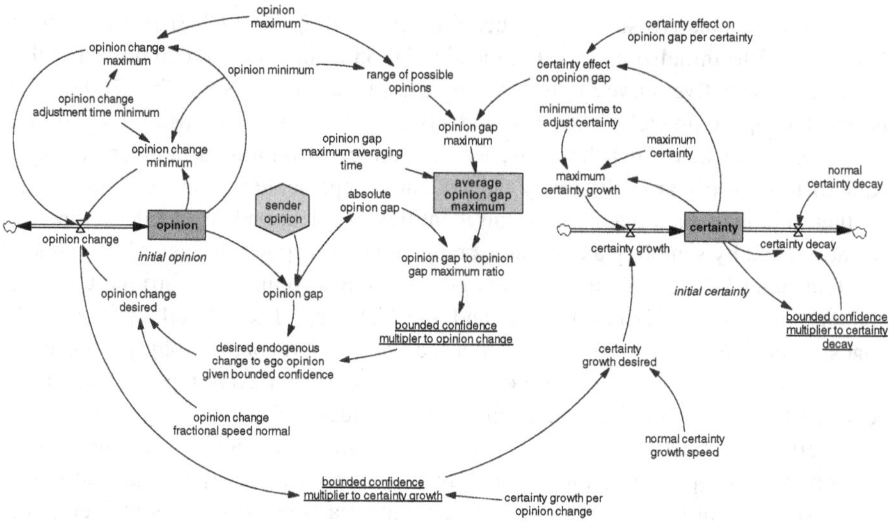

Fig. 1. A model of bounded confidence opinion change

Consider the following examples in Fig. 2. Suppose a sender agent's opinion is held constant at 0.25, the amount the receiver agent is willing to close the gap is dependent upon its certainty and the size of the gap between them. With no starting certainty and an absolute gap of 0.5 resulting from the receiver's initial opinion of -0.25, over time, the agent is able to close the gap completely. With a level of moderate certainty (0.5), the agent will close only part of the gap before the growth of its certainty prevents further changes. When certainty is high or the gap is too large, then the agent will not close any portion of the gap, no matter how long they receive the same signal from the sending agent.

In the agent simulation, the signal received from a sender agent within its network or in response to a given event is not held constant as it is in the system dynamics model, but rather dependent upon discrete interactions. These interactions occur at

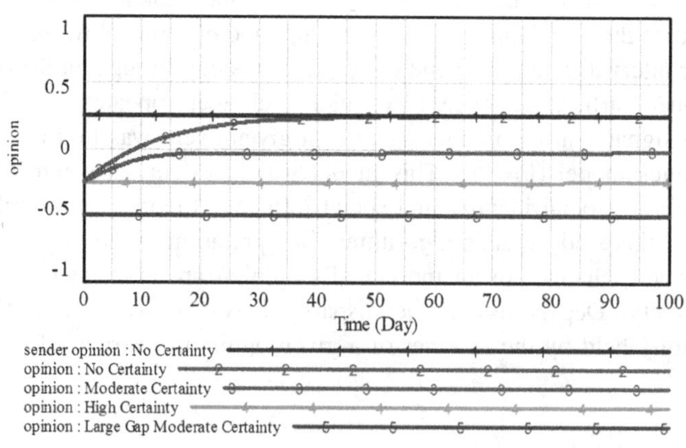

Fig. 2. Simulations of opinion change under a variety of bounded confidence conditions

different intervals that are scheduled based on a distribution of contact frequency. Given repeated interactions between two particular agents and no others, the same general behaviors can be created, but since each agent has a diverse network and because there are external events influencing the agent population, the overall dynamics have a variety of the emergent features that are characteristic of agent-based models. For this reason, we run Monte Carlo simulations on the various parameters affecting initialization and opinion changes in order to estimate the sensitivity and statistical significance of any population scale trends produced by the model.

2.2 A Model for Analyzing Unstructured Text

We introduce external stimulants to the agent-based model by turning online news and social media into a series of events. To extract significant events from unstructured text, we use a variety of preprocessing and statistical techniques. One of the important preprocessing stages is the recognition of location names. To complete this, we utilized the GeoNames (geonames.org) database of place names to extract locations referred to in documents. These locations were then used as metadata in the processing of text to create a Dirichlet-multinomial regression (DMR) model [5] of a corpus compiled from the Afghanistan News Center archive of articles. DMR is an unsupervised technique for extracting a set of topics that are conditional on metadata, such as a region in this case. Topics are distributions of words that represent a central concept. For example, one topic extracted from this dataset represents the concept of "road reconstruction," and the top words include project, road, and reconstruction.

For a given topic, we then set thresholds for the temporal probability (also referred to as the topic prevalence) in order to automatically determine when an event has occurred. When a topic moves above a threshold (one standard deviation above the mean) we note the start of an event and then set the duration once the topic prevalence drops below the threshold again. Events that go above a higher threshold (two standard deviations above the mean) denote more significant events that have stronger reaction distributions associated with them. The resulting stream of events then becomes the stimulus for the agent-based model described in the previous section.

3 Results

We present results on each of the four opinions by comparing the distributions from the 2005 survey, the 2006 survey, and the simulation. In particular, we wish to determine the extent to which the simulation can capture significant changes in opinion between the two surveys. In addition to plotting the three distributions, we also perform a series of statistical significance tests (specifically, the Mann-Whitney test, given the ordinal variables). Specifically, we first compare the 2005 and 2006 surveys to assess whether there was a statistically significant change in the real data. Second, we compare the 2005 survey to the simulation result to determine if the model produced any significant change. Finally, we compare the simulation result to the 2006 survey to see how well the simulated distribution compares with the "ground truth" distribution. Note if the second test reveals a significant difference in the correct direction (increasing or decreasing support for each opinion), this illustrates that the simulation is better than relying on the 2005 survey alone, assuming there significant movement in the 2006 survey. This is true even if the simulation does not capture the full magnitude of the change present in the 2006 survey, since the 2005 survey is obviously static.

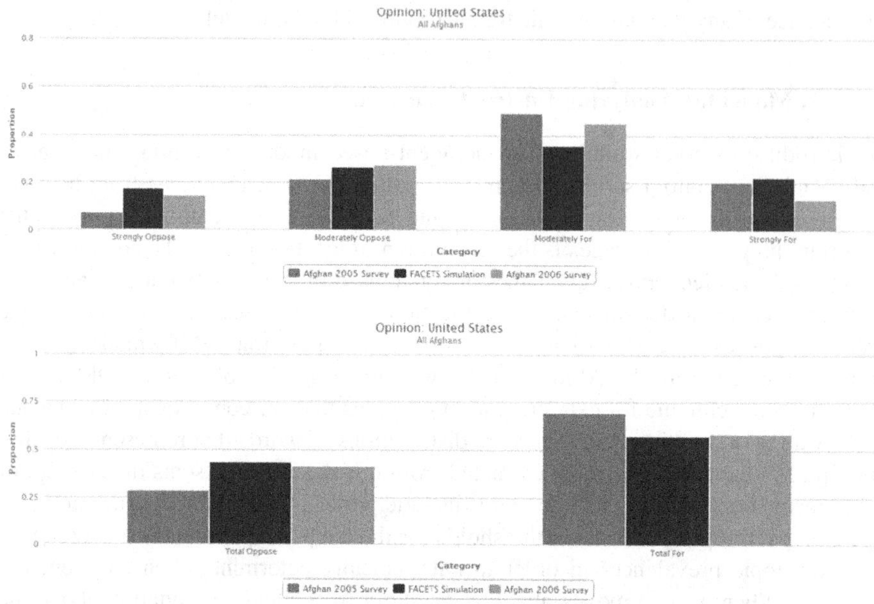

Fig. 3. Opinion toward the U.S.

In Fig. 3, we present the distribution of opinion toward the U.S. The upper panel shows the distribution for all four opinion categories. The light blue bars represent the 2005 survey; the dark blue bars represent the simulation; and the green bars represent the 2006 survey. There is a significant drop in opinion toward the U.S. between the 2005 and 2006 surveys ($p < .001$). The simulation captures this drop in support from the 2005 survey ($p < .001$) and is not significantly different from the 2006 survey ($p = 0.067$). Nevertheless, the simulation does not capture the decline in the "strongly for" category, but rather shows the decline in the "moderately for" category. To examine changes at grossest level – support or opposition – we combine the "strongly oppose" and "moderately oppose" categories and the "moderately for" and "strongly for" categories. The result is shown in the bottom panel of Fig. 3, which indicates a much closer fit between the simulation and the 2006 survey, and the difference is not statistically significant ($p = .611$). The simulation predicts a 12% drop in total support for the U.S., almost exactly the same as the 11% drop present in the 2006 data.

We repeat the exercise for opinion toward the Afghan government (Fig. 4; full distributions not shown). Once again, we see a significant drop in support between the two surveys ($p < 0.001$). The simulation also shows a significant drop in support compared with the 2005 survey ($p < 0.001$) but is significantly different than the 2006 survey ($p < .002$). When the categories are combined, however, the simulation result and the 2006 survey do not show a significant difference ($p = .967$), and the simulation exactly captures the 16% drop in total support between the 2005 and 2006 data.

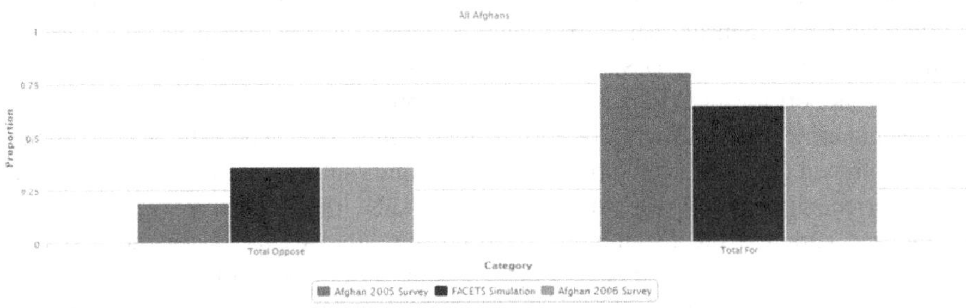

Fig. 4. Opinion toward the Afghan Government

In Fig. 5, we show the results for opinion toward Hamid Karzai. There is a steep drop in support between the 2005 and 2006 surveys ($p < .001$). The simulation indicates a drop in support ($p < .001$) but not nearly as much as indicated in the 2006 data ($p < .001$). When examining the combined categories, the difference between the simulation and the 2006 data is attenuated but still significantly different ($p < 0.001$). The simulation captures roughly half the magnitude of the decline (about 8% compared with 15% in the 2006 data).

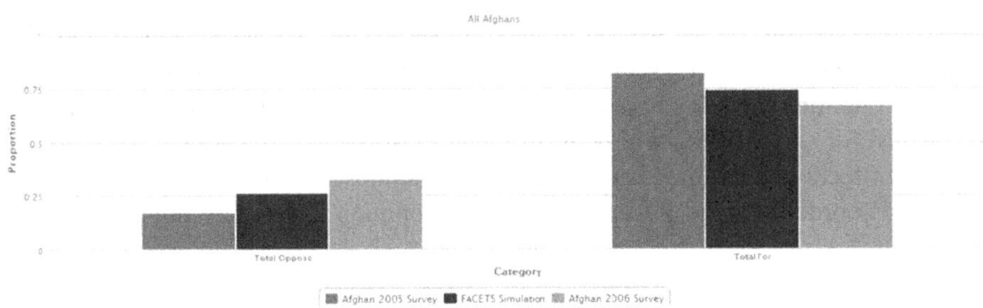

Fig. 5. Opinion toward Hamid Karzai

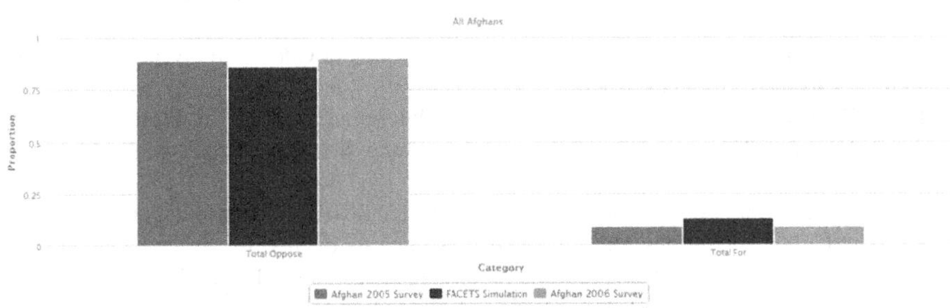

Fig. 6. Opinion toward the Taliban

Finally, in Fig. 6, we present the results for opinion toward the Taliban. In this case, there is no significant difference between the 2005 and 2006 surveys ($p = 0.715$), but the simulation detects a slight drop in opposition ($p < 0.001$). The simulated drop in opposition is approximately 3% when looking at the combined categories (bottom panel of the figure). While statistically significant, the magnitude is relatively small, approximately one quarter as large as the predicted (and correct) decline in support for the U.S. (approximately 12%) and one fifth as large as the decline in support for the Afghan government (approximately 16%).

To summarize, the simulation performed reasonably well, especially when looking at the combined categories for support or opposition. The simulation indicated change in the correct direction for opinion towards the U.S., the Afghan government, and Karzai, and captured the magnitude of the change almost exactly for the first two opinions. The simulation predicted a slight decline in opposition to the Taliban that was not present in the real data, but was much smaller than the predicted changes on the other opinions. This suggests the model is not simply detecting a significant drop in support across all opinions, but rather can detect nuances in the patterns present in the real data. It is worth emphasizing that we are forecasting one year into the "future," during a time of significant change on the ground. In short, the simulation results demonstrate significant improvement over simply relying on the 2005 survey by itself, and performed well in comparison to the actual changes in the 2006 data.

4 Conclusion

The intention of this paper was to provide an overview of a new methodology for combining text analytics of social and news media with an agent-based simulation model of opinion dynamics based on polling data. It is clear that traditional surveys and analysis of social media will be insufficient in isolation in accurately representing the state of local population. As such, methods like those discussed in this paper will be needed in order to help fill a critical gap in the opinion research community.

References

1. Asur, S., Huberman, B.A.: Predicting the Future with Social Media. In: Proceedings of the 2010 IEEE/WIC/ACM International Conference, pp. 492–499 (2010)
2. Mitchell, A., Hitlin, P.: Twitter Reaction to Events Often at Odds with Overall Public Opinion. Pew Research Center (2013)
3. Tajfel, H., Turner, J.C.: An Integrative Theory of Intergroup Conflict. In: Worchel, S., Austin, W.G. (eds.) The Social Psychology of Intergroup Relations. Brooks/Cole, Monterey (1979)
4. Grier, R.A., Skarin, B., Lubyansky, A., Wolpert, L.: SCIPR: A Computational Model to Simulate Cultural Identities for Predicting Reactions to Events. In: Second International Conference on Computational Cultural Dynamics, College Park, MD (2008)
5. Mimno, D., McCallum, A.: Topic models conditioned on arbitrary features with dirichlet-multinomial regression. In: Proceedings of the Twenty-Fourth Conference on Uncertainty in Artificial Intelligence (2008)

mFingerprint: Privacy-Preserving User Modeling with Multimodal Mobile Device Footprints

Haipeng Zhang[1,*], Zhixian Yan[2], Jun Yang[2], Emmanuel Munguia Tapia[2], and David J. Crandall[1]

[1] School of Informatics & Computing, Indiana University, Bloomington, IN, USA
{zhanhaip,djcran}@indiana.edu
[2] Samsung Research America, San Jose, CA, USA
{zhixian.yan,j3.yang,e.tapia}@samsung.com

Abstract. Mobile devices collect a variety of information about their environments, recording "digital footprints" about the locations and activities of their human owners. These footprints come from physical sensors such as GPS, WiFi, and Bluetooth, as well as social behavior logs like phone calls, application usage, etc. Existing studies analyze mobile device footprints to infer daily activities like driving/running/walking, etc. and social contexts such as personality traits and emotional states. In this paper, we propose a different approach that uses multimodal mobile sensor and log data to build a novel user modeling framework called mFingerprint that can effectively and uniquely depict users. mFingerprint does not expose raw sensitive information from the mobile device, e.g., the exact location, WiFi access points, or apps installed, but computes privacy-preserving statistical features to model the user. These descriptive features obscure sensitive information, and thus can be shared, transmitted, and reused with fewer privacy concerns. By testing on 22 users' mobile phone data collected over 2 months, we demonstrate the effectiveness of mFingerprint in user modeling and identification, with our proposed statistics achieving 81% accuracy across 22 users over 10-day intervals.

1 Introduction

Mobile devices such as smartphones and tablets have become powerful people-centric sensing devices thanks to embedded sensors such as GPS, Bluetooth, WiFi, accelerometer, touch, light, and many others. The devices have also advanced significantly in terms of computational capacity, memory and storage. These improvements have stimulated people-centric mobile applications, ranging from inferring and sharing real-time contexts such as location and activities [3,8] to identifying heterogeneous social behaviors of mobile users [6,9,11,15]. Most of these studies focus on using phone data to infer physical and social contexts for a particular user at a specific point in time. Additional studies have concentrated on analyzing long-term data from mobile devices to monitor trends and to establish predictive models of location [7] and app usage [12].

* This work was done while this author was a research intern at Samsung Research America - Silicon Valley.

W.G. Kennedy, N. Agarwal, and S.J. Yang (Eds.): SBP 2014, LNCS 8393, pp. 195–203, 2014.

In this paper, we take a significantly different view of mobile device data. Instead of focusing on real-time inference of contexts or social behaviors from phone sensors, we analyze multimodal mobile usage data to extract simple yet effective statistics that can uniquely represent mobile users. We construct a novel user modeling framework called 'mFingerprint' to define 'fingerprints' that try to uniquely identify users from mobile device data. Our experiments show the effectiveness of mFingerprint in modeling users and identifying them uniquely. In contrast to many user identification studies that try to identify users based on raw sensor data such as touch screen [2] and cell towers [5], mFingerprint computes high-level statistical features that do not disclose sensitive information from the phone, such as raw location, browser history, and application names. Therefore, applications can share, transmit, and use these descriptive privacy-preserving feature vectors to enable personalized services on mobile devices with fewer privacy concerns.

Research Challenges. It is non-trivial to build a system to identify users based on high level statistics of their mobile usage data, due to the following challenges: (1) Mobile devices collect a variety of multimodal data including logs from physical sensors such as GPS and accelerometer and application data like app usage and web browser history; how to choose sources to construct effective mobile fingerprints remains a question. (2) Mobile data is generated under complex real-life settings, introducing significant noise that demands robustness in processing. (3) In contrast to most existing offline data analysis approaches deployed on the server side, our mFingerprint framework focuses on designing lightweight energy-efficient algorithms that are able to run on mobile devices directly. (4) mFingerprint must avoid disclosing sensitive mobile data, such as geo locations from GPS, the applications installed, and URLs of website visited, to protect user's privacy.

Main Contributions. To address these challenges, this paper presents a novel approach for user modeling and identification based on digital footprints from mobile devices, with the following key contributions: (1) We build mFingerprint, a novel framework to analyze data from mobile devices and model users via digital footprints; (2) mFingerprint generates fingerprints from heterogeneous hardware sensors such as GPS, WiFi, and Bluetooth and soft sensors including app usage logs; (3) By designing a discriminative set of statistical features to capture mobile footprints, mFingerprint is able to identify users while preserving their privacy.

2 Related Work

Recently, user's digital footprints from mobile device have gained significant attention in various research areas such as mobile computing, data mining and social analysis.

In mobile computing and data mining, several studies build mobile systems to infer context offline or in real-time, such as detecting semantic locations (home and office) from GPS [14], identifying physical activities from accelerometer data [8], or estimating user's environmental properties such as crowdedness from Bluetooth [13] and noise-level [10]. These studies however focus on analyzing only single physical sensors to infer specific types of daily contexts and activities.

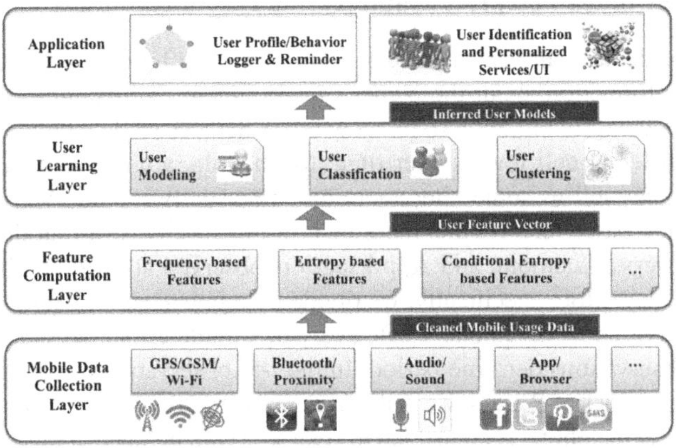

Fig. 1. The mFingerprint framework

For social and behavioral analysis on mobile data, the recent focus is mainly on continuously monitoring user's daily social contexts, such as inferring emotions from audio [11], detecting mood from communication history and application usage patterns [9], predicting user's personality using various phone usage data [4,6], and estimating user's profile and sociability level using Bluetooth crowdedness [15]. The mFingerprint framework further extends these studies to analyze multimodal mobile footprints (including both soft usage data and physical sensor data) by extracting discriminative statistical features as a fingerprint to model the user and provide accurate user identification with privacy preservation.

3 mFingerprint System Overview

Fig. 1 shows the four layers of the mFingerprint framework. The bottom layer collects various sensor readings using hardware sensors that trace location (e.g., GPS, WiFi, and cell tower), proximity (Bluetooth and microphone). Furthermore, soft sensor data such as application usage and web browser history are also recorded. The second layer of mFingerprint computes a set of privacy-preserving statistical features from these digital footprints. In this paper, we particularly focus on designing frequency and entropy based statistical features to capture mobile device usage patterns. Such features flow up to the third layer for user learning, which includes building user models via the feature vectors, identifying users via classification methods, and grouping users into meaningful clusters via unsupervised learning. The forth layer is the application layer where various applications can be established, such as inferring user profile, logging mobile behaviors, and creating personalized services and user interfaces.

4 Footprint Feature Computation and User Identification

We focus on designing simple frequency and entropy-based statistical features to create users' "fingerprints" and evaluate the features' performance in user identification.

4.1 Frequency Based Footprint Features

The number of devices and cell towers that are observed by a phone throughout the day provides information about the owner's environment. For example, a phone in a busy public place will likely observe many wireless devices while a phone in a moving car observes different cell towers over time. Meanwhile, a user's app usage patterns throughout the day tell us something about his or her daily routine. We thus propose simple frequency-based features that measure how much activity of four different types (WiFi, Cell towers, Bluetooth, and App usage) is observed at different time intervals throughout the day. More specifically, we divide time into T-minute time periods, and make observations about the phone's state every M minutes, with $M < T$ so that there are multiple observations per time period. In the i-th observation of time period t, we record: (1) the number of WiFi devices that are observed ($W^{t,i}$), (2) the number of cell phone towers that the phone is connected to ($C^{t,i}$), (3) the number of bluetooth devices that are seen ($B^{t,i}$), and the number of unique apps that have been used over the last m minutes ($A^{t,i}$). We then aggregate each of these observation types to produce four features in each time period:

$$F_W^t = \sum_i W^{t,i}, \qquad F_C^t = \sum_i C^{t,i}, \qquad F_B^t = \sum_i B^{t,i}, \qquad \text{and} \qquad F_A^t = \sum_i A^{t,i}.$$

A feature incorporating all of these features is simply the vector $F^t = [F_W^t \; F_C^t \; F_B^t \; F_A^t]$.

4.2 Entropy Based Footprint Features

While the simple frequency features above give some insight into the environment of the phone, they ignore important evidence like the distribution of this activity. For example, in some environments a phone may see the same WiFi hotspot repeatedly through the day, while other environments may have an ever-changing set of nearby WiFi networks. To illustrate this, Fig. 2 compares observed frequency versus anonymized device IDs for two users across each of the four observation types, for a period of 10 days. We can observe that User 2 is less active in WiFi and cell mobility compared to User 1, but has more Bluetooth encounters and uses more diverse apps.

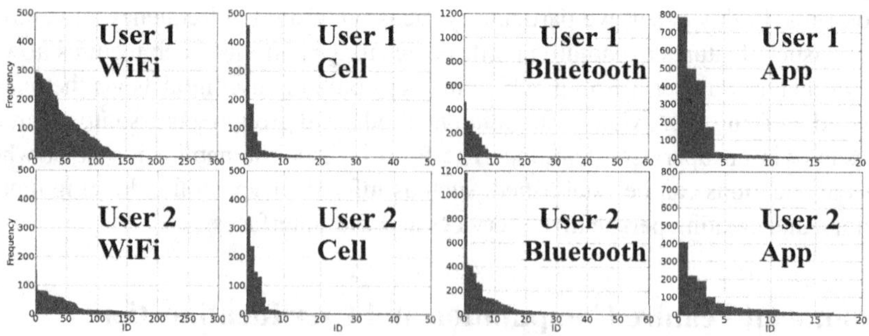

Fig. 2. Comparison of activity histograms for 2 users over 10 days. Y-axes are frequencies; X-axes are WiFi, Cell, Bluetooth, and App IDs.

We thus propose using entropy of these distributions as an additional feature of our user fingerprints. The entropy feature summarizes the distribution over device IDs, but in a coarse way such that privacy concerns are minimized. For WiFi, let W_j^t denote the number of times we observe wifi hotspot j during time period t. Then we define the WiFi entropy during time period t as,

$$E_W^t = -\sum_j \frac{W_j^t}{F_W^t} \log \frac{W_j^t}{F_W^t}.$$

Entropy features for Cell towers, Bluetooth, and Apps (E_C^t, E_B^t, and E_A^t, respectively) are computed in the same way, and we define a multimodal entropy feature vector $E^t = [E_W^t \ E_C^t \ E_B^t \ E_A^t]$, which incorporates all four perspectives.

4.3 Conditional Entropy and Frequency Based Footprint Features

In our system, we calculate the above entropy and frequency features conditioned on time and location. Intuitively at different times and at different locations, users have different patterns of application usage and surrounding devices (Bluetooth, WiFi, cell, etc.). For example, two users might have similar overall apps usage but one user always uses apps in the mornings, while the other uses them only in the afternoon. Two other users might have similar overall Bluetooth entropies but one might have more surrounding devices at work while the other observes the variety at a coffeeshop. Conditioning on time and space is thus useful to better differentiate users.

Conditional Features on Time. For the frequencies and entropies conditioned on time, we differentiate on time of a day and day of a week. Currently we distinguish between three fixed daily time intervals, mornings (0:00 - 8:59), working hours (9:00 - 17:59) and evenings (18:00 - 23:59), and two types of days, weekdays (Mon through Fri) and weekends (Sat and Sun). This gives five time periods over which we compute the conditional features. Future work might explore adaptive intervals instead.

Conditional Features on Location. We also compute frequency and entropy features conditioned on location. For each user, we filter and cluster their geo-locations in order to identify the top-k significant locations. From data collected at these k locations, we compute the conditional entropies and frequencies. There are two steps in finding significant locations: *Segmentation* and *Clustering*. In the segmentation step, we find periods of time when the phone appears to be stationary, by looking for time intervals when the IDs of surrounding devices are stable. In particular, we divide the data streams into 10-minute time frames and for each time frame, we record the IDs of the WiFi, Bluetooth and Cell towers. For adjacent time frames, we compute Jaccard similarity of the corresponding sets of IDs, $J(S_1, S_2) = |S_1 \cap S_2|/|S_1 \cup S_2|$, where S_1 and S_2 are sets of device IDs. If the similarity is larger than a threshold, we say that the device is stationary during the two time frames. Fig. 3(a) illustrates finding stationary and non-stationary periods according to WiFi readings. Similarly, we perform this on Bluetooth and Cell readings and take the union of all stationary time periods.

We then apply the DBSCAN clustering algorithm to the stationary segments in order to identify important locations. Fig. 3(b) shows the clustering results on the same

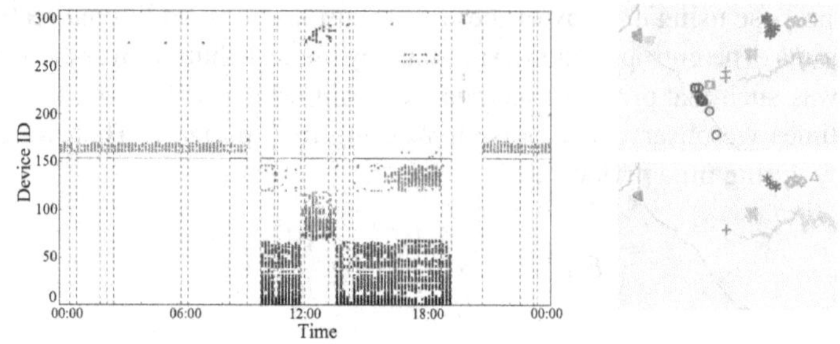

(a) (b)

Fig. 3. Time segmentation and location clustering: (a) Finding stationary times using WiFi. (b) Location clustering using all (top) and only stationary (bottom) data.

data with and without non-stationary points. We see that noisy signals (e.g., location points moving along highways) have been removed by keeping only stationary data, which generates better and fewer clusters. Note that our mFingerprint system computes location based conditional features using anonymized cluster and device IDs, not the original locations, to help preserve privacy. We choose $k = 2$ and statistics from the ith location will be compared with that from the location of the same rank across users.

5 Evaluation

We define a user identification problem in order to test whether mFingerprint can uniquely characterize users. We pose this as a classification problem: *can we build a multi-class classifier trained on the entropy and frequency fingerprint features labeled with user IDs, such that it tells which user a certain fingerprint vector belongs to?*

5.1 Data Collection and Experimental Settings

To collect data for our evaluation, we deployed an Android app called EasyTrack based on the Funf Open Sensing Framework [1]. This app has a customizable configuration with 17 data types, including WiFi, Bluetooth, cell tower, GPS, call log, app usage etc. We successfully recruited 22 users to install EasyTrack and collected their mobile footprints for about 2 months with some variation across users.

We first test the initial user identification performance using different time frame lengths. We uniformly sample the instances to make sure that the same number of instances is used to build the classifier for each time frame length. As shown in Fig. 4a, when the length of time frame increases, the classification accuracy generally improves, despite possible variations caused by weekday/weekend patterns. This suggests that in this range, longer time frames better capture the uniqueness. Since the data collection time span is about two months, longer time frames decrease the total number of time periods and thus there are fewer features for training the classifiers. In the following experiments, we fix the time period at 10 days.

With a 10-day time frame, we reach 107 time frames in total from 22 users. On average, each user has 4.86 time frames. The range is [2, 8] and the standard deviation is 2.2.

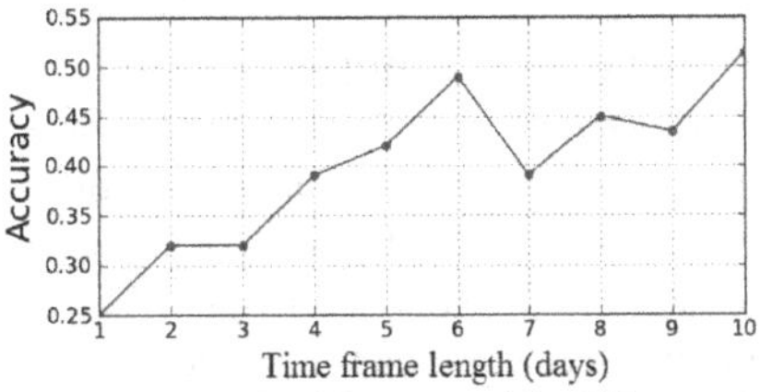

(a) Accuracy for 22 users with different time frame lengths, with the basic entropy features.

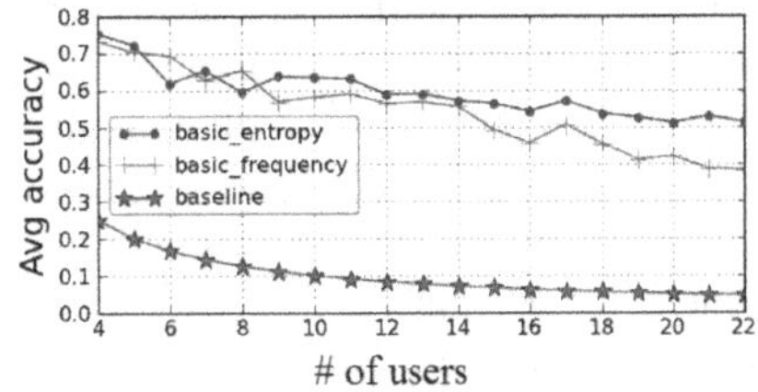

(b) Classification accuracy with different # of users for basic frequency and entropy features.

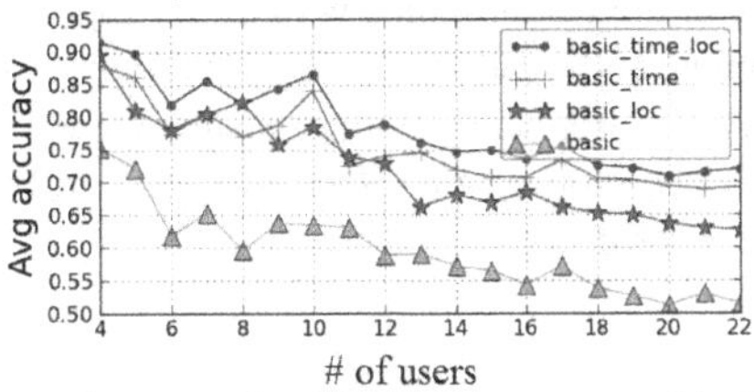

(c) Average classification accuracy with different number of users for entropy feature combinations.

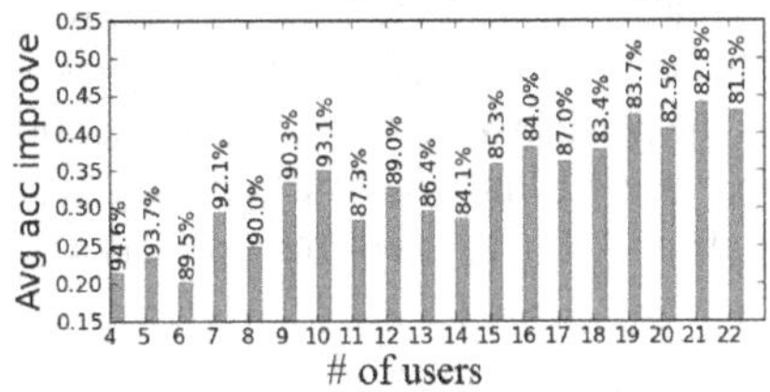

(d) Avg accuracy improvement for all features, compared to baseline frequency. Absolute avg accuracy is marked on the top of each bar.

Fig. 4. User identification classification results

In total, we have 64 features in mFingerprint. We test combinations of features (multimodal entropies/frequencies, conditional entropies/frequencies) on multiple classifiers including Naive Bayes, decision tree, SVM and Multilayer Perceptron. We report the results from the Multilayer Perceptron, which were best. The learning rate is 0.3, momentum is 0.2, the number of hidden layers is set to $\frac{\#features+\#classes}{2}$ where $\#features$ is 64 in mFingerpint and $\#classes$ is 22, which is the number of users. The number of epochs is set to 500. We use 10-fold cross validation for training and testing. We use accuracy as a performance measurement which is defined as $\frac{\#of\ correct\ predictions}{\#of\ instances}$.

5.2 User Identification Performance

We tested on varying numbers of users from 4 to 22, and observe that as the number of users increases, the average accuracy drops but is still significantly better than random guessing (see Fig. 4b and Fig. 4c). In these experiments, we randomly sample 10 from the $\binom{22}{n}$ possible combinations (where n is the number of users being tested), apply 10 fold cross validation on each of the samples, calculate the accuracy and then get the average over the 10 samples. On average, each sample group has 63 instances.

Performance of Standalone Frequency and Entropy Features. For basic multimodal frequency and entropy features, each vector has 4 dimensions, corresponding to WiFi, cell tower, Bluetooth, and apps, respectively. We compare their classification results as shown in Fig. 4b. Both frequency and entropy features outperform the random baseline significantly. Entropies have better performance for large numbers of users compared to frequencies: mean accuracy with entropies drops 22 percentage points from 4 users to 22 users, versus a 35 point drop using basic frequencies.

Performance of Conditional Features. We also compared various types of conditional features (Fig. 4c). When features including basic multi-modal entropies, location entropies and time entropies are all combined, the performance is the best. Time features perform slightly better than location features, perhaps because there are only 2 location-conditioned features but 5 time-conditioned features. With all three kinds of features combined, the accuracy is 71.96% for 22 users versus 91.54% for 4 users and 86.59% for 10 users. Though more features improve accuracy, more computation is required as well, especially for the clustering required by location conditioning.

Performance of All Features. Finally, we compute the performance with all mFingerprint features including basic frequencies, entropies, and conditioned features. Fig. 4d shows the performance improvement against the basic frequency feature results shown in Fig. 4b, with the absolute accuracy marked on the top of each bar: 94.68% for 4 users, 93.14% for 10 users and remains 81.30% for 22 users.

6 Conclusion

We presented mFingerprint and showed that its statistics (frequencies and entropies) computed from the device usage data and sensor data can be used as a fingerprint for user identification while preserving privacy. This serves as the key idea of the proposed mFingerprint framework to collect multimodal mobile data, compute footprint features, build unique user models, and serve personalized applications. We tested on a user-identification task and achieved over 81% accuracy even when the number of users reaches 22. The feature computation is designed considering both simplicity and energy-efficiency, and thus can naturally be fit into the on-device framework.

References

1. The Funf Open Sensing Framework, `http://www.funf.org/`
2. Angulo, J., Wästlund, E.: Exploring touch-screen biometrics for user identification on smart phones. In: Camenisch, J., Crispo, B., Fischer-Hübner, S., Leenes, R., Russello, G. (eds.) Privacy and Identity Management for Life. IFIP AICT, vol. 375, pp. 130–143. Springer, Heidelberg (2012)
3. Campbell, A.T., Eisenman, S.B., Lane, N.D., Miluzzo, E., Peterson, R.A., Lu, H., Zheng, X., Musolesi, M., Fodor, K., Ahn, G.-S.: The rise of people-centric sensing. IEEE Internet Computing 12(4), 12–21 (2008)
4. Chittaranjan, G., Blom, J., Gatica-Perez, D.: Who's who with big-five: Analyzing and classifying personality traits with smartphones. In: ISWC, pp. 29–36 (2011)
5. de Montjoye, Y.-A., Hidalgo, C.A., Verleysen, M., Blondel, V.D.: Unique in the crowd: The privacy bounds of human mobility. Scientific Reports 3 (2013)
6. de Montjoye, Y.-A., Quoidbach, J., Robic, F., Pentland, A(S.): Predicting personality using novel mobile phone-based metrics. In: Greenberg, A.M., Kennedy, W.G., Bos, N.D. (eds.) SBP 2013. LNCS, vol. 7812, pp. 48–55. Springer, Heidelberg (2013)
7. Do, T.M.T., Gatica-Perez, D.: Contextual conditional models for smartphone-based human mobility prediction. In: Ubicomp, pp. 163–172 (2012)
8. Kwapisz, J.R., Weiss, G.M., Moore, S.A.: Activity recognition using cell phone accelerometers. ACM SIGKDD Explorations Newsletter 12(2), 74–82 (2011)

9. LiKamWa, R., Liu, Y., Lane, N.D., Zhong, L.: Moodscope: building a mood sensor from smartphone usage pattern. In: Mobisys (2013)
10. Lu, H., Pan, W., Lane, N.D., Choudhury, T., Campbell, A.T.: Soundsense: scalable sound sensing for people-centric applications on mobile phones. In: MobiSys. ACM (2009)
11. Rachuri, K.K., Musolesi, M., Mascolo, C., Rentfrow, P.J., Longworth, C., Aucinas, A.: Emotionsense: a mobile phones based adaptive platform for experimental social psychology research. In: Ubicomp, pp. 281–290 (2010)
12. Shin, C., Hong, J.-H., Dey, A.K.: Understanding and prediction of mobile application usage for smart phones. In: Ubicomp, pp. 173–182 (2012)
13. Weppner, J., Lukowicz, P.: Collaborative crowd density estimation with mobile phones. In: SenSys (2011)
14. Yan, Z., Chakraborty, D., Parent, C., Spaccapietra, S., Aberer, K.: Semitri: a framework for semantic annotation of heterogeneous trajectories. In: EDBT, pp. 259–270 (2011)
15. Yan, Z., Yang, J., Tapia, E.M.: Smartphone bluetooth based social sensing. In: Ubicomp, pp. 95–98 (2013)

Poster Presentations

Using Trust Model for Detecting Malicious Activities in Twitter*

Mohini Agarwal and Bin Zhou

Department of Information Systems
University of Maryland, Baltimore County
Baltimore, MD 21250
{mohini1,bzhou}@umbc.edu

Abstract. Online social networks such as Twitter have become a major type of information sources in recent years. However, this new public social media provides new gateways for malicious users to achieve various malicious purposes. In this paper, we introduce an extended trust model for detecting malicious activities in online social networks. The major insight is to conduct a trust propagation process over a novel heterogeneous social graph which is able to model different social activities. We develop two trustworthiness measures and evaluate their performance of detecting malicious activities using a real Twitter data set. The results revealed that the F-1 measure of detecting malicious activities in Twitter can achieve higher than 0.9 using our proposed method.

Keywords: cybercrime, Twitter, heterogeneous social graph, trust model.

1 Introduction

Online social networks (OSNs) such as Twitter nowadays have become one of the major information sources for millions of users [7]. These online social networks mainly serve for two purposes: one as new web-based public media for information dissemination and the other as online platforms for social communication and information sharing. Along with the booming of many online social networks in recent years, however, are increasing concerns about the trustworthiness of information disseminated throughout the social networks and the privacy breaching threats of participants' private information. In contrast to traditional websites where people are only limited to the passive viewing of information posted on the websites, online social networks allow users to interact and collaborate with each other as creators of tremendous amounts of user-generated information. This unique characteristic of OSNs, along with the large population base of users, make them hotbeds of many cybercrimes (e.g., generating spams, rumors, fake messages).

* A preliminary version of this work was presented as a poster at the 2013 IEEE/WIC/ACM International Conference on Web Intelligence (WI 2013).

W.G. Kennedy, N. Agarwal, and S.J. Yang (Eds.): SBP 2014, LNCS 8393, pp. 207–214, 2014.

The commonly used method for social network companies to fight malicious activities is manual evaluation [19,20]. Users are able to report any suspicious activities to social network companies. Those user accounts are suspended if their malicious activities are verified by domain experts. This approach has several drawbacks, such as lack of efficiency and the incapability of detecting malicious activities in a timely fashion. In the past several years, several machine learning-based approaches analyzed features of social network users and adopted classification methods to detect untrustworthy and spam information [17,2,8,6,18,5,14,21,13,12,11]. However, the accuracy of those methods is still far from perfect.

The work in this paper is inspired by the concept of "trustworthiness" measure, an extended version of trust model [1], which intends to evaluate how trustful a user is in online social networks and how creditable the information published by the user is. The trustworthiness of a user is a numerical score calculated for each user in online social networks. The measurement of trustworthiness needs to take into consideration many factors, such as the user's activities, social connections, user profiles, and many more. The lower the trustworthiness of a user, the more likely that the information disseminated by the user is not reliable.

In online social networks, users participate in various social activities. A robust model of trustworthiness needs to be constructed based on different social activities. As these activities are not isolated, we propose a novel heterogeneous social graph representation to integrate different social activities using a unified model. In this paper, we focus on Twitter, which is a popular online social network platform. The proposed heterogeneous social graph captures different social activities such as posting a tweet, posting other people's tweets (also called *retweet*), and sending directed messages in Twitter. We develop two different trustworthiness measures on the proposed heterogeneous social graph of Twitter and adopt a simple threshold-based method to detect malicious activities.

The remaining of the paper is organized as follows. In Section 2, we review some recent studies related to our work. In Section 3, we present the novel heterogeneous social graph representation of Twitter. Two different trustworthiness measures are discussed in Section 4. In Section 5, we provide some empirical results conducted on a real Twitter data set. Section 6 concludes this paper.

2 Related Work

Online social networks are comparatively inexpensive and accessible to enable anyone, even private individuals, to publish or access information. Inevitably, this new public social media provides new gateways for malicious users to achieve various malicious purposes, such as generating spams and disseminating rumors.

Spam generally refers to "unsolicited, unwanted message intended to be delivered to an indiscriminate target, directly or indirectly, notwithstanding measures to prevent its delivery [3]." There are different strategies for spam detection. Many current techniques of social spam detection largely depend on the set of features

Table 1. Different social activities in Twitter

ID	Involved Entities	Description
(a)	User u_1 and User u_2	u_1 follows u_2
(b)	User u_1 and Tweet t_1	u_1 posts t_1
(c)	Tweet t_1 and Hashtag Topic h_1	t_1 covers h_1
(d)	Tweet t_1 and Tweet t_2	t_2 is a retweet of t_1
(e)	Tweet t_1 and User u_2	t_1 is a tweet directly mentioned (sent) to u_2

extracted from user behaviors and social interactions[17,2,8,6,18,5,14,21,13,12,11]. For example, Lee *et al.* [14] classified Twitter users into polluter and legitimate users using 18 profiles-based factors such as number of followings and followers. However, existing studies do not consider different social activities as an integrative unit in online social networks.

In addition to spam detection in social networks, rumor identification in social networks also attracts much attention. A rumor is usually defined as "a statement whose true value is unverifiable or deliberately false [15]. Online social networks provide new ways for information dissemination and sharing. Ratkiewicz *et al.* [16] studied the patterns of rumor generation and spreading in online social networks, and the results revealed new ways for rumors to be created and spread through online social networks. Several existing automated detection approaches of rumors in online social networks are based on traditional NLP sentiment analysis technique [4]. Different supervised learning models, such as Markov model [9] and SVM classifier [22], have been applied for identifying rumors in social networks. These methods more or less rely on some selected features for building the detection model, including content-based features, network-based features, and social network specific Memes [15].

3 A Heterogeneous Social Graph Representation of Twitter

In Twitter, users are the major entities, along with their social relationships (e.g., *Follower* and *Following* relationships) and social activities (e.g., posting *Tweets* and *Hashtag Topics*[1]). Some typical social activities in Twitter are summarized in Table 1.

To differentiate malicious activities from legitimate ones, intuitively, if we can measure how trustful each social activity is, we can classify those activities into corresponding categories. This idea is related to the research of trust models [1] which have been extensively studied in psychology and social science. Our idea is to extend the traditional trust model to measure how trustful a social activity is in online social networks.

Intuitively, every entity (user/tweet/hashtag topic) is calculated a trustworthiness score. The lower the trustworthiness of an entity, the more likely that the

[1] In Twitter, a special hashtag symbol # is used to denote a topic in a tweet (e.g., a tweet of "... #topic ...").

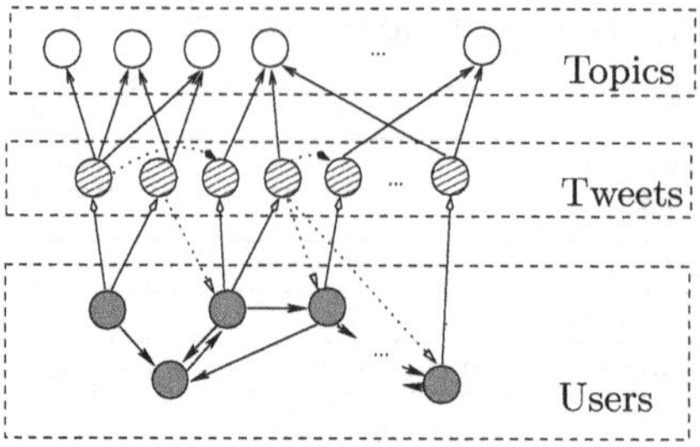

Fig. 1. A unified heterogeneous graph representation of Twitter

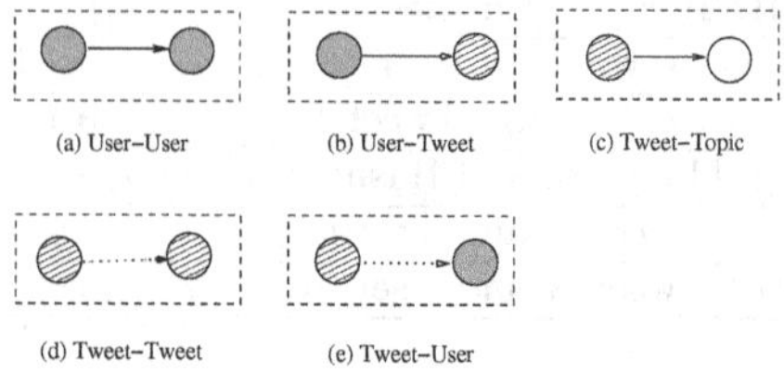

Fig. 2. Different types of edges in the heterogeneous social graph

associated social activities are not reliable. A typical trust model involves an information propagation process. Thus, to formalize the trustworthiness measure in online social networks, one challenge is to model different social activities, as described in Table 1, in a unified way. Our proposal is illustrated in Figure 1, which considers all the five social activities using a unified heterogeneous graph representation.

In general, our proposed model is a heterogeneous graph representation. There are three types of vertices in the graph which correspond to three major entities in online social networks (e.g., users, tweets, and hashtag topics). Directed edges connecting vertices in Figure 1 represent different types of social activities. Each type of the edges is illustrated in Figure 2. First, an edge from user u_i to user u_j means that u_i relates to u_j in the network (e.g., u_i is following u_j in Twitter). Second, an edge from user u_i to tweet t_j indicates that u_i is the author of t_j (e.g., u_i posts a tweet t_j in Twitter). Third, an edge from tweet t_i to topic h_j represents that h_j is one of the topics covered in t_i (e.g., h_j is a hashtag topic in a tweet t_i). In addition, there are two more types of directed edges in the graph. One edge starts from tweet t_i and points to another tweet t_j. This represents that t_j is a retweet of t_i. Another type of edges connects a tweet t_i and a user u_j. This specifically captures the *mention* function in Twitter.

4 Backward Propagation of Trustworthiness

The heterogeneous graph representation shown in Figure 1 can be used to model the propagation of trustworthiness in online social networks. The idea is similar to PageRank in web search engines: a directed edge from one vertex v_i to another vertex v_j represents an endorsement from v_i to v_j. If one vertex receives many endorsements, this vertex is considered to be trustful.

A similar idea is applied over the heterogeneous social graph. At the initialization step, every vertex in the graph is assigned the same trustworthiness score 0.5 (e.g., whether the entity is trustful or not is undecided). Then, we start an iterative trustworthiness propagation process using backward propagation. That is, if there is a direct edge from vertex v_i to vertex v_j in the graph, the trustworthiness of v_j is propagated back to v_i. The usage of backward propagation

Algorithm 1. Detecting malicious activities in online social networks

Input: a heterogeneous graph representation $G(V, E)$, a trustworthiness threshold θ;
Output: a set of malicious activities Mal;

initialize a trustworthiness score of 0.5 to each v in G;
perform the iterative trustworthiness propagation process on G until reaching a stable status; each vertex v is calculated a trustworthiness score $T(v)$;
initialize Mal to be $\emptyset$;
for every $v \in V$ **do**
    if $(T(v) < \theta)$ **then**
        let $Mal = Mal \cup \{v\}$;
    end if
end for
return Mal;

is consistent with the representation of many social activities in Figure 1. For example, consider the scenario that u_i follows u_j (i.e., a direct edge from u_i to u_j), if u_j is a malicious user, it is likely that u_i is not a legitimate user either. The similar analysis can be applied on other types of social activities as well. The trustworthiness propagation process is iterated until a stable status of vertices in Figure 1 is achieved. Every vertex (i.e., user, tweet, or trending topic) will be calculated a numerical score in the range [0,1]. We name this score as **Normal Trustworthiness Score**. The higher the trustworthiness score, the more likely that associated social activities are legitimate.

Based on the concept of trustworthiness score, a straight-forward solution to detect malicious activities in online social networks is described in Algorithm 1.

In Algorithm 1, the most important step is to calculate the trustworthiness score for every node in the heterogeneous graph representation of the social networks. If a trustworthiness threshold θ is given, any vertices having a trustworthiness score less than θ shall be classified as malicious rather than legitimate. In practice, the threshold could be set manually. Some practical solutions such as the Simulated Annealing algorithm can also be adopted to recommend a threshold value which could lead to highest performance in terms of detection accuracy.

In the original model of trustworthiness propagation, we treat every node in the graph representation as the same. In reality, we may already have some background knowledge about malicious activities in online social networks. For example, Twitter maintains a blacklist for actual malicious user accounts. There is also a list of spammers in Twitter. These user accounts are regarded as 100% malicious users. An interesting idea is whether these background knowledge can be integrated into the trustworthiness measure and detection methods so as to improved the detection performance.

We also develop a biased trustworthiness propagation in the heterogeneous graph representation. At the very beginning, the background knowledge is well captured in the graph. For example, if some nodes are considered to be actual malicious users, their initial trustworthiness scores will be set to 0. On the other hand, if some nodes are considered to be actual legitimate nodes, their initialized trustworthiness scores will be set to 1. After the biased initialization step, the remaining detection method is similar to Algorithm 1. The same backward propagation process is conducted until a stable status is reached. We refer to this as **Biased Trustworthiness Score**.

The biased trustworthiness propagation method shares some similar characteristics of topical-sensitive PageRank [10]. An important question in the biased trustworthiness propagation method is the selection of blacklists and whitelists. The decision could have great impact in the detection performance. When the blacklits and whitelists are not available, an alternative solution is to assign initial scores using previous studies (e.g., [14]): we can use a set of features associated with each user account and build a classification model. The probability score can be used as the initial score.

5 Evaluation

We applied the two backward propagation processes on a real Twitter data set crawled during the period from 2/1/2012 to 4/30/2012. The crawled data mainly contain the Follower/Following graph, the public profiles of crawled users, and the tweets. The resulting Twitter data set contains 5.5 million users and 12 million tweets.

To collect the ground-truth information, 12 volunteers were invited to manually label a sampled set of data. Whether an entity is malicious is labeled by assigning 0 point to each vote of "malicious", 0.5 point to each vote of "borderline", and 1 point to each vote of "legitimate". The final label is determined by the average of points from all votes: an average of over 0.5 point is "legitimate", an average of less than 0.5 point is "malicious", and an average of 0.5 point is "undecided". As a result, we obtained a labeled data set of 10K users (1,704 are malicious and 8,296 are legitimate), 20K tweets, and 3K hashtag topics.

We examined the performance using the two trustworthiness measures for detecting malicious users. If the calculated trustworthiness score is equal to or greater than a selected threshold θ, the user is classified as "legitimate"; otherwise, the user is classified as "malicious". For calculating the biased trustworthiness score, only the obtained average scores for tweets and trending topics are used as initialized values. The purpose is to make a fair comparison for the two measures regarding the performance of detecting malicious users.

In Figure 3, we plot the F-1 measure when the threshold θ varies from 0.3 to 0.7. The performance using the biased trustworthiness score outperforms that using the normal trustworthiness score. In addition, when θ is set to be 0.4, the F-1 measure is the highest.

We also considered another strategy for detecting malicious users. We simply treat the obtained trustworthiness score as a feature and build a classification model using the trustworthiness score along with other obtained profile-based features, such as the number of followers/followings, locations, languages, etc. A decision tree model in *weka* package is used as the classification model.

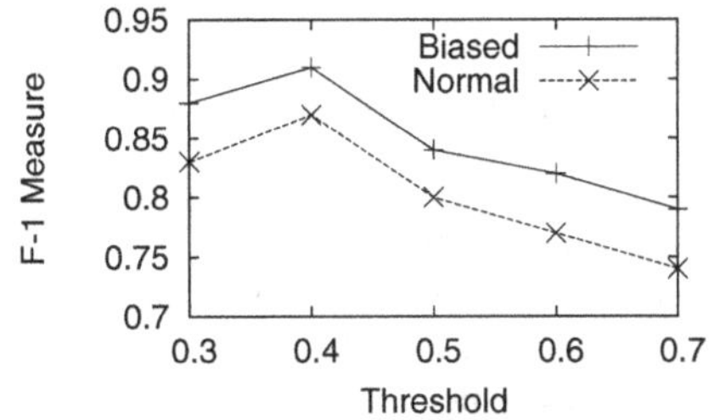

Fig. 3. F-1 measure of detecting malicious users

Table 2. F-1 measure of detecting malicious users using classifiers

Classifier (profile only)	0.8954
Classifier (profile + normal)	0.9468
Classifier (profile + biased)	0.9584

The F-1 measure of classifiers using different feature sets is shown in Table 2. The feature of trustworthiness score is very useful for detecting malicious users. The F-1 measure is improved from 0.8954 to 0.9468 (normal) 0.9584 (biased) and when the feature of trustworthiness score is considered in the classification.

6 Conclusion

In this paper, we developed a heterogeneous social graph for Twitter and presented an extended trust model on the graph to propagate trustworthiness in the graph. The proposed trustworthiness measure can be used to detect malicious activities in Twitter with high accuracy. As future studies, we plan to consider different types of trustworthiness propagation strategies to evaluate their performance.

Acknowledgment. The authors are very grateful for the valuable comments from anonymous reviewers. This research is supported in part by a Samsung Global Research Outreach grant and a UMBC Special Research Assistantship/Initiative Support grant. All opinions, findings, conclusions and recommendations in this paper are those of the authors and do not necessarily reflect the views of the funding agency.

References

1. Alfarez Abdul-Rahman, S.H.: A distributed trust model (1997)
2. Benevenuto, F., Rodrigues, T., Cha, M., Almeida, V.: Characterizing user behavior in online social networks. In: Proceedings of the 9th ACM SIGCOMM Conference on Internet Measurement Conference, pp. 49–62. ACM, New York (2009)
3. Cormack, G.V.: Email spam filtering: A systematic review. Foundations and Trends in Information Retrieval 1(4), 335–455 (2008)

4. Duric, A., Song, F.: Feature selection for sentiment analysis based on content and syntax models. In: Proceedings of the 2nd Workshop on Computational Approaches to Subjectivity and Sentiment Analysis, Stroudsburg, USA, pp. 96–103 (2011)
5. Gao, H., Chen, Y., Lee, K., Palsetia, D., Choudhary, A.: Poster: online spam filtering in social networks. In: Proceedings of the 18th ACM Conference on Computer and Communications Security, pp. 769–772. ACM, New York (2011)
6. Gao, H., Hu, J., Wilson, C., Li, Z., Chen, Y., Zhao, B.Y.: Detecting and characterizing social spam campaigns. In: Proceedings of the 10th ACM SIGCOMM Conference on Internet Measurement, pp. 35–47. ACM, New York (2010)
7. Grant, R.: Social media around the world 2012 (October 2012)
8. Grier, C., Thomas, K., Paxson, V., Zhang, M.: @spam: the underground on 140 characters or less. In: Proceedings of the 17th ACM Conference on Computer and Communications Security, pp. 27–37. ACM, New York (2010)
9. Hassan, A., Qazvinian, V., Radev, D.: What's with the attitude?: identifying sentences with attitude in online discussions. In: Proceedings of the 2010 Conference on Empirical Methods in Natural Language Processing, Stroudsburg, USA, pp. 1245–1255 (2010)
10. Haveliwala, T.: Topic-sensitive pagerank. In: Proceedings of the 11st International World Wide Web Conference (WWW 2002), Honolulu, Hawaii, pp. 784–796. ACM (2002)
11. Heymann, P., Koutrika, G., Garcia-Molina, H.: Fighting spam on social web sites: A survey of approaches and future challenges 11(6), 36–45 (2007)
12. Irani, D., Webb, S., Pu, C.: Study of static classification of social spam profiles in myspace. In: Proceedings of the Fourth International Conference on Weblogs and Social Media (ICWSM 2010). The AAAI Press (2010)
13. Jin, X., Lin, C.X., Luo, J., Han, J.: Socialspamguard: A data mining-based spam detection system for social media networks. PVLDB 4(12), 1458–1461 (2011)
14. Lee, K., Caverlee, J., Webb, S.: The social honeypot project: protecting online communities from spammers. In: Proceedings of the 19th International Conference on World Wide Web, WWW 2010, pp. 1139–1140. ACM, New York (2010)
15. Qazvinian, V., Rosengren, E., Radev, D.R., Mei, Q.: Rumor has it: identifying misinformation in microblogs. In: Proceedings of the Conference on Empirical Methods in Natural Language Processing, Stroudsburg, USA, pp. 1589–1599 (2011)
16. Ratkiewicz, J., Conover, M., Meiss, M., Gonçalves, B., Patil, S., Flammini, A., Menczer, F.: Truthy: mapping the spread of astroturf in microblog streams. In: Proceedings of the 20th International Conference Companion on World Wide Web, pp. 249–252. ACM, New York (2011)
17. Stringhini, G., Kruegel, C., Vigna, G.: Detecting spammers on social networks. In: Annual Computer Security Applications Conference (2010)
18. Thomas, K., Grier, C., Song, D., Paxson, V.: Suspended accounts in retrospect: an analysis of twitter spam. In: Proceedings of the 2011 ACM SIGCOMM Conference on Internet Measurement Conference, pp. 243–258. ACM, New York (2011)
19. `http://support.twitter.com/articles/64986-how-to-report-spam-on-twitter`
20. `http://www.facebook.com/help/?page=204546626249212`
21. Xie, Y., Yu, F., Achan, K., Panigrahy, R., Hulten, G., Osipkov, I.: Spamming botnets: signatures and characteristics. In: Proceedings of the ACM SIGCOMM 2008 Conference on Data Communication, pp. 171–182. ACM, New York (2008)
22. Yang, F., Liu, Y., Yu, X., Yang, M.: Automatic detection of rumor on sina weibo. In: Proceedings of the ACM SIGKDD Workshop on Mining Data Semantics, pp. 13:1–13:7. ACM, New York (2012)

Predicting Guild Membership in Massively Multiplayer Online Games

Hamidreza Alvari[1], Kiran Lakkaraju[2], Gita Sukthankar[1], and Jon Whetzel[2]

[1] University of Central Florida, Orlando, Florida, USA
[2] Sandia National Labs, Albuquerque, New Mexico, USA
halvari@eecs.ucf.edu, {klakkar,jhwhetz}@sandia.gov, gitars@eecs.ucf.edu

Abstract. Massively multiplayer online games (MMOGs) offer a unique laboratory for examining large-scale patterns of human behavior. In particular, the study of guilds in MMOGs has yielded insights about the forces driving the formation of human groups. In this paper, we present a computational model for predicting guild membership in MMOGs and evaluate the relative contribution of 1) social ties, 2) attribute homophily, and 3) existing guild membership toward the accuracy of the predictive model. Our results indicate that existing guild membership is the best predictor of future membership; moreover knowing the identity of a few influential members, as measured by network centrality, is a more powerful predictor than a larger number of less influential members. Based on these results, we propose that community detection algorithms for virtual worlds should exploit publicly available knowledge of guild membership from sources such as profiles, bulletin boards, and chat groups.

Keywords: group formation, MMOGs, community detection, homophily.

1 Introduction

Guilds in massively multiplayer online games (MMOGs) have been shown to parallel real-world social structures such as work teams and friendship networks [1, 2]. Previous work has leveraged data from MMOGs and virtual worlds to conduct large-scale studies of group formation [3–5]. In the real-world, group membership can be a gateway to increased social capital. However, membership benefits are often more tangible in the virtual world, where guilds can confer direct social, economic, and military benefits to the players in the form of privileged communication channels, shared assets, control of physical territories, and tacit mutual defense agreements [6]. Note that in many MMOGs guild membership is an exclusive relationship, in which players can only belong to a single guild. Since the players are forced to choose, an examination of guild membership in these games can be highly revealing of the players' internal assessment of the relative advantages of different social situations.

In this paper, we analyze the composition of guilds in Game X, a browser-based exploration MMOG [7]. Game X features player-led guilds who vie for physical and economic control of the fictional game world. Although nation-level conflict exists in Game X, unlike games such as Everquest (EQ) and World

W.G. Kennedy, N. Agarwal, and S.J. Yang (Eds.): SBP 2014, LNCS 8393, pp. 215–222, 2014.

of Warcraft (WoW), guilds are not primarily an enabler for player vs. player raiding activities. Our aim is to develop a computational model of the processes driving group formation and evolution within Game X and their interaction with in-game conflicts.

In this paper, we present results from the first part of our research agenda—the creation of an agent-based simulation for modeling group formation in Game X. The simulated players make guild membership decisions in a game-theoretic way by calculating the relative utility of joining vs. switching guilds. To evaluate our model, we compare the output of the simulation to the ground truth guild membership from a stable time period prior to the first nation-level conflict. Seeding our algorithm with a small number of known guild members provides the largest performance boost, particularly when the players have a high centrality. Since the leadership of guild is highly predictive of guild membership in Game X, we suggest that semi-supervised community detection approaches to community detection are likely to be particularly fruitful in this domain for a static analysis of network structure.

2 Related Work

To use virtual worlds for social science research, it is necessary to validate the "mapping" of group behavior in virtual worlds to real-world analogs [2, 8]; a topic of key importance is understanding how models of MMOG guild membership relate to group formation in the real world. A second question is whether these models generalize to guilds in a different MMOG setting. In this paper, we use Game X, a turn-based massively multiplayer online game, as our research testbed; Game X requires players to strategize how to make effective use of limited actions rather than encouraging "grinding" gameplay in which the players perform repetitive activities to gain wealth and experience in a low-risk way. Guilds in Game X provide one way of overcoming the action limitations, since multiple players can coordinate their action budgets toward the same mission.

A clear dichotomy between research approaches is whether they seek to recover the static guild membership structure [9, 10] or create a process-oriented model of group formation [3, 5]. For instance, Shah and Sukthankar showed that even transient groups from different regions in Second Life possess a group "fingerprint" that can be recognized by examining a combination of network and topic features [9]. Static models of community detection based on network structure have been employed within virtual worlds without attempting to understand the process by which these communities were created [10]. Our approach is based on a modification of a stochastic community-detection algorithm, NGGAME [11]; since NGGAME separately models the decision-making actions of individual agents rather than optimizing a measure of network partitions, it is well suited for modeling the dynamics of group formation and evolution. Similar to our work, the group formation model proposed by Ahmad et al. [5] also uses a stochastic optimization process by which agents join and leave guilds according to a combination of network and player attributes. The aim of their work is to model changes in the number and size of guilds rather than the membership. Chen et al. [12] note that WoW guilds follow a life cycle in which guilds are formed from a small group of low level members and transition between different stages such as small, elite, unstable, newbie, and big. Only a few guilds succeed—most disband after failing to solve guild management problems and some become unstable because of the strong level disparity between the founding members and a revolving membership of lower-level members.

Game X does not have a concept of character level, but more experienced players have more wealth and higher average skills. Previous work has examined the role of avatar skills in guild composition [5]. Skills are often correlated with player behavior since to advance in a skill the player either has to spend time developing the skill or has to allocate limited character design resources towards acquiring it. This would indicate that guilds may exhibit skill homophily. However, in cases where guilds must operate like teams to accomplish missions, cultivating *skill diversity* can be advantageous. In this paper, we evaluate the use of both skill homophily and diversity for predicting guild membership, in combination with network structure and existing membership.

3 Method

To predict the guild membership, our agent-based simulation seeks to optimize each players' utility through a stochastic search process. Following work described in [13], we treat the process of guild formation as an iterative game performed in a multi-agent environment in which each node of the underlying network graph is a rational agent who decides to maximize its total utility. In this paper, we examine the contribution of two factors, network similarity (C_{ij}) and skill diversity (D_{ij}), toward predicting guild membership. Our code is freely available at: `https://github.com/hamidalvari/GuildDetection`.

Suppose that we have a graph $G = (V, E)$, with $n = |V|$ vertices and $m = |E|$ edges representing the network data. During the simulation, each agent can select an action (*join*, *switch* and *no operation*) to modify or retain its guild membership. In our method, each node of G represents a selfish agent which has a utility function denoted by u_i which is a linear function of two parts:

$$u_i(S) = \frac{1}{m}\sum_{j\in s_i}(\alpha C_{ij} + (1-\alpha)D_{ij}), \tag{1}$$

where s_i is the label of the guild that agent i belongs to and $\alpha \in [0, 1]$.

An agent periodically decides to choose one of the operators according to the player's current utility. The set of all such guilds is denoted by $[k] = 1, 2, \ldots, n$. We also define a strategy profile S = (s_1, s_2, ..., s_n) which represents the set of all strategies of all agents, where $s_i \subseteq [k]$ denotes the strategy of agent i. In our framework, the best response strategy of an agent i with respect to strategies S_{-i}

of other agents is calculated as: $\arg\max_{s'_i \subseteq [k]} u_i(S_{-i}, s'_i)$. The strategy profile S forms a pure Nash equilibrium of the community formation game if all agents play their best strategies.

One can compute the similarities between each pair of vertices in G with respect to some local or global properties, regardless of whether the nodes are directly connected or not. When it is possible to embed the graph vertices in the Euclidean space, the most commonly used measures are Euclidean distance, Manhattan distance and cosine similarity, but when a graph cannot be embedded in space, adjacency relationships between vertices are used [14]. In this work we use separate similarity measures for the two halves of the utility function. For the first half, we use neighborhood similarity [11] to quantify structural similarity between users:

$$C_{ij} = \begin{cases} w_{ij}(1 - \frac{d_i d_j}{2m}) & A_{ij} = 1, w_{ij} >= 1 \\ \frac{w_{ij}}{n} & A_{ij} = 0, w_{ij} >= 1 \\ \frac{d_i d_j}{2m} & A_{ij} = 1, w_{ij} = 0 \\ -\frac{d_i d_j}{2m} & A_{ij} = 0, w_{ij} = 0 \end{cases} \tag{2}$$

where w_{ij} is the number of common neighbors node i and j have. C_{ij} assumes its highest value when two nodes have at least one common neighbor and are also directly connected, i.e. $A_{ij} = 1$.

To evaluate the predictive value of skill diversity between users, we calculate the L_2 norm over the t skills using:

$$D_{ij} = \frac{1}{t}\sqrt{\sum_{d=0}^{t} \frac{1}{l_{max}^d}(l_i^d - l_j^d)^2} \tag{3}$$

where vector $l_i = \{l_i^1, l_i^2, ..., l_i^t\}$ and $l_j = \{l_j^1, l_j^2, ..., l_j^t\}$ are skill vectors for player i and player j, respectively. Using this measure, we can evaluate whether skill diversity or homophily (the inverse) are more predictive. Figure 1 shows our proposed framework. After calculating network similarities between each pair of agents (Equation 2) and skill diversity (Equation 3), the multi-agent game commences and the agents have the option of switching guilds. The community structure of the network emerges after agents reach the local equilibrium. One research question of interest is to examine how the proposed method benefits from including partial knowledge of the guild membership. To do so, we select M members from each guild to seed our agent-based simulation using one of the following centrality measures calculated with the JUNG package[1]: degree, betweenness, and eigenvector centrality.

4 Game X Guild Dataset

In Game X, players can communicate with each other through in-game personal messages, public forum posts and in chat rooms. To create our dataset, we

[1] `http://jung.sourceforge.net/`

1. Calculate network similarity between each pair of users
2. Calculate homophily between each pair of skill vectors
3. Select a subset of the initial guild structure
4. Repeat until convergence in the agents' utilities
   (a) Iterate over agents
   (b) Iterate over actions (join, switch, no action)
      i. Calculate the change in agent utility resulting from the action
      ii. If the change exceeds a threshold, execute action.
      iii. Update guild membership

Fig. 1. Agent-based simulation

selected a relative stable period before the first nation-level conflict, during which time relatively few new users joined the game. The network structure was created by examining all personal messages sent during the time period from day 350 to 400; edges were added if the players exchanged 25 or more messages during the time period. Players who were unaffiliated with a guild were removed from the time period, but singleton guilds (with one member) were left in the dataset. The Guild X dataset contains 1150 nodes, 1936 edges, and 135 guilds; it has a clustering coefficient of 0.135. Figure 2 shows a histogram of guild sizes and node degree distribution.

Players skills are attributes of a player that can impact their ability to collect resources and successfully attack and defend. Skills can increase and decrease based on the activities of the player. Players who focus on gathering resources will have high resource gathering skills, while players who focus on combat will have high combat skills. Since players have a limited amount of turns, they may not be able to excel in both. There are 10 skills that all players can have, divided into three groups: 1) *Combat* skills, which help player attack and defend; 2) *Gathering* skills, which help players gather resources more effectively; and 3) *Movement* and *Hiding* skills, which allow a player to move and hide better. There is an additional skill that is oriented toward repair.

5 Results

In this paper, we evaluate the predictive power of our method for identifying guild membership from the data. First we must define criteria to measure how similar the discovered partition is to the partition we wish to recover (the ground truth guild membership). Here, we evaluated our results with respect to a well-defined metric, normalized mutual information (NMI). In the best condition, our proposed method for predicting guild composition scores an NMI of 0.9, assuming that we use the best utility measure and initialize the model with the ten players with the highest centrality measures in each guild. If we initialize the model with 4 players per guild, the NMI is 0.7.

To understand effect of the initialization options, we compare the performance of selecting the seed group using different centrality measures (degree,

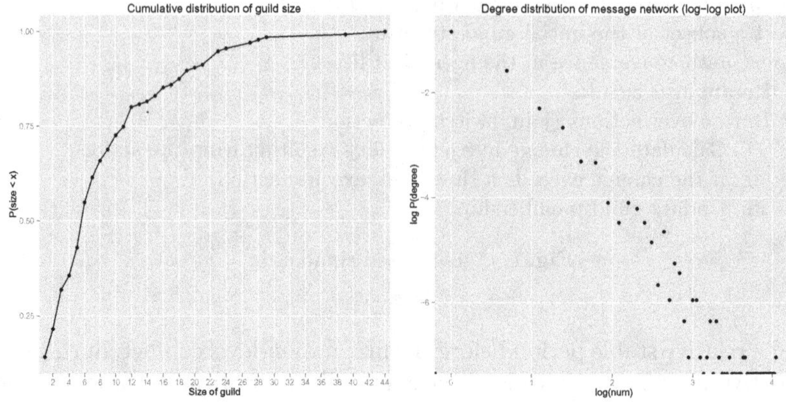

(a) (b)

Fig. 2. (a) Histogram of guild sizes (b) Degree distribution for the dataset

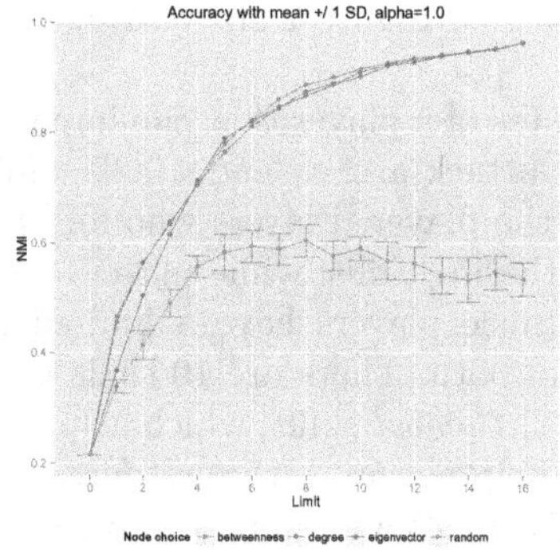

Fig. 3. Effects of exploiting different node choice options (network-based utility only). There is a clear separation between random (pink) and centrality-based (other) selection criteria on asymptotic NMI performance.

betweenness, eigenvector) vs. a random benchmark. There is a clear performance distinction between random selection of members, which reaches a maximum NMI of 0.6 even with a large number of seed members, and centrality-based which can achieve a near perfect NMI.

To tease apart the relative contribution of the network-based and skill-based utility measures, we vary the parameter α. When α is 1, only the network-based utility is used and when α is 0, the utility measure relies entirely on the player skill attributes. If we look at the unsupervised case where we provide no seed information to the algorithm, α=1 (network only) is the clear winner. Skill diversity (higher L_2 distance) slightly outperforms skill homophily ($1/L_2$) in the unsupervised case. Interestingly, when a seed set is provided, the best performing

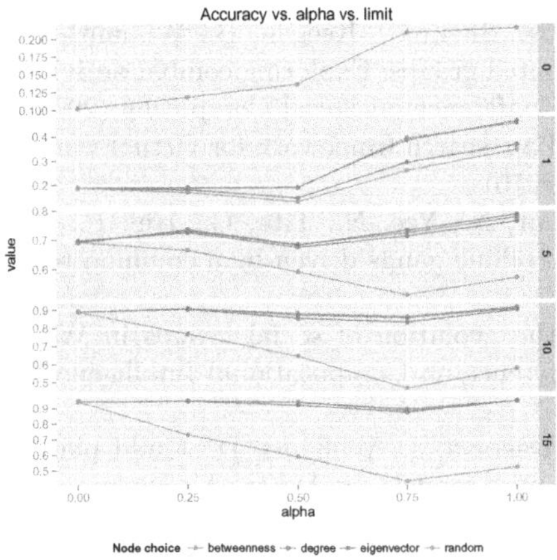

(a)

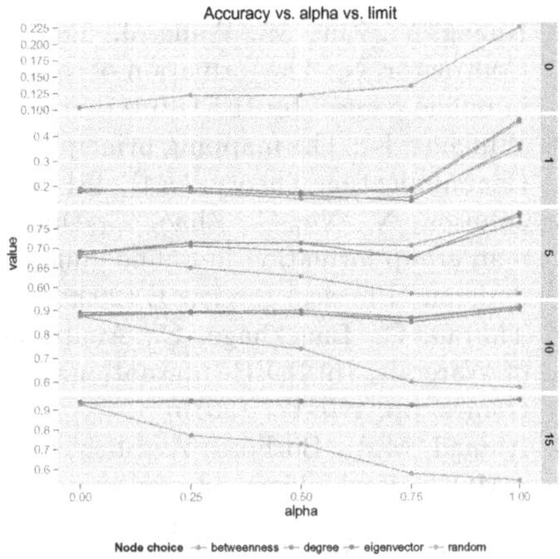

(b)

Fig. 4. Accuracy vs. α vs. different subset sizes. (a) Skill diversity (b) Skill homophily.

α value is dependent on the number of seed members provided. With a high number of seed members, the algorithm is relatively insensitive to the α value and both skill diversity and homophily are useful; Figure 4 shows the relative performance of the algorithm with different seed set sizes.

6 Conclusion and Future Work

This paper presents a predictive model of guild membership in the Game X MMOG that incorporates social ties, skill attributes, and existing guild membership. Our results show that differences in gameplay between Game X and more raiding-centric MMOGs create an environment in which skill attributes are a less important consideration than network structure. Knowledge of existing guild members has the highest predictive power, particularly if the members have a high centrality measure. Currently, we are exploring other methods for selecting influential players as the seed, including the use of "privileged" guild members and members with high average skill attributes. One strength of our approach is that it is relatively simple to extend the utility measure to account for economic and military factors that affect player decisions. In future work, we plan to replicate this study using economic utility measures, combined with the trade network.

Acknowledgments. Research at University of Central Florida was supported by NSF award IIS-0845159. Sandia National Laboratories is a multi-program laboratory managed and operated by Sandia Corporation, a wholly owned subsidiary of Lockheed Martin Corporation, for the U.S. Department of Energy's National Nuclear Security Administration under contract DE-AC04-94AL85000.

References

1. Huang, Y., Zhu, M., Wang, J., Pathak, N., Shen, C., Keegan, B., Williams, D., Contractor, N.: The formation of task-oriented groups: Exploring combat activities in online games. In: IEEE International Conference on Social Computing (2009)
2. Williams, D.: The mapping principle, and a research framework for virtual worlds. Communication Theory 20(4), 451–470 (2010)
3. Johnson, N., Xu, C., Zhao, Z., Duchenaut, N., Yee, N., Tita, G., Hui, P.: Human group formation in online guilds and offline gangs driven by a common team dynamic. Physical Review E (2009)
4. Thurau, C., Bauckhage, C.: Analyzing the evolution of social groups in World of Warcraft. In: IEEE International Conference on Computational Intelligence in Games, pp. 170–177 (2010)
5. Ahmad, M.A., Borbora, Z., Shen, C., Srivastava, J., Williams, D.: Guild play in mMOGs: Rethinking common group dynamics models. In: Datta, A., Shulman, S., Zheng, B., Lin, S.-D., Sun, A., Lim, E.-P. (eds.) SocInfo 2011. LNCS, vol. 6984, pp. 145–152. Springer, Heidelberg (2011)
6. Williams, D., Ducheneaut, N., Xiong, L., Zhang, Y., Yee, N., Nickell, E.: From tree house to barracks: The social life of guilds in World of Warcraft. Games and Culture 1(4), 338–361 (2006)
7. Lakkaraju, K., Whetzel, J.: Group roles in massively multiplayer online games. In: Proceedings of the Workshop on Collaborative Online Organizations at the 14th International Conference on Autonomous Agents and Multi-Agent Systems (2013)
8. Castronova, E., Williams, D., Shen, C., Ratan, R., Xiong, L., Huang, Y., Keegan, B.: As real as real? macroeconomic behavior in a large-scale virtual world. New Media and Society 11(5), 685–707 (2009)
9. Shah, F., Sukthankar, G.: Using network structure to identify groups in virtual worlds. In: Proceedings of the International AAAI Conference on Weblogs and Social Media, Barcelona, Spain, pp. 614–617 (July 2011)
10. Pang, S., Chen, C.: Community analysis of social networks in MMOG. Communications, Network and System Sciences 3, 133–139 (2010)
11. Alvari, H., Hashemi, S., Hamzeh, A.: Detecting overlapping communities in social networks by game theory and structural equivalence concept. In: Deng, H., Miao, D., Lei, J., Wang, F.L. (eds.) AICI 2011, Part II. LNCS, vol. 7003, pp. 620–630. Springer, Heidelberg (2011)
12. Chen, C., Sun, C., Hsieh, J.: Player guild dynamics and evolution in massively multiplayer online games. Cyber Psychology and Behavior 11(3) (2008)
13. Alvari, H., Hashemi, S., Hamzeh, A.: Discovering overlapping communities in social networks: a novel game-theoretic approach. AI Communications 36(2), 161–177 (2013)
14. Wasserman, S.: Social Network Analysis: Methods and Applications. Cambridge University Press (1994)

Emoticon and Text Production in First and Second Languages in Informal Text Communication

Cecilia R. Aragon[1], Nan-Chen Chen[1], Judith F. Kroll,[2] and Laurie Beth Feldman[3,4]

[1] Human Centered Design & Engineering Dept., University of Washington, Seattle, WA, USA
{aragon,nanchen}@uw.edu
[2] Department of Psychology & Center for Language Science, Pennsylvania State University, University Park, PA, USA
jfk7@psu.edu
[3] Department of Psychology, The University at Albany-SUNY, Albany, USA
[4] Haskins Laboratories, New Haven, CT, USA
lfeldman@albany.edu

Abstract. Most of the recent research on online text communication has been conducted in social contexts with diverse groups of users. Here we examine a stable group of adult scientists as they chat about their work. Some scientists communicated in their first language (L1) and others communicated either in their L1 or in a second (L2) language. We analyze the production in English of emoticons and of lines of text and compare measures in L1 and L2 speakers. L1 and L2 speakers differed significantly along multiple measures. English L1 speakers used more lines of text per message. English L2 (French L1) speakers used more emoticons per message. Patterns suggest compensatory emoticon/text productivity. In future analyses we will undertake a more fine-grained analysis of how emoticon use varies across social and linguistic settings. Computer-mediated communication is often viewed as impoverished, but even our initial research provides hints that users repurpose the technology according to social dynamics previously associated only with face-to-face communication.

Keywords: informal text communication, emoticons, bilingualism.

1 Introduction

Written communication poses special problems for understanding that do not arise when speaking and many of these problems recur in online communication. Nonetheless, advancing technology means that online communication is becoming more common not only in social but also in work domains. Many of these work contexts require people to talk in a language other than their native or first language (L1). Understanding how bilingual speakers communicate online may become crucial to productive cross-cultural team interactions and decision-making across remote workplaces. Effective social interaction encompasses understanding not only the transfer of information from others but also their emotions. That challenge is exacerbated in what has previously been considered an impoverished computer-mediated environment [1]. The insertion of emoticons is one option to convey emotion in

W.G. Kennedy, N. Agarwal, and S.J. Yang (Eds.): SBP 2014, LNCS 8393, pp. 223–228, 2014.

online text communication. Emoticons appear to function similarly to facial cues, tone of voice, and body language in face-to-face communication. We hypothesize that the use of emoticons in informal text communication (ITC) is becoming systematic and may now echo many of the patterns observed in face-to-face communication.

We focus here on emoticon use in a bilingual, cross-cultural scientific collaboration where chat serves as the primary method for coordinating scientific tasks for periods of many hours a day. The data set consists of nearly half a million lines of chat collected over a four-year collaboration and includes emoticons as well as text. The initial findings that we report here are novel in several respects. The text communication derives from a task shared by a quasi-permanent group of adult scientists. About half of them communicate in both the first (L1) and the second (L2) language and the remainder only in their L1. Therefore, we can compare L1 and L2 communication in the same social and collaborative work environment. Gesture and other nonverbal behaviors enrich the text they accompany. Adopting a psycholinguistic framework, we hypothesize a parallel function for emoticons in text chat [2] as has been postulated for the role of gesture, and other nonverbal behavior in the context of face-to-face conversation. It has been asserted that gestures are less common in the presence of shared knowledge [3]. In this case, speakers may be less likely to gesture when they engage in a shared task. However, gestures may be compensatory for a lack of high proficiency in which case speakers may be less likely to gesture when proficiency is high. Our assumption is that text communication engages similar mechanisms to spoken discourse and that many of the same factors that shape spoken communication, including the use of gesture, will be present. If so, then emoticon use should be modulated by the same factors that affect displays of emotion in spoken discourse.

2 Prior Research

2.1 Bilingual Speakers

There is consensus among researchers that bilingual speakers activate both of their languages when they read, listen, or talk in either one of their languages [4]. Because both languages are active concurrently, the cognitive functioning of bilinguals is not analogous to that of monolinguals. In fact, differences between bilinguals and monolinguals are evident in many domains. For example, the tendency for a bilingual to express emotion is greater when the L2 is the ambient language. Many claim that relative to the L1, communicating in the L2 makes self-disclosure feel less threatening and allows the speaker to remain more distant [6]. Childhood memories for events from immigrants that are described in the L2 tend to have less detail and be emotionally less charged than those in the L1 [7]. Not only are bilinguals more likely to speak freely about emotionally charged topics in their L2 than L1, but at a physiological level, taboo words spoken in L2 elicit smaller electrodermal changes from baseline in bilingual speakers than do those in the L1 [8].

In the present study, we examine the consequences of bilingualism for online text communication and emoticon insertion in a professional work context with a shared task. The virtual work environment that is enabled by current technology provides a new opportunity for testing and extending models of language processing.

2.2 Emoticon Use in Informal Text Communication

Text-based computer-mediated communication that is spontaneously generated without editing or rewriting constitutes what we call "informal text communication" (ITC). It is more similar to spontaneous speech than to formal writing. Examples include chat, instant messaging, microblogs such as Twitter, and certain forum posts/message boards. Note that e-mail, online profiles, product reviews, blog posts, and webpages are excluded because they resemble written discourse. ITC may be synchronous or asynchronous; essential is that it serves to communicate with other humans without substantial reflection, similar to spontaneous speech.

Even a perusal of the literature [9, 10] indicates that the options when communicating in text-based chat environments are becoming more elaborated and less prescriptive, and that humans are adapting available technologies to meet social needs such as to convey emotion in communication. We hypothesize that the use of emoticons in ITC parallel many of the practices observed in face-to-face communication.

3 Background on the Scientific Collaboration and Chat Dataset

The chat dataset was produced by an international astrophysics collaboration consisting of about 30 members; about half of the scientists work at several different locations in the U.S. and the other half in three research institutes and universities in France. All the French scientists also speak English, and English is the official language of the collaboration. Collaboration members use English in the chat whenever an English speaker is present; French speakers may revert to French whenever they are alone in the chat.

The astronomers' task is complex and requires coordination on telescope observation especially when working under time pressure. The primary means of communication during remote telescope observation are AIM (AOL Instant Messenger) chat (augmented by a virtual assistant) and VNC (virtual network computing).

4 Analysis and Results

4.1 Data

Description of the Chat Corpus. The corpus consists of a total of 485,045 chat messages. The logs include 1,319 days (nearly four years from 2004 to 2008), and cover approximately 12-hour sessions during which observations from a remote telescope were coordinated. A line of chat refers to a single message. Messages are posted as soon as the user hits return in the chat client.

The chat data included only participants who contributed over 2000 lines of chat and who used more than 30 emoticons over the period of data collection. The resulting corpus included the productions of 8 native English speakers and 10 native French speakers, with a total of 259,256 lines of chat and 3343 emoticons.

Participants. A total of 18 astrophysicists in the United States and France who were members of the collaboration formed the sample. There were 8 American scientists and 10 French scientists. Typically 5-6 people were present in the chat at any one time.

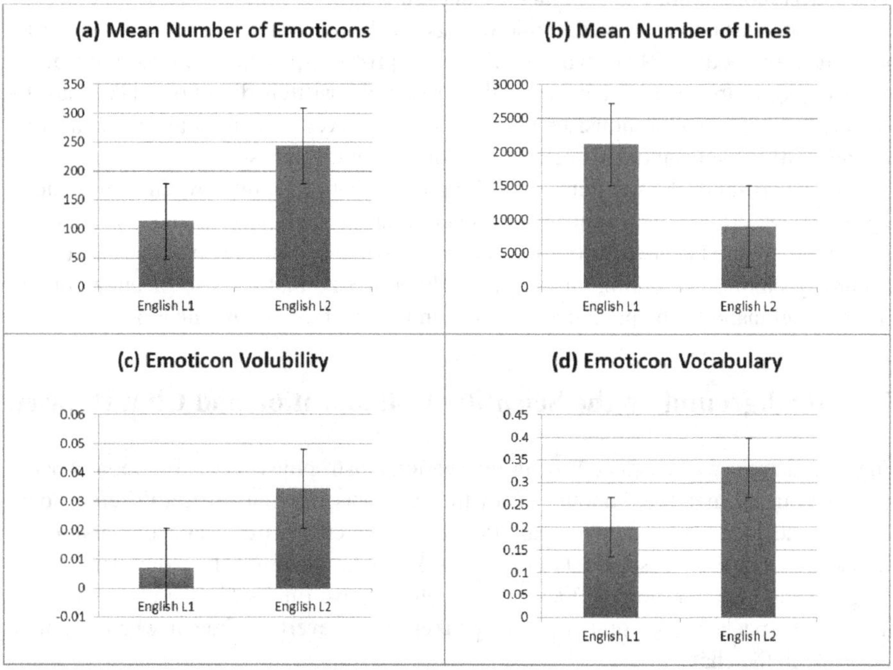

Fig. 1. (a) English L1 speakers (American) tended to insert more emoticons into their text than English L2 speakers did. (b) English L1 speakers (American) on average produced more lines of text than did English L2 speakers (French). (c) English L2 speakers (French) used more emoticons per text message (higher emoticon volubility) than did English L1 speakers (American). (d) English L2 speakers (French) had larger emoticon vocabulary size than English L1 speakers (American). Whiskers denote standard error (SE).

4.2 Results

Token Based Measures of Emoticon Use. Results of an ANOVA contrasting L1 and L2 speakers showed that English L2 speakers tended to insert more emoticons into their text messages than did English L1 speakers [$F(1, 16) = 2.859$, $p < .11$]. This is depicted in Fig. 1a. Results are consistent with the claim that displays of emotion are more frequent and easier to produce in one's L2. Even if reliable, an alternative account based on overall proficiency with emoticons cannot be dismissed from these data alone. Conversely, English L1 speakers tended to write longer text messages than did English L2 speakers [$F(1, 16) = 3.354$, $p < .086$]. This is depicted in Fig. 1b. Most importantly, English L2 speakers used more emoticons per text message than did English L1 speakers [$F(1, 16) = 7.319$, $p < .016$]. We define production of emoticons per lines of text as *volubility*. It is depicted in Fig. 1c. Ongoing analyses further examine the potential compensatory relation between emoticon and text usage.

Type Based Measures of Emoticon Use. In a second set of analyses we examined *vocabulary size*, or the number of different emoticons that L1 and L2 speakers used. Results are expressed in terms of number of different emoticons relative to the total number of emoticon types documented in the corpus (total=64). Here the effect of L1 [$F(1, 16) = 3.95$, $p < .06$] was marginally significant. It suggested that English L2 speakers are using more different emoticons over all. This is depicted in Fig. 1d.

5 Discussion and Conclusions

This work contrasts emoticon production in ITC in a work setting by American speakers of English as the L1 and French speakers of English as the L2. We introduced four measures of production. First we examined mean number of emoticons and mean number of lines of text in each message. Then we examined the mean number of emoticons per lines of chat. We defined this token-based measure of emoticon production as *volubility*. Most important is that the *production rate* based on number of emoticons per lines of chat was higher for participants communicating in an L2 than in an L1 English. Finally, we looked at emoticon *vocabulary size*, a type measure of emoticon production that reflects the mean number of unique emoticons produced by an individual divided by the number of unique emoticons present in the entire chat (64). Vocabulary size was higher in L2 than L1 English.

Most novel in our findings is the greater volubility and vocabulary of emoticons produced by scientists communicating in their L2 than in their L1. Results are consistent with the claim that communicating in the L2 makes it easier to convey emotion because it introduces emotional distance and reduces any sense of vulnerability relative to communicating in the L1 [6, 8].

Not only do L2 chatters have higher volubility than L1 chatters, they also produce fewer lines of text. Consistent with the claim that emoticons in chat serve a purpose similar to gesture and facial expressions in face-to-face communication the increased emoticon volubility among scientists communicating in their L2 could reflect a type of compensatory behavior. Accordingly, the reduced accessibility of words to express an intended meaning in L2 could be causally related to the increased production of non-verbal elaborations such as emoticons. This pattern warrants further investigation.

6 Limitations and Future Work

We have only begun to examine chat behavior and how it varies for communication in an L1 as compared to an L2 and our sample consists of a relatively limited number of individuals. However, number of observations per person over a four-year period is large, adding reliability to our analyses.

More important is that status of English as either a first or second language could not be manipulated experimentally. Therefore, it remains possible that the difference between groups (emoticons per lines of text; number of different emoticons) reflects skill differences in emoticon use rather than constraints on communicating in one's L1 or L2. To attenuate the contribution of uncontrolled differences across groups, subsequent analyses will examine interactions between L1/L2 speakers and various language environments defined by the composition of L1 and L2 speakers. We ask whether productions in L2 vary more with language environment than do those in L1. It is unlikely that these interactions reflect group differences.

Acknowledgments. The writing of this paper was supported in part by NIH Grant HD053146 and NSF Grants BCS-0955090 and OISE-0968369 to J.F. Kroll; by NIH Grant HD053146 to Haskins Laboratories.

References

1. Gill, A.J., French, R.M., Gergle, D., Oberlander, J.: The language of emotion in short blog texts. In: CSCW, vol. 8, pp. 299–302 (2008)
2. Gullberg, M., McCafferty, S.G.: Introduction to gesture and SLA: Toward an integrated approach. Studies in Second Language Acquisition 30, 133 (2008)
3. Holler, J., Wilkin, K.: Communicating common ground: How mutually shared knowledge influences speech and gesture in a narrative task. Language and Cognitive Processes 24, 267–289 (2009)
4. Kroll, J.F., Bobb, S.C., Wodniecka, Z.: Language selectivity is the exception, not the rule: Arguments against a fixed locus of language selection in bilingual speech. Bilingualism: Language and Cognition 9, 119–135 (2006)
5. Zhang, S., Morris, M.W., Cheng, C.-Y., Yap, A.J.: Heritage-culture images disrupt immigrants' second-language processing through triggering first-language interference. Proceedings of the National Academy of Sciences (2013)
6. Pavlenko, A.: Emotions and Multilingualism. Cambridge University Press (2007)
7. Schrauf, R.W., Hoffman, L.: The effects of revisionism on remembered emotion: the valence of older, voluntary immigrants' pre - migration autobiographical memories. Applied Cognitive Psychology 21, 895–913 (2007)
8. Harris, C.L., Aycicegi, A., Gleason, J.B.: Taboo words and reprimands elicit greater autonomic reactivity in a first language than in a second language. Applied Psycholinguistics 24, 561–579 (2003)
9. Brooks, M., Kuksenok, K., Torkildson, M.K., Perry, D., Robinson, J.J., Scott, T.J., Anicello, O., Zukowski, A., Harris, P., Aragon, C.R.: Statistical affect detection in collaborative chat. In: Proceedings of the 2013 Conference on Computer Supported Cooperative Work, pp. 317–328. ACM (2013)
10. Park, J., Barash, V., Fink, C., Cha, M.: Emoticon Style: Interpreting Differences in Emoticons Across Cultures. In: Proceedings of the 7th International AAAI Conference on Weblogs and Social Media (2013)

The Needs of Metaphor

David B. Bracewell

Language Computer Corporation
Richardson, TX USA
david@languagecomputer.com

Abstract. In this paper we semi-automatically construct a multilingual lexicon for Maslow's seven categories of needs. We then use the semi-automatically constructed lexicons and a metaphor recognition system to analyze the change in rate of the expression of needs in the presence of metaphor. We examine four languages, English, Farsi, Russian, and Spanish, and focus on metaphors whose target concept is related to poverty or taxation.

1 Introduction

Metaphor is a pervasive literary mechanism allowing individuals to make comparisons of seemingly unrelated concepts. The most prominent theory of metaphor is Lakoff's Contemporary Theory of Metaphor [1] in which concepts are mapped from an abstract *target* domain into a concrete *source* domain (e.g. "Time is Money" or "Love is War.") allowing for the target to be discussed, understood, and affect assessed through the source. While some mappings of targets to sources appear to be universal, e.g. orientational metaphors tend to be more similar across cultures, others vary between cultures, e.g. in some cultures the future is in the front of us whereas in others it is in the back [2].

By analyzing the metaphors expressed by a culture, a greater understanding of the culture can be gained. Moreover, understanding the differences in the expression of metaphors between cultures can facilitate better communication and negotiation. The differences found in the expression of metaphor are born out of such factors as socio-economic conditions, religious and traditional teachings, and social norms. One way in which the differences are apparent is through the lexical choices employed by individuals when metaphorically describing a target concept, its cause, and its impact on the individual or the whole of society.

In this paper, we examine these lexical choices through the lens of human motivation and needs utilizing a framework proposed by Abraham Maslow [3,4]. Needs are related to individuals' instincts and are an important factor in the motivation of human behavior. They facilitate an exploration of the affect and culture associated with metaphor. We hypothesize that expression of needs will vary among cultures with regards to a target concept and in the face of metaphor. We compare and contrast the expression of needs in the presence of metaphor in *English*, *Farsi*, *Russian*, and *Spanish* with a focus on the United States, Iran,

W.G. Kennedy, N. Agarwal, and S.J. Yang (Eds.): SBP 2014, LNCS 8393, pp. 229–236, 2014.

Russian Federation, and Mexico respectively. Furthermore, we focus on the target concepts of *poverty* and *taxation* to constrain the analysis. We chose poverty and taxation as they are sub-concepts of Economic Inequality which we are performing a larger analysis of metaphor on.

2 Background and Related Work

In his 1943 Psychological Review paper entitled "A Theory of Human Motivation," Maslow proposed a hierarchical framework for human motivation which has become simply known as Maslow's hierarchy of needs. Maslow's framework is based on his assumption that there are multiple motivational systems which are independent and fundamental and that these motivations form a hierarchy based on the priority of needing to be fulfilled. Maslow originally outlined a five stage model which was later extended with Cognitive and Aesthetic needs. The seven stage model (listed in order from lowest level to highest level) along with definitions are as follows:

1. **Physiological:** The basic needs of human survival, including food, water, and air.
2. **Safety:** The need for protection from forces such as the elements and war and the desire for security, order, and stability.
3. **Belongingness and Love:** Interpersonal needs revolving around family, friendship, and intimacy.
4. **Esteem:** The need to be respected and valued by others.
5. **Cognitive:** The need to learn, discover, create, and explore in order to gain knowledge and increase intelligence.
6. **Aesthetic:** The need for a feeling of intimacy with nature and beauty.
7. **Self-Actualization:** The need to realize ones full potential.

Maslow believed the motivations in his hierarchical model to be universal features of human nature. Assuming that Maslow's belief of universality is true, the needs can serve as an indicator to cultural difference based on variations in usage. Moreover, the lack, fulfillment, or overabundance of a need in relation to a metaphor may be an indicator of its affect in terms of polarity or modifications to the intensity associated with the polarity. In this paper we focus only on the usage and leave studying the fulfillment or lack thereof for future work.

Research on metaphor has been performed in many fields, including, psychology linguistics, sociology, anthropology, and computational linguistics. A number of theories of metaphor have been proposed, the most prominent of which is the Contemporary Theory of Metaphor [1]. Automated methodologies for processing metaphor can be broken down into two main categories: recognition and interpretation. Interpretation of metaphor involves determining the intended, or literal, meaning of a metaphor [5]. The recognition of metaphor entails determining if an expression is literal or figurative. Work on automated metaphor recognition dates back to the early 1990's with the work of Fass [6] based on selectional preference violations and more recently with the work of Shutova [7] using noun-verb clustering.

3 Identifying Maslow's Needs

Identification of Maslow's needs is done using lexical matching against a set of semi-automatically constructed lexicons. This type of identification is similar to other work in the area, such as [8] who used custom lexicons for the identification of moral foundations. Construction of our need lexicons was inspired by the approach presented in [9] for sentiment. We first constructed initial lexicons by assigning English WordNet [10] senses to each of the seven category of needs. A total of 248 WordNet senses were manually assigned among the seven categories of needs. Attempts were made to choose senses whose associated synset was as culturally neutral as possible, i.e. slang and metaphoric senses found in WordNet were omitted. Table 1 gives examples of WordNet senses assigned to each of the seven needs.

Table 1. Example WordNet senses for each of the seven categories of needs

Need	Example WordNet Senses
Physiological	food#N#1, nourishment#N#1, shelter#N#1
Safety	safety#N#1, security#N#1, wellness#N#1
Belongingness and Love	family_unit#N#1, friendship#N#1, love#N#1
Esteem	esteem#N#1, clout#N#2, admiration#N#1
Cognitive	knowledge#N#1, insight#N#2, cognition#N#1
Aesthetic	beauty#N#1, shape#N#8, balance#N#2
Self-Actualization	morality#N#1, creativeness#N#1, self-realization#N#1

We automatically mapped the manually defined English WordNet senses into equivalent senses in the Farsi, Russian, and Spanish WordNets using the available interlingua links. We used FarsNet [11] for Farsi, RussNet [12] for Russian, and the Multilingual Central Repository [13] for Spanish.

The set of manually identified and automatically translated senses were then expanded via the synonymy, hyponymy, and derivationally related form relationships. The expansion of the senses serves to increase the extraction coverage of each of the seven needs as well as providing possible cultural specific variations in lexical items by using the language specific relations. In lieu of attempting the difficult task of word sense disambiguation we discarded the sense level information and instead used the surface form of the senses as the entries in the lexicons.

4 Data Collection

We collected a corpus of general text for each of the four languages from the web to use as a representation of general language usage. Collection was performed using a general web crawl procedure with each language-specific crawl beginning with a set of seed domains (news, blog and forums) known to be related to the given language. In total, we collected approximately 20,000 documents for English, 7,000 documents for Farsi, 6,000 documents for Russian, and 10,000 documents for Spanish.

We additionally constructed *poverty* and *taxation* specific corpora, see [14] for one method for such collection, for each of the four languages. We used the Yahoo! Boss API[1] to locate relevant pages given a set of keywords related to the concepts. In total, we collected 7,229 English documents, 2,246 Farsi documents, 1,863 Russian documents, and 3,439 Spanish documents for poverty and 4,365 English documents, 1,158 Farsi documents, 1,266 Russian documents, and 1,458 Spanish documents for taxation.

For analyzing needs in the presence of metaphor, we extracted sentences from the poverty and taxation specific corpora which contained a metaphoric expression related to poverty or taxation. Metaphor identification was performed using a combination of selectional preference violations, semantic and syntactic mismatch, imageability, and comparison to known metaphors, some of the details are covered in [15] and [16]. The method achieves approximately 70% F-measure for English, 65% for Spanish, 65% for Russian, and 69% for Farsi. The details of the method are left for future publication.

5 Analysis

Our analysis focuses on comparing and contrasting metaphoric expressions relating to poverty and taxation as evidenced through the rate of usage of the needs. We calculate the rate of usage as needs as the number of sentences containing a lexical item related to a need, i.e. found in a lexicon. By comparing the rate of usage, the bias of the lexicons can be minimized. Bias in the lexicons occurs when a lexicon entry is highly polysemous and thus matches more than it should. An important note is that this analysis does not make any claim to clear cultural differentiations, but only reports on what is manifested in the data.

We compare the rate of usage of needs between:

- Sentences containing a metaphoric expression related to a target domain **and** the target domain specific corpus.
- Sentences containing a metaphoric expression related to a target domain **and** the general corpus

To infer the influence of metaphor and domain on the expression of needs, we looked for statistically significant changes in occurrence rates. Statistical significance was determined using G^2, which is commonly used when comparing frequencies for words across corpora.

We first analyze the difference in rate of usage for Poverty, shown in Tables 2. When discussing the concept of poverty[2], it would be expected, at least from an

[1] `http://developer.yahoo.com/boss/search/`

[2] All four countries are listed as having low human poverty in the United Nation's Human Poverty Index `http://hdr.undp.org/en/statistics/data/hd_map/hpi/`

Anglo-American perspective, that the more basic needs, i.e. those lower in the hierarchy, would have increased usage. This increased usage would stem from the fact that those who are impoverished have real needs for basic necessities such as, food, water, and shelter. Likewise, it would be expected to see a decreased usage of higher level needs as those in need of food often are not worried about self-actualization.

Table 2. Change in rate of need expressions between metaphors pertaining to the target concept of poverty and a poverty specific corpus and general corpus. (items marked with an * have $p < 0.001$).

Poverty	Change in rate of usage compared to the poverty corpus				Change in rate of usage compared to the general corpus			
	English	Farsi	Russian	Spanish	English	Farsi	Russian	Spanish
Physiological	-2.4%*	7.6%	-4.0%	-0.4%	-0.4%	10.9%*	4.2%	4.2%*
Safety	-2.1%	8.8%*	-0.5%	-1.9%	-9.2%*	6.3%	-0.8%	-1.1%
Belongingness and Love	6.4%*	8.9%*	9.1%*	7.1%*	14.9%*	10.3%*	14.4%*	12.0%*
Esteem	-2.0%	-13.3%*	-3.8%	-1.3%	-6.6%*	-16.1%*	-18.6%*	-5.9%*
Cognitive	0.2%	-10.3%*	0.7%	-2.9%*	1.7%*	-10.6%*	1.8%	-8.9%*
Aesthetic	-0.2%	0.0%	-0.8%	-0.7%	-0.4%	0.0%	-0.1%	-0.3%
Self-Actualization	0.1%	-1.7%	-0.6%	0.0%	0.0%	-0.9%	-0.9%	0.0%

As can be seen in Table 2 this trend of increased usage of lower level needs seem to be present across all four of the languages. Examining the usage of the lower level needs (Physiological, Safety, and Belongingness and Love) in the presence of metaphor, we found a number of interesting differences. First, there were significant increases in the expression of Physiological needs over the general corpora in Farsi and Spanish and a significant decrease over the poverty corpora for English. This suggests that Physiological needs are less mentioned in normal conversation in Farsi and Spanish. The decrease in usage over the poverty corpus in English suggests that poverty metaphors highlight other aspects of poverty or are associated with concepts not directly related to needs. Similarly, there was a significant decrease in the expression of Safety needs in English compared to the general corpus, which suggests that either Safety is not a crucial concern in regards to poverty or that Safety needs are a more general concern. There was an increase in the expression of Safety needs in Farsi compared to the poverty corpus suggesting that in Farsi speaking countries Safety is an important concern for those in poverty or the discussion of poverty. Interestingly Belongingness and Love saw increases over the poverty corpora and general corpora in all four languages. Examining the data we found that often children and family are emphasized in the context of poverty, which are associated with Belongingness and Love.

Expression of the higher level needs was decreased in the presence of poverty metaphor across all four languages. The expression of Esteem needs say significant decreases compared to the general corpus in all four languages. There was a further significant decrease compared to the poverty corpus in Farsi. This suggests, as one would expect, that discussions on poverty do not manifest themselves in such a way that the need of Esteem (clout, admiration, etc.) is important. Likewise, the expression of Cognitive needs were decreased among all languages but Russian (no significant change) and English (significant increase) when comparing to the general corpora and further decreased compared to the corresponding poverty corpus in Farsi and Spanish. Examining the English and Russian data revealed that they both often talk about Poverty causing or being caused by the lack of education.

We next analyzed the change in rates of usages for needs in the presence of taxation metaphors, listed in Table 3. As a point of comparison the percentage of GDP from tax revenue, according to the Heritage Foundation[3] for the four countries of interest are 26.9% for the United States, 6.1% for Iran, 36.9% for the Russian Federation, and 29.7% for Mexico.

Table 3. Change in rate of need expressions between metaphors pertaining to the target concept of taxation and a taxation specific corpus and general corpus. (items marked with an * have $p < 0.001$).

Taxation	Change in rate of usage from taxation corpus				Change in rate of usage from general corpus			
	English	Farsi	Russian	Spanish	English	Farsi	Russian	Spanish
Physiological	26.4%*	-6.5%*	-11.9%*	4.0%*	27.2%*	-10.3%*	-10.2%*	5.8%*
Safety	-6.3%*	2.5%	-1.9%	5.4%*	-12.3%*	-3.9%	-1.2%	9.6%*
Belongingness and Love	-8.3%*	5.4%*	-4.3%	-3.8%*	-3.9%*	4.6%*	-0.4%	-0.7%
Esteem	-9.7%*	4.0%	21.7%*	-1.7%	-9.7%*	16.9%*	14.6%*	-3.9%
Cognitive	-2.0%*	-4.0%	-0.9%	-3.6%*	-1.7%*	-4.3%	-0.7%	-10.8%*
Aesthetic	0.0%	0.00%	-2.3%	0.0%	0.7%	0.0%	-1.2%	0.4%
Self-Actualization	-0.1%	-1.5%	-0.4%	-0.3%	-0.3%	-3.0%	-0.9%	-0.3%

As can be seen in Table 3 the most striking changes in the lower level needs between languages comes in the expression of Physiological needs in English, Russian and Spanish. English and Spanish both had increased rates of usage with English increasing by 27.2% and Spanish by 4.0% over their respective taxation corpora. An example English metaphor related to taxation found in the corpus is "high taxes are taking the air out of the economy", in which the target, "taxes", is causing "air", a Physiological need, to be depleted. In contrast, Russian, which has the highest percentage of GDP from tax revenue, saw a decrease of 10.2% over the general corpus and a further decrease of 11.9% over the taxation corpus.

[3] http://www.heritage.org/index/ranking

Significant changes are seen in the expression of Safety for Spanish (increase compared taxation and general corpora) and English (decrease compared to taxation and general corpora). The increase in Spanish is seen through metaphors relating to physical harm and burden, e.g. "carga de impuestos" (burden of taxes), "robo de impuestos" (tax theft). The expression of Belongingness and love had significantly increased usage in Farsi for compared to both the general and taxation corpora and significantly decreased for English (general and taxation corpora) and Spanish (taxation corpus).

Expression of the higher level needs was decreased in the presence of taxation metaphor except for the expression of Esteem in Russian and Farsi. Farsi and Russian had significant increases (16.9% and 14.6% respectively) in the rate of Esteem needs compared to their respective general corpus. There was an even higher and significant change, 21.7%, compared to the taxation corpus. The only other significant changes were for the rates of Cognitive needs in English and Spanish, which both were decreased for the their respective general corpus and taxation specific corpus.

6 Conclusion

In this paper we presented the construction of dictionaries covering Maslow's seven categories of needs in four languages (English, Farsi, Russian, and Spanish). Using the constructed dictionaries we analyzed the expression of needs, in terms of rate of occurrence, for a general web corpus, domain specific corpora for the domains of poverty and taxation, and sentences containing metaphors pertaining to poverty and taxation. Through this analysis we are able to show the differences in the expression of needs as related to poverty and taxation and in particular in the presence of metaphor.

The goal of the analysis presented in this paper is to ultimately use the expression of needs as signal to the affective properties of metaphor. In order to accomplish this goal we first studied the relationship between metaphor and needs and whether or not these relationships are culturally different. As was seen there are cultural differences in the expression of needs in the presence of metaphor, which is encouraging for our future work. The analysis in this paper only scratched the surface of what cultural comparisons can be made through the expression of needs and metaphor.

Acknowledgments. This research is supported by the Intelligence Advanced Research Projects Activity (IARPA) via Department of Defense US Army Research Laboratory contract number W911NF-12-C-0025. The U.S. Government is authorized to reproduce and distribute reprints for Governmental purposes notwithstanding any copyright annotation thereon. Disclaimer: The views and conclusions contained herein are those of the authors and should not be interpreted as necessarily representing the official policies or endorsements, either expressed or implied, of IARPA, DoD/ARL, or the U.S. Government.

References

1. Lakoff, G., et al.: The contemporary theory of metaphor. Metaphor and Thought 2, 202–251 (1993)
2. Lakoff, G., Johnson, M.: Metaphors we live by, Chicago London, vol. 111 (1980)
3. Maslow, A.: A theory of human motivation. Psychological Review 50, 370–396 (1943)
4. Maslow, A.: Motivation and personality. Harper (1954)
5. Shutova, E.: Models of metaphor in nlp. In: Proceedings of the 48th Annual Meeting of the Association for Computational Linguistics, pp. 688–697. Association for Computational Linguistics (2010)
6. Fass, D.: met*: a method for discriminating metonymy and metaphor by computer. Comput. Linguist. 17(1), 49–90 (1991)
7. Shutova, E., Sun, L., Korhonen, A.: Metaphor identification using verb and noun clustering. In: Proceedings of the 23rd International Conference on Computational Linguistics, COLING 2010, pp. 1002–1010. Association for Computational Linguistics, Stroudsburg (2010)
8. Graham, J., Haidt, J., Nosek, B.A.: Liberals and conservatives rely on different sets of moral foundations. Journal of Personality and Social Psychology 96(5), 1029 (2009)
9. Bracewell, D.B.: Semi-automatic creation of an emotion dictionary using WordNet and its evaluation. In: 2008 IEEE Conference on Cybernetics and Intelligent Systems, pp. 1385–1389 (2008)
10. Fellbaum, C.: Wordnet: An electronic lexical database (1998), WordNet is available from, `http://www.cogsci.princeton.edu/wn`
11. Shamsfard, M., Hesabi, A., Fadaei, H., Mansoory, N., Famian, A., Bagherbeigi, S., Fekri, E., Monshizadeh, M., Assi, S.: Semi automatic development of farsnet; the persian wordnet. In: Proceedings of 5th Global WordNet Conference, Mumbai, India (2010)
12. Azarova, I., Mitrofanova, O., Sinopalnikova, A., Yavorskaya, M., Oparin, I.: Russnet: Building a lexical database for the russian language. In: Proceedings of Workshop on Wordnet Structures and Standardisation and How this Affect Wordnet Applications and Evaluation, Las Palmas, pp. 60–64 (2002)
13. Atserias, J., Villarejo, L., Rigau, G., Agirre, E., Carroll, J., Magnini, B., Vossen, P.: The MEANING Multilingual Central Repository. In: Proceedings of the 2nd Global WordNet Conference (GWC), Brno, Czech Republic (January 2004)
14. Bracewell, D.B., Ren, F., Kuroiwa, S.: Mining News Sites to Create Special Domain News Collections. Computational Intelligence 4(1), 56–63 (2007)
15. Bracewell, D.B., Tomlinson, M.T., Mohler, M.: Determining the conceptual space of metaphoric expressions. In: Gelbukh, A. (ed.) CICLing 2013, Part I. LNCS, vol. 7816, pp. 487–500. Springer, Heidelberg (2013)
16. Mohler, M., Bracewell, D., Hinote, D., Tomlinson, M.: Semantic signatures for example-based linguistic metaphor detection 27 (2013)

Behavior in the Time of Cholera: Evidence from the 2008-2009 Cholera Outbreak in Zimbabwe

Anne Carpenter*

Department of Economics
University of California Irvine

Abstract. Despite the potential benefits of investments in water and sanitation, individual level water treatment remains low in many developing countries. This paper explores the dynamic relationship between water transmitted infectious disease and water treatment behavior. Using evolutionary game theory, I endogenize water treatment decisions in a mathematical model of cholera. I calibrate the model for the '08-'09 cholera outbreak in Zimbabwe. I show that prevalence dependent water treatment behavior is a factor contributing to endemic cholera. Additionally, I find that in absence of WHO interventions in Zimbabwe, the share of the population treating their water would have converged to a level that would have enabled cholera to persist in the population.

Keywords: Cholera, Infectious Disease, Water Treatment, Prevalence Dependent Behavior, Economic Development.

1 Introduction

Water transmitted diseases are the fifth leading cause of global mortality. According to World Health Organization (WHO) estimates, 94% of water transmitted diseases could be prevented with access to clean water and sanitation [1]. Despite the promising returns to investment in clean water and sanitation, individual use of disease prevention products, like water chlorination tablets, is low in developing countries [2]. If on average individuals in developing countries were forced to spend their entire income on subsistence, then this would explain the low investment in disease prevention products. However, individuals living on less than \$1 a day spend an average of 10%-18% of their income on alcohol, tobacco, and local festivals [3]. This suggests that even the very poor have some discretion in the way they spend their income. So, why is investment in water treatment products so low despite the potential for water transmitted disease? While many factors likely explain low water treatment levels, some researchers

* I would like to thank Garret Ridinger, Stergios Skaperdas, Michael McBride, and Dan Bogart for helpful comments and suggestions. I acknowledge and appreciate financial support from the Economics Department at the University of California, Irvine.

W.G. Kennedy, N. Agarwal, and S.J. Yang (Eds.): SBP 2014, LNCS 8393, pp. 237–244, 2014.

have indicated that individuals may exhibit prevalence dependent behavior, and therefore, may base decisions to treat water on the incidence of water transmitted diseases in the environment [4,5].

This paper explores the dynamic relationship between cholera incidence and water treatment behavior. The hypothesis is that if individuals exhibit prevalence dependent water treatment behavior, then there should be an increase in the share of the population treating their water during a cholera outbreak. However, theoretical research on prevalence dependent behavior suggests that outbreak induced behavior change will not be sufficient to eliminate an outbreak [4,5]. The key finding of this paper is that prevalence dependent water treatment behavior is a factor contributing to endemic cholera. Furthermore, in absence of the WHO interventions during the 2008-2009 cholera outbreak in Zimbabwe, the share of the population treating their water would have converged to a level that would have allowed a high cholera incidence to persist. Therefore, these results support theoretical model predictions regarding the inability of individuals exhibiting prevalence dependent behavior to coordinate their individual actions to eliminate disease.

2 Related Work

Microeconomic research has focused on the individual level barriers to water treatment that may explain low investment in water treatment products in developing countries. Through the use of randomized controlled trials factors such as information, credit constraints, and product pricing have been varied to explore the role of such factors in individual water treatment decisions [6,7,8]. However, this research typically ignores the fact that water treatment decisions take place in a dynamic disease environment. The idea that individuals might exhibit prevalence dependent behavior with regard to prevention of certain diseases has been demonstrated in several empirical studies [5]. However, these studies do not explore the role of prevalence dependent behavior in the decision to treat water. Although some mathematical models have incorporated behavioral responses to disease [9,10,11,12,13,14], the primary focus of these models is sexually transmitted diseases, with a few exceptions. This model is the first of my knowledge to explore the dynamic relationship between water treatment behavior and cholera incidence.

Furthermore epidemiologists have developed mathematical models of cholera that incorporate biological realism in an effort to provide more accurate forecasts of epidemic magnitudes [15,16,17]. These models are used to explore potential epidemics both in the absence of and in the presence of potential interventions in order to determine the best policy response to an outbreak. However, these models assume that individual decisions, including water treatment decisions, are constant over time and exogenous to the disease environment.

This paper contributes to the literature in both economics and mathematical epidemiology by combining the behavioral focus of economic models with the biological focus of the mathematical epidemiology models. By modeling the decision to treat water as a function of the share of the cholera infected population,

I endogenize the decision to treat water. This enables me to explore the influence of the dynamic disease environment on water treatment decisions and the ability of individuals to coordinate prevention measures to eliminate an outbreak.

3 Model

3.1 Water Treatment Game

To model changes in population water treatment behavior, I use an evolutionary game [18]. In the game, agents have two possible strategies: treat water (T) or do not treat their water (NT). Table 1 provides the game.

Table 1. Water Treatment Game

	T	NT
T	$\beta - C$	$\beta - C(I(t))$
NT	$\beta - C(I(t))$	$-C(I(t))$

In this evolutionary game, α represents the share of the population not treating their water, and $(1 - \alpha)$ represents the share of the population treating their water. Here, β is the benefit from treating water, and C is the cost of treating water. The variable $I(t)$ represents the number of infected individuals which changes over time, t. Additionally, $C(I(t))$ is the cost of not treating water and is an increasing function of $I(t)$. Therefore, the cost of not treating water increases as the number of cholera infected individuals increases.

Because individuals' payoffs depend upon the aggregate share of individuals making each choice, this model has the nice property that it can account for the role that externalities might play in behavior change. In table 1, it is easy to see that if $C > C(I(t))$, then the game is a hawk-dove game. In this case, a portion of the population will free-ride on the water treating group in equilibrium. When cholera infection is unlikely, a larger portion of the population will not treat water in equilibrium. However, if $C < C(I(t))$, then the game is a coordination game where the dominant strategy is to treat water. Therefore, at higher infection levels, the positive externality imposed by the water treating group is not sufficient to allow free-riding to exist in the population.

The replicator dynamic presented in equation 1 captures the evolution of the population water treatment over time, t.

$$\frac{d\alpha(t)}{dt} = \alpha \left[\Pi(NT) - \bar{\Pi} \right] . \tag{1}$$

Here, α represents the share of the population not treating their water, $\Pi(NT)$ is the payoff from not treating water, and $\bar{\Pi}$ is the average payoff for all strategies. Equation 1 demonstrates that the change in the share of the population not treating their water moves according to the changes in the underlying payoffs.

If the payoff from not treating water is larger than the average payoff, then the share of the population not treating their water will grow. Conversely, if the average payoff is larger than the payoff for not treating water, then the share of the population not treating their water will shrink. Equation 2 provides the final derivation for the change in the share of the population not treating their water (proof see S.I. 1.1).

$$\frac{d\alpha(t)}{dt} = \alpha(1-\alpha)\left\{(1-\alpha)\left[C - C(I)\right] - \alpha\beta\right\}. \tag{2}$$

3.2 Cholera Outbreak System

Following standard epidemiological practice, I use an SIR model to model the cholera outbreak [15,16,17]. The population can be divided into 3 categories: susceptible individuals, $S(t)$; infected individuals, $I(t)$; and recovered individuals with temporary cholera immunity, $R(t)$. The change in the cholera susceptible population is

$$\frac{dS(t)}{dt} = \mu N + \omega R(t) - q\alpha(t)S(t)\frac{B(t)}{k+B(t)} - \mu S(t). \tag{3}$$

Here, N represents the entire population, μ represents the birth and deaths rates, μN represents the number of individuals born into the population, and $\mu S(t)$ the number of deaths in the susceptible population unrelated to cholera. In this model, the number of recovered individuals with temporary cholera immunity is $R(t)$, ω is the rate at which people lose their temporary immunity, and $\omega R(t)$ is the number of individuals that lose their temporary immunity and again become susceptible to cholera. Here, k is the level of bacteria in the environment necessary for a 50% probability of cholera infection, $B(t)$ is the level of bacteria that exists in the aquatic environment, and $\frac{B(t)}{k+B(t)}$ is the probability that consumption of cholera contaminated water leads to cholera infection [19].

Standard models of cholera outbreak assume the fraction of the population that consumes contaminated water, $q\alpha(t)$, to be a constant. To incorporate changes in water treatment behavior, I break up the consumption of cholera contaminated water into the share of the population not treating their water, $\alpha(t)$, times the probability that untreated water is cholera contaminated, q. This model assumes that the probability that untreated water is cholera contaminated is constant and exogenous. Although q is assumed to be constant, it is possible that the infection rate influences the probability that untreated water is cholera contaminated. Unfortunately, no data exists at this time to explore this relationship. In this model, the assumption of a constant and exogenous q provides a considerable simplification of the model analysis; however, future research should seek to determine how the infection rate influences the probability that untreated water is cholera contaminated. It follows from the variable definitions that the product $q\alpha(t)S(t)\frac{B(t)}{k+B(t)}$ is the number of susceptible individuals that become cholera infected in time, t. The change in the population infected with cholera is

$$\frac{dI(t)}{dt} = \alpha(t)qS(t)\frac{B(t)}{k + B(t)} - (\mu_c + \mu + \gamma)I(t) \tag{4}$$

Here, μ_c is the cholera death rate, γ is the cholera recovery rate, $\mu_c I(t)$ is the number of infected individuals that die from cholera, $\gamma I(t)$ is the number of infected individuals that recover and receive temporary immunity to cholera, and $\mu I(t)$ is the number of infected individuals that die from causes unrelated to cholera. The change in the recovered and temporarily cholera immune population is

$$\frac{dR(t)}{dt} = \gamma I(t) - (\mu + \omega)R(t). \tag{5}$$

The change in V. cholerae bacteria in the aquatic reservoir is

$$\frac{dB(t)}{dt} = \theta I(t) - \delta B(t). \tag{6}$$

Here, θ is the rate at which infected individuals shed V. cholerae bacteria into the environment through defecation, δ is the death rate of V. cholerae bacteria in the aquatic environment, $\theta I(t)$ is the number of bacteria that are shed into the aquatic environment, and $\delta B(t)$ is the number of bacteria in the aquatic environment that die. The water treatment behavior discussed in section 3.1 is the new dynamic being explored in this paper. Equation 2 provides the change in the population share not treating their water. For simplicity, I assume that $C(I(t)) = \lambda\frac{I(t)}{N}$, a linear function of the infection rate. Here, $\lambda\epsilon[0, \infty)$ is a scalar that can be interpreted as the intensity with which the outbreak is perceived. Therefore, the change in the population share not treating their water is

$$\frac{d\alpha(t)}{dt} = \alpha(1-\alpha)\left\{(1-\alpha)\left[C - \lambda\frac{I(t)}{N}\right] - \alpha\beta\right\}. \tag{7}$$

4 Selected Calibration

To test the ability of the model to fit water treatment behavior during a cholera outbreak, I calibrated the model to the 2008-2009 cholera outbreak in Zimbabwe. To do so I use data on cholera incidence, outbreak induced changes in water treatment behavior, and cholera biology (see S.I. 2.1 for data description).

The calibrations for Mashonaland West illustrate the ability of the model to fit observe cholera incidence and water treatment behavior (see S.I. 2.2 for all provinces). The first graph provides the observed weekly cholera incidence and the calibrated incidence for the same period. The second graph demonstrates that the model can capture changes from the pre-outbreak behavior to the 2010 post-outbreak behavior.

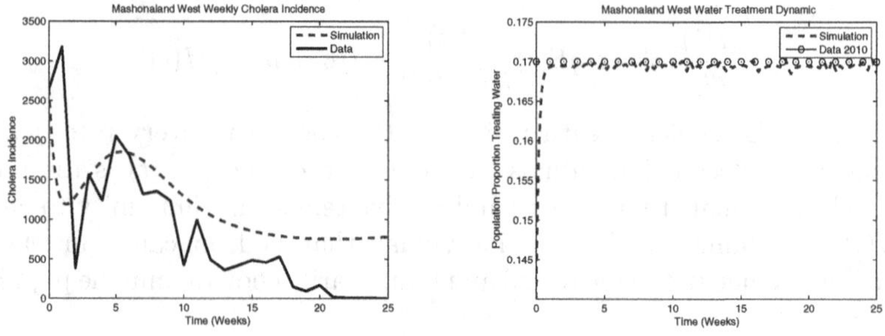

To calibrate the model, I chose the parameters β, C, and λ that best fit the observed data (for parameters see S.I. 2.2). I vary the costs and benefits for each province separately to determine calibration sensitivity to parameters chosen (see S.I. 3). I find that when the costs of water treatment are higher than the perceived benefits, the share of the population treating their water converges to a level that allows cholera to persist in the environment. However, when the costs of water treatment are sufficiently low relative to the benefits, cholera will be eliminated due to a greater increase in water treatment within the population. Prior research has explored the ability of V. cholerae bacteria to survive in the aquatic reservoir as an explanation for endemic cholera. This paper demonstrates that even when incorporating the persistence of bacteria in the aquatic environment, a cholera outbreak will be minimal if the costs of water treatment are sufficiently low. Thus, an explanation of endemic cholera should include prevalence dependent water treatment behavior as a contributing factor.

A drop in cholera incidence occurs in the data around week 7 that is not matched by the model predictions. However, around week 7, the WHO had partnered with multiple organizations working in Zimbabwe to provide clean water, water purification tablets, and cholera prevention information free of charge [20,21]. This intervention effectively reduced the cost of water treatment to zero, which allowed benefits of water treatment to outweigh costs of treatment thereby ending the outbreak. Thus, the model prediction should be viewed as the level of cholera incidence that would have persisted in absence of the intervention. This model suggests that because the costs of water treatment were higher than the perceived benefits, the increases in cholera incidence would not have induced a sufficient share of the population to treat their water. Therefore the outbreak would have persisted. This finding is consistent with theoretical research that suggests that if individuals exhibit prevalence dependent behavior, then they will be unable to coordinate individual actions to eliminate outbreaks [4].

5 Discussion

This paper explores the issue of low water treatment in developing countries considering the disease environment in which these choices take place. I find that high water treatment costs and low perceived water treatment benefits cause the share of the population treating their water to be low despite the persistence of cholera.

Model calibrations of the 2008-2009 cholera outbreak in Zimbabwe demonstrate two important implications of this modeling approach. First, cholera incidence can be eliminated with low water treatment costs relative to perceived benefits despite the presence of V. cholerae bacteria in the aquatic environment. Second, without the WHO interventions in Zimbabwe during the 2008-2009 cholera outbreak, the share of the population treating their water would have converged to a level that would have enabled a high cholera incidence to persist.

Previous research has focused on the ability of V. cholerae bacteria to survive in the aquatic environment as an explanatory factor in endemic cholera [19]. However, this paper demonstrates that cholera incidence can be eliminated when the costs of water treatment are sufficiently low relative to the benefits of water treatment regardless of the level of V. cholerae bacteria in the aquatic environment. Thus, prevalence dependent water treatment behavior coupled with high costs of water treatment can allow the population to converge to a level of water treatment that enables cholera to persist. Therefore, policy focused on eliminating endemic cholera should not only consider persistence of V. cholerae bacteria in the aquatic environment, but also should acknowledge that changes in water treatment behavior are likely prevalence dependent. Furthermore, policy makers should also consider that it may be very difficult, if not impossible, for individuals acting independently to coordinate actions to eliminate water transmitted diseases.

During the 2008-2009 cholera outbreak in Zimbabwe, the Zimbabwean government was reluctant to acknowledge or act to end the cholera epidemic [21]. Fortunately, the WHO acted to coordinate actions to provide clean water, water purification tablets, and cholera prevention information free of charge to individuals in affected regions [21]. This model demonstrates that although individuals may respond to outbreaks by changing outbreak related behaviors, the changes in behavior may not be sufficient to eliminate the outbreak. In the case of Zimbabwe, without the interventions by the WHO, the share of the population treating their water would have converged to a level that would have enabled a high cholera incidence to persist. Therefore, policy makers should not rely on individual behavioral responses to end epidemics. Instead, there is significant scope for policy makers to aid individuals in coordinating actions to eliminate outbreaks.

While this model provides a first pass at incorporating outbreak related behavior into a model of cholera outbreak, it does not provide a full biological analysis of a cholera epidemic. Future research should seek to incorporate more biological realness into the model of cholera incidence to provide better analysis of potential epidemic magnitudes. Specifically, incorporating the highly infectious cholera state and the associated health behavior, post-defecation hand-washing, will provide a more thorough analysis of the impacts of behavior change on cholera epidemics.

References

1. World Health Organization: Combating waterborne disease at the household level (2007), `http://www.who.int/water_sanitation_health/publications/combating_diseasepart1lowres.pdf`

2. Dupas, P.: Health behavior in developing countries. Annual Review of Economics 3, 425–499 (2011)
3. Banerjee, A.V., Duflo, E.: The economic lives of the poor. Journal of Economic Perspectives 21(1), 141–168 (2007)
4. Geoffard, P.Y., Philipson, T.: Disease eradication: Private versus public vaccination. The American Economic Review 87(1), 222–230 (1997)
5. Philipson, T.: Chapter 33 economic epidemiology and infectious diseases. Handbook of Health Economics 1, 1761–1799 (2000)
6. Madajewicz, M., Pfaff, A., Geen, A.V., Graziano, J., Hussein, I., Momotaj, H., Sylvi, R., Ahsan, H.: Can information alone change behavior? response to arsenic contamination of groundwater in bangladesh. Journal of Development Economics 84, 731–754 (2007)
7. Devota, F., Duflo, E., Dupas, P., Pariente, W., Pons, V.: Happiness on tap: Piped water adoption in urban morocco. American Economic Journal: Economic Policy 4(4), 68–99 (2012)
8. Ashraf, N., Berry, J., Shapiro, J.M.: Can higher prices stimulate product use? evidence from a field experiment in zambia. NBER Working Paper 13247, pp. 1–55 (July 2007)
9. Geoffard, P.Y., Philipson, T.: Rational epidemics and their public control. International Economic Review 37(3), 603–624 (1996)
10. Hyman, J.M., Li, J.: Behavior changes in sis std models with selective mixing. SIAM Journal of Applied Mathematics 57(4), 1082–1094 (1997)
11. Klein, E., Laxminarayan, R., Smith, D.L.: Economic incentives and mathematical models of disease. Environment and Development Economics 12, 707–732 (2007)
12. Kremer, M.: Integrating behavioral choice into epidemiological models of aids. Quarterly Journal of Economics 111(2), 549–573 (1996)
13. Auld, M.C.: Choices, beliefs, and infectious disease dynamics. Journal of Health Economics 22, 361–377 (2003)
14. Valle, S.D., Hethcote, H., Hyman, J., Castillo-Chavez, C.: Effects of behavioral changes in a smallpox attack model. Mathematical Biosciences 195, 228–251 (2005)
15. Anderson, R.M., May, R.M.: Infectious Diseases of Humans: Dynamics and Control. Oxford University Press (1991)
16. Andrews, J.R., Basu, S.: Transmission dynamics and control of cholera in haiti: An epidemic model. Lancet 377, 1248–12455 (2011)
17. Mukandavire, Z., Liao, S., Wang, J., Gaff, H., Smith, D.L., Morris Jr., J.G.: Estimating the reproductive numbers for the 2008-2009 cholera outbreaks in zimbabwe. PNAS 108(21), 8767–8772 (2011)
18. Sandholm, W.H.: Population Games and Evolutionary Dynamics. The MIT Press (2010)
19. Codeco, C.T.: Endemic and epidemic dynamics of cholera: The role of the aquatic reservoir. BMC Infectious Diseases 1 (2001)
20. UN OCHA: Weekly situation report on cholera in zimbabwe no. 10 (January 21, 2009), http://reliefweb.int/sites/reliefweb.int/files/resources/A40A3859537F4E9385257546006E5B01-Full_Report.pdf (January 2009)
21. UN OCHA: Evaluation of the wash response to the 2008-2009 zimbabwe cholera epidemic and preparedness planning for future outbreaks (November 2009), https://zw.humanitarianresponse.info/document/evaluation-wash-response-2008-2009-zimbabwe-cholera-epidemic-and-preparedness-planning

Mutual Information Technique in Assessing Crosstalk through a Random-Pairing Bootstrap Method

Xinguang (Jim) Chen[1] and Ding-Geng Chen[2]

[1] University of Florida Department of Epidemiology, Gainesville, FL 32611, USA
Wayne State University School of Medicine, Detroit, MI 48201, USA
Jimax168chen@gmail.com

[2] School of Nursing & Department of Biostatistics and Computational Science
University of Rochester Medical Center, Rochester, NY 14642, USA
Din_Chen@URMC.Rochester.edu

Abstract. Crosstalk plays a critical role in prevention research to promote purposeful behavior change through randomized controlled trials. However, two challenges prevent researchers from assessing crosstalk between subjects in the intervention and the control conditions that may contaminate an intervention trial. First, it is very hard if not impossible to identify who in the intervention group have talked with whom in the control group; therefore the crosstalk effect cannot be statistically evaluated. Second no method is readily available to quantify crosstalk even if we know who has talked with whom. To overcome the challenges, we devised the random-pairing bootstrap (RPB) method based on statistical principles and adapted the mutual information (MI) technique from the information sciences. The established RPB method provides a novel approach for researchers to identify participants in the intervention and the control groups who might have talked with each other; the MI itself is an analytical method capable of quantifying both linear and nonlinear relationships on a variable between two groups of subjects who might have experienced information exchange. An MI measure therefore provides evidence supporting the effect from crosstalk on a target variable with data generated through RPB. To establish the PRB-MI methodology, we first conducted a systematic test with simulated data. We then analyzed empirical data from a randomized controlled trial (n=1360) funded by the National Institute of Health. Analytical results with simulated data indicate that RBP-MI method can effectively detect a known crosstalk effect with different effect sizes. Analytical results with empirical data show that effects from within-group crosstalk are greater than those of between-group crosstalk, which is within our expectation. These findings suggest the validity and utility of the RBP-MI method in behavioral intervention research. Further research is needed to improve the method.

Keywords: Randomized pairing; Crosstalk, Mutual Information, Informational Correlation, Behavioral Intervention research, Randomized controlled trials.

1 Introduction

1.1 Crosstalk and Its Significance in Health Sciences

Our health, to a great extent, is determined by our behavior. Purposeful behavior change thus emerges as the top priority for modern health promotion and disease

W.G. Kennedy, N. Agarwal, and S.J. Yang (Eds.): SBP 2014, LNCS 8393, pp. 245–252, 2014.

prevention. To promote better health and to prevent diseases, intervention programs are devised and tested to assist people in adapting health-enhancing behavior and to reduce health risk behavior. Numerous such programs have been developed and tested and a number of them showed a "significant effect" in promoting purposeful behavior change through strict randomized controlled trials. Many of these programs are recognized by the Centers for Disease Control and Prevention (CDC) and are packed up for use through the Diffusion of Effective Intervention (DEBI) project (http://www.effectiveinterventions.org/en/Abou tDebi.aspx). Typical programs include "Focus On Youth with Informed Parents and Children Together" for HIV prevention among adolescents [1] and "Mpowerment" for young gay and bisexual men to engage in safer sex [2].

1.2 Challenges for Evaluating Crosstalk

To establish a behavioral intervention program, the standard randomized controlled trial (RCT) design for drug testing has been adopted. Despite many strengths of RCT, a critical research issue remains: contamination due to crosstalk between subjects in the intervention group and subjects in the control group [3-4]. When an intervention subject is in contact and communicates with a control subject, he/she may incidentally or purposefully pass the intervention information to a control subject. Such crosstalk may interfere with the research efforts to correctly evaluate an intervention program [5-8] due to at least two opposite mechanisms. (1) *Diluting effect*: Crosstalk may result in the acceptance of the intervention by some control subjects, leading to behavior change. For example, it is likely that after receiving a smoking prevention education, some intervention students may talk with control students about it, leading to reductions in tobacco use among the control students. Obviously, any diluting effect will reduce the statistical power to detect a program that eventually works. (2) *Enhancing effect*: Instead of persuading control subjects, crosstalk may increase the resistance of some control subjects to the intervention, widening the differences between the two comparison groups. If this "magnifying effect" is not considered, researchers will not be able to rule out programs that do not work.

In addition to between-group crosstalk, within-group communication is also likely and such crosstalk may also have significant effect on purposeful behavior change. Within-group crosstalk may lead to idea synchronization among those who agree with the intervention, enhancing purposeful behavior change. On the contrary, within-group communication may also lead to chaos among those who disagree with the intervention, attenuating the effect of a devised program to achieve purposeful behavior change. Up to date, no reported research can be found that has investigated this within-group crosstalk in RCT behavioral trials.

Although the concept of crosstalk is nothing new, no valid method is readily available to evaluate its effect in RCT. The lack of appropriate quantification methodology prevents researchers from assessing the effect of crosstalk in conducting a behavioral intervention trial. However, in communication and information sciences, methodologies, including linear and nonlinear models have been used in quantifying the effect of communication between paired individuals across groups through experimental designs [9-11] and in nature settings (e.g., married couples, therapist-client) [6, 10,

12-13]. However, none of these methods can be directly used to assess crosstalk for behavioral intervention trials with a RCT design, mainly because of the challenges to identifying the subjects who talked with each other.

1.3 Purpose of This Study

The purposes of this study are two folds: 1) To report the random pairing and bootstrap method we invented to detect pairs of subjects who may have crosstalked with each other and 2) to utilize the mutual information method in quantifying the potential effect, if it exists, from crosstalk. We achieved the goal by first testing the methodology using simulated data, followed by an analysis of empirical data.

2 Methods

2.1 Random Pairing-Bootstrap (RPB)

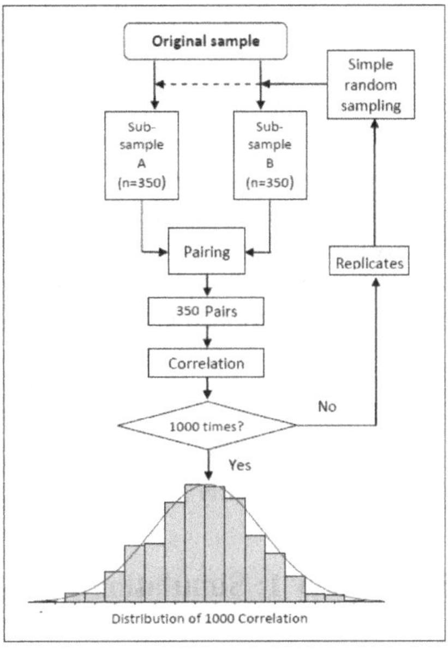

Although it is practically hard to accurately identify who crosstalked with whom in a behavioral intervention trial with a RCT design; however, as long as such crosstalk results in idea/behavior synchronization, it is likely to detect the effect without explicitly identifying the individual participants who talked with each other. This can be achieved using the random-pairing bootstrap (RPB) method (the figure on the right) in four steps. (1) Randomly select n pairs of subjects from the study sample (size=N), n < N and then pair the selected individuals; (2) compute the between-pair correlation index for the variable of interest, (3) repeat steps 1 and 2 a large number of times (e.g., 1000 to 5000) to obtain a large number of the correlation indexes; and (4) to obtain the mean and 95% confidence interval like the bootstrap methodology commonly used in statistical analysis.

The essence of the RPB is that random pairing makes it possible to match those who might have talked with each other that rendered idea synchronizations/behavioral changes. If the estimated coefficients for all paired samples approach zero, it suggests neither crosstalk, nor a crosstalk effect. If the estimated coefficients showed a normal distribution, the mean of these coefficients will be an estimate of the crosstalk effect.

2.2 Correlation and Mutual Information

Pearson correlation R has been used in assessing the relationship between two variables X and Y, or the same variable for subjects in Group X paired with subjects in Group Y:

$$R = \frac{\text{cov}(X,Y)}{\sigma_x \sigma_y} \quad \text{.......................................} (1)$$

However, if the synchronization effect through crosstalk is nonlinear, the Pearson correlation method will no longer be adequate.

To more effectively quantify crosstalk, we emphasize the mutual information (MI) technique. MI is a measure of the dependence between two variables [14]. $MI = 0$ for two variables that are independent; MI increases if the dependence of the two variables increases such that one is a function of another. For two random variables X and Y, MI is mathematically defined as:

$$MI = \sum_{b \in B} \sum_{a \in A} p(x,y) * \log\left(\frac{p(x,y)}{p(x)p(y)}\right) \quad \text{.................} (2)$$

As an information-based measure, a primary strength of MI over R is that MI does not require a linear assumption. As long as two variables co-vary with each other, whether the relation is linear, exponential, or power, MI can be used to quantify the strength of the relationship. Different from R, which varies between -1 and 1, MI is non-negative from zero to infinitive. To standardize MI, we used the informational correlation coefficient R_I (Mathematics Encyclopedia: http://www.encyclopediaof math. org/index.php/Information_ correlation_ coefficient):

$$R_I = \sqrt{1 - e^{-2MI(x,y)}} \quad \text{..............................} (3)$$

The R_I defined above converts MI so that the results can be compared with R. It is easy to mathematically prove that for any normally distributed variables, $R_I = |R|$.

3 Study 1: Simulation Analysis to Assess the Validity of RPB

Following the RPB protocol described above, data (N=500) with a bivariate normal distribution were generated on a computer with known correlations (five levels). Random pairs with size n were selected and the R_I computed. Different pair sizes were also analyzed to evaluate the sensitivity of the method to sample size.

Results in Table 1 indicate that (1) the estimated R_I through the RPB method is close to the true linear correlation r. As expected, 95% confidence intervals of the estimated R_I decreases as the sample size increases, but still covers the true simulated value. (2) Larger sample sizes are needed for small rs. For behavioral intervention studies, a sample size n=100 pairs appears to be adequate to detect an underlying correlation with $r \geq 0.200$. We used this criterion for the next two studies.

Table 1. Estimated informational correlation coefficient R_I (95% CI) using RPB and computer-generated data with a known linear correlation

Random Pairs (n)	Computer generated data (N=500) with a known correlation				
	r = 0.000	r = 0.100	r = 0.200	r = 0.500	r = 0.800
50	0.000 (0.000, 0.380)	0.000 (0.000, 0.407)	0.136 (0.000, 0.442)	0.430 (0.000,0.645)	0.724 (0.570,0.831)
100	0.000 (0.000,0.354)	0.063 (0.000,0.358)	0.184 (0.000,0.415)	0.479 (0.250, 0.630)	0.776 (0.684, 0.844)
150	0.000 (0.000, 0.329)	0.066 (0.000, 0.335)	0.186 (0.000, 0.407)	0.491 (0.314, 0.613)	0.790 (0.714, 0.848)
200	0.000 (0.000, 0.305)	0.080 (0.000, 0.339)	0.189 (0.000, 0.386)	0.499 (0.342, 0.603)	0.796 (0.730, 0.847)
250	0.000 (0.000, 0.286)	0.095 (0.000, 0.312)	0.192 (0.000, 0.372)	0.501 (0.358, 0.604)	0.797 (0.744, 0.842)

4 Study 2: Simulation Analysis to Assess the Validity of MI

To test whether *MI* is valid in quantifying both linear and non-linear relations, we compared the estimated R_I with two known nonlinear relations, a power function defined as $y = x^b + \varepsilon$ and an exponential function defined as $y = e^{bx} + \varepsilon$, where $x \sim N(0, \sigma^2)$, and b = 2 and σ = 2 for both functions. Likewise, observations (N=500) were generated for variable *x* with its range set between -5 and 5; the observations for *y* were thus computed using the defined equations. With this setting, the Pearson correlation for the power function is $R_{x,y} = 0$ and 0.519 for the exponential function.

Following the RPB protocol described earlier, random samples were drawn for n = 50 ~ 250 respectively from computer-generated data. R_I coefficients were calculated for individual samples. The R_I estimate and its 95% CI were determined using 5000 repeating samples. For comparison purpose, the same procedure was used to estimate *R* and its 95% CI.

Results in Table 2 indicate that the estimated *R* for the power function data, centered at zero, with 95% CI = -0.007, 0.003. This is a correct assessment of the underlying linear correlation for power function. The same correlation for the exponential function varied between 0.521 and 0.523, slightly greater than but still very close to, its theoretical correlation of 0.518. Although the estimated *R* failed to detect the nonlinear power relationship, it extracted the linear portion of the exponential relationship. Likewise, the estimated R_Is were relatively stable when the sample size greater or equal to 100, supporting the result observed in Study 1.

Table 2. Assessment of various given effect of crosstalk using the informational correlation R_I in comparison with the Pearson correlation R with computer generated data (N=500)

Random Pairs (n)	Power Function		Exponential Function	
	Informational correlation(R_I)	**Pearson correlation(R)**	**Informational correlation(R_I)**	**Pearson correlation(R)**
50	0.381 (0.000, 0.577)	-0.002 (-0.330, 0.329)	0.636 (0.555, 0.703)	0.543 (0.428, 0.645)
100	0.635 (0.532, 0.726)	0.006 (-0.211, 0.225)	0.665 (0.612, 0.722)	0.525 (0.441, 0.602)
150	0.781 (0.729, 0.825)	-0.007 (-0.179, 0.159)	0.682 (0.638, 0.739)	0.523 (0.457, 0.573)
200	0.842 (0.809, 0.869)	0.003 (-0.137, 0.135)	0.697 (0.655, 0.740)	0.521 (0.468, 0.565)
250	0.875 (0.853, 0.896)	-0.001 (-0.118, 0.110)	0.708 (0.668, 0.746)	0.521 (0.478, 0.557)

5 Study 3: Analysis of Empirical Data

To empirically test the RPB-MI, we analyzed data we collected in a project conducted in the Bahamas [15-16]. Students (n=1360) were randomized into three groups: Group 1 with students and their parents receiving HIV prevention, Group 2 with only students receiving intervention, and Group 3 received environmental preservation education as control. A theory-guided program with 10 sessions was delivered in classroom settings after baseline assessment. Follow-up data were collected at a pre-determined time schedule. The program has been demonstrated to improve HIV knowledge, self-efficacy for condom use and condom skills [15-16]. Data collected at baseline, 12 and 24 months were analyzed. Three variables, HIV knowledge, self-efficacy and skills to use a condom were assessed using validated instruments with Cronbach alpha >0.80. The same RPB was used for sampling and the R_I and its 95% CI were determined by 5000 repeating samples.

Results in Table 3 show that the estimated R_Is all are greater than 0.000, suggesting active within-group crosstalk. The estimated R_Is were the highest for condom skills, lowest for HIV knowledge, with self-efficacy in between. In addition, relative to the other two variables, there appeared to be a larger increase in the estimated R_I regarding HIV knowledge at the 24-month follow up.

As expected, data in Table 4 show that the estimated R_Is for the crosstalk between any two of the three groups are smaller compared with the R_Is for the within-group crosstalk in Table 3. In addition, the R_Is between Group 1 and 2 appear to be greater than the R_Is between Group 1 and 3 and the R_Is between Group 2 and 3, which is also reasonable. Lastly, there is an increase in the estimated R_Is for HIV knowledge, which could be due to the effect of behavioral intervention, as previously observed [15].

Table 3. Estimated R_I (95% CI) for within-group crosstalk at baseline and follow-ups, the Random-Paring Bootstrap (RPB) method (n=100 random pairs)

Time/variable	Group 3 (Control) (N=497)	Group 1 (Interv. 1) (N=436)	Group 2 (Interv. 2) (N=427)
HIV knowledge			
Baseline	0.372(0.243, 0.472)	0.416(0.320, 0.508)	0.424(0.328, 0.517)
12 months	0.323(0.216, 0.428)	0.337(0.234, 0.446)	0.381(0.291, 0.470)
24 Months	0.444(0.349, 0.540)	0.575(0.460, 0.743)	0.663(0.463, 0.766)
Self-efficacy			
Baseline	0.475(0.372, 0.572)	0.488(0.389, 0.579)	0.450(0.350, 0.543)
12 months	0.407(0.312, 0.509)	0.418(0.314, 0.512)	0.422(0.329, 0.505)
24 Months	0.474(0.387, 0.555)	0.425(0.334, 0.522)	0.461(0.365, 0.558)
Condom skills			
Baseline	0.691(0.614, 0.759)	0.727(0.651, 0.793)	0.743(0.659, 0.816)
12 months	0.703(0.638, 0.773)	0.722(0.629, 0.822)	0.721(0.652, 0.805)
24 Months	0.684(0.592, 0.777)	0.793(0.707, 0.861)	0.735(0.638, 0.822)

Table 4. Estimated R_I (95% CI) for between-group crosstalk at baseline and follow-ups, the Random-Paring Bootstrap (RPB) method (n=100 random pairs)

Time/variable	Between Group 1 and Group 3	Between Group 2 and Group 3	Between Group 1 and Group 2
HIV knowledge			
Baseline	0.283(0.103, 0.414)	0.255(0,000, 0.400)	0.284(0.063, 0.422)
12 months	0.282(0.049, 0.421)	0.296(0.089, 0.434)	0.309(0.110, 0.466)
24 Months	0.418(0.231, 0.574)	0.395(0.209, 0.541)	0.485(0.290, 0.642)
Self-efficacy			
Baseline	0.391(0.242, 0.533)	0.357(0.211, 0.489)	0.368(0.214, 0.502)
12 months	0.335(0.178, 0.476)	0.355(0.209, 0.504)	0.337(0.182, 0.466)
24 Months	0.394(0.261, 0.517)	0.376(0.230, 0.497)	0.318(0.140, 0.445)
Condom skills			
Baseline	0.624(0.520, 0.740)	0.606(0.496, 0.711)	0.633(0.536, 0.749)
12 months	0.621(0.517, 0.770)	0.608(0.504, 0.735)	0.632(0.536, 0.764)
24 Months	0.618(0.511, 0.720)	0.584(0.467, 0.685)	0.638(0.490, 0.769)

6 Discussion and Conclusions

In this study, we introduced the RPB-MI method for assessing crosstalk in behavioral intervention research. The validity of the method was tested with simulated data and the utility of the method was assessed with empirical data. From the study findings we can conclude: (1) The RPB protocol is effective in capturing person-to-person communication even if the underlying relationship of such information exchange is nonlinear. (2) The MI-based R_I is more powerful than the Pearson correlation R in assessing the effect of crosstalk [10-12]. (3) The RPB-MI approach adds a new tool for behavioral intervention research [8]. Despite the advantages of the RPB-MI, several technical issues remain, such as the understanding of the informational correlation in assessing nonlinear correlation and sample size determination.

References

1. Stanton, B., Cole, M., Galbraith, J., Li, X., Pendleton, S., Cottrel, L., Marshall, S., Wu, Y., Kaljee, L.: A randomized trial of a parent intervention: Parents can make a difference in long-term adolescent risk behaviors, perceptions and knowledge. Archives of Pedi. Adol. Med. 158, 947–955 (2004)
2. Kegeles, S.M., Hays, R.B., Coates, T.J.: The Mpowerment Project: A community-level HIV prevention intervention for young gay men. Am. J. of Public Health 86, 1129–1136 (1996)
3. Howe, A., Keogh-Brown, M., Miles, S., Bachmann, M.: Expert consensus on contamination in educational trials elicited by a Delphi exercise. Med. Edu. 41, 196–204 (2007)
4. Torgerson, D.J.: Contamination in trials: is cluster randomisation the answer? BMJ 322, 355–357 (2001)
5. Fritz, K., McFarland, W., Wyrod, R., Chasakara, C., Makumbe, K., Chirowodza, A., et al.: Evaluation of a peer network-based sexual risk reduction intervention for men in beer halls in Zimbabwe: results from a randomized controlled trial. AIDS Behav. 15, 1732–1744 (2011)
6. Gottman, J.M.: Detecting cyclicity in social interaction. Psychological Bulletin 86, 338–348 (1979)
7. Keirse, N.C., Hanssens, M.: Control of error in randomized clinical trials. European J. Obst. Gyn. Reprod. Bio. 92, 67–74 (2000)
8. Lang, D.L., DiClemente, R.J., Hardin, J.W., Crosby, R.A., Salazar, L.F., Hertzberg, V.S.: Threats of cross-contamination on effects of a sexual risk reduction intervention: fact or fiction. Prev. Sci. 10, 270–275 (2009)
9. Gottman, J.M.: Marital Interaction: Experimental Investigations. Academic Press, New York (1979)
10. Gottman, J.M., Murray, J.D., Swanson, C.C., Tyson, R., Swanson, K.R.: The Mathematicals of Marriage: Dynamic Nonlinear Models. MIT Press, Cambridge (2002)
11. Guastello, S.J., Pincus, D., Gunderson, P.R.: Electrodermal arousal between participants in a conversation: nonlinear dynamics and linkage effects. Nonlinear Dynamics Psychol Life Sci. 10, 365–399 (2006)
12. Hartkamp, N., Schmitz, N.: Structures of introject and therapist-patient interaction in a single case study of inpatient psychotherapy. Psychotherapy Research 9, 199–215 (1999)
13. Nowak, A., Vallacher, P.R.: Dynamical Social Psychology. Guilford Press, New York (1998)
14. Shannon, C.E.: The mathematical theory of communication. Bell Syst. Techn. Journal 27, 379–423 (1948)
15. Chen, X., Stanton, B., Gomez, P., Lunn, S., Deveaux, L., Brathwaite, N., et al.: Effects on condom use of an HIV prevention programme 36 months postintervention: a cluster randomized controlled trial among Bahamian youth. Int J. STD AIDS 21(9), 622–630 (2010)
16. Chen, X., Lunn, S., Deveaux, L., Li, X., Brathwaite, N., Cottrell, L., et al.: A cluster randomized controlled trial of an adolescent HIV prevention program among Bahamian youth: effect at 12 months post-intervention. AIDS Behav. 13, 499–508 (2009)

New Methods of Mapping

The Application of Social Network Analysis to the Study of the Illegal Trade in Antiquities

Michelle D'Ippolito

Department of Anthropology, University of Maryland College Park,
College Park, MD, 20740, USA
mrdippolito@gmail.com

Abstract. This study examines the application of a human-agent based network to the illegal trade in antiquities. Specifically, this study tests whether the hierarchical pyramidal structure proposed by law enforcement in the case of Giacomo Medici's trafficking ring is accurate. The results of the analysis reveal discrepancies in perceptions of how antiquities trafficking networks are organized, how they operate, and how cultural patterns and representation of criminal activity influence the perception of such network structures.

Keywords: Illegal Antiquities Market, Social Network Analysis, Network Theory, Ucinet, Anthropology, Economics, Criminology, Law.

1 Introduction

How do we understand the nature, structure, and function of a phenomenon like the antiquities market?[1] Because it operates in both legal and illegal spheres of society, it poses a unique challenge to social scientists and law enforcement. The trade in illegal antiquities has been closely compared to other illegal trafficking markets and to organized crime. These comparisons, while providing insight into the motivations behind transnational criminal activity in general, have propagated the misconception that the illegal antiquities trade is hierarchically structured, similar to the mafia [1]. The illegal antiquities market is distinct from other forms of trafficking in that the cultural heritage being moved is a "finite resource that cannot be cultivated or manufactured…and artifacts must be laundered in order to appear legitimate" [2].

This study examines networks of individuals operating both legally and illegally in the antiquities market inductively through the application of social network analysis. Specifically, this study asserts that the hierarchical pyramidal structure most often

---

[1] The art market in general and the antiquities market are distinguished in this study. The art market includes contemporary art and antiquities. Here, the antiquities market is defined as dealing with archaeological and ethnographic objects. While there is no clear definition of what makes an object old enough to be an antique, the objects discussed in this study are from archaeological sites at least five hundred years old. As such in this study an object must be at least five hundred years old to be considered an "antiquity."

W.G. Kennedy, N. Agarwal, and S.J. Yang (Eds.): SBP 2014, LNCS 8393, pp. 253–260, 2014.

proposed by law enforcement is an accurate representation of the structure of an antiquities trafficking. This research is excerpted from a larger study and will only focus on one case study involving the dealer Giacomo Medici and the Italian carabinieri. Social network analysis is used to test two hypotheses: (1) when using the same set of data[2] the program Ucinet [3] will produce a non-hierarchical visualization of the network that differs from what was described by the Italian police; (2) Ucinet's representation of the network will more accurately describe the structure and dynamics of the network and its actors.

2 Methodology

The present study combines the fields of criminology, social network analysis, and the illegal trade in antiquities to examine the structure of an Italian network of smugglers first uncovered in the early 1990s with investigations continuing for over a decade. Specifically, it focuses on one branch, or cordata, of the network involving Giacomo Medici. The literature that informs the larger study from which this case study is abstracted pulls from the scholarship in criminology, anthropology, economics, and network theory. The literature covers transnational criminal organization [1,4], theories on influence and power in the illegal antiquities market [5], and the use of centrality and betweenness in determining network structure in corporate hierarchies and quasi-legal activities [6-9].

To test the first hypothesis, the network is first visualized based on the categories of the actors in the network (i.e., tombaroli, middleman, dealer, etc.). Each node represents a category of actor and each relationship encompasses buying, trading, and selling. The movement of objects through the network is used to validate the new model generated by social network analysis. Interpersonal relationships of the actors are used to analyze the nature of influence and power in the network. For the second hypothesis, Medici's ego-network is examined. Statistical measures developed by Linton Freeman are used to examine the centrality and prestige both at the actor and network level [8,9]. Four measures within centrality are used due to a lack of specific data on directed relations:

- Degree Centrality: Assesses prominence within the network as measured by the amount of "activity" a node experiences.
- Closeness Centrality: Uses proximity to determine the centrality of an actor. The faster an actor can interact with others, the more central it is to the operation and structure of the network [10].

[2] Every effort was made to use the same primary sources as were used by the Italian police in the original investigation. Italian legal policy states that all original documentation pertinent to a trial is not available to the public until the trial is completely resolved. Although Giacomo Medici's trial and appeal have concluded, the documents are not yet available. "Data" here refers to the information available in the organigram and associated qualitative open-source data available on the relationships between actors in the network.

- o Betweenness Centrality: Assumes that all paths are equally likely in the communication of information and looks at those actors who act as gateways and tollbooths for information in the network [8,11].
- o Information Centrality: Assumes that information may easily take a more circuitous route either by "random communication" or by being "channeled" through intermediaries to "shield" or "hide' information [12].

All four measures are calculated at the actor level or the network level. At the actor level, they indicate the actor's prominence in relation to the specific measure. At the network level, they indicate how variable or heterogeneous the actor centralities are in the network.

The Italian police and prosecutor asserted that Giacomo Medici was part of a pyramidal hierarchical network based on two pieces of evidence: an organigram[3] [13] found in the home of an actor in the network describing a rough hierarchy of relationships and (2) the use of the term "cordata" to describe the structure of the network by a looter (or tombarolo) [13].[4] The four centrality measures above were used to verify whether these assumptions about the structure of the network are accurate both in the specific case of Medici and in the Theoretical network (see Figure 2). They also informed how the perception of the network affected assumptions about its operations.

3 The Italian Network

3.1 Overview

The theft of several Melfi vases on January 20, 1994 triggered an investigation of an extensive underground network of antiquities smuggling by the Italian Art Squad led by Roberto Conforti.[5,6] The investigation quickly expanded to include prominent and influential dealers in Italy, including Giacomo Medici. Several of the central figures in the investigation, Medici included, maintained meticulous records of their involvement not only with the network's other branches' (or cordate's) dealers and

[3] The organigram is an organized chart of the structure of the network drawn from the perspective of one inside the network. The relationships were defined as primarily unidirectional on the bottom and bi-directional towards the top of the hierarchy. Based on this information, a relationship was defined as buying, selling, or trading. It was most likely created sometime between 1990 and 1993.

[4] Cordata refers to a mafia-like structure in Italian, which, as a classic structure in organized crime, is typically considered hierarchical.

[5] Roberto Conforti, a veteran carabinieri, has worked in drugs, homicides, and spent considerable time undercover working against "the most formidable, well-equipped, determined, and organized criminals that Italy has produced," [12]. At the time of the theft under investigation, Conforti had already been the head of the Art Crime Unit for four years and had expanded the unit to operate in Palermo, Florence, and Naples.

[6] The Melfi vases stolen from the museum were terracotta pots, each roughly 2,500 years old. They depicted stories from the Greek classics of goddesses, athletes, and scenes of dancing and feasting [12]. The condition and artistic quality of the vases made them very important culturally and historically to the Melfi region and very desirable to collectors and museums.

middlemen, but also with a periphery of museums, collectors, and scholars in both Europe and the United States. The expanse of the network required effective tracking of the movement and sale of objects. Along with the organigram, four archives of information were discovered during the course of the investigation. The broader network's activities were found to cover at least Italy, Switzerland, France, Germany, England, Japan, and the United States. These included looting, smuggling, theft, and laundering of objects to auction houses. By 2006 Giacomo Medici had been convicted in Italy, while other prominent members were awaiting trial in England and Egypt. As of 2012, Medici was serving an 8 year term and owed a €10 million fine to Italy for damage caused to cultural heritage [14].

3.2 Comparison of Network Structure

The results of the visual analysis supported the first hypothesis. Using the same network-level dataset on the relationships as defined in the organigram, the visualization produced by Ucinet featured a highly centralized network with two actors of equal interest: the dealer and the international dealer. The pyramidal structure proposed by Italian law enforcement places the international dealer underneath the buyers (auction house, collectors, and museums), suggesting that while the international dealer is important, the buyers would be most influential.

As shown in Figures 1 and 2, the relationships between the categories in both the pyramid and centralized networks do not change; however, their representations of the influence of actors within the networks differ. In order for the pyramidal structure to be an effective description of the structure of the network, participants would have to be subject to a monopoly, long-term agreements, and sanctions [2]. Based on the available data on Medici's branch of the larger trafficking network, no monopolies existed on the sources or buyers of antiquities. Nor did agreements to purchase or work together (formal or otherwise) exist between the tombaroli, capà zona, middlemen, and dealers [13]. Without the first two, it would have been impossible for sanctions to exist. The hierarchical nature of the pyramid, then, does not reflect the nature of influence or power in the network.

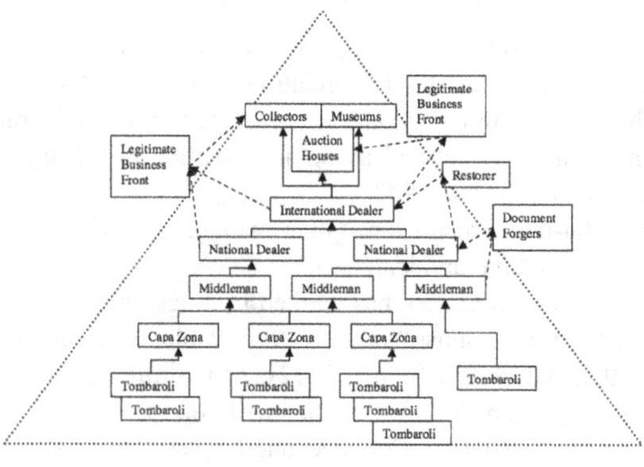

Fig. 1. Pyramidal Structure proposed by law enforcement as adapted from descriptions of the network by Watson and Todeschini and interpreted based on the organigram created by Pasquela Camera

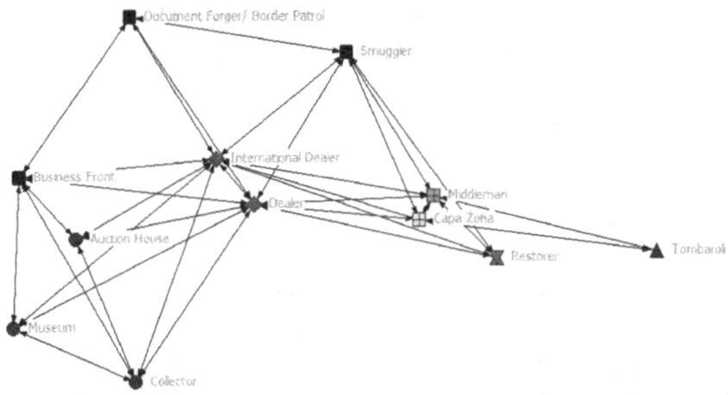

Fig. 2. Theoretical Model of Medici Network. Node color reflects the category of the actor. Node shapes correspond to function in the network: buyers are circular, laundering experts are square, diamonds are dealers, triangles are sources of looted objects, and boxes are intermediaries.

The centralized network represented in the Theoretical Model (Figure 2) more accurately represents the flexible nature of the relationships in the network and the movement of antiquities through it. The dealer and international dealer control the flow of objects from the source to the buyers in that they have the most direct relationships with the buyers.[7] Giacomo Medici, as a dealer, had significant influence over his connections in the network both to persuade buyers to consider objects of questionable provenance and work directly with peripheral actors to prepare the objects for transportation. Yet, there was opportunity for other actors in the network to work with others should they chose, even their influence over the network was not significant. The Theoretical Model is more realistic in reflecting how objects move through the network; the flexibility accounts for the unpredictability of movement that is sometimes seen in the movement of illegal antiquities.

3.3 Representation of Network- and Actor-Level Structure and Dynamics

The results of the statistical analyses support the second hypothesis. Analyses run on both the network level and the actor level datasets confirmed the central role of the dealer and international dealer in the network. They also suggested several patterns in the structure of power and influence in the network based in the behavioral patterns known to have occurred during the height of Medici's activity.

Overall, the results found that for the degree, betweenness, and closeness measures, the network-level results indicated a much more decentralized network than the

---

[7] It is interesting to note that here unidirectional relationships tend to be shown more on the peripheries than bidirectional relationships, despite the fact that in reality this is not always the case. While it is true that the buyers do not have a "relationship" with the dealer, they do know of him and are often closely associated with him. This shows one of the weaknesses of this type of visualization – whether a path exists between two actors depends on how "relationship" is defined. In this case it encompasses buying, trading, and selling. Because most buyers only buy and do not sell or trade, they are shown as further out in the network. One must add weights (e.g, closeness of relationship) to get a more realistic visualization.

results of the actor-level analyses.[8] With centrality indices ranging from 22 – 60 percent for these three measures, the network-level model (Figure 2) is only moderately centralized.[9] At the actor-level, results showed a highly centralized network with measure results ranging from 76 – 86 percent.

Table 1. Comparison of Summary Statistics for Theoretical Model (Network-Level) and Network Model (Actor-Level)

	Maximum	Minimum	Mean	Standard Deviation
Theoretical Model	3.316	1.364	2.463	0.504
Network Model	1.980	0.655	1.158	0.322

Of particular interest is the information centrality measure, as seen in Table 1 above. The results showed that both at the network and actor-level, there was a large amount of variation between actors that are highly influenced and influential.[10] Yet the variation was not consistent across categories of actors in the Network Model. The activities of the tombaroli were similar, yet not all of them were equally influenced by the capa zona and middlemen. The same is true across every category of actor (see Figure 3). This suggests that while the network may be centralized, assumptions of who has power and influence among the actors is fluid, which may be a reflection of the fluid participation of the actors in both legal and illegal activities [1].

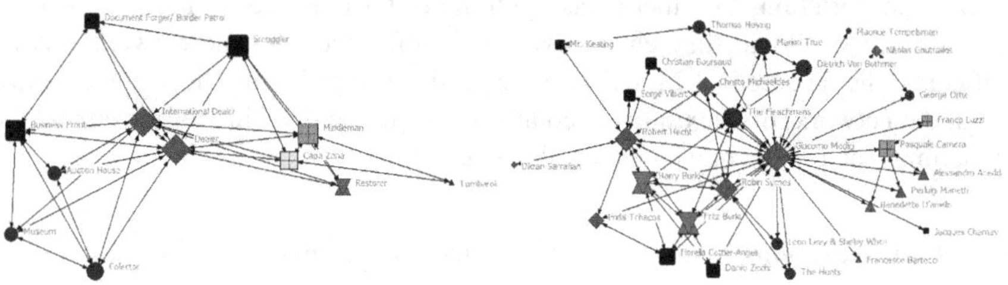

Fig. 3. Freeman Information Centrality measures calculated at the actor level for the Theoretical Model (top) and the Network Model (bottom). Node size indicates prominence of actors in the network; nodes color reflects the category of the actor (dealer, restorer, etc.). Node shapes correspond to their function in the network: buyers are circular, laundering experts are square, diamonds are dealers, triangles are sources of looted objects, and boxes are intermediaries.

8 Differences in the degree of centrality between the Theoretical and Network Models can in part be explained by the differences in the number of nodes in each graph.

9 The centrality measures for degree, closeness, and betweenness provided a network-level centralization index that aggregated the individual actor results. The degree index expresses the inequality, or variance, of the distribution of influence in the network as a percent out of 100. Closeness indicates variability in the data in terms of the proximity of the actors as a percent out of 100. the more valuable the data, the closer to100 the index will be. Betweenness has only one statistic for centralization out of 100 percent and includes all individual centralization measures, including those without intermediaries, so can be misleading.

10 Information centrality does not provide an index; it assumes that all paths are not equally likely, instead relying on choices of actors in the network.

4 Discussion

The results of the centrality measures suggest that Medici's network of actors, while highly centralized, also had the capacity to be extremely fluid and adaptable in terms of power and influence. This is contrary to the notion of a mafia-like hierarchical structure [1]. Such a network would have stronger relationships outside of business, possibly including familial relationships, of which Medici's network had very few. Instead, the fluidity of the influence and power can be attributed to the ease with which actors in the network are required to operate in both the legal and illegal markets [1]. Their influence over other actors at any one time would be determined by which sphere they are operating in. A tombaroli selling to Medici directly under a legal sale has equal authority and influence over the transaction, whereas the same situation in the illegal sphere would have a different balance of power.

Of all the centrality measures, information centrality portrays the complexity of the perception of power and influence best. The measure uses a system of weights to calculate the most probable path based on the strength of relationship and the frequency of interactions, what Stephenson and Zelen call "strength of signals" [12]. In the antiquities market, this is called "triangulation" [13] and from what studies have been able to show, is the most frequent and effective means of evading authorities [13]. As shown in Figure 3, with the exception of a few categories, most of the nodes are of roughly equivalent size, suggesting that there are a number of equally likely paths depending on the situation, which is an accurate reflection of the transportation methods used by Medici described in the data [13].

5 Conclusion

Ultimately the results of this study suggest that a network perspective on the structure of illegal antiquities trafficking is a more accurate reflection of the complexities of power and influence among the actors. It also can help to inform our understanding of the current perceptions of criminal organization that guide the work of law enforcement. Further, it suggests that the inductive method taken here can provide a more flexible framework from which generalizations across antiquities networks may be elucidated.

Power and influence in the illegal antiquities market is fluid and depends on the specific actors involved at any one time. This is contrary to the perception held by the Italian law enforcement. The notion of a single prominent actor controlling the activities of all others would not only be unsustainable but also unrealistic given the scale of operations crossing continents, like the larger network Medici was. This more nuanced view of the illegal antiquities network is more realistic and can provide deeper insight into the perception of criminal activities.

More broadly, this study suggests that a means of generalizing across different antiquities networks, while not necessarily possible at the actor-level due to difference in social norms and mores, may be possible at the methodological-level. The methodology used could provide one possible means of generalizing. Using both deductive and inductive research methods and a modeling method, like social network analysis, that can incorporate both allows for consistent analyses across different countries. One important feature of a generalized framework not considered here is the effect of social norms and mores of the society in which networks operate on the decisions and behaviors of both the criminals and law enforcement officials that pursue them. Most importantly, such a methodology would require time and additional data for new observations on the behavioral patterns of the participants in order to increase the validity and reliability of the model.

Obtaining data will always be a challenge for this kind of research due to the nature of illicit markets. However, using social network analysis to better understand the market based on the data that is available may reveal new sources information in existing records or better new ways of collecting data during investigations.

References

1. Williams, P.: Networks, Markets, and Hierarchies. In: Williams, P., Vlassis, D. (eds.) Combatting Transnational Crime: Concepts, Activities, and Responses, pp. 57–87. Frank Cass, London (2001)
2. Campbell, P.: The Illicit Antiquities Trade as a Transnational Criminal Network: Characterizing and Anticipating Trafficking in Cultural Heritage. International Journal of Cultural Property 20, 113–151 (2013)
3. Borgatti, S.P., Everett, M.G., Freeman, L.C.: Ucinet for Windows: Software for Social Network Analysis. Analytic Technologies, Harvard (2002)
4. Polk, K.: Whither Criminology in the Study of the Traffic in Illicit Antiquities? In: Green, S.M.A.P. (ed.) Criminology and Archaeology: Studies in Looted Antiquities. Oñati International Series in Law and Society. Hart Publishing, Oxford (2005)
5. Bruinsma, G., Bernasco, W.: Criminal Groups and Transnational Illegal Markets: A More Detailed Examination on the Basis of Social Network Theory. Crime Law & Social Change 41, 79–94 (2004)
6. Burnkrant, R.E., Cousineau, A.: Information and Normative Social Influence in Buyer Behavior. The Journal of Consumer Research 2, 206–215 (1975)
7. Dawes, P.L., Lee, D.Y., Dowling, G.R.: Information Control and Influence in Emergent Buying Centers. Journal of Marketing 62, 55–68 (1988)
8. Freeman, L.: A Set of Measures of Centrality Based on Betweenness. Sociometry 40, 35–41 (1977)
9. Freeman, L.C.: Centrality in Networks: I. Conceptual Clarification. Social Networks 1, 215–239 (1979)
10. Sabidussi, G.: The centrality index of a graph. Psychometrika 31, 581–603 (1966)
11. Anthonisse, J.M.: The Rush in a Graph. Mathematische Centrum, Amsterdam (1971)
12. Stephenson, K., Zelen, M.: Rethinking Centrality: Methods and Applications. Social Networks 11, 1–37 (1989)
13. Watson, P., Todeschini, C.: The Medici Conspiracy: The Illicit Journey of Looted Antiquities from Italy's Tomb Raiders to the World's Greatest Museums, pp. 16–18. Public Affairs, New York (2007)
14. Felch, J.: 'Quick takes', Los Angeles Times (January 20, 2012), `http://articles.latimes.com/2012/jan/20/entertainment/la-et-quick-20120120` (accessed November 15, 2013)

A New Paradigm for the Study of Corruption in Different Cultures

Ya'akov (Kobi) Gal[1], Avi Rosenfeld[2], Sarit Kraus[3,4], Michele Gelfand[4], Bo An[5], and Jun Lin[6]

[1] Department of Information Systems Engineering, Ben-Gurion University, Israel
[2] Department of Industrial Engineering, Jerusalem College of Technology, Israel
[3] Department of Computer Science, Bar-Ilan University, Israel
[4] Institute for Advanced Computer Studies, University of Maryland, USA
[5] Department of Engineering, Nanyang Technological University, Singapore
[6] Department of Engineering and Management, Beihang University, China
kobig@bgu.ac.il, rosenfa@jct.ac.il, sarit@cs.biu.ac.il, mjgelfand@umd.edu, boan@ntu.edu.sg, linjun@buaa.edu.cn

Abstract. Corruption frequently occurs in many aspects of multi-party interaction between private agencies and government employees. Past works studying corruption in a lab context have explicitly included covert or illegal activities in participants' strategy space or have relied on surveys like the Corruption Perception Index (CPI). This paper studies corruption in ecologically realistic settings in which corruption is not suggested to the players a priori but evolves during repeated interaction. We ran studies involving hundreds of subjects in three countries: China, Israel, and the United States. Subjects interacted using a four-player board game in which three bidders compete to win contracts by submitting bids in repeated auctions, and a single auctioneer determines the winner of each auction. The winning bid was paid to an external "government" entity, and was not distributed among the players. The game logs were analyzed posthoc for cases in which the auctioneer was bribed to choose a bidder who did not submit the highest bid. We found that although China exhibited the highest corruption level of the three countries, there were surprisingly more cases of corruption in the U.S. than in Israel, despite the higher PCI in Israel as compared to the U.S. We also found that bribes in the U.S. were at times excessively high, resulting in bribing players not being able to complete their winning contracts. We were able to predict the occurrence of corruption in the game using machine learning. The significance of this work is in providing a novel paradigm for investigating covert activities in the lab without priming subjects, and it represents a first step in the design of intelligent agents for detecting and reducing corruption activities in such settings.

1 Introduction

Corruption – the abuse of entrusted power for private gain, is a global phenomenon that severely diminishes economic growth, education opportunities,

W.G. Kennedy, N. Agarwal, and S.J. Yang (Eds.): SBP 2014, LNCS 8393, pp. 261–268, 2014.

availability of health and welfare services, and undermines the ability of governments to implement needed policies [1]. Because corruption is inherently a covert activity and difficult to prove, it is notoriously hard to measure, and convictions are few and far between, especially in countries where corruption is often viewed as permissible. Thus most assessment tools have relied on surveys, such as the well-established Corruption Perception Index (CPI) which ranks countries based on their perceived levels of corruption, as determined by expert assessments and opinion surveys [8].

Lab experiments have studied corruption as three-player games involving a briber, a bribee, and a third party which is negatively affected by the bribe [9,2]. Other works have studied how wages and detection probabilities affect corruption levels when a public "official" can make a unilateral decision about how much money to divert from public funds, at a risk of being detected and punished [5,6]. In all of these works, illegal activities and bribes are primed or explicitly stated in players' strategy space, which does not reflect the secretive and dynamic nature in which corruption develops. A notable exception is the work by Falk and Fischbacher [4] in which illegal activities are framed as "take" actions, and are shown to reciprocate over time in complete information settings. They did not simulate a repeated task setting but asked subjects to provide a complete description of their strategy. Our study is more realistic in that we did not make references to any covert activities, participants could not observe such activities in the game and interactions among participants were conducted in real-time.

The goal of this paper is to study corruption in a realistic way as an evolving process in which participants engage in repeated economic activities. We used a board game in which three bidders compete to win contracts by submitting bids in repeated auctions, and a single auctioneer determines the winner of each auction. Bidders earn bonus points in the game if their bids are accepted. The auctioneer's score is constant and does not depend on their bids. The winning bid is paid to an external "government" entity. At given times in the bidding process, participants could exchange private messages with each other.

We played 276 games in Israel, the U.S. and China spanning hundreds of subjects. We analyzed the logs that were generated by the games posthoc in each country. We measured two types of corrupt activities, one in which the auctioneer did not pick the bidder with the highest bid, and one in which there was a distinct bribe that was transferred from a bidder to the auctioneer in return for getting chosen.

The results showed that corruption occurred in about 33% of games played in the U.S., 29% of games played in Israel and 56% in the games played in China. These results partially follow the CPI, in that China was the country that exhibited the largest amount of corruption. However, we measured a lower level of corruption in Israel as compared to the U.S. despite the higher CPI level in Israel. We also found that corruption benefited the auctioneer when compared to the case in which there was no corruption. However, while the bribing players benefited from this activity in Israel and in China, they did not benefit from bribery in the U.S. Further analysis revealed that bribes in the U.S. were often

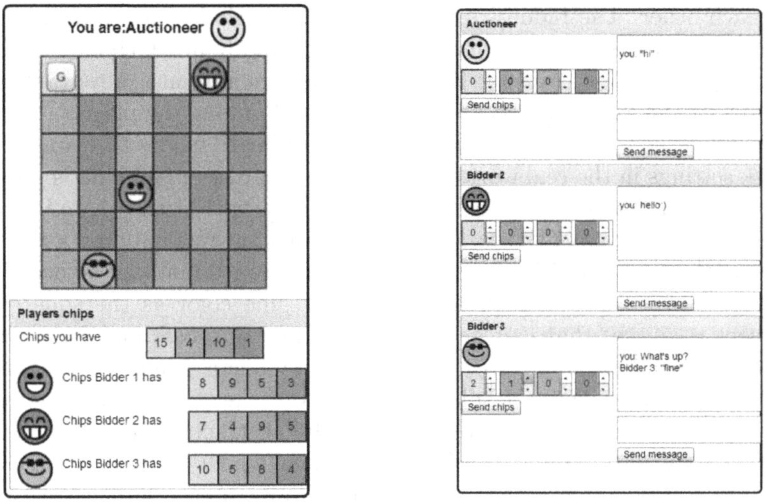

(a) Main board panel (b) Message panel

Fig. 1. Snapshots of Olympic Game GUI

excessively large, which prevented the winning bidders from completing their tasks. In all countries, we found that there was a significant increase in the ratio of corruption in the last round of play. Additionally, we were able to predict the occurrence of corrupt activities in all countries using supervised learning approaches using information that was publicly available during the game.

The contributions of this paper are threefold. First, we provide a new empirical framework to study corruption in a lab setting without priming subjects. Second, we show that corruption is endemic to people's interaction in different countries in a way that is not predicted by the CPI. Third, we provide a predictive model for the occurrence of corruption in the lab.

2 Implementation Using Colored Trails

Our empirical study was based on a test-bed called Colored Trails [7], which we adapted to a multi-round contract allocation game called the Olympic Game. In the Olympic Game, a city is preparing to host the Summer Olympics by creating the necessary infrastructure such as hotels, transportation and restaurants. There are four participants in the game: three *bidders* and one *auctioneer*. The game comprises a board game of colored squares which includes players' icons and a goal square. Players are allocated a set of chips of colors chosen from the same palette as the board square. All players have full view of the board game and their allocated chip sets, however bidders cannot observe the chips of other players.

Figure 1 shows the main game panel from the point of view of the auctioneer. At the heart of the game is players' abilities to exchange resources and messages with each other. The bidders send bids to the auctioneer in the form of chips. The auctioneer has full authority to make decisions about which bidder to choose for each project. The chips for the winning bid are automatically transferred to an external "government" and is not distributed among the players.

The Olympic Game is a variant of a family of board games that are analogous to task settings in the real world. Paths on the board from a bidder's position to the goal square represent carrying out a contract, with each square in the path corresponding to one of the tasks in the contract such as building a restaurant, hiring workers, and acquiring permits. To move an icon into an adjacent square a bidder player must turn in a chip of the same color as the square. Chips represent resources that can be used for bidding purposes and for completing the contracts. These chips directly translate to monetary units at the end of the game and are awarded to the subjects. The advantage of using the Colored Trails game is that it creates an environment which "abstracts away" domain specific details while still providing a task-like context in which decisions are made. The abstraction provided by Colored Trails is especially appropriate for investigating decision-making in different cultures because it avoids culturally-loaded contexts (e.g., religious connotations) that may confound people's behavior.

The Olympic Game is played repeatedly for an undisclosed number of rounds (four rounds in our study). In each round, a new contract is generated for a different project for the Olympic city. The goal for each round is to win the bid and advance to the goal square. Each round is played on a new board game, in which players' location on the board as well as their initial chip allocations may change. Players' roles do not change between rounds (i.e., the same auctioneer chooses the winning bid in all rounds).

The interaction in each round of the game proceeds in a sequence of phases with associated time limits. In the *strategy phase* (60 seconds), all players are given a chance to study the board and to think about possible strategies in the game. In the *exchange phase* (90 seconds), players may exchange resources and messages in free text with each other using the game interface. An example is shown in Figure 1 (bottom). The messages can be initiated by any of the participants, and are only visible to the message initiator and receiver. In the *bidding phase* (60 seconds), each of the bidders can choose to make a bid to win the contract for the round. The bid is the number of chips the bidder is willing to pay the government for the purpose of winning the contract. Bidders cannot see each others' bids. All of the bids are relaid to the auctioneer at the end of the bidding phase. At this point, the auctioneer can choose one bidder to win the contract. In the *execution phase* the bid of the winning bidder is deducted from the chips in its possession, and the winning bidder is advanced to the goal (automatically) given its available resources. In particular, if the winning bidder cannot get to the goal square, its icon is moved as close as possible to the square using its available chip set. Note that only the winning bidder is allowed to move towards the goal, and is deducted the chips used to move towards the goal.

At the end of each round, the score for each participant is computed as follows: One point is given for each chip in the possession of a participant at the end of the round; a 100 point bonus is given to a winning bidder who is able to reach the goal. Otherwise, one point is deducted for every square in the path from the final position of the winning bidder and the goal square; a 5 point bonus is given to an auctioneer for choosing a winning bidder that is able to reach the goal square. There are several reasons for choosing these parameters. First, they reflect the fact that reaching the goal is the most important component in the game. Reaching the goal is analogous to completing the contract for which a winning bidder is chosen using the resources that are available to the bidder at the execution phase. The auctioneer is an employee of the government, and receives a constant salary for each round, as represented by the chips that are allocated to the auctioneer at the onset of the round. As a government employee, it receives only a nominal bonus for choosing a bidder that wins the contract.

3 Empirical Methodology, Setup, and Results

We ran 101 instances of the game in the U.S., 91 instances in Israel and 84 instances in China, totaling 276 participants. Our analysis based on those game instances in which one of the bidders was chosen (and in which corruption or bribery had a chance to occur according to our definition).[1] In all countries subjects were students enrolled in a university or college degree program. Each participant was given an identical 30 minute tutorial on the Olympic game which consisted of a self-guided presentation. Participation in the study was contingent on passing an on-line quiz about the game. There were at least 8 subjects in the lab at any given session (playing two consecutive games), and participants could not see others' terminals or communicate to each other in any way outside of the computer interface. All participants were paid a constant sum for their participation (about \$10) plus a bonus that depended on their performance in the game (an additional \$2-\$7). The participants were randomly allocated to their respective roles in the game. All games comprised four rounds, but this information was not made public to the subjects.

We defined two classes of covert activities in the game. Corruption was defined as the case in which the auctioneer did not choose the highest bidder (or one of them in case of ties). Bribery was defined as a special case of corruption in which the auctioneer accepted resources from a bidder who was subsequently chosen to win the contract despite not submitting the highest bid. We hypothesized that the occurrence of corruption will closely follow the Corruption Perception Index level in each country. Thus, we expected China to exhibit the highest level of corruption and bribery in the game, followed by the Israel and the U.S. In addition, we hypothesized that the number of instances of corruption will increase with the number of rounds played in the same game. Lastly, we expected that corruption and bribery will result in diminished profit for the government, and increased revenue for the auctioneer and bidder. All results

[1] There was no bidder chosen for a contract only in a minority of cases, less than 8 games in each country.

Table 1. Corruption and bribery measures for all countries

	Bribery	Corruption
U.S.	19%	33%
Israel	14%	29%
China	**46%**	**56%**

reported in this section were verified for statistical significance in the $p < 0.05$ range using appropriate ANOVA, chi-square and t-tests.

To illustrate a corrupt exchange in one of the games, consider the following discourse fragment between the auctioneer (A) and bidders (B1, B2, B3) in one of the games played in the U.S: A to B1: "wanna send up some chips? B3 sent me some." B2 to A: "Do you send me some back?" A to B2: "I make the decision on the final choice of bidding, so no". B2 to A: "Or is this just a bribe? haha" In this game, B2 sent the auctioneer a bribe of 24 chips and was chosen as the winning bidder despite submitting the lowest bid.

Table 1 shows the percentage of corruption and bribery activities in the game. These results partially follow the Corruption Perception Index in that China was the country that exhibited the largest amount of corruption. However, we measured lower level of corruptions in Israel than in the U.S., despite the higher corruption level index in Israel.

There was no significant correlation between the number of rounds played and the occurrence of corruption in any of the countries. However, in all countries, the number of corruption cases increased significantly from round 1 to round 4. Specifically, the corruption ratio increased from 25% to 57% in the U.S.; from 21% to 46% in Israel; and from 37% to 67% in China.

Figure 2 compares the performance (measured by average score) of the government ("gvrn"), auctioneer ("acnr") and the winning bidder ("bdr") for those games that exhibit and do not exhibit corruption in the three countries. (The results for bribery activities followed the same trend and are not reported due to brevity concerns). As shown by the figure, the auctioneer significantly benefited from corruption in the U.S., but not in China and in Israel (there was no significant difference in the score to auctioneer with and without corruption in these countries). In addition, while winning players significantly benefited from corruption in Israel and in China, they did not benefit from bribery in the U.S.

To explain this result, we measured the average number of chips sent by the winning bidder, and the number of times the winning bidder defaulted on its contract (was not able to complete its task) when corruption existed. We found that the average number of chips sent in China (11.79) was highest, followed by the U.S. (7.14) and Israel (4.2). In addition, we found that there were 6 instances of failed contracts following corruption in the U.S. as opposed to 3 instances in China and only 1 instance in Israel. Such failed contracts occurred because the winning bidder did not retain enough chips to be able to complete its task.

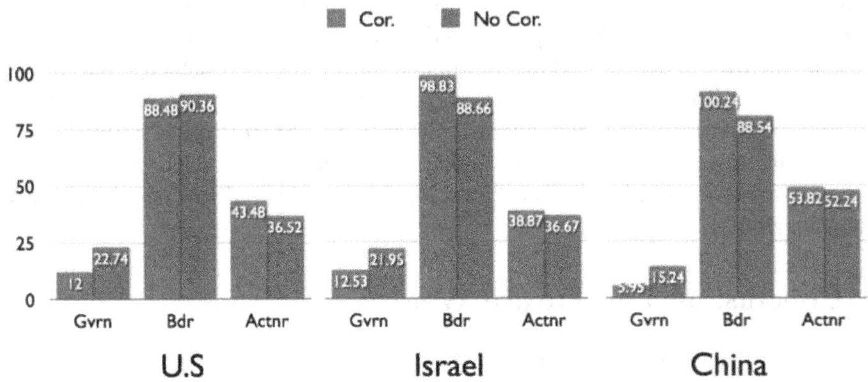

Fig. 2. Average scores for participants in the 3 countries

Table 2. Modeling corruption overall and across each of the three countries with all attributes and exclusively with the StateProfit value

	All – Acc.	Recall	China – Acc.	Recall	Israel – Acc.	Recall	U.S. – Acc.	Recall
Corr. (all)	77.54%	0.75	72.62%	0.83	75.82%	0.37	74.26%	0.59
Corr. (profit)	**80.07%**	**0.78**	**77.38%**	**0.89**	**79.12%**	**0.48**	**76.23%**	0.56

4 A Machine Learning Model of Corruption

One goal of this study was to quantify and predict when corruption occurs. In constructing our models, we considered all game specific data and general demographic attributes. Game specific information included: the initial position of each bidder, the round number, the board configuration, and the distance of that bidder from the goal. We also considered the known outcome of the game: how much the state profited from the winning goal. Demographic information included the bidders' age, sex, and country. We intentionally did not consider private information relating to messages and chip exchanges or identity of winning bidders. We employed a standard decsision tree classifier using ten-fold cross validation.

We first considered a cross-cultural model for corruption, the results of which are found in left-most section of Table 2. In constructing these results, we observed that the decision tree for all of the data in predicting corruption had the following rule: If StateProfit $<= 15$, then a series of three rules involving the distance of Bidder1 from the goal, the Auctioneer's gender, and Bidder2's age, otherwise, if StateProfit > 21 then there is no corruption, otherwise again a series of rules involving Bidder2's age and Bidder1's distance from goal.

We found that the country of origin attribute did not constitute a main attribute within the decision trees. This was noted from the absence of this attribution from the output of the decision trees, implying that the rules primarily based on StateProfit are independent of the 3 countries we considered.

However, as the previous section documented, we did observe that differences exist across cultures. Thus, we postulated that explicitly creating decision trees for each culture might yield additional insights, the results of which are found in columns 3–8 of Table 2. These results show, as was the case previously, that the accuracy of corruption models exclusively with the StateProfit attribute were more accurate and usually yielded higher recall than those with all attributes (these instances are bolded within the table). This again confirms the significance of this attribute.

5 Conclusions and Future Work

This paper presented a new empirical framework to investigate corruption. This novel setting allows us to study people's actions without biasing their actions through explicitly including corruption as a factor within the game, or merely noting their subjective feelings about corruption. This work has implications for both the social and computational sciences. It presents a new tool for studying corruption in the lab. We do not claim that corrupt behavior in the game predicts the occurrence of corruption in real life. We *do* claim that covert activities occur in a more realistic manner than in traditional studies. Our test-bed can be used to study the effects of different mechanisms and policies on the evolution of corruption over time. It opens the door for designing computer agents (whether by researchers or students [3]) that play the game and attempt to reduce corruption by adopting different strategies, based on machine learning methods.

Acknowledgements. This research is supported in part by ERC grant #267523.

References

1. United nations handbook on practical anti-corruption measures for prosecutors and investigators (2012)
2. Abbink, K., Irlenbusch, B., Renner, E.: An experimental bribery game. Journal of Law, Economics, and Organization 18(2), 428–454 (2002)
3. Chalamish, M., Sarne, D., Lin, R.: The effectiveness of peer-designed agents in agent-based simulations. Multiagent and Grid Systems 8(4), 349–372 (2012)
4. Falk, A., Fischbacher, U.: "Crime" in the lab-detecting social interaction. European Economic Review 46(4), 859–869 (2002)
5. Fehr, E., Gächter, S., Kirchsteiger, G.: Reciprocity as a contract enforcement device: Experimental evidence. Econometrica 65(4), 833–860 (1997)
6. Fehr, E., Kirchsteiger, G., Riedl, A.: Does fairness prevent market clearing? an experimental investigation. The Quarterly Journal of Economics 108(2), 437–459 (1993)
7. Gal, Y., Grosz, B., Kraus, S., Pfeffer, A., Shieber, S.: Agent decision-making in open mixed networks. Artificial Intelligence 174(18), 1460–1480 (2010)
8. Transparency international. Corruption perceptions index (2011)
9. Treisman, D.: The causes of corruption: a cross-national study. Journal of Public Economics 76(3), 399–457 (2000)

A Study of Mobile Information Exploration with Multi-touch Interactions

Shuguang Han[1], I-Han Hsiao[2], and Denis Parra[3]

[1] School of Information Sciences, University of Pittsburgh, Pittsburgh, PA, United Sates
shh69@pitt.edu
2 The EdLab, Teachers College Columbia University, New York, NY, United Sates
ih2240@columbia.edu
[3] School of Computer Science, PUC Chile, Chile
dparra@ing.puc.cl

Abstract. Compared to desktop interfaces, touch-enabled mobile devices allow richer user interaction with actions such as drag, pinch-in, pinch-out, and swipe. While these actions have been already used to improve the ranking of search results or lists of recommendations, in this paper we focus on understanding how these actions are used in exploration tasks performed over lists of items not sorted by relevance, such as news or social media posts. We conducted a user study on an exploratory task of academic information, and through behavioral analysis we uncovered patterns of actions that reveal user intention to navigate new information, to relocate interesting items already explored, and to analyze details of specific items. With further analysis we found that dragging direction, speed and position all implied users' judgment on their interests and they offer important signals to eventually learn user preferences.

Keywords: Multi-touch interactions, implicit relevance feedback, mobile information seeking behaviors.

1 Introduction

The massive user adoption and the rapid improvements of mobile technologies have attracted researchers from academia and industry to investigate how people use mobile devices compared to the traditional desktop or laptop computers. For instance, several studies comparing user behavior during web search tasks [1,3,4] have found differences between desktop and mobile devices in terms of query length, user click patterns, and search time distribution. Other researchers have leveraged location information - which is more easily captured in mobile devices - to investigate how people search for near restaurants or tourism landscapes [5,6,7]. Touch-enabled devices provide additional features that can be used to augment the user search experience, such as drag, pinch-in (zoom in), pinch-out (zoom out), and swipe.

Though search is an important activity, there are other important activities such as information exploration. In search tasks, users usually have specific information needs and the search results are ranked by their relevance. However, in other user

W.G. Kennedy, N. Agarwal, and S.J. Yang (Eds.): SBP 2014, LNCS 8393, pp. 269–276, 2014.

activities such as reading of news or social media posts users mostly lack a clear goal and the items are commonly displayed in lists sorted by chronological order. Therefore, we argue that in this scenario the multi-touch interactions can be exploited to improve the user experience but they cannot be interpreted in the same way as in traditional search tasks, where the top results are assumed to be the most relevant ones. In this article, we investigate how users behave in multi-touch enabled mobile devices while exploring information, how differently they behave compared to search tasks, and more importantly, whether we can find a relation between multi-touch interactions and users' interests.

2 Related Works

To display the same amount of information as in desktop computers, users in mobile devices need to perform additional actions (e.g., page scrolls, zooms) because of the limited screen size. Implicit user feedbacks can be utilized to improve information filtering. Page dwell time [9] and mouse cursor movements [8, 10] have been already identified as important implicit feedback on desktop devices. Besides them, multi-touch feedback has been also found to be able to generate significant improvements [2] on ranking search results. For example, the zooming-in behavior may suggest that users are interested in the targeted content block, whereas the fast swiping behavior may indicate that they are only "skimming" the non-relevant content. Implicit feedback is also used to generate recommendations. For example, Hu et al. [11] introduced the concept of "confidence" in a matrix factorization model for weighting the implicit feedback, and Pan et al. [12] further considered weighting both positive feedback and negative feedback, treating it as One Class Collaborative Filtering problem.

To the best of our knowledge, implicit feedback is mostly used in information retrieval or recommendation tasks where the items are ranked by relevance to the users' queries or interests. Moreover, the traditional reading patterns over listed results learned from information retrieval systems, e.g., the F-Shape [14], are no longer useful when analyzing exploration tasks over items not sorted by relevance but rather chronologically such as a list of news or social media posts. We argue that finding the most relevant items from a list of search or recommendation results is significantly different from reading a list of items for exploration. Therefore, in this paper, we are interested in studying further implicit feedback in mobile devices: understand user behavior during information exploration, interaction patterns while performing this activity, and how those patterns can be used to infer users' interests.

3 Research Design

3.1 Task and Data Collection

The task was designed to resemble the reading of news or social media posts. However, we chose a task of academic content by displaying a list of scientific publications because: i) news and social media posts are usually augmented with images, videos or external links, which introduce variables that will influence user's preference on our analysis; and ii) we believe that a mobile application that supports exploration in the academic domain can generate future impact on scientific communication while it is still rare recent years. In this study, we focused on the computer science domain. Since the academic conferences have evolved to be the major channel for the dissemination of research achievements in this domain, the publications for this user study were chosen from three conferences in Information Retrieval (SIGIR), World Wide Web (WWW) and Human Computer Interaction (CHI). To minimize the bias that some users may have already read those publications, we considered the publications of those conferences during 2013, which are available online only less than three months before the time of our study. We have 739 publications (392 from CHI, 137 from SIGIR and 210 from WWW). The title, abstract, author and affiliation information was displayed in the user interface.

3.2 The Experimental System: ConfReader

We developed a mobile web application called ConfReader[1]. Fig. 1 presents a screenshot of the article list page, which provides a list of articles with title and briefed abstracts up to 180 characters. By quickly swiping left or right, users can bookmark an article, as shown in Fig. 3. By tapping an article, users will see the detailed information with full title, abstract, author and affiliation information, shown in Fig. 2. All the pages have a navigation bar fixed as a footer in order to facilitation navigation to the article list page or check the number of articles being bookmarked.

ConfReader is able to log the following multi-touch interactions on a mobile device: hold, tap, double-tap, drag up, drag down, drag left, drag right, pinch in, pinch out. Each interaction is defined as starting from touching the screen until the user releases the fingers. We implemented the functionalities based on an open source Javascript library for multi-touch gestures, the hammer.js[2].

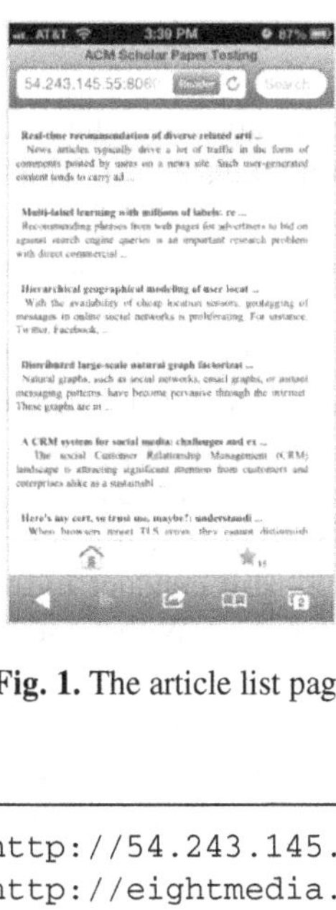

Fig. 1. The article list page

Fig. 2. The detail page

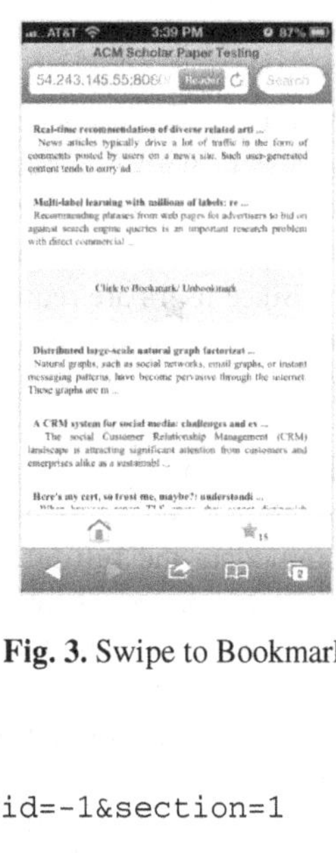

Fig. 3. Swipe to Bookmark

[1] http://54.243.145.55:8080/acmmw/index.jsp?uid=-1&section=1

[2] http://eightmedia.github.io/hammer.js

3.3 Participants and Procedure

Our target participants were PhD students majoring in computer or information science, considering that our data collection is from three conferences in computer science related domains. In total, 15 PhD students from 5 universities in China (2) and United States (3) were recruited. 9 are female and 6 are male.

The experiment began with an introduction to the study. Then, participants were given a 5-minute tutorial about the system, to continue with a training task around 10 minutes and to be familiar with the system on one of the 3 conferences. After the training, each participant was asked to finish three tasks in up to 10 minutes for each, each task corresponding to one conference. The order of those three tasks (conferences) was rotated based on the Graeco-Latin Square design. For each user task, we randomly assigned a list of 40 papers from one conference. They were asked to explore all those articles to interact with our experimental system, and to choose the top ten most relevant articles based on their personal interest. All of the four tasks were completed using our system with no interventions by the experimenters.

4 Result Analysis and Discussions

4.1 Descriptive Statistics on Multi-touch Interactions

In the study, we collected 3,519 multi-touch interactions (3,041 from article list pages and 478 from detailed pages) from 15 users. Within the given 30 minutes for exploration, the actual-time-on-task to interact with the article list pages spent 25.62±(18.83) minutes, which is double of the time spent on detailed pages, with only 13.44±(10.88) minutes. This is understandable due to the list pages had already provided a short snippet of the abstracts, users will tap into detailed pages only if they want to confirm. The action percentage of each page is presented in Table 1.

As for actions, users seldom used the pinch-in, pinch-out or double-tap. The result is consistent with the study of web search behaviors on mobile devices [2]. This is either due to the interface displayed proper font size and layouts resulted in smooth exploring process or because user has no preference on performing multi-touch behaviors. Since users are required to drag left/ right to bookmark articles in the article list pages, it was not surprising that dragging left/ right take high proportion in the article list page but not in the detailed pages. The actions are dominated by the tapping (39.12% in detail page) and dragging up/down (44.03% and 14.8% in the list page). In the detailed pages, users perform more tapping. In the detailed pages, users have more tapping. We think it may because most of the detailed pages are actually able to accommodate the full information in one page; however, users still conduct further interactions for reconfirming in case they missed some hidden content. Without scrollable content, the actions were counted as tapping. The dragging down behaviors in the article list page take around 15% of all actions which demonstrated a dynamic exploring process instead of linear browsing that goes from top to bottom. However, the deeper implications of dragging up and dragging down are still unclear, which are the focuscs discussed in the following sections.

Table 1. Percentage of users' multi-touch interactions on different pages

	Article list page	Detail page
Drag up	44.03%	40.17%
Drag down	14.80%	6.28%
Drag left	6.61%	5.02%
Drag right	22.73%	7.95%
Tap	11.05%	39.12%
Pinch in/ Pinch out/ Hold/ Double-tap	0%	0 %

4.2 Dragging Down vs. Dragging Up

In our system, each logged interaction consists of a set of metadata, such as the distance (distance has moved on screen), centerX/Y (the X/Y axis positions when a touch begins), velocityX/Y (the touch speed on X/Y axis). To obtain a deeper understanding of the dragging up/ down actions on both the article list pages and the detail pages, we analyzed those metadata. Table 2 shows the statistics of the metadata information and the comparison between dragging up and down. We found that when dragging down on the article list page, users tend to put fingers at a higher position (in Y axis) comparing to the dragging up. It makes sense because a higher position allows more distances for dragging. Indeed, the dragging up in article list page has a longer average distance, although no significance is found. Since the article list pages contain more content than in detailed pages, the centerY is always bigger. A fair comparison may use the relative position in Y axis; however, our system cannot log such information because the limited API functionality from the open-source library. The reported mean of centerX is near the third quarter position of the screen.

In terms of velocity, the dragging down are significantly faster than dragging up, which suggests that users are more likely to exploring information when dragging up. While in dragging down, they are only skimming or relocating information to what they recalled from memory. We think that the consistent slow dragging up gesture may demonstrate users' attentive behavior in identifying articles for their interests, which we will test in the next section. We didn't compare between two pages because there are only few explorations on detailed pages. For several users, there are even no dragging up/ down in detailed pages, from which we cannot make fair comparisons.

Table 2. The comparsion of gesture metadata information (with standard deviation) for dragging up/ down on different pages. * means significance at 0.1 level, ** means 0.05 level and *** means 0.01 level. Significance tests are based on Generalized Linear Models (GLMs).

	Article list page		Detailed page	
	Drag down	Drag up	Drag down	Drag up
centerX	571 ± (240)	554 ± (266)	583 ± (227)	537 ± (284)
centerY	**4676 ± (2861)***	4357 ± (2690)	917 ± (425)	982 ± (332)
velocityX	**0.382 ± (1.721)****	0.168 ± (0.322)	**0.319 ± (0.496)*****	0.170 ± (0.496)
velocityY	**1.294 ± (4.660)****	0.760 ± (1.623)	**0.899 ± (1.244)***	0.610 ± (1.354)
distance	65.78 ± (57.46)	60.51 ± (39.09)	51.49 ± (43.32)	53.70 ± (34.32)

4.3 Inferring Users' Interests from Dragging Up/Down

Dragging up/down is the dominating behavior in both article list and detail pages. In mobile devices, dragging up/down served the same functions as mouse scrolling up/down in desktop, except that users need to tap on a certain item to move up/down. When users drag on a specific item, we assume that there are N surrounding articles on users' reading zone. N is a small value considering the small screen size. As an initial step, we set N = 1 and assume that users will drag quickly if they are not interested in those articles; otherwise, they will slow down and read carefully. We would like to test whether there is a significant correlation between the dragging speed (on Y axis) and user interests.

Each user was required to bookmark articles on three different conferences, when studying users' interactions on one conference, the bookmarked articles on the other two are used to model users' real interest. User interest **I** is represented as a vector that aggregates bookmarked documents (**B**) using vector space model over all of the words in vocabulary, i.e. $\mathbf{I} = \langle w_1, w_2, \dots, w_{|V|} \rangle$. $|V|$ is vocabulary size. Weight for each word in each document w_{ij} is represented as Formula (1), which denotes the TF-IDF for word w_i in d_j. To test the assumption, we calculate the cosine similarity between $\mathbf{D_r}$ / $\mathbf{D_{r+1}}$ / $\mathbf{D_{r-1}}$ (i.e. the article users dragged on, the article above the dragged article and the article below) and user interest **I.** The similarity is used to measure the interestedness of the dragging area.

$$w_i = \sum_{d_j \in \mathbf{B}} \frac{w_{ij}}{|\mathbf{B}|} \quad (1)$$

Our system logged the dragging speed for each drag up/down action. Thus, we can compute the Pearson correlation between dragging speed (we use a log transformation because the speed distribution is highly skewed) and the interestedness of the dragged area. The correlation coefficients are -0.01 (no significance) for dragging down while **-0.1 (*p*<0.01)** for dragging up, which indeed provides evidence to our assumption that the dragging speed is reduced when users read content similar to their interest. In summary, this result can indicate that users mostly drag up to explore new items, whereas they drag down to relocate information relevant to their interest.

4.4 Predicting Users' Bookmarks Using Rich Interactions

With the goal of producing personalized recommendations, we aimed at evaluating whether identifying user interest based on touch interactions is comparable with identifying it based on bookmarks. In our study, each user needs to read a list of articles of 3 conferences and bookmark 10 for each. Using the logged actions at each conference, we performed an evaluation simulating an online recommender. To explain the protocol, let's assume a user **u1** is exploring WWW conference and at the moment of bookmarking each article, we produce recommendations based on the logged actions for CHI conference. The evaluation protocol is described as following:

- Given the user **u1** and the conference WWW, we follow the actions in sequential order and keep track of each article tapped in a vector **I_t**, and each article dragged up/down (and its immediate upper and lower articles in the list) in a vector **I_d**.

- When **u1** bookmarks a paper we add that paper to the vector **I_b**, and then we generate three sets of recommendations from the papers in the CHI conference based on their similarity to the user interests. The first recommended list **R_b** is generated using **I_b** (Pred_book), the second recommended list **R_t** using the tapped articles **I_t** (Pred_tap), and the third list **R_d** using the dragged articles **I_d** (Pred_drag). Since drag speed is negatively correlated with user interests, Pred_drag filters out those drags with high speed (speedY>2.0).
- The papers that the user **u1** bookmarked in CHI will be used as ground truth for evaluating the algorithms. We calculate the precision at 10 (P@10) and also Average Precision (AP) for each recommended list **R_b, R_t** and **R_d**.

We repeat the three previous steps for each article bookmarked at each conference, completing 10 evaluations, which are each of the ticks in the x-axis of Figure 4 and Figure 5. We also include a random prediction (Pred_Rand) by randomly guessing the top 10 recommendations. Since only 40 articles were displayed for each conference, the random baseline already achieved high performance. Comparing to the random baseline, the Pred_tap only improves the performance after the 3rd or 4th bookmarks, which is due to the data sparseness. Users may result in *judging-a-book-by-its-cover* even if they were interested in the article. Pred_drag has a smaller sparseness problem and works better than using the Pred_tap, which confirms the value of multi-touch interactions. The best performance is achieved by Pred_book, which considers users' explicit feedbacks. Though explicit feedback is indeed more valuable than implicit feedback, the former is usually scarcer in real systems. Given our results, we argue that multi-touch actions in mobile devices offer a promising alternative for non-invasive preference elicitation. We also observe that precision curves for both Pred_tap and Pred_drag have a convex shape, which indicates that implicit feedbacks might introduce noise when aggregated over time. Leveraging each type of implicit feedback and finding a better tradeoff among them are our next research focus.

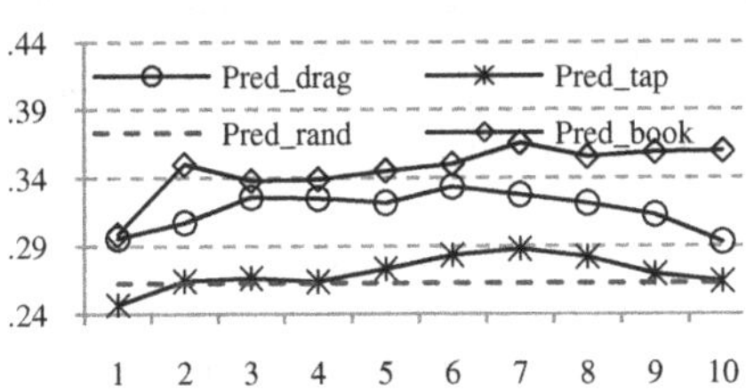

Fig. 4. The P@10 evaluation on the i-th (x-axis) bookmarked paper

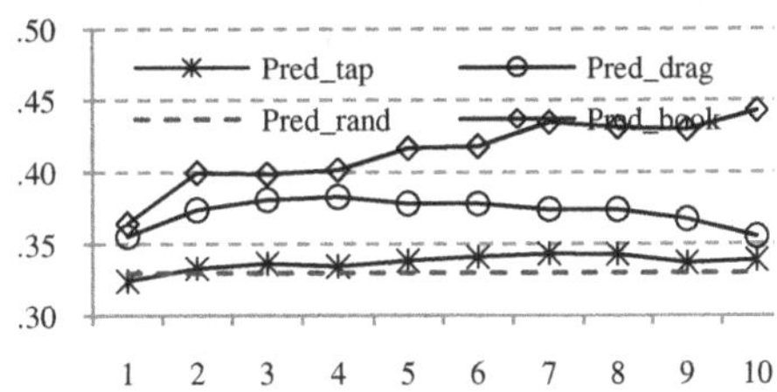

Fig. 5. The AP evaluation on the i-th (x-axis) bookmarked paper

5 Conclusion

In this paper, we studied the mobile information exploration behaviors through a controlled user study, in which users were asked to explore and bookmark a list of conference publications. Users tended to read and explore information when dragging up but relocating information when dragging down. We also found that dragging up/down actions' speed and position can be used to infer user interest over articles and also as additional feedback to predict users' future actions.

However, there are still several open issues to incorporate multi-touch interactions in predicting user interests. First of all, in this paper, we assumed that users were reading surrounding articles (the one being dragged, one upper and one lower). Follow up studies are needed for justifying this assumption, analyzing user's reading zones through eye-tracking. Secondly, we considered actions isolated. The drag down and tap actions could be combined with drag up actions for further improving the recommendation performance. For example, a slowly drag up with follow-up tapping on articles may further confirm user interest. Thirdly, with more exploration on the displayed articles, user's interests may change dynamically, particularly in the exploratory task where user lacks a clear goal. A possible solution is to consider weight decay on articles being explored based on temporal information, a recent explored article may receive more weight than the old one.

References

1. Song, Y., Ma, H., Wang, H., Wang, K.: Exploring and exploiting user search behavior on mobile and tablet devices to improve search relevance. In: WWW, pp. 1201–1212 (2013)
2. Guo, Q., Jin, H., Lagun, D., Yuan, S., Agichtein, E.: Mining touch interaction data on mobile devices to predict web search result relevance. In: SIGIR 2013, pp. 153–162 (2013)
3. Yi, J., Maghoul, F., Pedersen, J.: Deciphering mobile search patterns: a study of Yahoo! mobile search queries. In: WWW 2008, pp. 257–266. ACM (2008)
4. Kamvar, M., Baluja, S.: Deciphering trends in mobile search. Computer 40(8), 58–62 (2007)
5. Ricci, F.: Mobile recommender systems. Information Technology & Tourism 12(3), 205–231 (2010)
6. Brunato, M., Battiti, R.: Pilgrim: A location broker and mobility-aware recommendation system. In: PerCom 2011, pp. 5265–5272. IEEE Computer Society (2003)
7. Dean-Hall, A., Clarke, C., Kamps, J., Thomas, P., Voorhees, E.: Overview of the TREC 2012 Contextual Suggestion Track. In: Proceedings of the 21st NIST Text Retrieval Conference (2012)
8. Huang, J., White, R., Buscher, G., Wang, K.: Improving searcher models using mouse cursor activity. In: SIGIR 2012, pp. 195–204. ACM (2012)
9. Morita, M., Shinoda, Y.: Information filtering based on user behavior analysis and best match text retrieval. In: SIGIR 1994, pp. 272–281 (1994)
10. Kong, W., Aktolga, E., Allan, J.: Improving passage ranking with user behavior information. In: CIKM 2013, pp. 1999–2008. ACM, New York (2013)
11. Hu, Y., Koren, Y., Volinsky, C.: Collaborative filtering for implicit feedback datasets. In: Eighth IEEE International Conference on Data Mining, ICDM 2008. IEEE (2008)
12. Pan, R., Zhou, Y., Cao, B., Liu, N.N., Lukose, R., Scholz, M., Yang, Q.: One-class collaborative filtering. In: Data Mining, ICDM 2008, pp. 502–511 (2008)
13. Manning, C.D., Raghavan, P., Schütze, H.: An Introduction to Information Retrieval, p. 181. Cambridge University Press (2009)
14. Lorigo, L., Haridasan, M., Brynjarsdóttir, H., Xia, L., Joachims, T., Gay, G., Pan, B.: Eye tracking and online search: Lessons learned and challenges ahead. Journal of the American Society for Information Science and Technology 59(7), 1041–1052 (2008)

Dyadic Attribution: A Theoretical Model for Interpreting Online Words and Actions

Shuyuan Mary Ho*, Shashanka Surya Timmarajus, Mike Burmester, and Xiuwen Liu

Florida State University
smho@fsu.edu, st13f@my.fsu.edu, {burmeste,liux}@cs.fsu.edu

Abstract. This paper presents a theoretical model for interpreting the underlying meaning in virtual dialogue from computer-mediated communication (CMC). The objective is to develop a model for processing dialogues and understanding the meaning of online users' social interactions based on available information behavior. The methodology proposed in this paper is built on a demonstrated observation that humans – in analogy to "sensors" in social networks – can detect unusual or unexpected changes in humans' trustworthiness based on observed virtual behaviors. Even with limited resources such as email, blogs, online conversations, etc., humans "sensors" can infer meaning based on observed behaviors, and assign attributes to certain words or actions. The idiosyncratic nature of human observations can be arbitrated by an attribution mechanism that provides the basis for a systematic approach to measuring trustworthiness. In this paper, we discuss a particular trust scenario called the Leader's Dilemma with the objective of identifying how anomalous online behavior can be interpreted as untrustworthy. We adopt the dyadic attribution model to analyze how a human disposition can be systematically uncovered based on words and actions, as evidenced by information behavior. This model is better suited for computational analysis of attribution engines. The novel goal of this research is to design a sensor system with the ability to attribute meaning to virtual interactions as supported by computer-mediated technologies.

Keywords: Sociotechnical system, information behavior, psychological theory.

1 Introduction

In modern society, information is commonly exchanged and posted online. People communicate naturally with each other using computer-mediated technologies. In the physical world, communication and interaction among social actors primarily relies on facial expressions, body gestures and eye contacts, which reflect thoughts expressed through words. Online communication is often studied with fMRI through social neuroscience and psychology to examine intent and motivation by analyzing human brain reactions in trust and deceitful situations [2, 10, 11]. However, in physical interactions as well as in the virtualized environment, an individual's thoughts, intents and actions are expressed through their language, written or spoken.

* Corresponding author.

W.G. Kennedy, N. Agarwal, and S.J. Yang (Eds.): SBP 2014, LNCS 8393, pp. 277–284, 2014.

Simply put, language is a composite of words; and words are a vehicle that expresses human information behavior. Words, whether spoken or written - whether monolog or dialogue - can convey a person's thoughts and intents at a fine-grained level. In order to understand an individual's information behavior online, we need to first understand ways to interpret the words they use.

Words are always associated with meaning. We can often deduce a person's commitment and intentions based on words. In this paper, we examine the results of one game scenario; the Leader's Dilemma game [4], which was played out by a group of online participants. Small group interactions with a Team-Leader and four members in virtual team arrangements were encouraged using the learning management system (LMS) at a university. Participants were geographically dispersed across the United States. As part of the game, the Team-Leader encounters an ethical dilemma; an unexpected award, in the form of micropayment, is presented by the Game-Master, an authoritative trusted party. The Team-Leader has to decide whether to distribute this reward to the team, or keep it for his/her own personal gain. With the data that was collected during game interactions, we intend to dissect the nuances in the words and actions exchanged by participants. Below is one excerpt extracted from the team's interactions. Team members, inadvertently acting as human sensors, were carrying on conversations with the Team-Leader. The highlighted grey areas in Table 1 are further analyzed in Table 4.

Table 1. An excerpt of conversations from the online Leader's Dilemma game

3:52 PM: **Ricky Player**: dude just divide equally 3:52 PM: **Ricky Player**: and save the trouble 3:52 PM: **Ashley Player**: but team leader sometimes you don't submit answers 3:53 PM: **Ashley Player**: a mutiny! hahahaha 3:53 PM: **Ricky Player**: but TL you have a very rong idea 3:53 PM: **Team-Leader Dragon**: i do not submit them because of u 3:53 PM: **Ricky Player**: wat is tat supp to mean 3:53 PM: **Ricky Player**: u get the answers even if wrong u are supp to sub,it 3:54 PM: **Ashley Player**: yes 3:54 PM: **Team-Leader Dragon**: we run out of time then we miss it 3:54 PM: **Ashley Player**: no, sometimes you don't submit when we say to 3:54 PM: **Ricky Player**: v find answers ur job is to keep track of time	3:54 PM: **Ashley Player**: and you should lead 3:54 PM: **Ashley Player**: yes 3:54 PM: **Team-Leader Dragon**: see tomorrow 3:54 PM: **Ashley Player**: ok 3:54 PM: **Team-Leader Dragon** has left the room. 3:55 PM: **Ricky Player**: oh man he got pissed 3:55 PM: **Ashley Player**: yeah 3:55 PM: **Ricky Player**: so waddup 3:55 PM: **Ashley Player**: hahaha 3:55 PM: **Ashley Player**: are we supposed to revolt? 3:55 PM: **Ricky Player**: i don no but he did not make sense 3:55 PM: **Ashley Player**: take over the game! hhahahaha 3:55 PM: **Ricky Player**: wat with bad evaluation 3:55 PM: **Ricky Player**: tats wat they are meant for 3:55 PM: **Ashley Player**: how we evaluate is our own business 3:56 PM: **Ashley Player**: of course in a real world situation, we'd be screwed 3:56 PM: **Ricky Player**: poor chap u put lot of pressure on him ;)

In the conversation reflected in Table 1, Ricky Player initiated a discussion about the sharing of the reward points being equal across the team. Ashley Player adds to the debate by saying that Team-Leader doesn't submit the points equally among the team. The debate continues even after Team-Leader leaves the forum. While Ricky Player showed sympathy for Team-Leader, Ashley Player complained about the Team-Leader by using the words "mutiny" and "wrong idea". The words "do not" and "because of u" from the Team-Leader asserted his/her level of management confidence, while holding the team responsible for an unspoken, unspecified reason. The words "poor chap" by Ricky Player expressed sympathy for the Team-Leader. These conversations demonstrate that words used in a virtual environment are valuable information as behavioral evidence for aggregated evaluation of human emotions and consensus-building towards the Team-Leader.

In order to understand how human factors play out in the context of their online interactions, it is critical to model the computational structure of the linguistic patterns and the lexical meaning of words as they relate to actions [7]. In this paper, we seek to contribute to a design for a computational model that captures a social actor's words and actions from their virtual communications for trust attribution.

2 Words and Intention in Human Interactions

Austin [1] stated that speech is a form of action, indicating that language as a window can reveal one's intimate "feelings," "thoughts," or "intentions." The social actor's use of words in context gives us an understanding of their actions. Austin speaks about two types of social actors' utterances, which shape their actions. The first one, performative utterance is stated as "the issuing of the utterance" is equal to "the performing of an action." In the above example, when Ricky Player says, "v find answers ur job is to keep track of time", it presupposes that the players are tasked with finding the answers and argues that the Team-Leader is supposed to submit the answers. The argument here is an example of performative utterance. The second one, constative utterance refers to a descriptive statement. For example, when Ashley Player says "a mutiny! Hahahaha" this describes her interest in taking over the game.

Words indicate the degree of a speaker's commitment towards a receiver's expectation, and any associated actions represent translation of these words into reality [3]. Words are a driving force behind the understanding of a social actor. The grouping and phrasing of the words is also of prime concern in understanding an actor. Words are the essence of text-based computer-mediated communication. In a CMC-enabled interacting environment, actor A acts while observers B_i observe. Observers B_i usually observe the distinctive patterns of actor A's words, the consistency between actor A's words and actions, and other interactions. Over time, observers B_i develop a feeling of trust towards the actor A. In our previous example, the Team-Leader did not commit to submitting team's answers on time, which initiated a sense of distrust across the team members.

People's intent is often communicated through natural language. Language used among social actors formulates a complex network of linguistic and syntactic expressions. Sometimes it may be possible to determine an actor's trustworthiness by analyzing the relationship between words used and actions taken. Table 2 describes a set of simple rules regarding comparisons of actor A's words and actions in the virtual communication between actor A and observers B_i. We can consider an actor A, who communicates with other social actors B_i and these observers B_i draw a conclusion based on messages received from actor A. A consistent match between an actor's words and actions can then be used to infer that actor A is trustworthy. Similarly, if the actor's words and actions do not correspond over time, then he may not be considered trustworthy. The actor may not be considered trustworthy if he/she promises to do something, but does not follow through with associated action. In the case where the actor promises less than his/her actions, it does not provide sufficient information for observers B_i to make a judgment. An act can be considered positive or negative depending upon the circumstances and actions taken. Any single act of non-correspondence can be considered as distinctive from normal behavior. In our previous example, at the moment when the Team-Leader decides not to submit the answers provided by players, the perceived trustworthiness of the Team-Leader (as actor A) was reduced.

Nevertheless, there is no assurance as to how precisely human intent can be fully expressed in this confined world of language. Even when the intent can be interpreted to a certain degree, there is no definite pattern of communicative cues in the language that is statistically generalizable [8]. Based on this, we can say that it is very hard to know the behavior of an actor without knowing the actors true intent. Without knowing the true intent, it's always hard to narrow down why an actor acts in a certain way.

Table 2. A simple illustration of behavioral parameters determining trustworthiness in the Leader's Dilemma game

Scenarios				
Actor A's behavior	Words = Actions	Words ≠ Actions	Words > Actions	Words < Actions
Observer B's Perception	Trustworthy	Questionable/not trustworthy	Questionable/not trustworthy	Not enough indication (positive vs. negative)

3 Trustworthiness Attributions

Trust can be built among observers over time based on the words and actions of a social actor in virtual communications. When actor A performs, observers B_i can attribute actor A's trustworthiness. Trust is built up among observers based on the consistency between words and actions of a social actor in a virtual communication. Rotter [14] defined trust as a generalized expectancy—held by an individual or a group—that the communications of an individual or group can be relied on regardless of situational specificity. Trust depicts a type of mutually dependent relationship between two individuals, two parties, or two groups: a trustor B_i, who may (or may not) trust a trustee A.

Similarly, *trustworthiness* refers to the generalized expectancy, concerning a degree of congruence between communicated intentions and behavioral outcomes as observed and evaluated, which remain reliable, ethical and consistent, as long as any fluctuation between perceived intentions and actions does not exceed the observers' expectations over time. Ho [5] stated that, trustworthiness refers to the inner characteristics or disposition of a focal actor, thus critical contextual information (e.g., situational context, the initiating agent's emotional state toward the receiving agent, the baseline or historical knowledge of each agent's response from prior communication, and the character as a whole, etc.) needs to be considered in any trustworthiness assessment situation.

Assessment of an actor's trustworthiness can be discerned by looking at language used in the virtual communication. In a world that is interpreted through language, action is also reflected in language. One of the ways by which we can infer an actor's disposition is to look for keywords that are characteristics of an attribute [8]. Consider the example where a social actor uses words such as "sorry," "apologize," "pardon," etc., when making a mistake. These keywords help us understand that the social actor follows ethics and accepts error. Integrity is a character trait, and it is likely to be dispositional. So, with the use of keywords, linguistic patterns can be constructed to model the behavior of an actor.

3.1 Internal vs. External Factors

Lieberman [12] provided two general sets of trustworthiness factors: competence (external or situational cause) and integrity (internal or dispositional cause). In general, competence is an external factor referring to the effective application of learned behavior. For example, an actor (Team-Leader from Table 1) can acquire external management skills. Integrity, on the other hand, is an internal (dispositional) characteristic, which refers to a person's internal dispositional condition. For example, suppose an actor (Team-Leader) is willing to sacrifice his/her own reward for the welfare of others.

In the above example, if the team pushes the Team-Leader to distribute the rewards equitably, the Team-Leader has two choices - to distribute them or keep them. If the Team-Leader agrees (in words) to distribute the rewards, but doesn't distribute all of them (action). This might demonstrate greed (internal causality) or insufficient rewards available (external causality). In the following section, we discuss attribution mechanisms to further the analysis regarding how observers make attributions.

3.2 Attribution Mechanisms

Attribution theory attempts to provide casual explanations for human behavior [9, 13], and reactions to that behavior. Because all human beings are of the same species and born with similar features and functions, a person should "know" and can "sense" from her or his own perceptions, using judgment to determine how the world operates [9] despite the fact that sometimes these attributions are not accurate.

Among many attribution theories, Kelley's [9] ANOVA model is considered one of the best known, which describes how individuals infer causes to a focal actor's behavior based on multidimensional observations of the focal actor's (person) behavior at different times (time) and circumstances (entity). The dimensions *person* × *time* × *entity* can be used to deduce the behavior of a person and to infer a cause of that behavior.

Additionally, three information types *consistency*, *distinctiveness*, and *consensus* can be used for behavioral analysis. Generally, the behavior of an actor is expected to remain *consistent* across time. But consistency in a virtual environment is defined as the extent to which an actor's words are consistent with associated actions. *Distinctiveness* refers to the extent to which the actor's response to a stimulus is distinctive over his/her responses to other stimuli. In a virtual environment, distinctiveness is defined as the extent to which an actor's actions are different as compared to generalized expectancy of his/her spoken words or promises over time. In Kelley's definition [9], *consensus* refers to the behavior of an actor being interpreted similarly across different people under the same stimulus. However, in a virtual environment, *consensus* refers to the observation or opinion and the extent to which members of a group are in agreement about an actor's behavior under typical circumstances over time. This is also called *group consensus*. In sum, using the above dimensions and three types of information, one can make attribution decisions about a person's behavior.

4 Dyadic Attribution Model

Ho and Benbasat [8] propose a dyadic attribution model to infer causes to a social actor's information behavior. Using the constructs defined in Kelley's attribution theory—and the causality of the actor—trustworthiness attribution mechanism can be built. Table 3 depicts the behavioral causal relationships among constructs using the dyadic attribution model.

Table 3. Behavioral casual relationships using a dyadic attribution model

Trustworthiness attribution	Cause	Consistency	Distinctiveness	Group consensus
Competent (usual)	External	High	Low	High
Integrity (ethical)	Internal	High	Low	Low
Competent (expected)	External	High	High	High
Innovative (unexpected)	Internal	High	High	Low
Incompetent (usual)	External	Low	Low	High
Unreliability	Internal	Low	Low	Low
Mistake (accidental)	External	Low	High	High
Betrayal	Internal	Low	High	Low

4.1 Example of Tagging and Attribution

Words can be tagged and analyzed at a granular level. This can be achieved by analyzing verbs in a sentence to deduce constative and performative tags. By using these tags, we can derive attribution of the three information types: *consistency*, *distinctiveness*, and *group consensus* for a given sentence. The first part of tagging focuses at the word level, while the latter part focuses at the sentence level, yet both are mutually inclusive. Once the identification of three types of information is done, we can deduce the actor's trustworthiness attribution using dyadic attribution model.

Table 4 shows how tagging can be done for the words used in Table 1. It is important to note that the degree of the emotion behind each sentence should also be captured. Different tags can identify words associated with a particular meaning, which help to identify the observers' trustworthiness attribution of the focal actor (Team-Leader). The tagging of words over time gives us an in-depth understanding about the behavior of the actor, and their intentions.

The degree of emotion associated with a particular word or sentence is an important indicator [6]. For example, the constative word "mutiny" indicates strong negative connotation of disagreement. The performative word "take over" indicates a strong degree of action towards the actor's intent to replace the management. This degree of emotion helps us to understand the intentions much better, thereby making a better attribution.

Table 4. Example of tagging to identify trustworthiness attribution through behavioral causes (words)

	Word Level		Sentence Level		
Category	**Constative**	**Performative/Promise**	**Consistency**	**Distinctiveness**	**Consensus**
3:52 PM: **Ricky Player**: dude just divide equally	--	<per>divide</per>	0	1	0
3:52 PM: **Ricky Player**: and save the trouble	--	<per>save</per>	0	1	1
3:52 PM: **Ashley Player**: but team leader sometimes you don't submit answers	--	<per>don't submit</per>	0	1	0
3:53 PM: **Ashley Player**: a mutiny! Hahahaha	<con>mutiny</con>	--	0	1	1
3:53 PM: **Team-Leader Dragon**: i do not submit them because of u	--	<per>do not submit</per>	0	1	0
3:53 PM: **Ricky Player**: u get the answers even if wrong u are supp to sub,it	<con>u are supp to sub, it</con>	--	0	1	0
3:54 PM: **Team-Leader Dragon**: we run out of time then we miss it	--	<per>run out of time, miss it</per>	0	1	0
3:54 PM: **Ricky Player**: v find answers ur job is to keep track of time	--	<per>find answers, keep track</per>	0	1	0
3:54 PM: **Ricky Player**: v don expect u to find answers	<con>don't expect</con>	--	0	1	1
3:54 PM: **Team-Leader Dragon** has left the room.	<con>left the room.</con>	--	0	1	0
3:55 PM: **Ashley Player**: take over the game! hhahahaha	--	<per>take over</per>	0	1	0
3:56 PM: **Ricky Player**: poor chap u put lot of pressure on him ;)	<con>poor chap, pressure on him</con>	<per>put lot of pressure</per>	1	1	0
Strengths			1	12	3
Inference result: Internal cause; Trustworthiness Attribution result: Betrayal			**Low**	**High**	**Low**

4.2 Analysis and Discussion

The dyadic attribution model helps analyze observed behavior of a focal actor proactively [8]. For example, the Team-Leader is asked by the team members to distribute the rewards, and thus the Team-Leader faces an ethical dilemma to either distribute the rewards equally among the team (loyalty), or to just keep them all to himself (betrayal). Based on the analysis of Table 4, the Team-Leader was hesitant in distributing the rewards as promised, and this can be characterized as low consistency. The Team-Leader's purposeful failure to submit the answers on behalf of his team can be categorized as high distinctiveness. The resulting arguments among the team members about the unfair distribution of rewards can be characterized as low group consensus. Aggregating these data-points allows us to designate the cause of behavior of the actor (Team-Leader) as being attributed to internal disposition. Based on the dyadic attribution (Table 3), such behavior falls into the category of betrayal.

5 Conclusions and Future Work

The paper identifies a sociotechnical approach to computationally modeling a system that mimics how human "sensors" are able to detect shifts of trustworthiness in potential threat situations. The conceptual framework of this inference model seeks to capture intelligence in conversations and develop algorithms that could automatically calculate attributions of trustworthiness.

While we have identified a number of trust factors pertaining to a single online situation, our methodology can further generalize the perceptions and identification of virtual disposition based on individual and group online conversations. We will use the resources of the "crowd," such as social media games, to simulate situations in which trustworthiness attribution can be exhibited. This is to collect public intelligence about various trust situations, group dynamics, and examine how a crowd attributes suspicious behavior. Then, we will use online games as research methods to formulate various situations. Online games are created using social media (e.g., Google+ Hangout, etc.) to collect public intelligence and study a series of trust-related scenarios. Using online gaming methodologies, one can engage people to contribute their knowledge and perceptions of trust. Finally, we will computationally model an attribution engine that can infer a high level of human attributes based on observed behaviors. The sensor computing system will have the ability to assign a probable subjective rating and assertions of actor's disposition at the individual level, and summarize a group's activities for that virtual interaction at the group level.

Acknowledgements. The authors thank the National Science Foundation for the support of Secure and Trustworthy Cyberspace EAGER awards #1347113 and #1347120. The first author also wishes to thank Conrad Metcalfe for his editing assistance.

References

1. Austin, J.L.: How to do things with words. Oxford University Press, New York (1962)
2. Emonds, G., Declerck, C.H., Boone, C., Seurinck, R., Achten, R.: Establishing cooperation in a mixed-motive social dilemma. An fMRI study investigating the role of social value orientation and dispositional trust. Social Neuroscience 9(1), 10–22 (2013)
3. Ho, S.M.: Attribution-Based Anomaly Detection: Trustworthiness in an Online Community. In: Liu, H., Salerno, J.J., Young, M.J. (eds.) Social Computing, Behavioral Modeling, and Prediction, pp. 129–140. Springer, Tempe (2008)
4. Ho, S.M.: Behavioral Anomaly Detection: A Socio-Technical Study of Trustworthiness in Virtual Organizations. In: School of Information Studies 2009, pp. 1–437. Syracuse, Syracuse University (2009)
5. Ho, S.M.: A Socio-Technical Approach to Understanding Perceptions of Trustworthiness in Virtual Organizations. In: Liu, H., Salerno, J.J., Young, M.J. (eds.) Social Computing, Behavioral Modeling, and Prediction, pp. 113–122. Springer, Tempe (2009)
6. Ho, S.M., Lee, H.: A Thief Among Us: The Use of Finite-State Machines to Dissect Virtual Betrayal in Computer-Mediated Communications. Journal of Wireless Mobile Networks, Ubiquitous Computing, and Dependable Applications (JoWUA): Special Issue of Frontiers in Insider Threats and Data Leakage Prevention 3(1/2), 82–98 (2012)
7. Ho, S.M., Katukoori, R.R.: Agent-based modeling to visualize trustworthiness: A socio-technical framework. Int'l Journal of Mobile Network Design and Innovation, Special Issue of Detecting and Mitigating Information Security Threats for Mobile Networks 5(1), 17–27 (2013)
8. Ho, S.M., Benbasat, I.: Dyadic attribution model: A mechanism to assess trustworthiness in virtual organizations. Journal of the American Society for Information Science and Technology (forthcoming, 2014)
9. Kelley, H.H., Holmes, J.G., Kerr, N.L., Reis, H.T., Rusbult, C.E., Van Lange, P.A.M.: The Process of Causal Attribution. American Psychology 28(2), 107–128 (1973)
10. Krueger, F., McCabe, K., Moll, J., Kriegeskorte, N., Zahn, R., Strenzlok, M., Heinecke, A., Grafman, J.: Neural correlates of trust. In: Proceedings of the National Academy of Sciences of the United States of America (PNAS) (2007)
11. Krueger, F., Hoffman, M., Walter, H., Grafman, J.: An fMRI investigation of the effects of belief in free will on third-party punishment. Social Cognitive and Affective Neuroscience Advance Access (2013)
12. Lieberman, J.K.: The litigious society. Basic Books, New York (1981)
13. Martinko, M.J., Thomson, N.F.: A synthesis and extension of the Weiner and Kelley attribution models. Basic and Applied Social Psychology 20(4), 271–284 (1998)
14. Rotter, J.B.: Interpersonal Trust, Trustworthiness and Gullibility. American Psychologist 35(1), 1–7 (1980)

Modeling Impact of Social Stratification on the Basis of Time Allocation Heuristics in Society

Michal Kakol, Radoslaw Nielek, and Adam Wierzbicki

Polish-Japanese Institute of Information Technology
Koszykowa 86, Warsaw, Poland
{michal.kakol,nielek,adamw}@pjwstk.edu.pl

Abstract. This paper describes a computational model of time allocation in a social network. It particularly focuses on the impact of inequalities or so called stratification process on the efficiency of resources allocation. The presented model is a multi-agent system with implemented network evolution mechanism. The stratification is modeled by the order in which the agents' actions are executed. The model was tested against as well different initial network parameters as a selection of time allocation heuristics. Efficiency of the system is measured using best applicable measures, e.g. agents schedule usage. By comparing the efficiency of the stratified system against baseline egalitarian system in our simulations we look for the answer whether social stratification brings benefits only for those who are on the top of the pyramid but decreases summarized performance of a society (in comparison to non-stratified systems).

1 Introduction

In fact, all modern developed societies are stratified and the only examples of the societies with no visible social hierarchy based on wealth or status are hunter-gatherer societies. Classification of people into different social hierarchy levels, e.g. based on race, wealth or caste, are said to have negative impact on the society. On the contrary, an egalitarian society is referred to not only as the fairest but also as the most efficient one.

Social stratification is a complex concept and manifests in the society in social, political or economic dimensions. Stratification exists in societies as well as in communities (even small ones) or within project groups (e.g. at work). In some social systems stratification is very rigid and cannot be changed by neither an individual nor a group (e.g. casts in India); in others it is much more flexible and individuals may improve their positions by themselves (picturesque American Dream).

Inequalities of time allocation potential differing among individuals of different status may be one of the social stratification consequences. People being at the top of social ladder generally do not wait. If you want to meet CEO of a big company, Prime Minister or a famous artist you have to adapt to their schedule. If they want to meet you, you probably reschedule your entire calendar immediately. More efficient use of limited resources (in this case time) is a way to develop prosperity.

In a very simple social system, where the only resource is time, the level in the social hierarchy would be determined by the capability to efficiently manage one's own

W.G. Kennedy, N. Agarwal, and S.J. Yang (Eds.): SBP 2014, LNCS 8393, pp. 285–292, 2014.

schedule. A good approximation of the control over one's own schedule can be an order in which agents arrange their meetings and fill up schedules – the person who arranges the meeting has the advantage of choosing the time (under the condition that the person invited to the meeting has to accept proposed date/time). Adding a mechanism of shifting the members' position in the execution queue accordingly to some social features can simulate flexibility of the hierarchy.

A multi-agent system of a society has been developed to verify following hypotheses: (1) *egalitarian societies and communities are in average more efficient in terms of distributed resource allocation based on interests of an individual, than stratified ones;* (2) *applying of advanced procedures of time allocation on agent level do not overcome the consequences of stratification; (3) stratification is always beneficial for privileged agents.*

The rest of the paper is organized as follows. Short theoretical introduction to sociological aspects and other papers concerning time allocation modeling can be found in section 2. Sections 3 and 4 describe the assumptions of the modeled world and mechanism by which the simulation is driven. Last two sections contain results presentation and interpretation, including conclusions and summarization.

2 Related Work

Stratified society is a system based on a socioeconomic hierarchy of its members who are divided into privileged or not privileged groups. [2] However, stratification does not need to be binary as it was in the early industrial class system where existence of three layers is assumed: upper, middle and lower class [15]. Despite the possibility of existing different hierarchy structures [6], we use one-dimensional abstract hierarchy for the sake of the modeling. Weberian stratification assumes that factors influencing hierarchy are wealth, prestige and power [11]. We believe that capability of efficient time management constitutes an expression of such factors.

It is widely agreed that differentiation of social groups into hierarchized layers is present in developed societies. These are worldwide phenomena and presumably ideally egalitarian society with equal opportunities seems to be a myth. A hugely influential book "Spirit Level" with a self-explaining subtitle: "Why more equal societies almost always do better" is aimed at raising the awareness of links between inequalities and various social problems (e.g. obesity, homicides, higher infant mortality etc.) [10]. Despite the existence of a sound evidence confirming that link, there is an ongoing discussion about the methodology as well as other critique followed this publication [16], [14] In contrast to the mentioned findings, Rowlingson and Karen [13] try to prove that correlation between income equality and social problems does not exist. Atkinson [1] goes even further and claims that inequality fuels economic growth. Although the issue of social stratification and inequalities was a subject to extensive research and debate over many years, the relation between inequality and prosperity still remains unsettled.

There are several works concerning simulation of time-use or time allocation. Fischer et al. [5] describe a model of time-use behaviors. The model was fed with empirical data-set gathered in a time-use diaries survey but ignores the social aspect of time-allocation and the role of interaction between the agents. Models where interactions between agents are used for time allocation have been researched by Wawer and Nielek but for different purposes, namely emergence of time-usage patterns [18] and infection spreading [9].

Empirical studies confirm an intuitive conclusion that all relationships decay over time, if they are not refreshed [12]. On the other hand the pace of decay may vary according to the closeness level. For example, lack of interaction on the link characterized by high strength value should not have greater impact on the relation strength – as both ends of high strength vertex can be referred to as "good" friends. [17]

3 The Model

The model described in this paper is a multi-agent system implemented in NetLogo[1] and based purely on social network approach (no spatial dimension). The size of the agents' population is constant over simulation. Removing or introducing new agents is not implemented in this model. A single agent corresponds to a person and the edge between agents represents an acquaintance relation. Edges are symmetrical. Agents can communicate only with agents they are directly connected. The relations of the actors can potentially be represented as vectors, and the network as a summation of these vectors, i.e., as a two dimensional matrix [8].

According to Social Brain Hypothesis and findings of Dunbar, human relationships are characterized by different level of intensity [3, 4], thus edges in the model have assigned strength attribute. The strength of the relationship between agents is dependent on the frequency of interaction between them and has been modeled according to Hays study [7]. The procedure of edge strength modification is described in detail in the section "Network evolution".

On a population level exists a finite set of all possible topics/interests (let call it *culture*). The size of *culture* is a parameter in the model. Every agent has a set of interests, which is a subset (the same size for all agents) of culture. Probability of being selected from *culture* follows a normal distribution. Therefore, some subjects are quite rare in population (e.g. Persian poetry) and some very popular (e.g. football). Additionally, every agent posses for each subject a willingness attribute, which is assigned randomly from 0 to 1 with uniform distribution for every interest. Willingness represents agent's inclination for a given subject.

Finally, every agent is equipped with the same number of time slots used in order to arrange meetings.

3.1 Network Evolution

The evolution of the network is shaped by three procedures that are invoked after every turn:

- *Relations strength updating* – basically this procedure look up all edges in the model and update their strength according to the activity in the last turn; namely, strength of every edge that has been used for successful

[1] Flowcharts and source code: `http://users.pjwstk.edu.pl/~s8811/papers/stratification/`

meeting arrangement is increased and strengths of unused edges are decreased; two separate updating functions have been implemented – linear and logarithmic, which binds the pace of relation change with their strength,

- *New relationship creations* – according to the simulation parameters new edges are created by two heuristics: *"my friend's friend"* approach where with some probability new edge appears between two agents who have at least one common friend and *"random meeting"*, which is simulated by the creation of an edge between two random agents in population;
- *Edge removing* – edges that are not used frequently enough (i.e. their strength reach zero) are removed.

3.2 Meeting Scheduling Procedures

Every turn for every agent simulation loop invokes a scheduling procedure that is responsible for finding partners for meetings and filling agent's time slots. Meeting takes place when two agents agree for a particular time slot (which has to be free) and subject. Every scheduled meeting consumes a time slot (is set as "in use" and will not be reused in the same turn) and willingness to discuss about a particular subject. Let W_{s1}^{A} and W_{s1}^{B} be willingness of discussing for subject 1 for agent A and B respectively, then a remaining willingness for agent A after the meeting may be calculated with formula presented below:

$$W'^{A}_{s1} = W'^{A}_{s1} - min(W_{s1}^{A}, W_{s1}^{B}) \quad (1)$$

It is worth noticing that the results of eq. 1 for one of two agents taking part in the meeting will always equal to 0. A meeting takes a single time slot, which number is limited and can be depicted as days of the week planned ahead by each agent.

The egoistic goal of the agent is to maximize one's own willingness usage. It is a combinatory problem and can me solved in many ways but finding optimal solution requires algorithms with exponential complexity. As users have only limited resources, they can only relay on approximation algorithms (heuristics). Two heuristics have been implemented and tested:

Simple – agent A choose a random agent B from their acquaintances89 and if they share a free time slot, selects a subject S, so that he maximizes $min(W_s{}^{A}, W_s{}^{B})$; every agent repeats this procedure few times (model's parameter),

Intelligent - agent A iterates over the list of its' subjects sorted in descending order by willingness to talk; for each of the iterated subjects all days of the week are used to find a partner B with maximal willingness to talk; this heuristic can be summarized as "trying until success".

3.3 Stratifying the System

The stratification in the proposed model manifests in three schedule filling initialization orders, which simulate different levels of inequality and rigidity [6]. The *baseline* used in the model is the *random order* of executing the agents and is treated as a metaphor of an egalitarian system with equal opportunities to make appointments; the inequalities and rigidity in such system are low. The *fixed order* method in which each agent in the system has constant place in the hierarchy, thus the agents are executed in the same order throughout the whole model run, in every turn. This can be referred to as a high level of inequality and rigidity in the system. In *weighted degree* differentiation of the agents occurs in every turn. In each turn, the agents (nodes) are executed according to their degree weighted by their relations' strengths – medium level of inequality and rigidity in the system.

4 Measures

All measurements were computed at the end of each turn, after all agents' interactions and network evolution. Next to the well-known network metrics (e.g. network density, average degree etc.) two measures have been used to express the efficiency of different stratification procedures and heuristics for arranging meetings (all used measurements were normalized to 1):

Schedule usage represents the percentage of all days in the schedules used in one turn and is presented on eq. 1,

$$Schedule_{Usage} = \frac{\sum_{i=1}^{N} T_i}{N * weekdays} \tag{2}$$

N denotes number of agents; T_i is the number of used time slots of i-th agent.

Willingness usage represents the percentage of "consumed" willingness of all agents in one turn in relation to all available willingness and is calculated with help of eq. 2,

$$Willingness_{Usage} = \frac{\sum_{i=1}^{N} SU_i}{\sum_{i=1}^{N} S_i} \tag{3}$$

N denotes number of the agents, SU_i is the sum of used willingness points in i-th agent schedule, S_i is the sum of willingness points of all subjects in the agent's interests set of i-th agent.

5 Results

The described model was tested for a wide range of initial parameters (among them variety of length of schedule, size of culture, relation update methods and decay/increase strength). Every set of parameters has been tested for three stratification methods (i.e. none, fixed and by degree) and two meetings scheduling procedures (i.e. simple and intelligent). Total number of combinations was 288. The number of agents was constant at 1000 and the initial network was always random of a given density. Every simulation has been repeated 30 times.

The model runs of certain parameter combinations were compared using four moments in time (250, 300, 500 and 1000 turn), which reflect initial warm up phase, the middle of the run and the final phase when stability is achieved. The Bonferroni method was used for multiple comparisons of the efficiency measure values.

Referring to the hypothesis no. 1 - *egalitarian societies and communities are in average more efficient in term of distributed self-seeking resource allocation than stratified.* The results of the 288 combinations prove that "None" execution method, which is the equivalent of egalitarian system, results in higher willingness usage in majority of cases. That is, in 81% of tested runs and time moments the "None" achieved higher willingness usage mean. Most importantly, the pairwise comparison of all three stratification methods in all the runs showed statistically significant differences in the means (p<0.0001). Although in 16,5% cases some of the compared pairs of mean values were not significantly different, these were mostly the pairs "Fixed" and "Degree" and among those cases (72%) still "None" showed better efficiency. The situations in which the "None" did not show better efficiency are mostly located in the first 300 turns of the model, these amounted to about 30% of runs for that moments in time. For the results from 1000th turn more than 97% of cases are showing better efficiency of the "None" method, which proves more efficient resource allocation in egalitarian systems.

Let us discuss a sample yet typical set of starting parameters: *Agents=1000; Culture size=500; Weekdays=15; Interests=15; New random relation strength=0.12.*

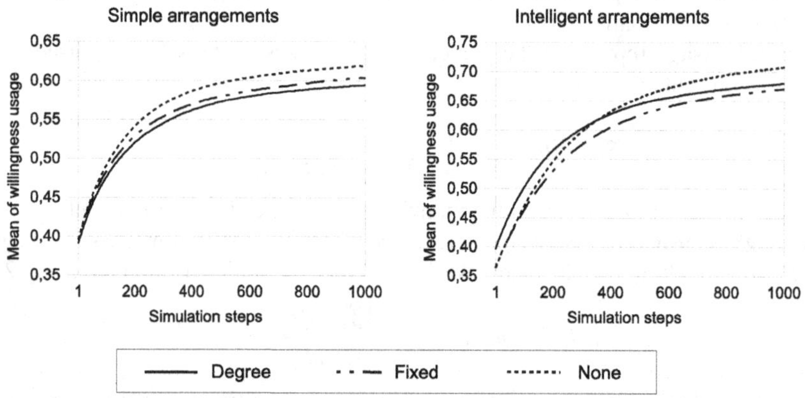

Fig. 1. Visualization of willingness usage throughout whole model run for a sample set of parameters. "Simple" time allocation heuristic on the left, "Intelligent" on the right.

Table 1. Significant results for different turns of the model run depicted in figure 1

			Means of willingness usage		
Arrangements	**Turn**	**Pr. > F**	**None**	**Degree**	**Fixed**
Intelligent	250	<.0001	(2) 0.574	(1) 0.555	(3) 0.586
Intelligent	300	<.0001	(2) 0.598	(1) 0.575	(3) 0.602
Intelligent	500	<.0001	(1) 0.657	(2) 0.626	(3) 0.645
Intelligent	1000	<.0001	(1) 0.708	(2) 0.670	(3) 0.680
Simple	250	<.0001	(1) 0.556	(3) 0.543	(2) 0.533
Simple	300	<.0001	(1) 0.570	(3) 0.553	(2) 0.545
Simple	500	<.0001	(1) 0.597	(3) 0.580	(2) 0.573
Simple	1000	<.0001	(1) 0.620	(3) 0.603	(2) 0.595

First 400 turns can be perceived as warm up, while stability is achieved on average after 1000th turn. The figure 1 depicts the model willingness usage output for two time allocation strategies. For both methods the non-stratified system shows better models efficiency in the area of model stability. In the "Intelligent" time allocation case however, the stratified execution method **by weighted degree** has the advantage over the **random** execution. This effect is apparent only and concerns the warm up phase of the model, which means that in approximately 30% of model runs during warm up the stratified methods are better than egalitarian ones, but in the area of stability, this percentage drops to roughly 2%. What proves the second hypothesis - *applying of advanced procedures of time allocation on agent level does not overcome the consequences of stratification.*

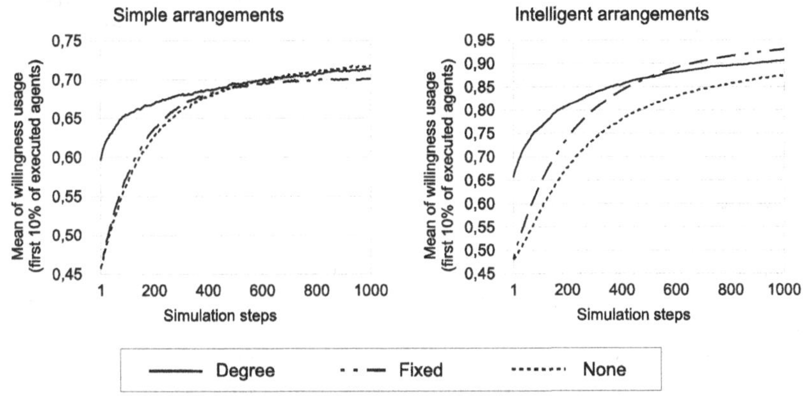

Fig. 2. Visualization of willingness usage for 10% of first executed agents throughout whole model run for a sample set of parameters. "Simple" on the left, "Intelligent" on the right.

The Figure 2 depicts the willingness usage for the first 10% of executed agents for both time allocation methods "Simple" and "Intelligent". The increase of efficiency on average for those first agents compared to the whole system amounts to 19% in case of "none", 26% for "fixed" and 29% for "degree" (calculated for all tested runs). In the Figure 2 the advantage of first to be executed agents for stratified execution methods over non-stratified ones is clearly visible and is present in about 92% of all model runs; this effect is not significant only for 8% of runs. In terms of tested model parameters, the agents privileged by the execution in the first place are most beneficial, compared to the whole system.

6 Conclusions

As the results of the carried out simulation show, the egalitarian systems are apparently on average more efficient in terms of distributed self-seeking resource allocation than stratified systems with unequal access to the resources. This effect is robust and is significantly visible in the majority of cases, especially when the model achieves stability. The advantage of stratified system over non-stratified one may exist during the initial phases of the model evolution and in longer term disappears in favor of non-stratified systems advantage.

Apart from simple time allocation heuristics with relatively high limitations an intelligent heuristic giving the agents more flexibility during their schedule filling terms was tested. The more intelligent heuristic was introduced to simulate the influence of the technological progress in communications, which allows people in general to communicate and manage their time more efficiently. The introduction of the intelligent time allocation in the model improves in fact the overall model efficiency; however, it still does not revert the previously described effect and the advantage of non-stratified egalitarian systems remain stable in long term.

It appears that stratification brings benefits to the privileged individuals, but this effect is temporary and these inequalities lead to the decrease of overall efficiency. This conclusion remains in line with the rather mainstream belief that egalitarianism is superior to the stratification and is more beneficial for all society members.

Acknowledgements. This work is supported by Polish National Science Centre grant 2012/05/B/ST6/03364

References

1. Atkinson, A.: Public Economics in Action. Oxford University Press, Oxford (1997)
2. Barker, C.: Cultural Studies: Theory and Practice, p. 436. Sage, London (ISBN 0-7619-4156-8)
3. Dunbar, R.I.M.: Determinants of group size in primates: a general model. In: Proceedings of the British Academy, vol. 88, Oxford University Press (1996)
4. Dunbar, R.I.M.: The social brain hypothesis. Brain 9, 10 (1998)
5. Fischer, I., Sullivan, O.: Evolutionary modeling of time-use vectors. J. of Economic Behavior & Organization 62(1), 120–143 (2007)
6. Grusky, D.B.: The Contours of Social Stratification. Social Stratification in Sociological Perspective (1994)
7. Hays, R.B.: The day-to-day functioning of close versus casual friendships. J. of Social and Personal Relationships 6(1), 21–37 (1989)
8. Leydesdorff, L.: The evolution of communication systems, arXiv:1003.2886 (2010)
9. Nielek, R., Wawer, A., Kotowski, R.: Two Antigenically Indis-tinguishable Viruses in a Population. In: Proceedings of the World Congress on Engineering, vol. 3 (2008)
10. Pickett, K., Wilkinson, R.: The Spirit Level: Why more equal societies almost always do better. Allen Lane, London (2009)
11. Ritzi-Messner, N., Veldhoen, C., Fassnacht, L.: The distribution of power within the community: Classes, Stände, Parties. J. of Classical Sociology 10(2), 137–152
12. Roberts, S.G., Dunbar, R.I.: The costs of family and friends: An 18-month longitudinal study of relationship maintenance and decay. Evolution and Human Behavior 32(3), 186–197 (2011)
13. Rowlingson, K.: Does income inequality cause health and social problems? (2011)
14. Saunders, P.: Beware False Prophets: Equality, the Good Society and The Spirit Level. Policy Exchange, London (2010)
15. Saunders, P.: Social Class and Stratification, Routledge (1990) ISBN 9780415041256
16. Snowdon, C.: The Spirit Level Delusion: Fact-checking the Left's New Theory of Everything. Little Dice (2010)
17. Sutcliffe, A., Wang, D.: Computational modelling of trust and social relationships. J. of Artificial Societies and Social Simulation 15(1), 3 (2012)
18. Wawer, A., Nielek, R., Kotowski, R.: Patterns in Time: Modelling Schedules in Social Networks. In: International Workshop on SOCINFO 2009. IEEE (2009)

The Social Aspect of Voting for Useful Reviews

Asher Levi[1] and Osnat Mokryn[2]

[1] Department of Information Systems, University of Haifa, Israel
[2] School of Computer Science, Tel Aviv Yaffo College, Israel

Abstract. Word-of-mouth is being replaced by online reviews on products and services. To identify the most useful reviews, many web sites enable readers to vote on which reviews they find useful. In this work we use three hypotheses to predict which reviews will be voted useful. The first is that useful reviews induce feelings. The second is that useful reviews are both informative and expressive, thus contain less adjectives while being longer. The third hypothesis is that the reviewer's history can be used as a predictor. We devise impact metrics similar to the scientific metrics for assessing the impact of a scholar, namely *h-index*, i_5*-index*. We analyze the performance of our hypotheses over three datasets collected from Yelp and Amazon. Our surprising and robust results show that the only good predictor to the usefulness of a review is the reviewer's impact metrics score. We further devise a regression model that predicts the usefulness rating of each review. To further understand these results we characterize reviewers with high impact metrics scores and show that they write reviews frequently, and that their impact scores increase with time, on average. We suggest the term *local celebs* for these reviewers, and analyze the conditions for becoming local celebs on sites.

1 Introduction

In today's Internet, reviews for products and services have become an online form of word-of-mouth. Reviewers contribute their time and energy to express their opinions and share their experience. Shoppers, as well as online service consumers, read these reviews as a first step in their decision-making process. With the growth in popularity of online reviews in popular commerce sites like Amazon, numerous sites have been created for the sole reason of becoming review portals: Yelp (businesses and services reviews), TripAdvisor (hotel reviews), and more.

Browsing through the plethora of reviews, readers find it hard to navigate and extract the useful information. Hence, a system that enables one to determine which reviews are more useful, or helpful, is needed. Sites like Amazon, Yelp and IMDb thus rely on a form of crowdvoting, in which the readers specify which reviews they find useful, or helpful. Specifically, Amazon customers and IMDb users vote whether a review is helpful or not; In Yelp, the users up-vote to indicate they find a review useful.

The purpose of this paper is to characterize which reviews are voted useful [1]. The benefits of automating this selection process are numerous. It enables the promotion

[1] The term useful refers also to helpful reviews. Throughout this paper, these terms are used interchangeably.

W.G. Kennedy, N. Agarwal, and S.J. Yang (Eds.): SBP 2014, LNCS 8393, pp. 293–300, 2014.

of potential useful reviews thus presenting users with current useful reviews upon their arrival, to improve the shopping / browsing experience; Useful reviews can be detected early, enabling sites to keep an up-to-date and current review system, without compromising quality; Useful reviews for low-traffic or new items can be detected, while today they might not gather enough votes to be noticed. Research in this area in the last years focused on classifying which reviews are voted useful or ranking useful reviews [1,2,3,4], taking into account textual characteristics, review sentiment, or product and user features. We continue to suggest three novel hypotheses for why reviews are voted useful.

Feelings have shown to influence decisions [5,6]. Physiologists have long demonstrated that people can induce feelings in others [7]. For example, imagine you enter a room in which a friend converses happily, laughing. Experience shows that you are likely to find yourself smiling at ease. Does a similar effect exist in text? Can reviews induce feelings and influence a reader's decision and therefore be perceived as more useful? We then come to our first hypothesis: (1) Useful reviews are more likely to express feelings.

Useful reviews are perceived as carrying important information to readers. Informative text contains more nouns, while descriptive text contains more adjectives. Additionally, longer text carries more information and might therefore be perceived as more useful. Our second hypothesis is hence: (2) Useful reviews are longer and more informative yet expressive (e.g., high ratio of nouns per adjectives accompanied with punctuation marks).

Profiling the reviewers reveals that some people tend to have a large number of useful reviews. As writing a review might be seen as a skill, the question is can skilled reviewers be identified, and are they more likely to keep producing useful reviews. To that end we use the average useful count a reviewer received. However, as the different sites employ different voting and counting systems, we proceed to create a set of unified metrics for capturing a reviewers "useful" counts for previous reviews. We thus devise a common metric for profiling a reviewer, borrowing from the scientific metrics for measuring a scholar's impact: the famous *h-index* and i_x*-index*. Our hypothesis is then: (3) A reviewer's impact score is a good predictor for the perceived usefulness of her future reviews.

We evaluate our hypotheses on three different datasets. Two datasets are from Yelp and one is from Amazon. All three data sets are cross domain, and are detailed in Section 3. Our results (detailed in Section 4) are conclusive, showing that the reviewer impact score is an excellent predictor, outperforming all others. Specifically, a SVM classifier with these users features gives an outstanding 97.8 accuracy (with AUC of 0.97) for the Amazon datasets, and almost as good results for the Yelp datasets. Moreover, our exact vote count prediction R-square values range between 0.561 and 0.871 across all evaluated datasets, with an RMSE of less then 1.7% of the predicted values. We further checked the statistical significance of each feature across the three datasets.

In Section 5 we explore the origin of these findings. We find a tendency for reviewers with high *h-index* to author many reviews. We further found that new reviewers are less likely to receive a high number of useful votes. Specifically, later reviews of the same author would on average be perceived as more useful and would be up-voted more as useful. We suggest the following explanation: Research on online engagement has demonstrated the importance of trust. An up-vote is an expression of trust, as the voter would only indicate as useful reviews she trusts their content to be truthful and accurate. It has been demonstrated that familiarity is a fundamental perquisite for trust, especially online [8,9]. Reviewers investing time and energy in writing reviews become known in online communities, in what we term *local celebs*, and their reviews are therefore perceived as useful.

2 Modeling and Hypotheses

A common review system is typically comprised of three different types of entities: a set $I = i_1, ..., i_N$ of N items (businesses, products or services); a set $U = u_1, ..., u_M$ of M users (or reviewers) that use the system and write reviews; and finally, a set $R = r_1, ..., r_T$ of T reviews that were written by the users, expressing their opinion on these items. Each review r written by reviewer r on item i contains the following information: users rating, review text, and the number of useful votes v. Formally, given a set of reviews R_i for a particular item i (business, product or service), we try to: Classify whether a review is useful or not; Predict the count of useful votes v for each review. The formal definition for a useful review is given in Section 3.

Our experience with reviews [10] has demonstrated that the text of the reviews contains additional information beyond the rating given by the reviewer. Consequently, it seems natural to assume that reviews perceived as useful would contain information found important by users. A survey of useful reviews led to the following observations:

1. Reviews expressing emotions are more likely to induce feelings, and hence be perceived as useful.
2. Useful reviews tend to be more informative (for example, have a higher nouns/adj. relation), and often contain expressive text expressions.
3. Users who historically have written useful reviews write more useful reviews.

An explanation of the text emotions and sentiment hypotheses can be found in the extended version [11], we concentrate here on modeling reviewer characteristics. Our observations have led to the conclusion that it is likely that the reviewer's information may also be a predictor for whether a review will be voted useful and how many votes it will acquire. It is reasonable to assume that some people are better than others in writing reviews, and hence their reviews will be perceived as more useful. Indeed, many sites display the reviewer information in a prominent way. In Yelp, for example, "Elite members" (top reviewer) are elected each year, and earn a badge that is emphasized online. In Amazon the "Top-1000" reviewers have a special tag displayed next to their name. The challenge ahead is to examine whether the reviewer's history can be used to predict the useful votes of her future reviews. To that end, a metric for describing the history of a reviewer needs to be formalized.

Reviewer Impact Metric. A reviewer profile usually consists of the following: number of reviews, average useful votes, sum of useful votes, star rating, and user average star rating.

To capture the *perceived impact* of a reviewer we take a detour to known methodologies to assess the impact of scientists and scholars. The *h-index* is a widely known and heavily used metric in science and academia. From Wiki: *The index is based on the distribution of citations received by a given researcher's publications, ... and reflects both the number of publications and the number of citations per publication.* Google Scholar further calculates a second metric, i_{10}-*index*, which describes the number of papers with 10 or more citations the scholar has published. Based on the above metrics, we define two variables to capture a reviewer's history: *h-index*, i_5-*index*.

Reviewer *h-index:* Reviewer has index h if h of her N_p reviews have at least h useful votes each, and the other $(N_p - h)$ reviews have no more than h useful votes each. Hence, an *h-index* of 10 means the reviewer has at least 10 reviews which were voted useful by at least 10 other people, and the rest of her reviews have less than 10 useful votes each.

Reviewer i_5-*index:* Reviewer has index i if she has i reviews with at least 5 useful votes each.

Fig. 1 shows the average number of useful votes users with the same *h-index* and i_5 *index* receive. Clearly, the better is a reviewer's impact, the more useful votes her reviews will receive on average. The complete list of features can be found here [].

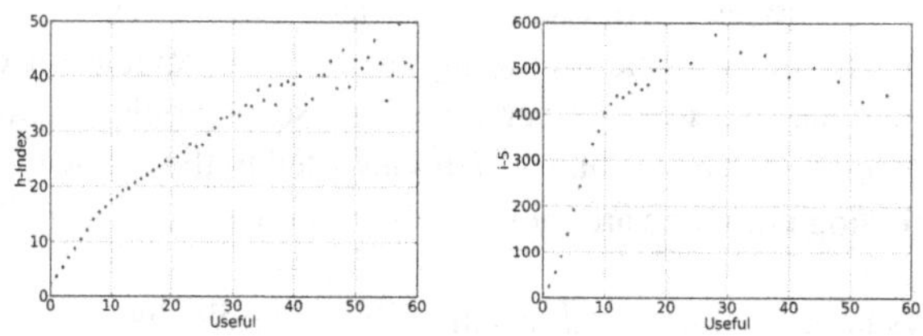

(a) *h-index* vs. useful vote (b) i_5 *index* vs. useful vote

Fig. 1. Average Useful Votes vs. Average Reviewer Metrics

3 Data

To evaluate our hypotheses we use three different datasets, as depicted in Table 1. Two data sets contain reviews about businesses from YelpThe first Yelp data set, referred to as *Yelp Bay Area*, was collected from Yelp in the Bay area, California, during May and June 2011. The second data set, referred to as *Yelp challenge*, was introduced in early 2013 by Yelp. It includes businesses' reviews from the greater Phoenix metropolitan area. The third data set, referred to as *Amazon*, is a set of product reviews obtained from Amazon during April 2010. Yelp users vote whether a review is useful, while Amazon users vote whether the review was helpful or not. (see Figures 2a, 2b). Using these votes, the community can filter relevant opinions more efficiently.

First, we need to convert the continuous variable "Useful" in Yelp, or "Helpful" in Amazon into a binary one. For the Amazon dataset, *Helpful* is a vote h out of v, where h is number of helpful votes, and v is the total number of votes. A review is marked as useful when h/v is greater then the threshold γ, and h is greater then threshold ζ.

Table 1. Summary of statistics for each of the data-set

Data Set	Users	Items	Reviews	Useful	%Useful
Amazon	21087	2103	23868	12877	53.9%
Yelp Bay Area	95296	66803	488805	63074	12.9%
Yelp challenge	43873	11537	229907	14422	6.2%

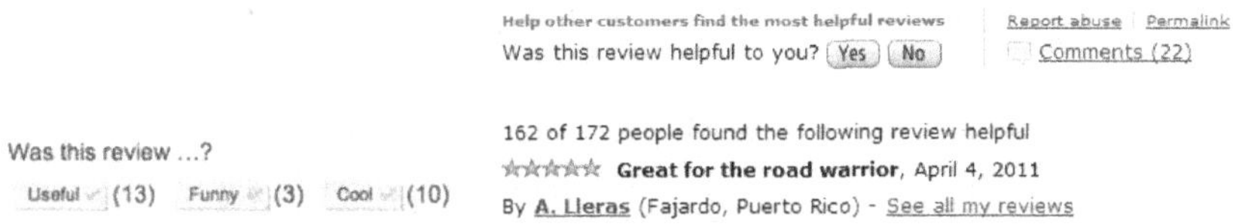

(a) Yelp review useful voting (b) Amazon review helpful voting

Fig. 2. Useful and Helpful Votes in Sites

[1] showed that the best threshold γ is 0.6. We chose γ as 0.6 and ζ as 5. In other words, if a review has more then 5 votes indicating that the review is helpful, and more than 60% of the votes indicate that the review is helpful, then we classify the review as "useful". Otherwise, we classify the review as "non-useful".

For the Yelp datasets *useful* is a vote h, where h is number of useful votes. Review is marked as useful when h is greater then the threshold ζ. The same ζ was chosen as for the Amazon dataset. In other words, if the review has more then 5 votes indicating that the review is useful, we classify the review as "useful". Otherwise, we classify the review as "non-useful".

4 Experiments and Results

We have conducted two large scale experiments to explore the validity of our hypotheses, as detailed in Section 2. A classification, and a useful score prediction. In both experiments, a review r is represented as an f-dimensional real vector x over a feature space F constructed from the information in r. Our methodology is based on a 6-fold cross validation procedure. To classify whether a review is useful or not, we use a supervised learning paradigm, based on a training set containing labeled data. The test set contains reviews represented as multi-dimensional feature vectors. We employ Support Vector Machines (SVM), using as kernel the radial basis function (RBF). The evaluation metrics are the classification accuracy and the area under the ROC curve (AUC).

We performed different series of binary classification experiments for evaluating our different hypotheses separately and combined, as detailed here. A detailed features list can be found in our extended version [11]. The first experiment, referred to as *All*, uses all the features; We then proceed to validate each one of our hypotheses as detailed. For evaluating our hypothesis on induction of emotions, referred to as *Emotions*, we used all the emotions features. For the text hypothesis, referred to as *Text*, we use the text features; We evaluate the combination of the two hypotheses above in an experiment referred to as *Emotions-Text*, using the combined set of features from *Emotions, Text*. Our third hypothesis, which takes into account the reviewer's impact, is referred to as

Table 2. Accuracy and AUC for the Usefulness Classifiers

Experiment	Yelp Bay Area		Amazon		Yelp Challenge	
	Accuracy	AUC	Accuracy	AUC	Accuracy	AUC
All	92.5%	0.79	97.7%	0.97	95.3%	0.68
Emotions	87.1%	0.5	59.5%	0.58	93%	0.5
Text	87.1%	0.5	60.1%	0.59	93.7%	0.5
Emotions-Text	87.1%	0.5	61.4%	0.61	93.7%	0.5
Reviewer	**92.3%**	**0.8**	**97.8%**	**0.97**	**95.2%**	**0.69**

Table 3. R-squared and RMSE for the Prediction Model

Dataset	R-squared	RMSE	Useful Range	RMSE(%)
Yelp Bay Area	0.607	3.02	0 – 180	1.67%
Amazon	0.871	7.56	0 – 635	1.19%
Yelp Challenge	0.561	1.47	0 – 120	1.22%

Reviewer. Table 2 shows the results for the experiments over all the datasets described in the previous section.

The first observation noted is that the experiments *Emotions*, *Text*, and *Emotions-Text* show low performance. The prediction accuracy for those experiments is high, while the AUC is low; Further analysis reveals that the precision of the classifier is high but the recall is low; This, with the fact that there is a low rate of useful reviews in the Yelp datasets (12.9% and 6.2%), may indicate that the prediction only succeeds in predicting "non-useful" reviews but fails in predicting the useful ones. The results for these experiments on the Amazon dataset support this observation: As the share of useful reviews in the Amazon dataset is much higher (53.9%), we see that indeed both the accuracy and the AUC are low. Hence we conclude that the results don't support our two first hypotheses.

The Reviewer Characteristics features set achieved the best performance across all datasets, if we take into account both the accuracy and AUC. Specifically, it outperformed all other feature combinations on the Amazon dataset, where there is a larger percentage of useful reviews, and performed similar to All Features experiment on both Yelp datasets. Hence, Reviewer Characteristics show to be a high quality and excellent predictor for the usefulness of a review, supporting our third hypothesis.

Our experiments also indicate that the predictive power increases when the "useful" percentage in the dataset increases; The size of the set seems less influencing; The Amazon data set contains 53.9% useful reviews, and even though the Yelp data sets are much larger, the performance using the Amazon dataset is significantly better. Yelp datasets support this observation; The Challenge dataset has only 6.2% of "useful" reviews, while the Bay Area has 12.9%. Indeed, the performance of the classification on the Bay Area dataset is better than that for the Challenge dataset. To predict the count of useful votes per review, we use the linear regression model (LR). The evaluation metrics used are the correlation coefficient (R-squared) of the linear regression model and the Root Mean Squared Error (RMSE) as a measure of the prediction error. Table 3 details the results. The R-squared values of our models range between 0.561 and 0.871 across the Yelp and Amazon datasets; This high values of correlated R-squared suggest that there is a high correlation between the features we choose as predictors, and the count of useful votes for each review. The RMSE is supporting the accuracy of the results, the values of the RMSE are less than 1.7% of the values that the model predicts.

We proceed to check the statistical significance of each feature across all three datasets, and find contributing features. A contributing feature has a statistical significant effect on the extent by which a review is perceived useful. Specifically, we include here only features for which $p < 1\%$. For the Yelp and Yelp Challenge datasets, 20 and 18 features, respectively, are found statistically significant out of the 36 checked. For the Amazon dataset only six features are found statistically significant. Moreover, out of the four features found highly significant for all datasets, two behave differently for the Yelp datasets and for the Amazon dataset. For example, lower product rating relates to higher useful votes for the Yelp datasets, while an opposite relation exists in the Amazon dataset, i.e., higher product rating has a positive relationship with the useful vote received. Other significant results are the following: In all datasets, the higher the average useful vote a reviewer has, the higher is the number of useful votes the review receives. Similarly, the more disgust emotion is expressed, the more the review is likely to gain more useful votes over all sets. In the Yelp datasets, the number of adjectives has a negative correlation with the useful vote, while the number of punctuations and question marks has a positive relation.

5 How to Become a Local Celeb

In this Section we search for the root cause for our results. Initial explanations are obvious ones: First, it could be that experts simply shine. These reviewers manage to capture the essence of what would be important to others and explain it vividly. The second explanation is stems from an underlying mechanism in social networks - preferential attachment [12]. Accordingly, the text of reviewers who are considered influential (i.e., earn Elite badges on Yelp or high rank badges in Amazon) is up-voted because the reviewer has acquired votes previously. Still, how does one become a Yelp Elite member or an Amazon top reviewer and acquires these votes?

Up-voting a review is a measure of trust: a reader cannot up-vote a review if she does not trust the reviewer to be truthful in expressing her opinion. Voting requires the reader to decide whether she finds the review useful. A positive decision demonstrates trust, and a fundamental requirement for that trust is familiarity [8]. People are more likely to trust the familiar, and familiarity obtained through frequent exposure has the potential to engender trust [9]. A high *h-index* captures users who write a lot of useful reviews. Indeed, the data shows that reviewers with a high average are quite prolific writers. Specifically, for the *h-index* range of $\{10, 70\}$, the number of reviews written on average was between 160 and 1000. Clearly, reviewers with a high *h-index* write a lot of reviews that people tend to like. When a reviewer writes frequently she becomes familiar, and hence trusted. Do people up-vote her reviews because they are useful or because they trust her judgment? Specifically, are reviewers writing a lot of reviews more likely to receive useful votes? Fig. 3 depicts the evolution of useful votes over time for reviewers. The reviewers are grouped in bins according to their average user vote and averaged, each bin containing at least five reviewers. An interesting observation is shown in Fig. 3a. The average number of useful votes a reviewer gets grows with the

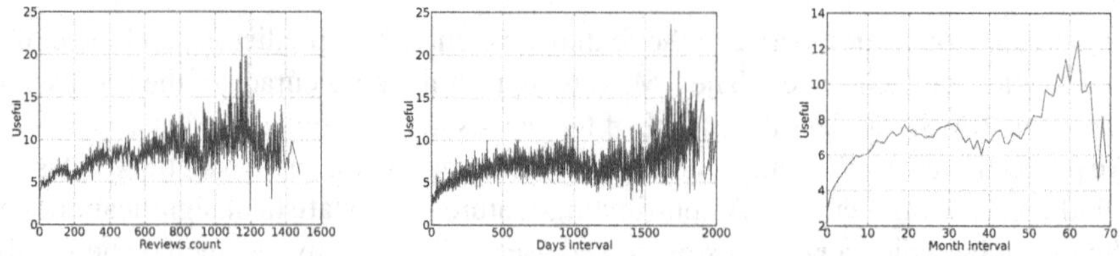

(a) Avg. useful per #reviews (b) Avg. useful per #days from the first review (c) Avg. useful per #months from first review

Fig. 3. Average useful votes with time/reviews# growing

number of reviews she writes. Similar results are shown in Fig. 3b and Fig. 3c: The average useful votes a reviewer gets grows with time passed from the first reviews she wrote. These local celebs do indeed work hard to become known, familiar and trusted online. The result of their hard work is the number of useful votes they receive.

References

1. Ghose, A., Ipeirotis, P.: Estimating the helpfulness and economic impact of product reviews: Mining text and reviewer characteristics. IEEE Transactions on Knowledge and Data Engineering 23(10), 1498–1512 (2011)
2. Kim, S.-M., Pantel, P., Chklovski, T., Pennacchiotti, M.: Automatically assessing review helpfulness. In: Proceedings of the 2006 Conference on Empirical MNLP, pp. 423–430 (2006)
3. Lu, Y., Tsaparas, P., Ntoulas, A., Polanyi, L.: Exploiting social context for review quality prediction. In: Proceedings of the 19th International Conference on WWW, pp. 691–700. ACM (2010)
4. Korfiatis, N., García-Bariocanal, E., Sánchez-Alonso, S.: Evaluating content quality and helpfulness of online product reviews: The interplay of review helpfulness vs. review content. Electronic Commerce Research and Applications 11(3), 205–217 (2012)
5. Bagozzi, R.P., Gopinath, M., Nyer, P.U.: The role of emotions in marketing. Journal of the Academy of Marketing Science 27(2), 184–206 (1999)
6. Griskevicius, V., Goldstein, N.J., Mortensen, C.R., Sundie, J.M., Cialdini, R.B., Kenrick, D.T.: Fear and loving in las vegas: Evolution, emotion, and persuasion. JMR, Journal of Marketing Research 46(3), 384 (2009)
7. Epstude, K., Mussweiler, T., et al.: What you feel is how you compare: How comparisons influence the social induction of affect. Emotion 9(1), 1 (2009)
8. Gefen, D.: E-commerce: the role of familiarity and trust. Omega 28(6), 725–737 (2000)
9. Siau, K., Shen, Z.: Building customer trust in mobile commerce. Communications of the ACM 46(4), 91–94 (2003)
10. Levi, A., Mokryn, O., Diot, C., Taft, N.: Finding a needle in a haystack of reviews: cold start context-based hotel recommender system. In: Proceedings of the Sixth ACM Conference on Recommender Systems, pp. 115–122. ACM (2012)
11. Levi, A., Mokryn, O.: The social aspect of voting for useful reviews (2014), `http://www.cs.mta.ac.il/~ossi/Pubs/LM2013.pdf`
12. Barabási, A.-L., Albert, R.: Emergence of scaling in random networks. Science 286(5439), 509–512 (1999)

Simulating International Energy Security

David Masad

Department of Computational Social Science
George Mason University
Fairfax, VA, USA

Abstract. Energy security is placed at risk by exogenous supply shocks, in particular political crises and conflicts that disrupt resource extraction and transportation. In this paper, a computational model of the security of international crude oil supplies is described, and its output analyzed. The model consists of country agents, linked geographically and by a data-derived oil trade network. Countries stochastically experience crises, with probabilities and durations drawn randomly from data-fitted distributions. The effect of these crises on secure oil supplies is measured globally and by country, and the effect of conflict contagion and spare production capacity are also estimated. The model indicates that Russia, Eastern Europe, and much of the Global South are at the greatest risk of supply shocks, while American producers are at greatest risk of demand shocks. It estimates that conflict contagion decreases energy security slightly, while spare capacity has minimal effect.

1 Introduction

Energy security is defined by the International Energy Agency as "the uninterrupted availability of energy sources at an affordable price"[1]. As the world's energy demand continues to expand, energy security is becoming increasingly critical not just to developed countries but to developing ones as well [2]. Much of the previous formal analysis of energy security has taken a portfolio-based approach, examining both the short-term risks[3,4] and long-term energy and environmental policy [5,6,7]. However, these approaches tend to aggregate political risk into a single indicator, and do not incorporate the complexity that characterizes international political in general, and conflict in particular [8,9]. In fact, oil supply shocks are often caused by domestic or international conflicts involving oil producers [10]. While the macroeconomic effects of oil shocks may be small [11], the price of oil is only an indirect measure of energy security. If a substantial volume of oil cannot be extracted or exported, and hence is unavailable for consumption, this poses an energy security risk regardless of the benchmark price.

In this paper, I present a novel model linking energy security with political instability. At the core of the model is the supply and demand for crude oil at the country level. Supply and demand are placed at risk when countries experience crises and conflicts. The model operates on the meso timescale, of months and

W.G. Kennedy, N. Agarwal, and S.J. Yang (Eds.): SBP 2014, LNCS 8393, pp. 301–308, 2014.

years rather than days or decades. It produces country-level and global time-series of oil security indicators with each iteration. By running multiple iterations with different parameters, I generate notional near-future scenarios of changes in international energy security. By incrementally enabling the possibility of conflict contagion, and of rapid increases of production to counterbalance supply shocks, I can estimate their impact on international energy security.

In the remainder of the paper, I describe the model in more detail, as well as its input data sources. I use the model to conduct several experiment and describe the output. Finally, I discuss my conclusions, both regarding international energy security and the challenges and opportunities of this type of modeling.

2 Model

2.1 Model Description

The model consists of countries, each of which has a set supply of and demand for crude oil[1]. Countries are linked geographically to their neighbors, and by a network of export-import relationships. Each monthly tick of the model, some countries enter crisis at random with probability based on their political instability. Crisis duration is randomly drawn from a power law distribution. The exports and imports of countries in crisis are considered *at risk*. At the end of each tick, the model measures several indicators of oil security. The Overall Ratio is the global ratio of secure demand to secure supply; a high ratio indicates a higher likelihood of a global supply shock, while a low ratio indicates a higher likelihood of a global demand shock. For each country, the model computes the ratio of total demand to secure (not at risk) supply (the Supply Ratio). The higher this ratio is, the more likely the country is to experience a supply shock. The model also computes the fraction of each country's total exports are going to countries not in crisis (Demand Ratio), estimating the risk of a demand shock.

If **Contagion** is enabled, when a country enters crisis it may trigger crises in neighboring countries based on their own instability. If **Assistance** is enabled, oil exporters may utilize their spare capacity and increase their production to balance a loss of secure oil by a major import partner, increasing the supply of oil to all of their trading partners.

2.2 Data Sources

The network of oil imports and exports comes from the United Nations-maintained COMTRADE database [12]. I extract all imports of crude and other unprocessed petroleum for 2012, the most recent year for which complete data is available.

For the majority of countries, I assume that supply and demand are equal to total exports and imports. For the top ten oil producers, I use US Energy Information Administration (EIA) estimates [13], country estimates [14,15], expert

[1] Full model code is available at `https://github.com/dmasad/riskworld`

reports [16], and media sources [17] for those countries' consumption of their own domestic production.

Spare capacity is particularly difficult to estimate, and varies over time and with the price of oil and other factors [18]. I use media sources [19] and subject-matter expert (SME) analysis [18] in order to estimate the spare capacity of major exporters. I also assign a default spare capacity of 10% based on SME input.

Political instability estimates are taken from the Economist Intelligence Unit's Political Index ratings [20]. These are in turn estimated based on the methodology developed by the Political Instability Task Force [21], which predicts the onset of internal instability in a two-year period with approximately 80% accuracy. Thus, these scores are normalized such that a rating of 10 (the maximum) is associated with an approximately 80% probability of instability within a 24-month timeframe, translating into a 6.5% monthly probability of crisis onset.

Crisis duration is drawn from a power-law distribution [22]. I calibrate the power law coefficient from two datasets: the UCDP/PRIO Armed Conflict Dataset [23] for military conflicts, and the Social Conflict in Africa Database [24], which includes both armed and lower-level social conflicts. I subset and combine both datasets to generate as complete as possible a set of crises and conflicts that may be associated with oil production shocks in Africa. I find the duration of each event in months, and fit a power law to the resulting distribution, resulting in an estimated coefficient of $-\mathbf{1.37}$.

3 Model Results and Analysis

3.1 Global Outcomes

Outcome Range: I run the model for 250 iterations for each permutation of the Contagion and Assistance parameters, 1,000 iterations in total; each iteration was run for 60 ticks, or 5 simulated years.

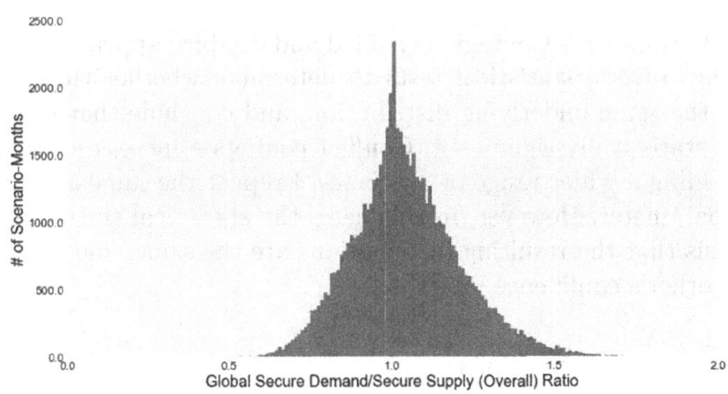

Fig. 1. Distribution of overall ratios across all scenario-months

Figure 1 shows a histogram of the ratio between total global secure demand and supply (overall ratio) across all iteration-months. It is centered very close to 1 (a balance between supply and demand) but with a longer tail to the right; extreme supply shocks are more likely than extreme demand shocks.

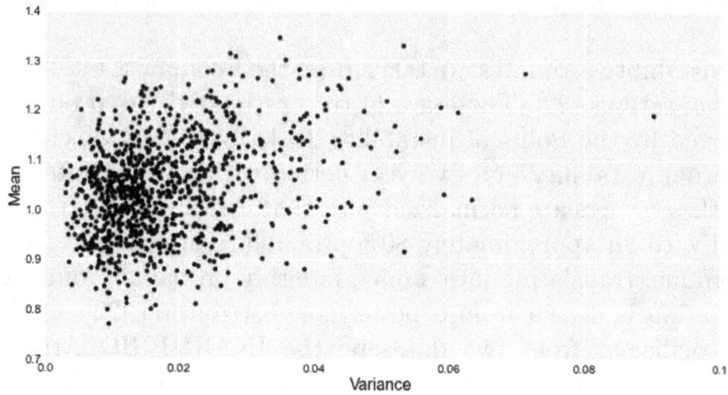

Fig. 2. Variance and mean of overall ratio for each model iteration

I characterize each model iteration by the mean and variance of the overall ratio, as shown in Figure 2. Correlation between mean and variance is 0.33, indicating that they are not necessarily linearly related to one another. A high mean ratio may indicate higher oil insecurity, but if the variance is low, it is 'stably' insecure; in contrast, the mean may be close to 1 or even below it, but high variance would indicate severe uncertainty and fluctuation – a distinct type of insecurity.

Contagion and Assistance: In order to disaggregate the effects of the Contagion and Assistance model parameters, I examine their results separately. Figure 3 shows probability densities of the Overall Ratio for the scenario-months for Contagion and Assistance, respectively.

The distributions with Contagion enabled and disabled appear extremely similar. I conduct several statistical tests to determine whether they are in fact drawn from the same underlying distribution, and conclude that the difference is small but statistically significant. Conflict contagion increases the mean and variance, yielding a wider range of outcomes. I repeat the same analysis for the Assistance parameter. However, in this case, the statistical tests cannot reject the hypothesis that the resulting distributions are the same, and the means are within each other's confidence intervals.

3.2 Country-Level Analysis

In addition to the aggregate global indicators, the model also tracks the Supply and Demand ratios for each country. We can analyze this output in order to

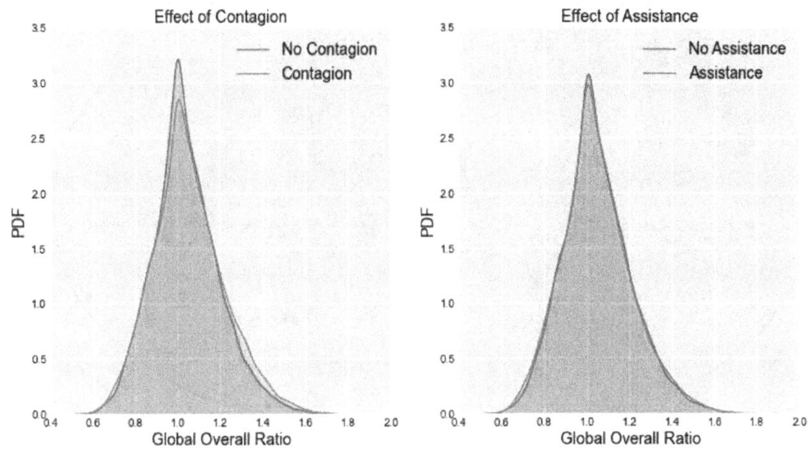

Fig. 3. Effects of Contagion and Assistance

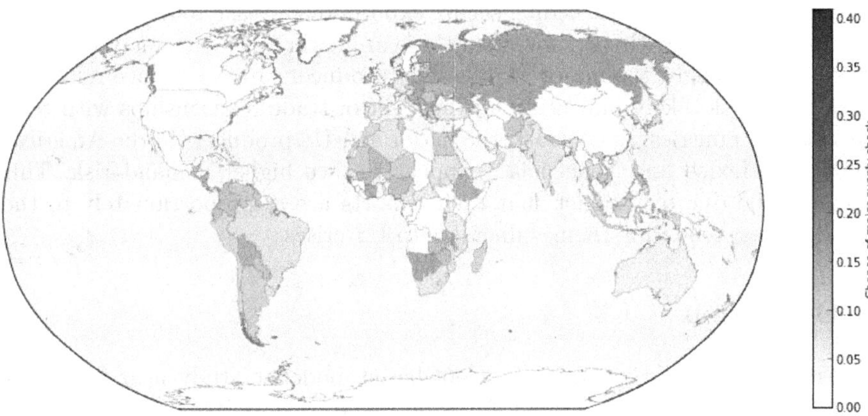

Fig. 4. Risk of a major supply shock (Supply Ratio > 2)

understand how countries' risks differ from one another, and to identify countries at high or low risk.

Figure 4 shows a map of the overall risk of a major supply crisis, with the supply ratio rising above 2 (that is, demand exceeds secure supply by at least 100%). Several things immediately jump out: major oil producers appear to be at the least risk, as their consumption of domestic production reduces their exposure to import disruption. Russia stands out as the one major producer to face a significant supply risk. The countries of the so-called 'Global South' appears to face higher risks than the more developed world, likely indicating a lower diversity of import sources.

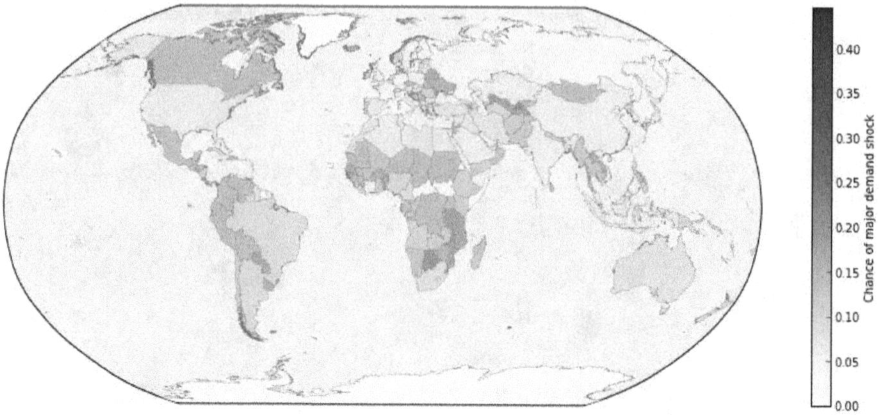

Fig. 5. Risk of a major demand shock (Demand Ratio < 0.5)

Figure 5 shows a similar map for demand risk, and indicates the likelihood of total possible exports exceeding secure exports by at least 100%. While there are fewer countries with very low risk, there are fewer with extremely high risk as well. Interestingly, the major Middle East producers appear to face relatively low demand risk, likely indicating a robust set of trade relationships with relatively stable countries. In contrast, the major non-US producers in the Americas – Canada, Mexico and Venezuela – appear to face higher demand risk. This appears to be due to the fact that their exports are disproportionately to the United States, rendering them vulnerable to US crises.

4 Discussion

In this paper, I have described an agent-based model to study near-future international energy security, with a focus on crude oil imports and exports. The model is data-driven, ties energy security to political crises and conflicts, and can study the effects of conflict contagion and of short-term oil production increases. We can draw two sets of conclusions from the model and its outputs: on energy security itself, as well as on the challenges of modeling it, and modeling international systems more broadly.

In terms of energy security, the model does not predict extreme shocks to be likely. Nevertheless, it shows that such extreme shocks may occur within the current configuration of the international oil system, and provides an estimate for their probability. The model indicates that consumption of domestic oil production provides a significant improvement in energy security, even in the absence of total energy independence. It also highlights the importance of diverse import and export partners. The model identifies the countries facing the greatest risks to their energy security, as well as making patterns in the risk visible – particularly the high risk in the countries of the 'Global South'.

Analysis of the Contagion and Assistance parameters yields additional interesting insight. The effect of contagion is statistically significant but small – perhaps suggesting the difficulty facing the ongoing debate over the phenomenon in political science, but also that the debate's conclusions may not be very important to policy. More counterintuitively, perhaps, the Assistance parameter does not have a significant effect. Here is somewhere the model can generate insights that are not otherwise obvious – specifically, that even if major oil producers can rapidly increase their output in response to crises, it is not enough to counterbalance the effects of those crises.

From a technical perspective, the model presented here is extremely simple. The interaction between agents is minimal and largely indirect, and with both Contagion and Assistance turned off, the model is essentially stochastically sampling from the space of possible crises and durations. Nevertheless, it is sufficient to capture the core aspects of the international oil system and produces useful results, demonstrating that the approach is viable. Future work can build on this framework, adding in additional submodels, dynamics and behaviors and measuring their effect against the baseline. Such additions may include explicit oil transportation and refining accounting for systemic choke-points; allowing countries to dynamically change their trading partners, supply and demand; incorporating the risks of non-political disasters both natural (e.g. Hurricane Katrina) and manmade (such as the Deepwater Horizon spill); and more.

However, as we have seen, even the simple model presented here had extensive input requirements, with data coming from various sources of varying completeness and reliability. Expanding the model would significantly increase the data requirements. Adding dynamics and behavior pose an even greater challenge. There is no complete theory of states' energy decisionmaking – in fact, such decisionmaking is almost certainly embedded within a broader policy framework, tied to both external interests and internal issues. It may be the case that adding additional elements will make the model more brittle without adding validity or increasing predictive power.

Finally, the model presented here may have another application, as a decision-support tool for policymakers. The model's user interface (both via input files and a graphical user interface) allows specific scenarios to be initiated and run, and their consequences and outputs evaluated. This may range from simply testing the effect of a particular events (e.g. a 12-month internal crisis in Saudi Arabia) to estimating the broader consequences of structural change (what would be the consequences if China were to become 10% less stable, or if Iran were to begin to export to Europe). By quickly setting up and observing a 'state of the world', a policymaker can augment her intuition, test hypotheses, and identify interesting issues or outliers for further qualitative or quantitative analysis.

References

1. International Energy Agency: Topic: Energy security (2013)
2. Yergin, D.: Ensuring energy security. Foreign Affairs 85(2), 69–82 (2006)

3. Wu, G., Liu, L.C., Wei, Y.M.: Comparison of china's oil import risk: Results based on portfolio theory and a diversification index approach. Energy Policy 37(9), 3557–3565 (2009)
4. Jewell, J.: International Energy Agency: The IEA model of short-term energy security. Technical report, International Energy Agency (2011)
5. Stirling, A.: Multicriteria diversity analysis: A novel heuristic framework for appraising energy portfolios. Energy Policy 38(4), 1622–1634 (2010)
6. Skea, J.: Valuing diversity in energy supply. Energy Policy 38(7), 3608–3621 (2010)
7. Jacobson, M.Z.: Review of solutions to global warming, air pollution, and energy security. Energy & Environmental Science 2(2), 148–173 (2009)
8. Cederman, L.E.: Emergent Actors in World Politics. Princeton University Press (1997)
9. Geller, A.: The use of complexity-based models in international relations: a technical overview and discussion of prospects and challenges. Cambridge Review of International Affairs 24(1), 63–80 (2011)
10. Hamilton, J.D.: What is an oil shock? Journal of Econometrics 113(2), 363–398 (2003)
11. Kilian, L.: Exogenous oil supply shocks: How big are they and how much do they matter for the U.S. economy? Review of Economics and Statistics 90(2), 216–240 (2008)
12. United Nations Statistical Division: COMTRADE (2013)
13. United States Energy Information Administration: Countries (2013)
14. Natural Resources Canada: Energy sector demand and consumption (January 2009)
15. United States Energy Information Administration: How much of the oil produced in the United States is consumed in the United States? (March 2013)
16. Mohamedi, F.: Iran primer: The oil and gas industry. Technical report, United States Institute of Peace (December 2010)
17. Rasmi, A.: The risks of Saudi oil consumption. The Daily Star Newspaper - Lebanon (June 2013)
18. Mearns, E.: Oil watch - OPEC crude oil production (November 2012)
19. Daya, A.: Saudi Arabia can raise output 25% if needed, Naimi says (March 2012)
20. Economist Intelligence Unit: Social unrest. Technical report (2013)
21. Goldstone, J.A.: Political Instability Task Force: A global forecasting model of political instability. Political Instability Task Force (2005)
22. Cioffi-Revilla, C., Midlarsky, M.I.: Power laws, scaling, and fractals in the most lethal international and civil wars. In: Diehl, P.F. (ed.) The Scourge of War: New Extensions on an Old Problem (2004)
23. Themnr, L., Wallensteen, P.: Armed conflicts, 1946-2012. Journal of Peace Research 50(4), 509–521 (2013)
24. Hendrix, C.S., Salehyan, I.: Social conflict in Africa database (SCAD) (2013)

Team PsychoSocial Assessment via Discourse Analysis: Power and Comfort/Routine

Christopher Miller, Jeffrey Rye, Peggy Wu, Sonja Schmer-Galunder, and Tammy Ott

Smart Information Flow Technologies (SIFT), 211 First St. N. #300, Minneapolis, MN USA 55401
{cmiller,jrye,pwu,sgalunder,tott}@sift.net

Abstract. We describe the theory, implementation and initial validation of a tool to assess interpersonal relationships and team states through non-intrusive discourse analysis. ADMIRE was developed to assess power/leadership relationships through scoring politeness behaviors in textual dialog to compute a graph of power asserted vs. afforded between individuals—an org chart—trackable over time. ADMIRE was formally tested in a military exercise where it proved 100% successful at deriving power relationships in 3 military chat rooms. Subsequent work extended ADMIRE's core approach with novel linguistic behaviors to identify "Team Comfort/Routine" (C/R) indicating when a team is performing in a well-understood, relaxed task context. Initial validation is provided by discriminating the disastrous Apollo 13 mission from others.

Keywords: Team dynamics, Leadership, Power, Stress, Routine, Psychosocial State Assessment, discourse analysis, politeness, textual analysis.

1 Introduction

The vast majority of human information exchange is accomplished through language, but language use is far more complex than simple semantic and syntactic content. We are exploring computational, model-based analyses of discourse using minimal linguistic processing and emphasizing social uses of linguistic forms. As others [e.g., 1-4], we are developing automated text processing to identify attitudes, emotions, intents and performance-related variables. Such automated analysis has great potential for gleaning knowledge about those involved, but it has previously either focused on personal texts such as diaries, journals, and blogs (e.g., [4]), or has tracked social networks [5] without delving into relationships between those who are interacting. The work we report here uses interactive discourse and relates it to computational models of language use to detect aspects of the views and relationships that each participant holds about the others, or about the task they are engaged in.

2 Inferring Relationships from Dialog: Power and Politeness

We previously developed a computational model of politeness in human interactions, based on Brown and Levinson [7]. Politeness behaviors are how we signal beliefs

W.G. Kennedy, N. Agarwal, and S.J. Yang (Eds.): SBP 2014, LNCS 8393, pp. 309–316, 2014.

about relationships and manipulate the others' beliefs. Brown and Levinson proposed that the function of politeness is to redress the face threat inherent in social interaction. In our implementation of their model [8], social interaction invariably threatens "face" and politeness "value" must offset the threat if the social status quo is to be maintained. The degree of face threat is a function of the power and social distance/ familiarity and of the degree of imposition of the interaction. Polite "redressive strategies" (such as the familiar behaviors we think of as "polite": "please", "thank you", honorifics, apologies, etc.) are used to offset the threat.

In prior work [7], we demonstrated a computational implementation of this model in culture and language training simulations. Recently, however, we have used observed politeness behaviors between individuals to infer their social relationships and attitudes toward each other. In principle, this is what an outside observer does naturally. If I see A calling B by an honorific, using formal language, making requests vs. demands, etc., I infer that A has (or believes s/he has) less power than B, that they are not close friends and/or that A is imposing on B to a degree. Our challenge was to enable a software algorithm to make similar inferences in an automated fashion.

Our approach, called ADMIRE (for Assessment of Discourse Media Indicators of Relative Esteem), parses and scores a corpus of textual data (email, IM Chat, discussion forum/blog, transcribed phone messages). The corpus is processed to identify speakers and addressees and eliminate misleading aspects such as quoted text from prior emails. The resulting interaction exchange set is run through our algorithm to recognize and score instances of politeness behaviors indicative of specific relationships between interactants, as described below.

2.1 ADMIRE's Architecture and Reasoning Approach

ADMIRE's algorithms and mathematical scoring techniques are described in depth in [8] and will only be summarized here. ADMIRE uses multiple techniques to create an interaction network by segmenting and assigning the input text stream into conversations (utterances and responses) from an actor to one or more others. It then uses a set of *Politeness Behavior Recognizers* (PBRs) to detect redressive actions in these utterances which accord or assert power, using strategies described in Brown and Levinson [7]. PBRs are simple linguistic recognizers which detect verbal patterns reflecting the speaker's perception about a social relationship. Generally, these could be written as simple rules using regular expressions and Boolean logic with a few key words, but some use deeper natural language processing and part of speech tagging.

Once recognized, behaviors are scored. Those which are highly indicative of a relationships (e.g., power) are weighted highly, whereas more common and "noisy" ones are given smaller weights. Scoring may be based on behavior frequency (linear), on presence given the number of opportunities (rate-- e.g., a greeting can generally be used only once per conversation) and, occasionally, only on absence (e.g., failing to respond to a greeting), with linear scoring most common. Once scored and weighted, the results for recognized PBRs are summed on a per-utterance basis. For power/leadership detection (described below), PBRs can be either positive (indicative of according power) or negative (indicative of asserting power) and the net value of each utterance is accorded from speaker to each hearer addressed.

This set of PBRs and their scoring can be reconfigured for different analyses—e.g., different PBRs or weights for different subcultures. The use of "Please" or gratitude expressions in many military chat environments distinctly suggests formality and that speaker has power over hearers, while in "normal" dialog they weakly suggest the opposite. Using the same basic algorithm but replacing PBR sets wholecloth converts from, for example, computing power relationships to, in our recent work, the degree of "comfort" or "routine" the team is experiencing.

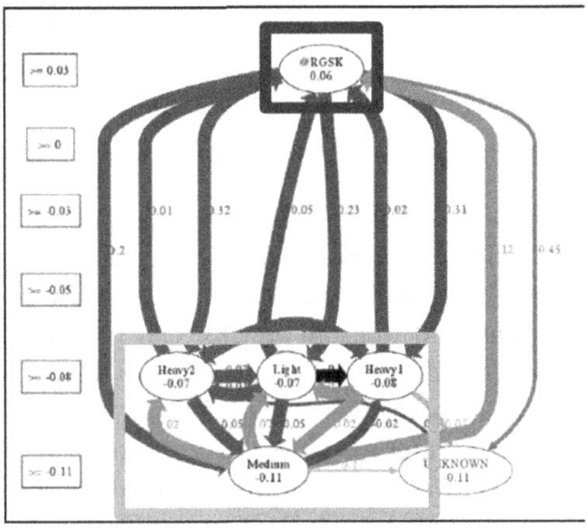

Fig. 1. Network graph of power scores and relationships derived for one chat room over a full week

Fig. 1 illustrates both a PBR set and their relative contributions to power analysis for chat room exchanges during a military exercise [8]. The PBR strategy "Leading Directive" (use of a directive verb first in an utterance—e.g., "Zoom in" or "Scan left") was the most common strategy (by count) and contributed the most value to the resulting analysis though, once weighted, that value was negative (power asserting). Other common strategies included "OutGroup" (use of second person pronouns and verbs—"you", "your", etc.) which were also weighted negatively, and of "InGroup" (first person plural pronouns and verbs—"we", "our", etc.) which were weighted in a positive direction (power according).

To graph power relationships—an "org chart"— we use a physics simulation to treat each link as a spring with a constant based on the number of interactions for that node and an equilibrium distance. The force-directed layout treats links with more messages as "stiffer"; greater effect on the results. Layout is one-dimensional and springs are directed, which makes positive links go up and negative links go down. Initially, the force-directed spring simulation sets each rank score at the mean of the politeness shown to that participant. The ranking algorithm then iteratively simulates spring physics until the system is stable using the "force" of politeness to pull or push a node based on the strength of politeness

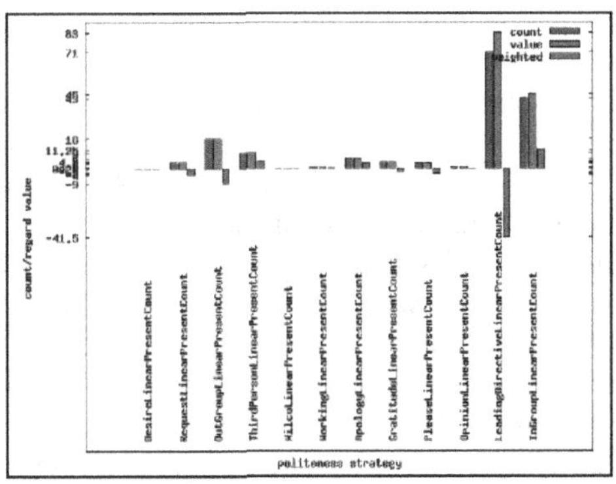

Fig. 2. Count, value and weight of behaviors from chat room text during a military exercise

behaviors directed to and from that participant. The result is a graph such as that shown in Fig. 2.

2.2 Power Assessment Application 1—Military Chat

ADMIRE was initially demonstrated on military IRC chat generated during a week-long military exercise where forces manned advanced equipment to gather intelligence on simulated communities (played by other military members) to identify insurgents and forestall attacks. Our participants coordinated sensor and analysis activities in 3 different IRC chat rooms during ~2 hr "vignettes" throughout the week. Each chat room involved 8-13 separate vignettes and 5-25 individuals. ADMIRE's goal was to infer the power structure in each room. We had a priori "ground truth" for the key power relationships expected for each room through predefined command structures. This knowledge was used for evaluation only, not in automated analysis of messages. For example, for the room in Fig. 2, formal command structures dictated that "RGSK," the unit commander, oversaw sensor operations by four subordinates (Heavy1, Heavy2, Medium and Light) and our evaluation criterion was whether or not ADMIRE would detect this relationship—as, in fact, it did (see Fig. 2).

Taken over the full week (13 sessions for one chat room, 8 for each of the others), ADMIRE's results were 100% successful at detecting the core "ground truth" relationships. Even when applied to individual vignettes (1-2 hours of data), ADMIRE's conclusions averaged 84% accurate. Results provided hints of even greater sensitivity: identifying individuals who exhibited subtle variations in power, and sessions during which command structures were inverted or in which one individual exerted more directive authority than was typical, etc.

2.3 Power Assessment Application 2: Apollo Radio Communications

We were subsequently awarded a NASA Research Agreement to investigate non-intrusive psycho-social state assessment using ADMIRE and other language analysis tools. Especially in long duration, deep-space missions, NASA needs awareness of crew psychological health and team relations. Current paths to such knowledge are all intrusive on either the astronauts' time (surveys, interviews) or physical state (wearable sensors) or both. Non-intrusive analysis of existing data streams (video, audio, textual) has potential to augment or replace more intrusive methods.

In this research, we used ADMIRE (as configured for military chat) to analyze transcribed radio communications from each of the 10 manned Apollo missions. While the PBRs used were not explicitly optimized for this environment, one goal was to explore whether PBR configurations from one specialized "culture" would work for a somewhat similar one. Another challenge was that, while there is a formal command structure in NASA operations, it is frequently loosely applied. NASA needs its astronauts to have great expertise and to take initiative when necessary. Thus, "leadership" ("directing the action" as opposed to formal command structures) tends to come from whoever has the most pertinent expertise and/or knowledge. While this makes our verbal assessment of power relationships are arguably more accurate to

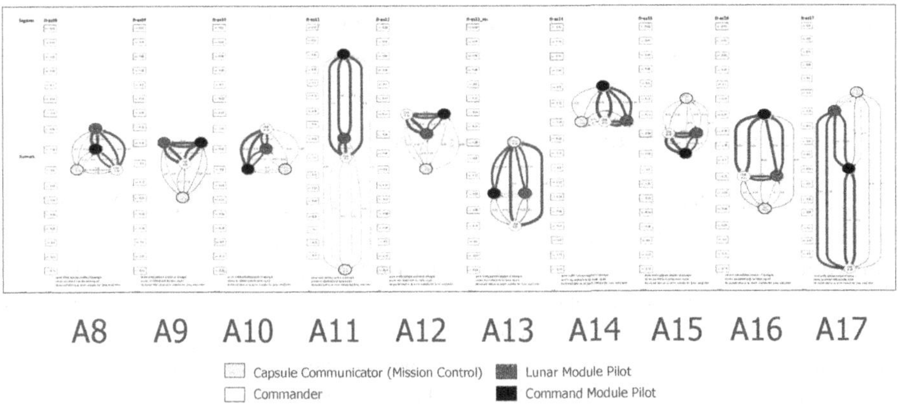

Fig. 3. Networked power/leadership graphs from communications for 10 Apollo missions

actual shifts in leadership throughout a mission, it also means that accurately assessing ground truth to evaluate our assessment success is more difficult.

Fig. 3 shows power/leadership assessments from the transcribed audio for each multi-day mission. In all transcripts, there were 4 primary speakers by role:

- *CapCom*—the ground-based "Capsule Communicator" who coordinates and generally delivers communications from all ground staff. CapCom is not the ground-based mission leader (the Flight Director's role), but rather a communication coordinator who relays messages to and from ground and space crew.
- *Commander (Cdr)*—in overall command of the space mission
- *Command Module Pilot (CMP)*—the astronaut who remains on the Command Module in space while the Cdr and LMP landed on the moon.
- *Lunar Module Pilot (LMP)*—the pilot of the Lunar Module and generally the lowest ranking individual on board. Both CMP and LMP had tasks before and after separation and were on flights where there was no separation or no Lunar Module.

While we can provide only qualitative analysis, one finding was the variety in leadership across crews. Another is that in missions focusing on specific units (e.g., the lunar module), the pilot for that unit tended to exhibit more power/leadership– e.g., Apollo 8 and 9 both focused on the command and lunar modules and the LMP takes on a higher leadership role than in other missions. Apollo 13, the disastrous mission where an Oxygen tank explosion after liftoff caused mission abort and threatened the crew's safe return, provides an interesting "natural experiment". Here, CapCom took on a strong leadership role that was different than any other mission. Though CapCom also exhibited high leadership in 15 and 17, the frequency of communications (as signaled by line width) was substantially lower–indicating lack of robustness for this power assessment. But in Apollo 13, CapCom exhibited more leadership over a longer interval than any other mission. In an hourly analysis, CapCom was ranked first or second during a critical 50 hr interval following the explosion in all but one hour. This corresponds to CapCom's increased role in relaying repair instructions (directives) to the spacecraft. By contrast, in the more typical Apollo 12, CapCom ranked first or second in less than 20% of intervals with no consistent pattern.

3 Comfort/Routine—Developing a New Assessment Technique

We have recently had additional success developing an assessor for the "Comfort" or "Routine" a team exhibits in communication during task performance—evidence of familiarity and lack of stress. In Apollo transcripts, other NASA domains, and from the literature, we saw indicators of a substantially different verbal behaviors when a team was "comfortable" with their task vs. not (for whatever reason). Comfort/Routine (C/R) is most typical when following a well understood routine without emotional, cognitive or physical stress– where everyone knows their job/role and things are going smoothly. We then sought automatically recognizable verbal behaviors indicative of positive vs. negative C/R. The resulting metric combines recognition for unfamiliar contexts (e.g., "cognitive" language indicative of causal or complex thought and "exclusionary" language indicative of contrasting alternate ideas) and recognition of adverse, cognitively or emotionally stressful environments (anger or anxiety terms) [4]. Increased use of past tense verbs indicates non-C/R because it is used more frequently during task performance to blame, diagnosis or simply for reviewing prior actions, all inconsistent with routine behaviors. In-group markers ("we" vs. "I" and "you") were seen as inversely related to C/R based on [11]. Filler terms ("uh," "er", "um") are rare in text interaction but when used, frequently indicate hesitancy or lack of agreement—and hence non-C/R. Finally, other candidates from the literature were tried and proved either not to be predictive of C/R within our implementation (e.g., length of utterance/words and prepositions use—suggested by [4] as indicating complex thought and therefore of non-C/R) or were impossible to reliably discriminate with the simple linguistic tools we were using (e.g., implicit task utterances like "now" vs. explicit ones like "push the red button now" have been found by [3, 12] to increase in skilled teams experience time pressure and stress).

Multiple behavior recognizers were used in the algorithm described. C/R could range from positive to negative infinity based recognized behavior count and then weighted positively or negatively. In practice, however, all behaviors we used were negatively weighted (cf. Fig. 4 below) and, therefore, most observed C/R scores ranged from zero to negative values. For C/R, we were less interested in pairwise relationships (though these can be graphed via the same spring algorithm), than in the overall summed or averaged score for the team. The result is an indication of the C/R value for the team in the context where the textual interactions were generated.

Fig. 4 illustrates behaviors recognized, scored and weighted by our C/R analyzer in data from a team interaction experiment where three subjects participated in a multi-hour planetary exploration simulation, coordinating their activities via textual chat [13]. We later had human raters not trained in our techniques read and score the communications for the degree of "team functionality" exhibited. In a resulting correlation analysis, human raters and our C/R analyzer agreed (R^2= .329, df= 15, p=.016).

We also applied the C/R analyzer to the same Apollo radio data above. While there is no objective measure of how "comfortable" or "routine" each Apollo mission was, it was reasonable that the Apollo 13 mission was the least "routine". In fact (Fig. 5), this was the result, with the computed score for C/R nearly 70% more negative than that for the other missions' average.

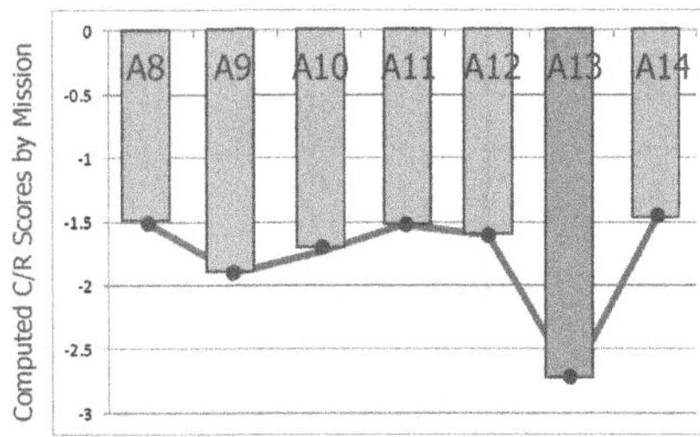

Fig. 4. Comfort/Routine scores for communications in each of 7 Apollo missions

We also computed C/R values for communications from each mission hour. We expected generally routine scores (comparatively high C/R) until the O^2 explosion, then a precipitous drop, followed by a climb back to "normal" C/R as the emergency was resolved. Instead, we saw much more variance (Fig. 6). In retrospect, this is fully understandable. NASA has an extraordinarily rich procedure environment and with the initial explosion, the crew began execution of familiar and well-drilled procedures associated with that type of problem in space. It was only when it became apparent that many subsystems were damaged, that they would have to abort the mission, that safe return to earth was uncertain, and especially that there was no clear procedure on how to react in this situation, that C/R decreased substantially. Even then, there were a series of problems to be solved and new problems discovered, with hourly triumph or defeat depending on outcomes, which continued to perturb C/R throughout the mission– exactly as our graph showed.

4 Future Work

This work is ongoing and novel data are being generated in individual and team experiments where concurrent survey data is being collected to serve as ground truth for validating our assessment techniques. Initial results, as described here, are promising and we are expanding the range of techniques we can bring to bear on NASA psychosocial state assessments. These results are already being delivered at rates far faster than the alternative: human mark up and coding of recorded speech or text data. Thus, it seems fair to conclude that these non-intrusive approaches to automated analysis of dialog to obtain team psychosocial states hold promise for future use in, at least, space exploration domains.

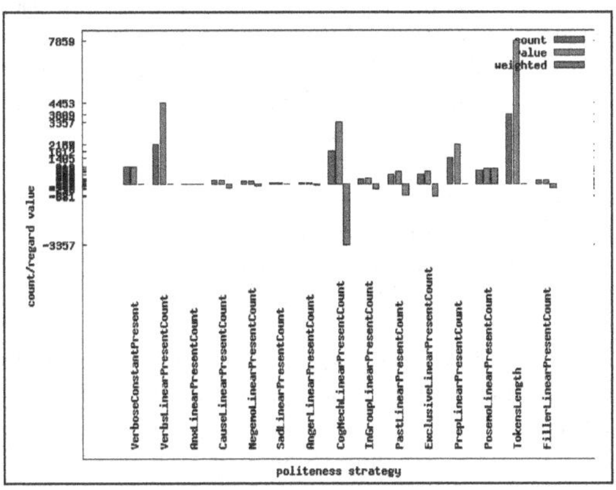

Fig. 5. Application of the C/R analyzer to team interaction chat from a planetary exploration exercise (left) and correlation between human ratings of that chat and the C/R scores.

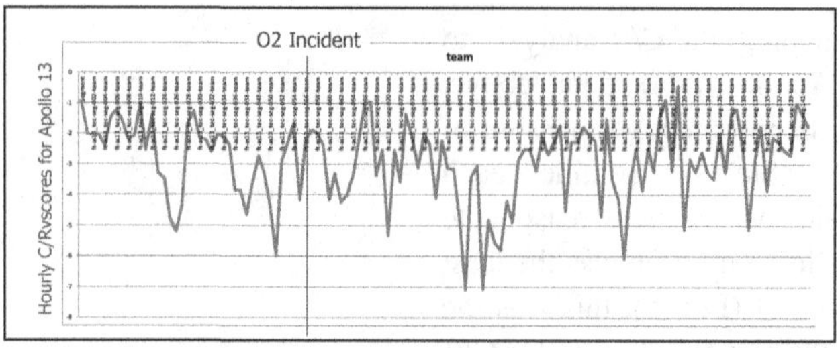

Fig. 6. Hourly C/R scores for Apollo 13 communications

Acknowledgments. This work was sponsored initially by The U.S. Office of Naval Research under contract #N00014-09-C-0264, with subsequent support from NASA's Human Research Program under contract #NNX12AB40G.

References

1. Bantum, E., Owen, J.: Evaluating the Validity of Computerized Content Analysis Programs for Identification of Emotional Expression in Cancer Narratives. Psych. Assessment 21, 79–88 (2009)
2. Landauer, T., Foltz, P., Laham, D.: Introduction to Latent Semantic Analysis. Discourse Processes 25, 259–284 (1998)
3. Shah, J., Breazeal, C.: Empirical Analysis of Team Coordination and Action Planning With Application to Human-Robot Teaming. Human Factors 52(2), 234–245 (2010)
4. Tausczik, Y., Pennebaker, J.: The Psychological Meaning of Words: LIWC and Computerized Text Analysis Methods. J. of Language and Social Psych. 29(1), 24–54 (2010)
5. Scott, J., Carrington, P. (eds.): The SAGE Handbook of Social Network Analysis. SAGE Publications (2011)
6. Brown, P., Levinson, S.: Politeness: Some Universals in Language Usage. Cambridge Univ. Press, Cambridge (1987)
7. Miller, C., Wu, P., Funk, H., Johnson, L., Viljalmsson, H.: A Computational Approach to Etiquette and Politeness: An "Etiquette Engine™" for Cultural Interaction Training. Proc. Behavior Rep. In: Modeling and Sim., Norfolk, VA, March 26-29, pp. 26–29 (2007)
8. Miller, C., Rye, J.: Power and Politeness in Interactions; ADMIRE—A Tool for Deriving the Former from the Latter. In: Proc. ASE Conf. on Social Informatics, pp. 177–184. IEEE, New York (2012)
9. Gonzales, A.L., Hancock, J.T., Pennebaker, J.W.: Language Style Matching as a Predictor of Social Dynamics in Small Groups. Communication Research 37(1), 3–19 (2010)
10. Serfaty, D., Entin, E., Deckert, J.: Team Adaptation to Stress in Decision Making and Coordination with Implications for CIC Team Training (Report No. TR-564, Vol. 1/2). ALPHATECH, Burlington (1993)
11. Gonzales, A.L., Hancock, J.T., Pennebaker, J.W.: Language Style Matching as a Predictor of Social Dynamics in Small Groups. Communication Research 37(1), 3–19 (2010)
12. Serfaty, D., Entin, E., Deckert, J.: Team Adaptation to Stress in Decision Making and Coordination with Implications for CIC Team Training (Report No. TR-564, Vol. 1/2). ALPHATECH, Burlington (1993)
13. Roma, P., Hursh, S., Hienz, R., Emurian, H., Gasior, E., Brinson, Z., Brady, J.: Behavioral and Biological Effects of Autonomous vs Scheduled Mission Management in Simulated Space - dwelling Groups. Acta Astronautica 68, 1581–1588 (2011)

Social TV and the Social Soundtrack: Significance of Second Screen Interaction during Television Viewing

Partha Mukherjee and Bernard J. Jansen

College of Information Science and Technology, Pennsylvania State University, University Park, PA, USA
pom5109@ist.psu.edu, jjansen@acm.org

Abstract. The presence of social networks and mobile technology in form of secondary screens used in conjunction with television plays a significant role in the shift from traditional television to social television (TV). In this research, we investigate user interactions with secondary screens during live and non-live transmission of TV programs. We further explore the role of handheld devices in this second screen interaction. We perform statistical tests on more than 418,000 tweets from second screens for three popular TV shows. The results identify significant differences in the social element of second screen usage whilst the show is on air versus after the live broadcast, with the usage of handheld devices differing significantly in terms of the number of tweets during live telecast of TV shows. Desktop devices as second screens are also significant mediums for communication. This research identifies the change in users' interaction habits in terms of information sharing and interaction for TV broadcasts via the presence of a computing device as a second screen.

Keywords: second screen, multiple screen interaction, social television, social sound track, social media.

1 Introduction

With the advent of Internet technology and the emergence of online social networking, the social possibility of TV has greatly expanded, as the merging of technologies now allow a number of social activities and conversation concerning TV content via social networks (e.g., Facebook, Twitter, Weibo etc.). This combination of TV and online social networks has forged a social TV that imparts feelings of togetherness and communication among people in dispersed locations. The social network has embedded itself within the modern TV culture and it acts as a social soundtrack for TV content.

The social soundtrack is an interesting communication interactivity that can be both real time (i.e., during the live broadcast) and non-real time (i.e., not during the live broadcast) based around TV shows. This social commentary happens on social networks, especially on Twitter. The integration of Twitter (or other online social

W.G. Kennedy, N. Agarwal, and S.J. Yang (Eds.): SBP 2014, LNCS 8393, pp. 317–324, 2014.

network) as the interactive medium with televised broadcasts marks the emergence of a new phenomenon augmenting the prior social aspect of TV. This new usage phenomenon is referred to as second screen (TV and computing device), although there may be multiple screens involved (TV and several computing devices). With the second screen phenomenon, the TV is the base device while the secondary screen is the computer (desktop or laptop), tablet, or smartphone. The secondary screen allows the social soundtrack, the conversation with others, regarding the particular TV program.

In this research, we investigate the use of the secondary screen during the telecast of three popular TV programs, specifically examining if viewer interactions differ during and after the live broadcast. We also examine the extent to which handheld devices play significant roles during second screen interaction. This research is important as the degree of usage of secondary screens can facilitate the personalization of TV content and advertising with the design of mobile apps. Findings can also assist both the channel owners and advertisers to formulate new strategies for TV airing, launching apps for product ads to engage more viewers, promote sales, and earn revenues.

2 Related Work

2.1 Social TV (sTV)

There is existing literature concerning socializing aspect of TV. Since the emerging social TV (sTV) relates to interaction among people associated with program watching, sTV highlights a special form of socialization, where the TV content is accompanied by social interaction. sTV is an emerging communication technology and practice in the fields of interactivity, mass media, computing, and entertainment.

sTV refers to a broad spectrum of systems that support and investigate the social experiences of TV viewers. Researchers have developed a series of sTV prototypes. Abreu, Almeida and Branco [1] developed the 2BeOn sTV system with the goal of making the users go online to ensure interpersonal communication. Coppens, Trappeneirs and Gordon [2] introduced the AmigoTV prototype, which brings personal content, voice communication, and community support by offering real time interaction among a group of friends around a TV show. Regan and Todd developed Media Center Buddies [3] that permits multiple users to engage in instant messaging while watching TV together.

Though the proposed sTV systems enrich TV based socialization, their design goal was to allow a small group to share a feeling of sociability or togetherness. The degree of sociability increased significantly via the integration of online social networks services with TV viewing. The conversation interactions in the social soundtrack can be considered as a form of end-user enrichment of the TV content [4]. Such end user enrichment enhances the sTV socialization via the user generated content and converts sTV into a community [5].

2.2 Integration of Prior Work

Despite the prior literature in sTV, there has been limited research on second screen interactions. Leroy, Rocca, Mancus and Gosselin [6] analyzed users' second screen behavior concerning where and when people look at their TV. Zhao, Zhong, Wickramasuriya and Vasudevan [7] extracted viewers' sentiments about US National Football League teams by analyzing real time tweets. Lochrie and Coulton [8] showed that Twitter is increasingly used as real time audience communication channel for sharing TV experiences.

Though the aforementioned research speaks to the usage of social networks to make TV more social via viewer interactions with the technology, the studies regarding significance of social interaction using secondary screen is scarce. As such, there are several unanswered questions concerning the second screen phenomenon. How is social media used during live TV broadcasts? How is social media used after the live broadcasts? What is the effect of mobile devices on the social soundtrack? What are the discussion patterns of second screen usage? What are the interaction points between TV and social media? These are some of the questions that motivate our research.

3 Research Question

Social networks allow for TV programs to be accessed and shared by viewers in a variety of ways. The audience can join in discussion while watching the show and have their interactions be viewed and responded by other members. Such interaction may or may not be active in terms of real time discussion during live telecast of the programs. The second screen technologies, such as smartphones, tablets and laptops / desktops, facilitate these interactions to occur anytime, anywhere irrespective of airing times of the TV shows. Commercial and other organizations are increasingly keen to capitalize on this second screen technology interaction to measure the viewing habits and reactions of the audience. Participating in active discussion using second screens while experiencing live TV shows may lead to an index regarding change in user's TV viewing habit. This issue formulates our first research question: -

1. *Is the usage of secondary screen significant for social interactions during live transmission of TV shows?*

There are two perspectives that are highlighted by the aforementioned research question. First, the significance of a secondary screen during live broadcast of TV shows identifies the adaptation of social network sites as the driver of real time conversation from the perspective of the audience whilst the program is on air. We refer to this as real time interaction (i.e., during the live broadcast). Second, the active interaction via social networks might create commercial opportunities by the advertising agents at the intersection of the social networks, the TV, and the second screen. So, the first research question regarding social significance of secondary screens during live show time begets the following research hypothesis.

Hypothesis 01: There is a significant difference between real time and non-real time social interaction based on a TV show w.r.t usage of secondary screen.

We further explore the presence of second screen technology during such sTV phenomenon in terms of handheld device usage. The ease of portability, mobility, and usage has triggered the widespread use of mobile devices, particularly the "always on", "always there" and "always connected" attributes. Thus, place and time independence play pivotal roles in promoting the touchscreen devices such as smartphones (e.g., iPhones, Android, Blackberry etc.) or tablets (iPad, Booklet, GriDpad etc.) for interactions in social networks. The intuition that tactile experience provides more interaction during live telecast refers to the second research question:

2. *Is the usage of mobile devices as secondary screen significant for interaction during live transmission of TV shows?*

Investigation of this question is important as the growing focus of production and consumption of interactions in the form of micro-blogs, texts, and symbolic messages helps to make mobile devices unique platforms for the development of personalized apps for social commerce. The apps can better exploit mobile devices' native capabilities for a more responsive and seamless experience. Based on the second research question, hypothesis 2 is formed.

Hypothesis 02: There is a significant difference between real and non-real time social interactions among viewers during the course of a TV show w.r.t use of mobile devices.

Hypothesis 02 indirectly leads us to one more hypotheses. Hypothesis 03 refers to the comparison between mobile and non-mobile devices (e.g., laptop, desktop etc.) to be used as the second screen medium for interaction during a live TV program.

Hypothesis 03: There is a significant difference in usage of mobile and non-mobile devices as secondary screen for interactions among viewers during live telecast of a TV show.

4 Data Collection

We selected three popular TV shows and collected users' interactions in form of tweets from Twitter. The TV shows selected for this research are: 1) *Dancing with the Stars,* 2) *Mad Men*, and 3) *True Blood*. The shows belong to three different genres and the reason for considering a spectrum of genres is to deal with the generalizability issue for the significance of second screen in social interaction during and after live broadcast of the TV shows. The tweets for *Dancing with the Stars* were collected for two consecutive weeks starting from 13th May 2013 to 25th May 2013. These two weeks account for selection of finalists and champion for season six respectively. Regarding *Mad Men* and *True Blood,* we collected tweets for three successive weeks in the month of June. For both shows, it spans from 9th June to 29th June 2013. As 23rd June was the date for the season finale for *Mad Men,* we stopped collecting tweets for both *Mad Men* and *True Blood* the following week. For each show, the tweets posted in English texts were collected. The number of tweets for *Dancing with the Stars,*

Mad Men and *True Blood* are 46269, 152259 and 220390 respectively. Here the queries are the TV show names. The number of tweets for *Dancing with the stars* is less than that for other two TV shows as the version of Twitter API for *Dancing with the Stars* tweets was older (API 1.0) compared to that used (API 1.1) in tweet collection for other two shows.

For identification of the use of mobile technology for viewer communication, we leveraged the *"Source"* field of the tweets in Javascript Object Notation (JSON) format. We classified mobile tweets based on the existence of keywords within *"Source"* attribute. The keywords we used were *"iPhone", "iPad", "Android", "iOS", "Blackberry"* and *"Mobile Web"*. *"Mobile Web"* refers to access browser based internet services from handheld mobile devices. If the *"Source"* of a tweet contains any of these keywords, we classified it as a mobile tweet. Otherwise, we classified it as a non-mobile tweet. The keyword list seems to be exhaustive to classify mobile tweets from *"Source"* attribute.

We first focused on those tweets collected on the days of live telecast of the TV programs to measure the second screen effect on user interaction during the live broadcast. We segregated the count of tweets collected in 24x7 hours across the weeks for all three TV shows into fifteen minutes intervals and stored the time versus count data in three spreadsheets, each for one televised program. We also stored the numbers of mobile and non-mobile tweets with the intervals of fifteen minutes for each show in the time vs. count format. We annotated the timings of the tweets generated and categorize them as "real time second screen" (rtSS) and "non-real time second screen" (nrtSS) tweets w.r.t Eastern Daylight Time (EDT). The rtSS tweets indicate that the tweets are posted during live broadcasts. The nrtSS counterparts are the ones posted by the users while the TV shows are not on air. We need to focus on the tweets as rtSS tweets collected in hours shown in Table 1 combining the show timings of all six different US time zones (i.e., Eastern, Pacific, Central, Mountain, Alaska and Hawaii) considering the time differences w.r.t EDT. The airing time for all three TV shows is about 60 minutes each day except the week for champion selection for *Dancing with the Stars*. The airing time of *Dancing with the Stars* in final week is about two hours each day.

Table 1. Time in hour w.r.t EDT focusing collection of rtSS tweets for three TV shows

	Sun	Mon	Tue	Wed
Dancing with the Stars		8 PM, 9 PM,11 PM	9 PM, 10 PM	12 AM
Mad Men	10 PM	1 AM		
True Blood	9 PM	12 AM		

5 Result

We transformed the data via the Box-Cox transformation using log transformation function *log(variable + 1.0)* as the data is not normal. Using the log transformation, the data is successfully normalized, though a bit of skewness exists on the left (i.e., the histogram of transformed data nearly follows normal distribution). The Box-Cox transformation of the non-normalized data increases the efficiency of univariate t-tests [9]. In SPSS, we ran three separate t tests between relevant combinations of (rtSS, nrtSS) and (mobile, non-mobile) pairs to evaluate the hypotheses. Two tailed statistical t-tests were carried out at 95% confidence interval where t-statistic for each test is 1.96 at significance level $\alpha = 0.05$ for large degrees of freedom. $\alpha = 0.05$ is feasible standard to catch effects, large enough to be of scientific interest.

For hypothesis 01, the two-tailed t test result indicates that there is a significant difference between the rtSS and nrtSS categories of tweets for all three TV shows as observed in Table 2. So, hypothesis 01 is fully supported. The statistical significance of hypothesis 01 led us to explore the direction of the second screen effect. For one tailed two sample t test we have $t_{critical} = 1.645$ at $\alpha = 0.05$. From Table 2, it is evident that t-values of all three shows are significantly greater than $t_{critical}$. So, we conclude that social interactions in terms of number of postings during the live telecast of the shows are greater than the postings when the shows are not aired.

Table 2. T-statistic with p-values obtained from two-tailed t test between rtSS & nrtSS tweets for the TV shows

	T-value	p-value
Dancing with the Stars	11.152	0.00 < 0.05
Mad Men	8.017	0.00 < 0.05
True Blood	5.060	0.00 < 0.05

For Hypothesis 02, we perform t test between two independent samples: 1) tweets posted from handheld devices during live broadcast of the shows (rtSS) and 2) tweets posted from handheld devices when the shows are not live (nrtSS). The results in Table 3 show that there are significant differences of means of mobile tweets between rtSS and nrtSS categories of postings for all three TV shows. The result fully supports hypothesis 02. We carried out one tailed two sample t test to investigate the direction of second screen effect on mobile interactions. From Table 3, it is evident that t-values of all three shows are significantly greater than $t_{critical}$ (1.645). This infers that the mean of usage of mobile devices for real time sharing of TV experiences is greater than that for sharing of views when the programs are not being televised real time.

Table 3. T-statistic and p-values obtained from two tailed t test for rtSS & nrtSS mobile tweets

	T-value	p-value
Dancing with the stars	13.173	0.00 < 0.05
Mad Men	8.140	0.00 < 0.05
True Blood	4.956	0.00 < 0.05

For hypothesis 03, the result of the two tailed t-test, displayed in Table 4, does not infer that there is statistical significance between the usage of mobile and desktop devices for interaction when a TV show is airing.

Table 4. T-statistic with p-values obtained from two tailed t test for rtSS tweets posted by mobile & non-mobile devices

	T-value	p-value
Dancing with the Stars	1.465	0.145 > 0.05
Mad Men	1.887	0.064 > 0.05
True Blood	0.797	0.429 > 0.05

6 Discussion

In this research, we investigated two main research questions pertaining to second screen interaction highlighting the use of social networks in sharing conversations about TV shows. Both of these questions are examined from perspective of human information processing in terms of the volume of tweets posted about the TV media content. Intuitively, the proximity of second screens will increase the social interaction among the viewers about the TV shows. Our results show that the use of second screen during live transmission of TV occurs based on the rise in real time social communication. The findings are important as they indicate the changed of TV viewing habit of the users in terms of information sharing behavior and social interactivity. During live transmission of the TV programs, the presence of a second screen generates significantly more tweets than when the program is not live. It is noteworthy that usage of mobile devices gains considerable momentum during such second screen interaction when the TV show is on the air. Though there is rise in usage of mobile phones for sharing communication during live transmission of a TV show, the use of desktop devices for posting of tweets is still strong for second screen usage. It is interesting, from Table 4, that neither mobile nor desktop devices claims clear majority as a medium of second screen technology. Our assumption is that viewers may be watching the shows or reviewing the social soundtrack on their desktop computers. Thus, it is easier for them to share social media content via that hardware platform.

Regarding practical implication retailers can advertise their brands or services during a live TV show, targeting the users on social media as potential consumers. The growth of social interaction via second screen during live transmission of TV shows increases the possibility of purchase intent among the potential consumers of the brand. On the other hand this will help broadcasters to generate increased revenue from retailers by allowing them to air personalized ads during the TV shows.

7 Conclusion

Though our data set spans for two/three weeks for each TV show, the results regarding evaluating the interaction effect of the second screen phenomenon in this research indicates that the adjacency of second screen during live transmission of TV shows generates higher social interaction relative to that when the show is not airing live. In addition, higher engagement of mobile users is observed when the show is live. We believe that our research provides valuable novel contribution concerning user's behavior and interaction while viewing of mass media in a relatively new but emerging avenue of user behavior research and personalization of ad and TV content.

For future work, we will test the effect of the second screen phenomenon mechanism over a number of TV shows from different genres across several subcontinents to find whether the results translate to a spectrum of genres and non-US TV shows with large data set. We will further conduct content analysis of the social soundtrack to determine the content of the conversation occurring via user interactions with the second screen.

References

1. Abreu, J., Almeida, P., Branco, V.: 2BeOn-InteractiveTelevision Supporting Interpersonal Communication. In: Multimedia 2001, pp. 199–208. Springer (2002)
2. Coppens, T., Trappeniers, L., Godon, M.: AmigoTV: towards a social TV experience. In: Proceedings from the Second European Conference on Interactive Television "Enhancing the Experience", University of Brighton (2004)
3. Regan, T., Todd, I.: Media center buddies: instant messaging around a media center. In: Proceedings of the Third Nordic Conference on Human-Computer Interaction. ACM (2004)
4. Cesar, P., et al.: Enhancing social sharing of videos: fragment, annotate, enrich, and share. In: Proceedings of the 16th ACM International Conference on Multimedia. ACM (2008)
5. Alliez, D., France, N.: Adapt TV paradigms to UGC by importing social networks. In: Adjunct Proceedings of the EuroITV 2008 Conference (2008)
6. Leroy, J., et al.: Second screen interaction: an approach to infer tv watcher's interest using 3d head pose estimation. In: Proceedings of the 22nd International Conference on World Wide Web Companion (WWW) (2013)
7. Zhao, L., Wickramasuriya, J., Vasudevan, V.: Analyzing twitter for social tv: Sentiment extraction for sports. In: Proceedings of the 2nd International Workshop on Future of Television (2011)
8. Lochrie, M., Coulton, P.: Sharing the viewing experience through second screens. In: Proceedings of the 10th European Conference on Interactive TV and Video. ACM (2012)
9. Freeman, J., Modarres, R.: Efficiency of t-Test and Hotelling's T 2-Test After Box-Cox Transformation. Communications in Statistics-Theory and Methods 35(6), 1109–1122 (2006)

Social Network Structures of Primary Health Care Teams Associated with Health Outcomes in Alcohol Drinkers with Diabetes

Marlon P. Mundt and Larissa I. Zakletskaia

Department of Family Medicine, University of Wisconsin School of Medicine and Public Health, Madison, Wisconsin, USA
Marlon.mundt@fammed.wisc.edu

Abstract. This study evaluates if social network structures in primary care teams are related to biometric outcomes of diabetic alcohol drinkers. The study results show that primary care teams with less hierarchical face-to-face social networks (i.e. more connection density, more 3-tie closures and less network centrality) have better controlled HbA1c, LDL cholesterol and blood pressure among their diabetic alcohol drinking patients. Notably, more interconnected primary care teams, with members who engage others in face-to-face communication about patient care, who feel emotionally supported by their coworkers, and who feel like they work with friends, share the same goals and objectives for patient care and have better patient outcomes, as evidenced by the diabetic biometric measures of their team's patients.

Keywords: alcohol, social networks, primary care, diabetes.

1 Introduction

Diabetes is a pervasive public health problem. In the US, diabetes prevalence is 7% [1]. Diabetes is a risk factor for cardiovascular morbidity and mortality, with diabetes costs exceeding $91 billion per year [2]. Glycemic control (HbA1c) is a key modifiable predictor of diabetes complications. A one percent reduction in Hb1Ac is associated with a 21% reduction in risk for diabetes-related mortality and a 37% reduction in risk for microvascular complications [3].

Given the tremendous disease burden and the economic costs, it is essential to understand how to provide optimal care to diabetic patients who consume alcohol [4-10]. Primary care (PC) providers may be best suited to offer the most effective care to alcohol drinkers with diabetes due to their ongoing relationships with patients, their understanding of patient health concerns in the context of other medical problems, and more frequent opportunities to discuss self-care behavior [11, 12]. Close to 85% of diabetes patients are treated in primary care [13].

Primary care (PC) practitioners work as care teams. Notably, PC team social interactions, or social ties, may be of critical importance for delivering high quality care to diabetic alcohol drinkers [14-20]. Rigorous scientific investigation of the causal

W.G. Kennedy, N. Agarwal, and S.J. Yang (Eds.): SBP 2014, LNCS 8393, pp. 325–332, 2014.

pathways linking PC team social interactions to health care outcomes of alcohol-drinking diabetic patients is missing and is sorely needed.

To fill this gap, the present study uses a social network analysis approach to investigate if PC team social connections (or social networks) relate to the biometric patient outcomes, such as glycemic control (HbA1c), blood pressure (BP) and cholesterol (LDL), of alcohol drinkers with diabetes. PC team social connections can be defined as instrumental ties and expressive ties among team members [21-23]. Instrumental ties are information- and cognition-based, centered around formal working relationships in order to accomplish a task [24, 25]. Expressive ties are affect-based and are characterized by intimacy, trust, care, and a sense of belonging [26, 27].

Specifically, we aim to study the following research question:

Research Question: Do the structure of instrumental and expressive social ties in PC teams relate to effective control of HbA1c, BP and LDL in the PC teams' panels of diabetic alcohol drinkers?

2 Methods

2.1 Overview and Sample

This undertaking is a pilot feasibility study designed to investigate the study research question. It uses a convenience sample of 5 primary care clinics associated with a large Midwestern University. A total of 7 primary care clinics were invited to participate in the study and 5 clinics agreed. The Institutional Review Board of the University of Wisconsin approved the study.

All physicians, physician assistants, nurse practitioners, medical assistants, laboratory technicians, radiology technicians, and medical reception staff were invited to schedule the 30-minute face-to-face structured survey interview. Eligibility criteria included 18 years of age or older, ability to read and understand English, and employment at the study site in a patient care or patient interaction role. A total of 114 health professionals at the 5 study clinics were invited to participate and 110 (96%) completed the study survey. Study participants were 95% female and included 13 physicians, 3 nurse practitioners, 21 registered nurses, 21 medical assistants or licensed practice nurses, 27 medical receptionists, 18 laboratory or radiology technicians, 5 clinic managers, and 2 other clinic staff (Table 1). One fifth of study participants had worked at their practice for 1 year or less and just over a third worked part-time, defined as 75% time or less.

2.2 Survey Instrument

The study instrument was a 30-minute face-to-face structured interview survey administered by a trained research assistant. The survey examined social connections and team climate in the primary care teams. The social network measures were self-generated and pilot tested by the research team.

Table 1. Study Sample (n=110)

Female, n (%)	105 (95.4)
Position, n(%)	
Physician	13 (11.8)
NP/PA	3 (2.7)
Clinic Manager	5 (4.6)
RN	21 (19.1)
LPN/MA	21 (19.1)
Medical Receptionist	27 (24.5)
Lab/Radiology Tech	18 (16.4)
Other	2 (1.8)
Years at Clinic, n (%)	
1 year or less	21 (19.1)
>1 to 6 years	49 (44.5)
>6 years	40 (36.3)
% Full Time, n (%)	
75% or less	38 (34.6)
>75%	72 (65.6)

The survey included 5 social network questions asking respondents about their dyadic relations with every other staff member of the clinic. Each health professional identified their interactions in the past 6 months in terms of (1) face-to-face communication about patient care, (2) electronic communication about patient care, (3) friendship, (4) advice, and (5) emotional support. Respondents were given a complete clinic staff roster and asked to provide an answer to the social network question in relation to each name on the clinic staff list.

In addition, survey respondents were asked to complete the Team Climate Inventory (TCI-14) and a demographics section which included gender, job title, percentage of full-time employment, and years working at the clinic.

3 Measures

3.1 Definition of Care Team

Subjects were asked to indicate who is on their care team from a staff roster. The care team was defined as 'the smallest unit of individuals within the clinic that care for a specific patient panel.' For the analysis, care team membership included a lead clinician, either a physician, nurse practitioner, or physician assistant, and all clinic employees who either indicated on the survey that they belonged to that lead clinician's care team or who were named by the lead clinician as a care team member.

3.2 Social Network Parameters for Instrumental and Expressive Ties

Social network parameters (*density*, *transitivity*, *centralization*) were hypothesized to be related to diabetic patient outcomes. *Density* was calculated as the percentage of network ties present divided by the total network ties possible. *Density* measures overall connectedness within the care team. In a dense network, information can flow quickly between team members and social processes may result in positive intentions to use new information in daily practice.

Transitivity was calculated as the number of closed 3-person triads in the network divided by the number of closed 3-person triads plus the number of open 2-sided triads. Low *transitivity* corresponds to a more hierarchical care team network, whereas high *transitivity* corresponds to a more egalitarian care team network.

Network centralization measures organization around a single individual. *Centralization* was calculated as the sum of the differences in degree nominations between the highest degree node in the network and all other nodes, divided by the largest sum of differences possible in any network of the same size.

The *density*, *transitivity*, and *centralization* parameters of instrumental and expressive ties were used as independent predictors of diabetes biometric outcomes in the analysis. Instrumental tie social network questions included face-to-face communication, electronic communication, and advice. Expressive tie social network questions included friendship and emotional support.

3.3 Diabetic Patient Outcomes

The primary outcomes were biometric measures for diabetic patients of the care team who consumed alcohol in the past 12 months and who had at least one visit with the lead clinician in the past 12 months, and at least 2 visits in the past 36 months. Diabetes diagnoses were determined by the presence of 2 validated ICD-9 diabetes codes (250.00-250.03, 250.10-250.13, 250.20-250.23, 250.30-250.33, 250.40-250.43, 250.50-250.53, 250-60-250.63, 250.70-250.73, 250.80-250.83, 250.90-250.93, 357.2, 362.01, 362.02, 366.41) on 2 separate occasions within the past 3 years. Alcohol use was determined by a positive response to alcohol screening. The analysis calculated the percentage of diabetic alcohol drinkers who had effectively controlled hemoglobin A1c (<7.0%), LDL cholesterol (<100mg/dL), and BP blood pressure (<130/80 mm/Hg) in the past 12 months in the PC teams' patient panels. All biometric measures were derived from Electronic Health Records (EHR) and aggregated to the primary care team level.

3.4 Staff Turnover

Staff turnover for the care team was measured as the proportion of team members who had worked in the clinic for one year or less.

3.5 Team Climate

Team climate was measured with the 14-item short version of the Team Climate Inventory (TCI-14), using 5-point Likert scales [28]. The TCI-14 measures PC team members' perception of (1) focusing on clear and realistic goals to which the team members are committed (*shared vision*), (2) interactions between team members that are participatory and interpersonally non-threatening (*participatory safety*), (3) commitment to high standards of performance and appraisal of weaknesses (*task orientation*), and (4) support for innovation attempts including cooperation and development of new ideas (*innovation support*). For each TCI-14 subscale, items were coded from 1 to 5 and summed to produce an individual subscale score. Subscale scores were then averaged across team members to create a team measure for each TCI-14 subscale.

4 Statistical Analysis

The analysis first computed pairwise bivariate correlations between social network parameters, staff turnover, TCI-14 subscales, and patient biometric outcomes. Multivariate analysis then applied structural equation modeling (SEM) to test the path by which the social network parameters related to diabetic outcomes.

The analysis used Ucinet 6 for constructing networks and calculating network parameters, SAS Version 9.2 for correlations, and MPlus Version 7.11 for SEM.

5 Results

This study surveyed 20 primary care teams at 5 primary care clinics. Care team size ranged from 12 to 22 individuals, with a typical size of 17 to 18 team members.

Table 2. Unadjusted Correlations between Social Network Characteristics, Team Climate, and Biometric Outcomes for Alcohol-Drinking Diabetic Patients (n=20 care teams, N=551 patients)

		Team Climate					Diabetic Patient Panel		
Network Characteristic		Shared Vision	Participatory Safety	Task Orientation	Innovation Support	Staff Turnover	A1c	LDL	BP
	Mean (sd)	13.0 (0.5)	11.0 (1.2)	8.6 (0.5)	8.0 (0.7)	0.17 (0.08)	0.50 (0.08)	0.53 (0.12)	0.44 (0.14)
Instrumental Ties									
Face-to-Face Density	0.51 (0.09)	0.83***	-0.10	0.09	-0.71***	-0.58***	0.28	0.41	0.42*
Face-to-Face Transitivity	0.38 (0.05)	0.71***	0.08	0.15	-0.54***	-0.59***	0.22	0.41	0.34
Face-to-Face Centralization	0.24 (0.06)	-0.19	-0.68***	-0.71***	-0.03	0.06	-0.16	-0.25	0.01
Electronic Density	0.36 (0.05)	0.72***	-0.23	-0.11	-0.35	-0.11	-0.25	0.33	0.12
Electronic Transitivity	0.27 (0.04)	0.12	-0.22	-0.32	0.09	0.14	-0.58**	-0.06	-0.1
Electronic Centralization	0.24 (0.07)	-0.37	-0.67***	-0.75***	-0.03	-0.14	-0.11	-0.30	-0.18
Advice Density	0.65 (0.06)	0.02	0.78***	0.71***	0.62***	0.35	-0.08	0.12	-0.12
Advice Transitivity	0.50 (0.08)	-0.3	0.59***	0.58***	0.46**	0.14	0.22	0.05	0.06
Advice Centralization	0.33 (0.09)	-0.48**	-0.56**	-0.65***	0.05	0.02	-0.31	-0.42	-0.55**
Expressive Ties									
Friendship Density	0.66 (0.09)	0.47**	0.40*	0.43*	-0.21	-0.68***	0.48*	0.39	0.34
Friendship Transitivity	0.54 (0.12)	0.62***	0.27	0.36	-0.42*	-0.67***	0.46*	0.48*	0.43*
Friendship Centralization	0.27 (0.06)	-0.33	-0.32	-0.43*	0.30	0.45**	-0.30	-0.28	-0.41
Emotional Support Density	0.60 (0.07)	0.28	-0.01	-0.15	-0.03	-0.72***	0.4	0.11	-0.11
Emotional Support Transitivity	0.43 (0.07)	0.48**	0.10	0.03	-0.12	-0.70***	0.28	0.24	0.02
Emotional Support Centralization	0.33 (0.06)	-0.34	-0.22	-0.20	0.00	0.01	0.09	0.02	-0.21

*p<.10, **p<.05, ***p<.01

As seen in Table 2, mean density of face-to-face communication within the care team was 0.51, indicating that roughly half of all possible face-to-face communication ties possible were present. Staff turnover within the past year averaged 17 percent.

Higher density and higher transitivity of face-to-face communication, friendship, and emotional support were associated with lower staff turnover, higher shared team vision, and better hemoglobin A1c, LDL cholesterol, and blood pressure control for alcohol-drinking patients with diabetes (Table 2). Transitivity in electronic communication was negatively correlated with hemoglobin A1c.

Figure 1 presents SEM of biometric outcomes for diabetic alcohol drinkers. Panel A shows friendship transitivity in primary care teams is associated with a higher percentage of patients with controlled HbAlc (<7.0%), by way of lower staff turnover. In panel B, face-to-face communication transitivity is associated with a higher percentage of patients with controlled LDL cholesterol (<100mg/dL), by way of shared vision within the team. Finally, in panel C, face-to-face communication transitivity is linked with a higher percentage of patients with controlled blood pressure (<130/80 mm/Hg), by way of shared vision.

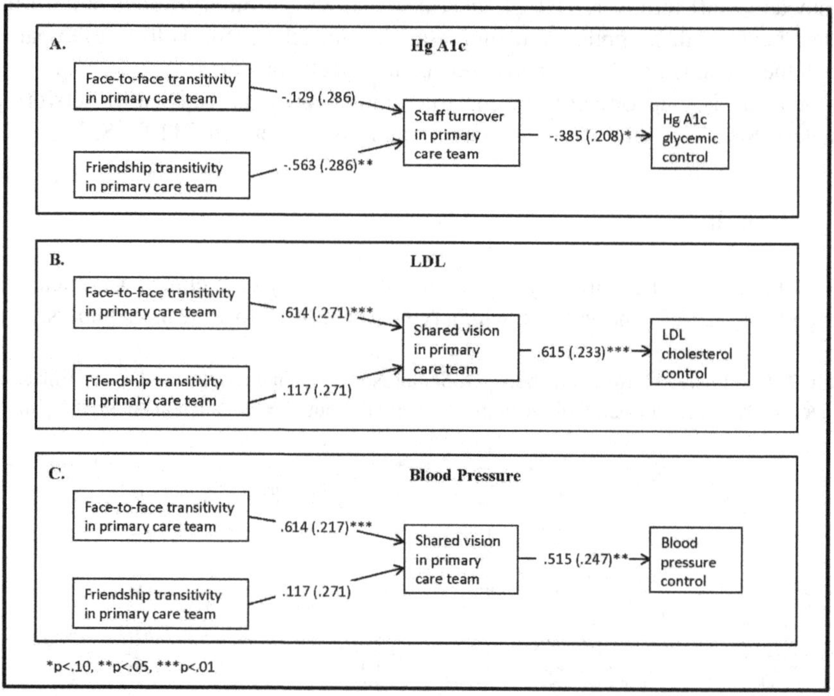

Fig. 1. Structural Equation Models of Social Network Ties in Primary Care Teams and Biometric Outcomes for Alcohol-Drinking Diabetic Patients (n=20 teams, N=551 patients)

6 Discussion

The study's objective is to evaluate an association between social network structures in PC teams and biometric outcomes of diabetic alcohol drinkers. Overall, our results show that PC teams with less hierarchical face-to-face communication ties (i.e. more connection density, more 3-tie closures and less network centrality) are more likely to have diabetic alcohol drinkers with effectively controlled HbA1c, LDL cholesterol and blood pressure. Interestingly, PC teams with highly interconnected electronic communication ties (via email or electronic medical records) are associated with higher likelihood of diabetic alcohol drinkers with poorly controlled HbA1c among diabetic alcohol drinkers. Our results suggest that lean electronic communication interactions about patient care may miss the finer understanding that comes with rich face-to-face communication interactions.

6.1 Limitations

Our diabetic health outcome measures are not adjusted for patient characteristics due to small sample size. Our results do not account for frequency of primary care use by diabetic drinking patients. Higher utilization of medical services may provide better diabetes outcomes, but it could also point to a finer understanding of patient health concerns and a more patient-centered approach to diabetic care which may be indicative of better functioning PC teams.

6.2 Conclusions

Less hierarchical, more emotionally supportive PC teams with more face-to-face communication ties, greater shared vision and low staff turnover are better suited for delivering high quality care to diabetic alcohol drinkers.

References

1. Mokdad, A.H., Ford, E.S., Bowman, B.A., et al.: Prevalence of obesity, diabetes, and obesity-related health risk factors. JAMA 2003 289(1), 76–79 (2001)
2. Hogan, P., Dall, T., Nikolov, P.: American Diabetes A. Economic costs of diabetes in the US in 2002. Diabetes Care 26(3), 917–932 (2003)
3. Stratton, I.M., Adler, A.I., Neil, H.A., et al.: Association of glycaemia with macrovascular and microvascular complications of type 2 diabetes (UKPDS 35): prospective observational study. BMJ (Clinical research ed) 321(7258), 405–412 (2000)
4. Ahmed, A.T., Karter, A.J., Warton, E.M., Doan, J.U., Weisner, C.M.: The relationship between alcohol consumption and glycemic control among patients with diabetes: the Kaiser Permanente Northern California Diabetes Registry. J. Gen. Intern. Med. 23(3), 275–282 (2008)
5. Chew, L.D., Nelson, K.M., Young, B.A., Bradley, K.A.: Association between alcohol consumption and diabetes preventive practices. Fam. Med. 37(8), 589–594 (2005)
6. Ahmed, A.T., Karter, A.J., Liu, J.: Alcohol consumption is inversely associated with adherence to diabetes self-care behaviours. Diabetic Med. 23(7), 795–802 (2006)
7. Engler, P.A., Ramsey, S.E., Smith, R.J.: Alcohol use of diabetes patients: the need for assessment and intervention. Acta Diabetol. 50(2), 93–99 (2013)
8. Cox, W.M., Blount, J.P., Crowe, P.A., Singh, S.P.: Diabetic patients' alcohol use and quality of life: relationships with prescribed treatment compliance among older males. Alcohol Clin. Exp. Res. 20(2), 327–331 (1996)
9. Johnson, K.H., Bazargan, M., Bing, E.G.: Alcohol consumption and compliance among inner-city minority patients with type 2 diabetes mellitus. Arch. Fam. Med. 9(10), 964–970 (2000)

10. Fleming, M., Brown, R., Brown, D.: The efficacy of a brief alcohol intervention combined with %CDT feedback in patients being treated for type 2 diabetes and/or hypertension. J. Stud. Alcohol 65(5), 631–637 (2004)
11. Ettner, S.L.: The relationship between continuity of care and the health behaviors of patients: does having a usual physician make a difference? Med. Care. 37(6), 547–555 (1999)
12. Cook, W.K., Cherpitel, C.J.: Access to health care and heavy drinking in patients with diabetes or hypertension: implications for alcohol interventions. Subst. Use Misuse 47(6), 726–733 (2012)
13. Dijkstra, R., Braspenning, J., Grol, R.: Implementing diabetes passports to focus practice reorganization on improving diabetes care. Int. J. Qual. Health C 20(1), 72–77 (2008)
14. Stevenson, K., Baker, R., Farooqi, A., Sorrie, R., Khunti, K.: Features of primary health care teams associated with successful quality improvement of diabetes care: a qualitative study. Fam. Pract. 18(1), 21–26 (2001)
15. Mohr, D.C., Benzer, J.K., Young, G.J.: Provider Workload and Quality of Care in Primary Care Settings Moderating Role of Relational Climate. Med. Care 51(1), 108–114 (2013)
16. Sparrowe, R.T., Liden, R.C., Wayne, S.J., Kraimer, M.L.: Social networks and the performance of individuals and groups. Acad. Manage. J. 44(2), 316–325 (2001)
17. Balkundi, P., Harrison, D.A.: Ties, leaders, and time in teams: Strong inference about network structure's effects on team viability and performance. Acad. Manage. J. 49(1), 49–68 (2006)
18. Chambers, D., Wilson, P., Thompson, C., Harden, M.: Social Network Analysis in Healthcare Settings: A Systematic Scoping Review. PLoS One 7(8) (2012)
19. Schmutz, J., Manser, T.: Do team processes really have an effect on clinical performance? A systematic literature review. Br. J. Anaesth. 110(4), 529–544 (2013)
20. Davidoff, F.: Heterogeneity Is Not Always Noise Lessons From Improvement. JAMA 302(23), 2580–2586 (2009)
21. Zhou, S.H., Siu, F., Wang, M.H.: Effects of social tie content on knowledge transfer. J. Knowl. Manag. 14(3), 449–463 (2010)
22. Lin, C.P.: To share or not to share: Modeling tacit knowledge sharing, its mediators and antecedents. J. Bus. Ethics 70(4), 411–428 (2007)
23. Zhong, X.P., Huang, Q., Davison, R.M., Yang, X., Chen, H.P.: Empowering teams through social network ties. Int. J. Inform. Manage. 32(3), 209–220 (2012)
24. Lee, D., Pae, J., Wong, Y.: A model of close business relationships in China (guanxi). Eur. J. Marketing 35(1/2), 51–69 (2001)
25. Ibarra, H., Andrews, S.B.: Power, Social-Influence, and Sense Making - Effects of Network Centrality and Proximity on Employee Perceptions. Adm. Sci. Q. 38(2), 277–303 (1993)
26. Berman, E.M., West, J.P., Richter, M.N.: Workplace relations: Friendship patterns and consequences (according to managers). Public Adm. Rev. 62(2), 217–230 (2002)
27. Gibbons, D.E.: Friendship and advice networks in the context of changing professional values. Adm. Sci. Q. 49(2), 238–262 (2004)
28. Anderson, N.R., West, M.A.: Measuring climate for work group innovation: development and validation of the team climate inventory. J. Organ. Behav. 19(3), 235–254 (1998)

Emerging Dynamics in Crowdfunding Campaigns

Huaming Rao[1,2], Anbang Xu[2], Xiao Yang[3], and Wai-Tat Fu[2]

[1]Dept. of Computer Sci & Tech, Nanjing University of Sci & Tech
[2]Dept. of Computer Science, University of Illinois at Urbana-Champaign
[3]Dept. of Automation, Tsinghua University

Abstract. Crowdfunding platforms are becoming more and more popular for fund-raising of entrepreneurial ventures, but the success rate of crowdfunding campaigns is found to be less than 50%. Recent research has shown that, in addition to the quality and representations of project ideas, dynamics of investment during a crowdfunding campaign also play an important role in determining its success. To further understand the role of investment dynamics, we did an exploratory analysis of the time series of money pledges to campaigns in *Kickstarter*[1] to investigate the extent to which simple inflows and first-order derivatives can predict the eventual success of campaigns. Using decision tree models, we found that there were discrete stages in money pledges that predicted the success of crowdfunding campaigns. Specifically, we found that, for the majority of projects that had the default campaign duration of one month in Kickstarter, money pledges inflow occurring in the initial 10% and 40-60%, and the first order derivative of inflow at 95-100% of the duration of the campaigns had the strongest impact on the success of campaigns. In addition, merely utilizing the initial 15% money inflows, which could be regarded as "seed money", to build a predictor can correctly predict 84% of the success of campaigns. Implication of current results to crowdfunding campaigns is also discussed.

Keywords: Crowdfunding, predictor, decision tree, dynamic.

1 Introduction

Crowdfunding platforms have provided a novel means for individuals to raise money for their creative works from the crowd. Although it is a relatively new phenomenon, many successful campaigns were able to solicit hundreds of thousands or even a million dollars. On the other hand, an incredible rush of projects are launched on these platforms every day, but more than half of them have failed. According to the statistics from Kickstarter, the largest crowdfunding platform, only about 44% projects meet their funding goal. On other platforms, the success rate might be even lower[2].

---

[1] `http://www.kickstarter.com`

[2] `http://www.theverge.com/2013/8/7/4594824/less-than-10-percent-of-projects-on-indiegogo-get-fully-funded`

W.G. Kennedy, N. Agarwal, and S.J. Yang (Eds.): SBP 2014, LNCS 8393, pp. 333–340, 2014.

Many online tools, such as *Kickspy*[3], *Kicktraq*[4] and *CanHeKickIt*[5], provide tracking tools and basic trend projection to predict the final result of a campaign. Research has shown that even in dynamic processes, in which past events impact future events, emergent patterns can also be found. Given that the crowdfunding campaign is a dynamic process, it is inherently interesting to see how the campaign goes over time and how the trajectory of its amount of pledged money is associated with its success or failure, and to what extent one can identify emergent patterns that are useful for understanding crowdfunding campaigns. The goal of this paper is to present preliminary results on modeling the time series of money pledges in crowdfunding campaigns to investigate what and how emergent patterns occur, and the extent to which these patterns can predict the success of campaigns.

To the best of our knowledge, our current study is one of the first models studying the dynamics of crowdfunding campaigns. We used the method of decision tree to analyze the time series data of the amount of pledged money over time. Our goal is to identify the significant periods that are more or less predictive of campaign success. To preview our results, we found that money pledges inflow that occurs in the initial 10% and 40-60%, and the first order derivative of inflow (i.e. the rate of increasing money pledges) at 95-100% of the duration of the campaigns had the strongest impact on the success of campaigns. In addition, merely utilizing the initial 15% money inflows, which could be regarded as "seed money", to build a predictor can correctly predict 84% of the success of campaigns.

2 Related Work

Many studies have been conducted to explore the factors that impact the successful fund-raising on the crowdfunding platforms. The incentive of participants of these platforms is one of the most important factors that many researchers [5][6][7][8] have been working on. Hui et al. [7] conducted interviews with project creators and bakers, trying to understand what kind of actions should be done in order to reach the funding goal after a campaign gets started. They found that successful projects need a lot of efforts not only online but also offline.

Some other researchers focus on the static representation of the project, such as the attributes related to the project page, or the initial settings of a campaign. According to their results, some useful suggestions have been provided to help the creators start their campaigns. For example, the presence of a video [4] and the quality of the video [11] are close related to the outcome of a campaign; Campaigns set with a smaller amount of funding goal tend to be successful [12]; The duration of a campaign is negatively related to the success of the campaign [11].

[3] `http://www.kickspy.com`
[4] `http://kicktraq.com`
[5] `http://canhekick.it`

Recently, there are more and more studies focused on the dynamic aspects of the campaigns. Mollick [11] offered a description of the underlying dynamics of success and failure among crowdfunding ventures, which provides insight into how the actions of the founders may affect their ability to receive entrepreneurial financing. Wash [13][14] studied the predictability of campaigns over time by analyzing the dataset from a charity crowdfunding site to understand how the act of completing a project can lead to larger donations. Etter et al. [3] proposed a method by combining direct information and social features to accurately predict the success of *Kickstarter* campaigns. Some researchers [11][9] investigated the content of project updates. It was found that campaigns with frequent updates more likely successful. Xu et al. [15] developed a taxonomy of the types of updates during the campaigns and studied how different types of updates and the representation of updates were associated with the success of campaigns. In addition, online social activities were found to play an important role in determining the outcomes of a campaign [1][11].

3 Data Collection and Pre-processing

We took advantage of the site *Kickspy* that is especially designed to help easily discover and research Kickstarter projects. More importantly, it keeps tracking those projects during their campaigns and presents "Funding Graph" of them. Therefore, we used a custom extractor to fetch the information about the amount of pledged money over time from this graph of each campaign. We crawled the pages in a reverse chronological order and in total we collected 8,529 campaigns from this site that started between March 19, 2013 and May 17, 2013. So for each campaign i, we obtained a series of data points that represented the current pledged money M for campaign i at a given time period t (See Fig 1), which is denoted by:

$$M_{t_1}^i, M_{t_2}^i, ..., M_{t_{l_i-1}}^i, M_{t_{l_i}}^i$$

Where $i \in I$ represents the ith campaign for the set of all campaigns (I), and l_i represents the total number of time periods for the ith campaign.

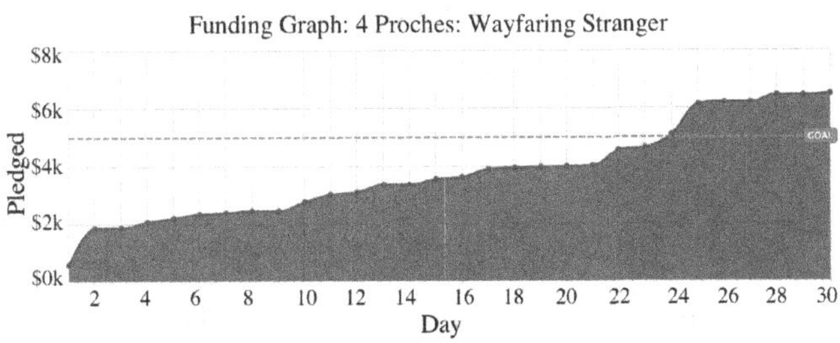

Fig. 1. A typical trajectory of money pledges of a campaign: downloaded from *Kickspy*

In addition, we also collected the corresponding basic information of those campaigns from *Kickstarter*, such as duration, funding goal, launched time, category, etc. To get an overview of how the duration were set among those samples, we drew the histogram of them as shown in Fig. 2. It was found that most of the projects set their duration around 1 month. And considering that the duration as a very important feature may impact the success of campaigns [11], to control this variable, we chose to use those projects of which the duration was set within 28 days to 31 days as our dataset for next analysis, which occupy the majority of the collected campaigns.

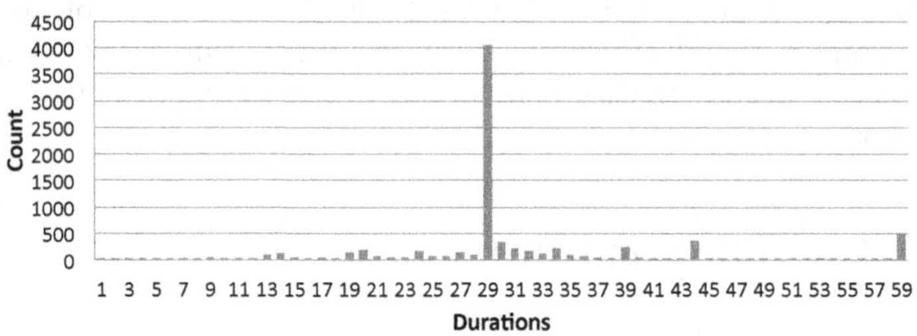

Fig. 2. Histogram of duration of collected campaigns: it shows that the duration of the majority of the collected campaigns was set around 1 month

Considering l_i varies among campaigns, we resampled those data points every 5% of the duration with respect to the launched time of each campaign. Then we obtained 20 states for every project at the corresponding relative times (5%, 10%, ..., 95%, 100%). To make it more comparable among campaigns, we normalized the current amount of pledged money M of each state by dividing it by its funding goal G:

$$m_j^i = \frac{M_j^i}{G^i}, j \in [1, 20], i \in I$$

To capture the dynamic change of the amount of pledged money, we calculated the increase m' in the normalized amount of pledged money at each relative time and its derivative m'':

$$m_j'^i = m_j^i - m_{j-1}^i$$

$$m_j''^i = m_j'^i - m_{j-1}'^i$$

In the following analysis, we would respectively use these two sets of values as input variables (i.e. $\{m_1'^i, m_2'^i, \ldots, m_j'^i\}$ and $\{m_1''^i, m_2''^i, \ldots, m_j''^i\}$, $j \in [1, 20], i \in I$) to train predictors.

4 Data Analysis

In order to explore which periods during campaigns are relatively important, for each relative time, we applied a decision tree model[6] to train predictors by using the series of values of either m' or m'' before that time as input variables. So as time increases, more variables were put into the model, each of which represents the state of a certain period before current time.

The reason to use a decision tree model rather than other models is partly due to its simplicity, and its general ability to capture the dynamics well. The method is also well accepted and is available from standard software packages. This makes the results easily replicable by others. The predictor from a decision tree corresponds to a structural tree and the root node indicates the most important variable that predicts the success of campaigns. We will use the same decision tree model to report accuracies and which variables are more predictive as captured by the model.

4.1 Method

The dataset was randomly partitioned into 10 equal size subsets, of which a single subset was retained as the validation data for testing the predictor, and the remaining 9 subsets were used as training data. Then for each relative time, the training process was repeated 10 times, with each of the 10 subsets used exactly once as the testing data. So we obtained 10 predictors for each state. Then we selected the most frequent relative time to which the root node variables belong as the most significant time before current time.

4.2 Results

The whole process above was repeated over the values of m' and its derivative m''. The prediction accuracies over time by using the series of m' before current time as the predictors in the decision tree model are shown in Fig. 3. In general, as expected, the performance of the predictors steadily improves, as more inflow periods (m') are included. This comes from the fact that, as the model has more data to predict success, its accuracy improves. The accuracies, on the other hand, are surprisingly high early on. In particularly, even with only the first 15% of the money inflows, the model can accurately predict whether a campaign will be successful or not with an accuracy of 0.84. Note that, this result is consistent with a recent study that just came out [3], in which they used a much more complex model to attain comparable performance.

Fig. 3 also shows the most important period (right y-axis) before current time that predicts the success of a campaign from the decision tree model. For example, as m' increases from 10% to 40% (x-axis), the most predictive period stays at 10%. In general, it is interesting to see that there are four such most predictive periods that are relatively stable (10%-40%, 45%-55%, 60%-80% and

[6] We used *ctree* in R package *party* to train predictors.

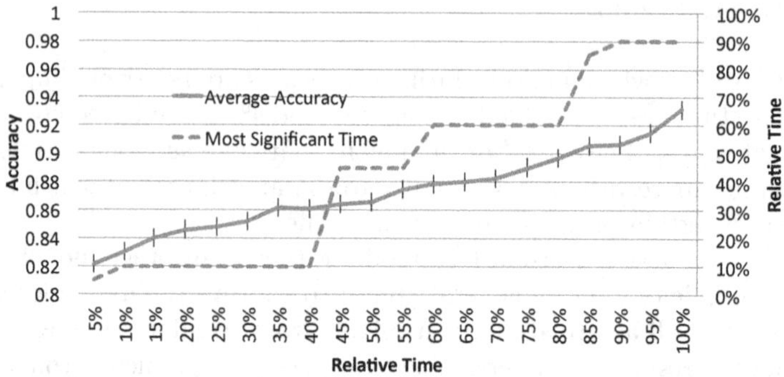

Fig. 3. Prediction accuracies over time by using the values of money inflows and the selected significant time before current time: The accuracies shown is the average over 10 runs and the error bars show the standard deviation over these runs; The most frequent relative time to which the root node variables of the 10 tress (predictors) belong was selected as the most significant time within that duration from beginning to current relative time.

90%-100%). What they imply is that during these periods, the most significant period stays the same even when more inflow data are included in the decision tree model. These patterns suggest that, while the model accuracies in general increase, the general structure of the decision tree stays relatively stable. On the other hand, the model suggests that around 40-60%, there are more changes in the structures of the money inflows, which makes the later period become more predictive than earlier ones. Looking at the first 80% of the duration of the campaigns, it seems that the most "active" periods could be around the first 10% as well as between the 40-60%. We will explain the last period (90-100%) next.

Fig. 4 shows the result of the prediction accuracies over time by using the values of m''. In contrary to Fig. 3, the performance of the predictors does not increase much until the very last stage of the duration of campaigns. The most important period also does not change until the end, jumping from 5% to 100%. This suggests that, though the predictor by using the values of the rate of the increasing money pledges can achieve an accuracy higher than 0.82, the use of more data of this information over time does not improve its predict power, except for those at the very last stage. One obvious reason is that during the last stage of the campaigns, the rapid increase of money inflows play a most more significant role in reaching the funding goal.

4.3 Discussions

As shown in Fig. 3, there are three major turn points (around 10%, 45% and 85% of the duration), which occur after the two relatively "stable" durations spanning from 10% to 40% and 60% to 80%. While in this exploratory analysis we do not have enough evidence to pinpoint exactly what contributes to this

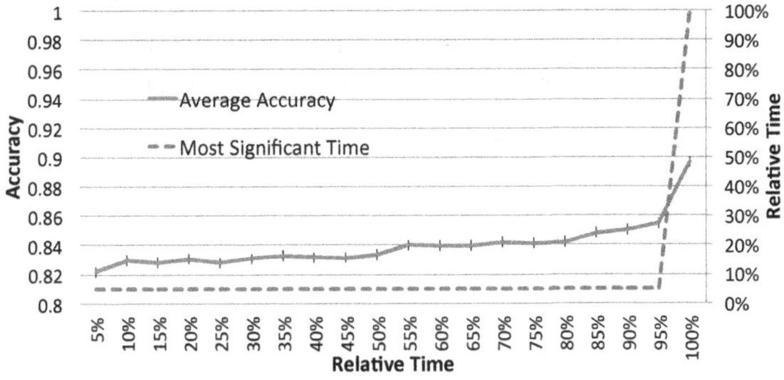

Fig. 4. Prediction accuracies over time by using the values of the derivative of money inflows and the selected significant time before current time: The accuracies shown is the average over 10 runs and the error bars show the standard deviation over these runs; The most frequent relative time to which the root node variables of the 10 tress (predictors) belong was selected as the most significant time within that duration from beginning to current relative time.

interesting patterns, we did find emergent patterns across this large set of funding campaigns. Most importantly, using merely the initial 15% money inflows, the decision tree model is already able to predict up to 84% accuracies of the success of campaigns. If regarded the initial 15% money inflows as "seed money", this result seems similar with what Andreoni [2] and List [10] found during their studies on fund-raising for threshold public goods that publicly announced "seed money" will increase charitable donations, though it is not yet clear whether they share a common theory.

Another possible reason for these emergent patterns is related to one of Xu et al.'s recent work [15], which investigated the dynamic use of different kinds of project updates during crowdfunding campaigns. They evenly divided the duration of a campaign into three phases: initial, middle, and final phase. They found that the number of project updates could significantly impact their success, and these project updates did tend to occur in batches. It is therefore possible that our current results reflect some systematic patterns of behavior by the crowdfunding participants. More analysis is needed to understand how current patterns are related to any patterns of the participants' behavior.

5 Conclusion

We conducted an exploratory study to analyze the time series of money pledges within Kickstarter to investigate the relationships between different periods during the campaign and its final result. It was found that the periods around 10% and 60% of the overall duration of the campaigns had a stronger impact on the success of campaigns. Although we did not yet have enough data to pinpoint the cause of these patterns, it did show that even with only money inflow and its derivative, the decision tree model could achieve a rather high accuracy in predicting a campaign's success. Our results, although preliminary, do seem promising for adopting this approach to improve the design of crowdfunding platforms that allow proposers to better monitor their money inflow and adjust their project description, updates, or use of social media to better advertise their campaigns so as to increase its likelihood of success.

References

1. Agrawal, A.K., Catalini, C., Goldfarb, A.: The Geography of Crowdfunding. SSRN Electronic Journal (2010)
2. Andreoni, J.: Toward a Theory of Charitable Fund Raising. Journal of Political Economy 106, 1186–1213 (1998)
3. Etter, V., Grossglauser, M., Thiran, P.: Launch Hard or Go Home!: Predicting the Success of Kickstarter Campaigns. In: The Proceedings of the First ACM Conference on Online Social Networks, New York, NY, USA (2013)
4. Greenberg, M.D., Pardo, B., Hariharan, K., Gerber, E.: Crowdfunding Support Tools: Predicting Success & Failure. In: The CHI 2013 Extended Abstracts on Human Factors in Computing Systems, New York, NY, USA (2013)
5. Gerber, E.M., Hui, J.: Crowdfunding: Motivations and Deterrents for Participation. ACM Transactions on Computer-Human Interaction (TOCHI) 20, 34:1–34:32 (2013)
6. Gerber, E.M., Hui, J.S., Kuo, P.Y.: Crowdfunding: Why people are motivated to post and fund projects on crowdfunding platforms. In: The ACMs Conference on Computer Supported Cooperative Work and Social Computing (Workshop) (2012)
7. Hui, J.S., Gerber, E., Greenberg, M.: Easy Money? The Demands of Crowdfunding Work. Segal Technical Report. 12-04 (2012)
8. Hui, J.S., Greenberg, M.D., Gerber, E.M.: Understanding the Role of Community in Crowdfunding Work. In: Accepted: The ACM's Conference on Computer Supported Cooperative Work and Social Computing, Baltimore, Maryland, USA (2014)
9. Kuppuswamy, V., Bayus, B.L.: Crowdfunding Creative Ideas: the Dynamics of Projects Backers in Kickstarter. SSRN Electronic Journal (2013)
10. List, J.A., Reiley, D.L.: The Effects of Seed Money and Refunds on Charitable Giving: Experimental Evidence from a University Capital Campaign. Journal of Political Economy 110, 215–233 (2002)
11. Mollick, E.: The dynamics of crowdfunding: An exploratory study. Journal of Business Venturing 29, 1–16 (2014)
12. Muller, M., Geyer, W., Soule, T., Daniels, S., Cheng, L.-T.: Crowdfunding Inside the Enterprise: Employee-initiatives for Innovation and Collaboration. In: The Proceedings of the SIGCHI Conference on Human Factors in Computing Systems, New York, NY, USA (2013)
13. Wash, R.: The value of completing crowdfunding projects. In: The 7th International AAAI Conference on Weblogs and Social Media, Boston, Massachusetts, USA (2013)
14. Wash, R., Solomon, J.: Coordinating Donors on Crowdfunding Websites. In: Accepted: The ACM's Conference on Computer Supported Cooperative Work and Social Computing, Baltimore, Maryland, USA (2014)
15. Xu, A., Yang, X., Rao, H., Huang, S.W., Fu, W.-T., Bailey, B.P.: Show me the Money! An Analysis of Project Updates during Crowdfunding Campaigns. In: Accepted: The Proceedings of the SIGCHI Conference on Human Factors in Computing Systems, Toronto, Canada (2014)

Incorporating Social Theories in Computational Behavioral Models

Eunice E. Santos [1], Eugene Santos Jr.[2], John Korah[1], Riya George[1], Qi Gu[2], Jacob Jurmain[2], Keumjoo Kim[2], Deqing Li[2], Jacob Russell[2], Suresh Subramanian[1], Jeremy E. Thompson[2], and Fei Yu[2]

[1] Computer Science Department, The University of Texas at El Paso, El Paso, TX, USA
{eesantos,jkorah}@utep.edu,
{rmgeorge,ssubramanian}@miners.utep.edu
[2] Thayer School of Engineering, Dartmouth College, Hanover, NH 03755, USA
{eugene.santos.jr,qi.gu,jacob.c.jurmain.th,keum.j.kim,
deqing.li.th,jacob.russell.th,jeremy.e.thompson,
fei.yu.gr}@dartmouth.edu

Abstract. Computational social science methodologies are increasingly being viewed as critical for modeling complex individual and organizational behaviors in dynamic, real world scenarios. However, many challenges for identifying, representing and incorporating appropriate socio-cultural behaviors remain. Social theories provide rules, which have strong theoretic underpinnings and have been empirically validated, for representing and analyzing individual and group interactions. The key insight in this paper is that social theories can be embedded into computational models as functional mappings based on underlying factors, structures and interactions in social systems. We describe a generic framework, called a Culturally Infused Social Network (CISN), which makes such mappings realizable with its abilities to incorporate multi-domain socio-cultural factors, model at multiple scales, and represent dynamic information. We explore the incorporation of different social theories for added rigor to modeling and analysis by analyzing the fall of the Islamic Courts Union (ICU) regime in Somalia during the latter half of 2006. Specifically, we incorporate the concepts of homophily and frustration to examine the strength of the ICU's alliances during its rise and fall. Additionally, we employ Affect Control Theory (ACT) to improve the resolution and detail of the model, and thus enhance the explanatory power of the CISN framework.

Keywords: Socio-cultural behavioral models, Computational social science, Social networks, Group stability, Somalia.

1 Introduction

Computational models are useful for analyzing behavior in social systems. What is lacking is a generalizable method for incorporating essential details from a real-world scenario into a computational model. The critical aspects of our social environment that affect the decisions and actions we take can be thought of as culture – any information or behavior learned from our social setting. Incorporating culture into a

W.G. Kennedy, N. Agarwal, and S.J. Yang (Eds.): SBP 2014, LNCS 8393, pp. 341–349, 2014.

computational model can represent the unseen motivations that dictate the actions we take, and thus add realism to actors' actions within that model. However, there are a number of challenges that need to be overcome. Most simulations apply social behavioral rules in an ad hoc manner, often coding a standard set of behavior rules into the model to be applied for all actors in all situations. This is an opportunity missed, as there are many empirically-proven social theories in the literature which could provide theoretic underpinnings to actor behavior. Developing a computational model with the exceptional ability to incorporate pragmatic social theories would contribute greatly to the field of computational social modeling. This is precisely what we endeavor to achieve in our latest research.

In this paper, we employ a computational framework called a Culturally Infused Social Network (CISN) [1] for incorporating both culture and social theories into a complex simulation. CISNs leverage a probabilistic reasoning framework to represent various socio-cultural factors and relationships. Actors and entities in a CISN are embedded in social networks that model social interactions. Bridging the gap that exists between social theoretic methods and computational models is critical. One of the key insights is to mathematically represent the culture-dependent behavior using a probabilistic framework. Such a framework should expose critical factors and relations in the form of random variables (rv_s) and probabilistic rules. Mapping social theories into the computational framework can then be concretely reduced to the formulation of these rv_s and rules. Social theories can also be used to inform mathematical methods for combining the effects of the variables and rules. Mapping of social theories also involves formulation and interpretation of networks that represent social interaction. CISNs overcome the challenge of incorporating social theories through a unique architecture with separate components to represent the two fundamental aspects of any social interaction between two social actors, namely their contact opportunity and their cultural affinity. Both of these aspects are represented using social networks and cultural fragments. A cultural fragment, which is represented using a probabilistic reasoning network framework called Bayesian Knowledge Base (BKB), models a specific behavior of an actor or group. The CISN architecture defines infusion points where interactions prescribed by social theories can be incorporated into the model using Bayesian fusion algorithms. CISNs also represent the multi-scale organization of social systems by including networks to represent social structures at different levels (group, community, nation, etc.). Bayesian fusion algorithms help to combine the individual behaviors to form behaviors of groups at higher levels in the social system. For model validation, we continue with the investigation of group stability of major players in the 2006 Somali civil war. The initial results for this scenario [2] utilized only homophily theory to provide analysis of the interactions between the different groups and its impact on group stability. In order to demonstrate the generality of our methodology, we extend the initial results by applying Affect Control Theory (ACT) [3] and landscape theory [4] along with homophily, and show how a more nuanced analysis of the scenario can be conducted. In the following section, we provide a background on foundational concepts, and then follow it with a description of the approach used to construct CISNs incorporating elements of culture and social theories. Finally, we detail the application of our framework to the Somali scenario.

2 Background

Homophily [5], a prominent social theory, hypothesizes that actors of similar beliefs and goals will be attracted to each other and inclined to cooperate. We employ homophily to model the cohesion of clans within alliances. On the other hand, we have the notion of frustration, which we adapted from Axelrod and Bennett's landscape theory of aggregation [4]. In our reinterpretation, frustration suggests that actors taking mutually threatening actions will align against each other. At first glance it seems that high homophily should imply low frustration. This will usually be correct, but much can depend on influential variables. Two actors could have mostly identical beliefs, goals, and reasoning rules, which suggests they should want to work together, but a single difference could disproportionately influence them to take opposing actions, driving them apart. Affect Control Theory (ACT) [3] is a widely applied social theory with a mathematical formulation which proposes that individuals shape their impressions about other entities through social interactions, where they conduct themselves in a way such that the generated feelings are appropriate to the situation. By considering that an individual's goals/intents vary with the social event, we apply ACT in our framework to tune it based on actions and reactions, thus better reflecting complex real-world scenarios. Existing microsimulation based methods such as MoSeS [6] provide frameworks for incorporating social theories. However, actor behavior is usually kept linear, making them ill-suited to simulating social theories with dependence on actors' thoughts and choices. This is a major drawback that we address in our methodology by using BKBs [7] to model nonlinear actor behavior.

3 Approach

Before going into a deeper discussion on how social theories can be incorporated into real world scenarios using a CISN, we discuss some of the underlying representations and behavioral concepts used to incorporate culture into our models.

Bayesian Knowledge Bases (BKBs), Intent and Culture: The unique advantage of BKBs, unlike Bayesian networks, is that they do not require a complete probability distribution for all rv_s. A BKB is essentially a set of conditional probability rules (CPRs) linking rv_s in an "if-then" fashion. BKBs can be graphically represented as directed graphs with two types of nodes: (1) instantiation-nodes (I-nodes), which represent the states of rv_s; and (2) support-nodes (S-nodes), which represent CPR probability values. Representing culture in our model is essential because it can impact intentions, decisions and actions [1]. For use in our model, analysts can generate cultural fragments, which are BKBs representing actor behaviors and their cultural influences. We connect underlying cultural factors to actor behavior using the notion of intent [8]. The instantiation of rv_s, as represented by I-nodes, is used to represent four essential components of the intent framework: 1) Beliefs (B) – beliefs about others; 2) Axioms (X) – beliefs about self; 3) Goals (G); and 4) Actions (A). Multiple cultural fragments can be combined into a single BKB using the Bayesian fusion algorithm [9]. This is critical for modeling dynamism, as fragments, representing real

-time changes, can be added during simulation. Reasoning over the BKBs can help in predictions (e.g. belief updating) and explanations (e.g. contribution analysis).

CISN Framework: CISN (Fig. 1) [2] has a flexible, plug-n-play architecture that allows for multi-scale modeling of actors, groups, social structures, and interactions. At each level, behavior can be incorporated in the form of cultural fragments. The components of the framework deal with various aspects of social systems, including physical interactions, social influences, and perceptions. The component-based architecture allows for plug-n-play of methodologies and representations. Functional mappings between components combine information and analyses across the various methods. One key insight of the architecture is that social influence is a function of not only the similarities of the actors, but also of their physical interactions. An ideology network represents the ideological proximity of actors. In contrast, a contact network represents their communication proximity.

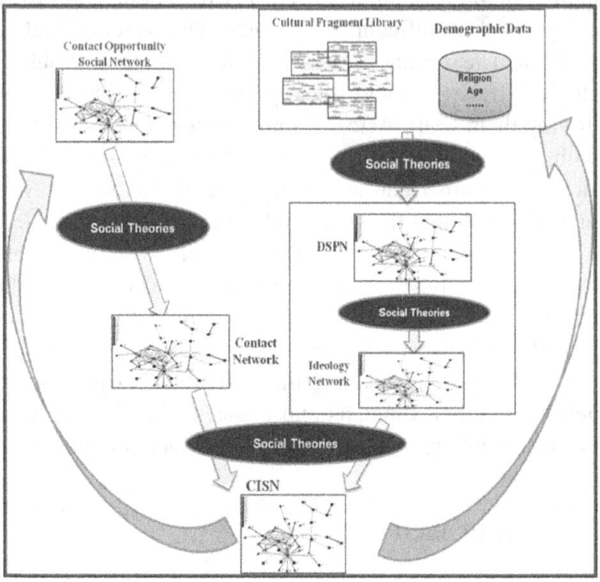

Fig. 1. CISN Architecture

An ideology network is comprised of nodes representing actors, goals, and actions. A weighted link between an actor and an action or goal indicates the calculated probability of the actor taking that action or setting that goal. Links between actors, indicating the similarity of the actors' culture and ideology, are calculated based on the actors' probability to pursue similar goals or actions. It is expected that actors with similar ideology will be more likely to collaborate, given the opportunity. Thus ideology networks help assess the desire or will to collaborate. Contact networks indicate the opportunity for actors to interact with other actors in the model. There can be a number of ways actors might interact: virtually, physically, financially, etc. Each of these interactions can be individually represented using contact opportunity social networks. A contact network is formed by combining all the contact opportunity social networks. The contact network and the ideology network are combined to form the CISN.

Changes in behavior can occur when an actor's perception of other actors changes [3]. Social perception has an impact on the role and status of an actor in a social system. This in turn will affect how he is treated or what social influence he has. The perception of the actors towards each other is represented in the Distributed Social Perception Network (DSPN), where the nodes represent actors, and edges represent perceptions. Note that the edges are bidirectional, as the sentiment that actor *a* has towards actor *b* may not be reciprocal. In a CISN, a DSPN provides input for generation of the ideology network. We can use social theories to inform the perception process. Fig. 1 indicates where social theories can be incorporated into the CISN.

4 Application

Table 1. Simulation Time Line

	Date	Major Events
1	08/16/2006	ICU seizes multiple ports that were supporting piracy.
2	09/30/2006	Minor skirmishes between ICU and Ethiopian troops. Some warlords defect to ICU.
3	10/10/2006	ICU captures complete annexation of Jubaland. More clans open negotiations with ICU.
4	10/26/2006	ICU declares Jihad against Ethiopian soldiers in Somalia.
5	11/01/2006	Puntland aligns with ICU against Somaliland.
6	11/26/2006	Ethiopian convoy in Baidoa is attacked by Pro-ICU forces.
7	12/02/2006	Multiple defections of groups both from and into ICU. ICU forces surround Baidoa and cuts off all support.
8	12/23/2006	Ethiopia deploys tanks and more soldiers near Baidoa.
9	12/25/2006	Ethiopia and TFG get the upper hand and push ICU back.
10	12/26/2006	ICU loses most of the territory gained since June. They are pushed back to Mogadishu.
11	12/27/2006	ICU surrenders most of the town without a fight. ICU leaders flee. TFG and Ethiopia captures Mogadishu.

For validation, we analyzed the conflict in Somalia during the latter half of 2006--a complex, rapidly-evolving, and well-documented scenario [10]. We explored the composition and dynamics of groups involved in the conflict, with the aim of explaining the Islamic Courts Union's (ICU's) sudden rise to, and even more sudden fall from, power. During this period, the main adversary of ICU was the Transitional Federal Government (TFG), which was devised during one of many peace conferences to stabilize the region. This conflict was framed by the ICU's successful occupation of Mogadishu in early June 2006 and the dissolution of the ICU in December of 2006. Since ICU was composed of multiple sub-groups, each with their own interests, goals and beliefs, understanding the stability of these groups is critical to determining the factors leading to the rise and fall of ICU. Most models in group stability focus on the cohesive forces of homophily and neglect the repulsive forces of frustration. We construct two competing models, based on: 1) only homophily; and 2) homophily and frustration combined. By comparing the stability analysis provided by the models with the actual events in the scenario, we validate our ability to embed social theories in computational frameworks and compare the efficacy of the two models. To further demonstrate the versatility of CISN, we analyze the same scenario through the lens of ACT. ACT has mathematical/statistical underpinnings. Applying ACT helps us understand the correlation between group stability and the change in other actors' perceptions of ICU.

Experimental Setup: For our simulation, we considered events during ICU's ascent and decline: June – Dec 2006. The major events included in our model are provided in Table 1. For each time step in Table 1, our simulation performs three major actions [2]: 1) Process Social Networks: We construct social networks to highlight relationships between actors affecting TFG and the ICU. We include both static and dynamic social networks, which are combined to form a single network using a weighted scheme. 2) Generate CISN: Actors are selected based on information extracted from the social networks. For each actor, relevant fragments are fused to form the actor's overall cultural fragment. Probability values for plausible actions are also calculated. The ideology network is constructed from all actors and plausible actions. Contact potentials from the various social networks are used to construct the final CISN. 3) Stability Analysis: As new alliances are made, the ideologies of the subgroups are fused with the core ICU and TFG ideologies to generate the overall group ideologies in the ideology network. Variance in the support among the subgroups is captured using an instability metric calculated from the deviation of each sub-group ideology from the group ideology. After the CISN is generated, we measure the instability metric for a selected target *rv*. A large instability value indicates that the group is likely to fragment.

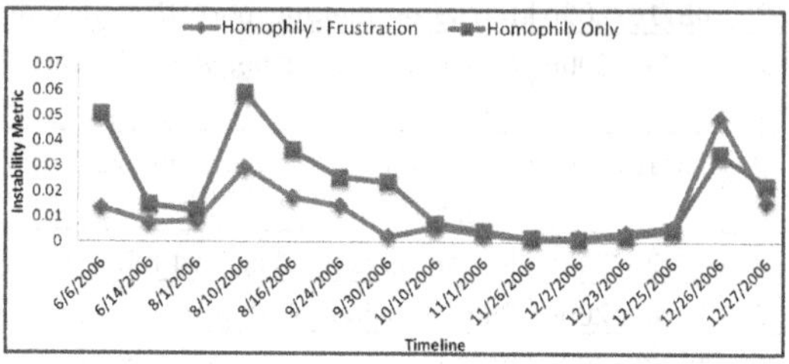

Fig. 2. Stability with Homophily and Frustration

The CISN is used to compute the degree to which each subactor influences the organization's overall decisions. A "super-fused" BKB, modeling the entire organization, is built by fusing all subactors' BKBs. Fragment contribution is used to calculate each subactor's contribution to the instantiations of a chosen target variable. In our simulations we chose *rv* "(A) Invade TFG territory" as the target variable. Finally, we find the variance of the contributions to the target variable across all the subactors by calculating the variance of the ratio of the contributions to each variable instantiation. This variance figure is our measure of instability. We employ ACT social theory to update individual perspectives based on unfolding events. Each sentiment is taken from an observer's point of view toward the participants involved. Three entities are considered for a situation: actors, events, and objects. An observer will assign *Evaluation-Potency-Activity* (EPA) values to the entities. The observer either finds the situation is consistent with cultural norms, or that it represents a deflection. In the case of deflection, the observer will modify its initial sentiment to conform to the situation. The EPA values are rated between [-3, 3] [11]. We use equations derived from ACT[1] to formulate the sentiment changes. To incorporate EPA values into the model, we

[1] http://www.indiana.edu/~socpsy/ACT/interact/subcultures/sex_averages.htm

select actors' goals, intentions, and behaviors that are related to each of the EPA components and build EPA fragments. The EPA fragments are fused with basic culture fragments to represent an actor's composite cultural influence. Since potency reflects how a region perceives the strength of ICU, the *potency* EPA value may serve as an indicator of the group instability.

Results and Analysis: Referring to Fig. 1, we can see that there is a social theory plug-in that feeds into the box containing the DSPN and ideology network. It is here that we applied Landscape Theory to gauge group stability by incorporating the concepts of homophily and frustration. Homophily contributes towards group cohesion, while frustration acts as a force of dissolution. In Fig. 2, we can observe the refinement provided by the incorporation of frustration. Without frustration, there is a maximum in the calculated instability of the group on 8/10/2006 that seems to indicate that the ICU should have fragmented and dissolved at this early stage. With the addition of frustration, we see that our results agree more closely with the actual timeline, where additional friction is apparent by a *local* maximum at 8/10/2006, but the timeline maximum now occurs properly at 12/26/2006. Thus we have a clear instance demonstrating how the incorporation of the social concept of frustration has greatly improved the accuracy of our results. It may seem counter-intuitive that adding frustration to the homophily model decreases instability. However, note that, mathematically, instability metric is measured using the variance of the contributions of the subactors to ICU's behavior, specifically contributions to the *rv* "(A) Invade TFG territory". Take for example 08/10/2006 when the strategic region of Beledweyne was incorporated into ICU. Since the start of this conflict Beledweyne, which borders the territories controlled by TFG and Ethiopia, was split between allies supporting ICU and TFG. Some extremist allies within ICU supported complete takeover of this region while moderate allies feared further escalation in violence. This variance is more pronounced when only homophily is considered. When we also include the frustration of the subactors with the ICU's efforts to attack TFG (from both moderate and extremist points of view), the variance of the contributions to ICU's behavior reduces, leading to lower instability values. In non-mathematical terms, the decrease in instability can be seen as disengagement of a set of subactors leading to lower conflicts within the ICU alliance.

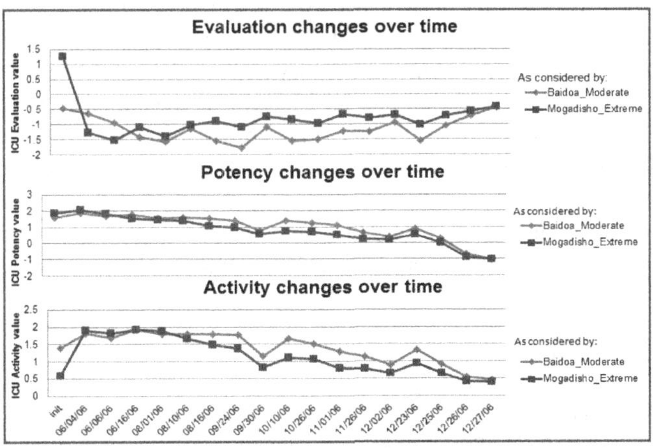

Fig. 3. Applying ACT to Stability Analysis – EPA Values

On examining the changes in the instability values in the homophily-frustration model, we see that the instability of ICU is high during its expansionist phase in the early part of the timeline. This can be explained by new groups with different socio-cultural make-up aligning themselves to ICU. However ICU's successful military operations contribute to its stabilization over time. This continues until Ethiopia initiates hostilities and starts gaining the upper hand. Military reversals increases the rift among the ICU allies, especially on the issues of resisting TFG, as indicated by the spike in ICU's instability metric around 12/26/2006. This eventually leads to the disintegration of ICU.

We model the important effects of perception and sentiment change using ACT. ACT argues that when presented with a conflict between an actor's EPA values and that actor's reactions to a particular event, an observer will respond by modifying his estimation of the actor's EPA values, either temporarily by associating modifiers for the actor (e.g. angry, eager, patient, etc.), or by assigning entirely new EPA values for the actor. At this point in our model, we explore the use of modifiers, and leave value reassignment for future research. Fig. 3 shows how EPA values toward ICU from two different groups, extremist and moderate, change over time. As we can see, the potency values for the ICU from both groups increase in the beginning as ICU seizes more regions and gains power, but decreases below zero after 12/25/06. This happens when Ethiopia and TFG get the upper hand, and ICU loses most of the territory they had gained. Thus, ACT provides additional explanation for the decline of ICU.

5 Conclusion

In this work, we provided key insights to incorporating social theories into computational models, and presented CISN as a generic framework that embodies these insights. We validated the viability of CISN by analyzing the stability of ICU during the 2006 conflict using landscape theory and ACT theory. We demonstrated the ability of CISN to compare social theories by showing the added accuracy provided by employing frustration and homophily for stability analysis. With ACT, we provided more details in the model in the form of actor perception and added explanation to the breakup of ICU. This work represents only the first step in embedding social theories in computational frameworks. Our next steps will be to embed different social theories and to pose a relevant categorization scheme for those theories.

References

1. Santos, E.E., Santos, Jr. E., Pan, L., Wilkinson, J.T.: Culturally Infused Social Network Analysis. In: Proceedings of the International Conference on Artificial Intelligence (IC-AI), pp. 449–455 (2008)
2. Santos, E.E., Santos Jr., E., Wilkinson, J.T., Korah, J., Kim, K., Li, D., Yu, F.: Modeling Complex Social Scenarios Using Culturally Infused Social Networks. In: Proceedings of the IEEE International Conference on Systems, Man, and Cybernetics, Anchorage, AK, pp. 3009–3016 (2011)

3. Heise, D.R.: Expressive Order: Confirming Sentiments in Social Actions. Springer, New York (2007)
4. Axelrod, R., Bennett, D.S.: A Landscape Theory of Aggregation. Br. J. Polit. Sci. 23, 211–233 (1993)
5. McPherson, M., Smith-Lovin, L., Cook, J.: Birds of a Feather: Homophily in Social Networks. Rev. Lit. Arts Am. 27, 415–444 (2008)
6. Wu, B., Birkin, M., Rees, P., Heppenstall, A., Turner, A., Clarke, M., Townend, P., Xu, J.: Moses: An Innovative Way to Model Heterogeneity in Complex Social Systems. In: 2010 Second Int. Conf. Comput. Model. Simul., pp. 366–370 (2010)
7. Santos Jr., E., Santos, E.S.: A Framework for Building Knowledge-Bases Under Uncertainty. J. Exp. Theor. Artif. Intell. 11, 265–286 (1999)
8. Santos, J. E., Zhao, Q.: Adversarial Models for Opponent Intent Inferencing. In: Kott, A., McEneaney, W. (eds.) Adversarial Reasoning: Computational Approaches to Reading the Opponents Mind, pp. 1–22. Chapman & Hall/CRC, Boca Raton (2006)
9. Santos, J. E., Wilkinson, J.T., Santos, E.E.: Bayesian Knowledge Fusion. In: Proceedings of the 22nd International FLAIRS Conference, pp. 559–564 (2009)
10. Barnes, C., Hassan, H.: The rise and fall of Mogadishu's Islamic courts. J. East. African Stud. 1, 151–160 (2007)
11. Heise, D.R.: Understanding events: affect and the construction of social action. Cambridge Univ. Press, New York (1979)

Contexts of Diffusion: Adoption of Research Synthesis in Social Work and Women's Studies

Laura Sheble and Annie T. Chen

School of Information & Library Science
University of North Carolina, Chapel Hill, NC, USA, 27599
sheble@live.unc.edu, atchen@email.unc.edu

Abstract. Texts reveal the subjects of interest in research fields, and the values, beliefs, and practices of researchers. In this study, texts are examined through bibliometric mapping and topic modeling to provide a bird's eye view of the social dynamics associated with the diffusion of research synthesis methods in the contexts of Social Work and Women's Studies. Research synthesis texts are especially revealing because the methods, which include meta-analysis and systematic review, are reliant on the availability of past research and data, sometimes idealized as objective, egalitarian approaches to research evaluation, fundamentally tied to past research practices, and performed with the goal informing future research and practice. This study highlights the co-influence of past and subsequent research within research fields; illustrates dynamics of the diffusion process; and provides insight into the cultural contexts of research in Social Work and Women's Studies. This study suggests the potential to further develop bibliometric mapping and topic modeling techniques to inform research problem selection and resource allocation.

Keywords: diffusion, social work, women's studies, research dynamics.

1 Introduction

Research synthesis methods (RSM), which include systematic reviews and meta-analysis, are among the most important recent methodological innovations in science, especially in applied fields. As RSM have diffused across science, the methods have been subject to adaptation and interpretation in the context of research fields. Because RSM typically require researchers to systematically engage with past research and empirically assess the relevance and strength of evidence presented in findings to integrate across studies, examination of how research synthesis is practiced in disciplines can provide insights about the nature of research, and values and beliefs of researchers within fields.

This study provides an overview of the dynamic processes associated with the diffusion of research synthesis in Social Work and Women's Studies. To an extent, the fields appear similar: each has a high proportion of women scholars; is aligned with interests of marginalized segments of society; and identified with the social sciences [1-2]. However, bibliometric mapping and topic modeling illustrate that the two fields

W.G. Kennedy, N. Agarwal, and S.J. Yang (Eds.): SBP 2014, LNCS 8393, pp. 351–358, 2014.

exist in very different cultural contexts, as reflected in different patterns of RSM use and non-use; engagement with the knowledge base; and prevalence of meta-research topics. The current study is part of a larger study designed to investigate how, and in what ways adoption of research synthesis methods reflects and is shaped by the cultural contexts of science fields; and the application of bibliometric mapping and text analysis to reveal field-level research dynamics.

1.1 Research Contexts and Research Synthesis

Research synthesis involves systematic, empirical integration of science knowledge across research studies. Methods of research synthesis require researchers to work within the framework of their own research project; and negotiate the methods, data, and reporting structures of the primary studies that comprise the data for the synthesis study. Adoption of research synthesis methods within a field is facilitated by common and relatively stable approaches to research design, methods, measures, and problem selection; and belief structures that associate value with accumulations of research-based knowledge [3]. Information resources that enable access to data and research literature, and organizations that support synthetic research methods contribute to adoption of the methods by facilitating access to data for synthetic studies [4-5], and contributing to visibility of syntheses and requisite tools and techniques to perform such studies. In fields associated with professional practice and/or policy, engagement with the evidence based practice (EBP) movement provides impetus to use research synthesis methods, which are an integral component of EBP.

2 Data, Analysis, and Visualization

Data for this study was extracted from the Thomson Science *Science* and *Social Science Citation Indexes* (*S/SCI*). To estimate the extent to which fields engage with RSM the S/SCI was searched with keyword and citing reference queries [6]. "Diverse" forms of research synthesis methods, those designed for synthesis of qualitative or interpretive research, were represented by a subset of these queries. A keyword search, refined with iterative searching and scanning, was used to estimate the prevalence of engagement with evidence-based practice. The "Social Work" and "Women's Studies" Web of Science categories were used to delimit research fields. Identified publications were rated "directly," "indirectly," or "not apparently" related to research synthesis methods based on titles, author keywords, and abstracts. Only directly or indirectly related items are included in bibliometric maps and topic models.

2.1 Bibliometric Mapping

Publications referenced by Social Work and Women's Studies RSM papers were overlaid on a global map of science to identify the knowledge base of each field. Cosine-normalized citation patterns across science reported in the 2010 *Journal Citation*

Reports (JCR), aggregated by Web of Science categories, are the basis of the global science network [7]. Citation patterns represent cognitive or socio-cognitive similarity between science fields, for which the categories are considered a proxy. The base map was visualized in Pajek [8] and overlaid with counts of references in Social Work and Women's Studies publications. Nodes represent fields, and are sized in proportion to the number of references observed. Shannon evenness and Rao-Stirling diversity [9-10] describe the distribution of references across science fields. Shannon evenness is a ratio of Shannon entropy, which measures abundance and evenness of entities across categories, to the maximum entropy possible. Rao-Stirling diversity accounts for distribution across fields, and the degree to which fields differ. Difference between fields is estimated with the citation matrix used to construct the global map of science.

2.2 Topic Modeling

Topic models were developed to summarize RSM-related publications using a variational Bayes implementation of Latent Dirichlet Allocation (LDA) [11-13]. Publications were represented by word co-occurrences in titles, author keywords, and abstracts using a 'bag of words' approach. Text preprocessing included removing labels from structured abstracts, applying the Porter stemmer, and identifying stop words and frequently occurring words. The number of topics selected was informed by perplexity scores for 5 to 30 topics, such that local perplexity minima were preferred. Topic labels consist of word stems most frequently associated with each topic. The document-topic matrix, which describes topic distributions across documents, was visualized in Gephi [14] as a bimodal network. Network partitions were identified by the Louvain algorithm [15] and are represented by color. Topic nodes were sized in proportion to the sum of associated document proportions, and edge thresholds were applied to reduce visual complexity. Overall, this approach can be described as a quantitatively guided qualitative overview of publication content.

3 Contexts: RSM in Social Work and Women's Studies

The extent to which researchers engage with past research, evidence based practice, and diverse forms of research synthesis methods (i.e., forms amenable to qualitative and/or interpretive research) likely influence RSM diffusion within fields. Social Work and Women's Studies differ markedly with respect to these factors (Table 1). Evidence-based practice has been important in the field of Social Work, but only marginally so in Women's Studies. About 1.5% of all Social Work publications that appeared in the twenty years after EBP was popularized in medicine [16] are associated with EBP. In Women's Studies, this figure is roughly one-tenth of that observed in Social Work. Similarly, based on S/SCI document type classifications, Social Work researchers are about ten times more likely to publish reviews in the journal literature. Between 1976 and 2011, reviews comprised 1.33% of all Social Work publications and 0.12% of Women's Studies publications. Social Work has engaged with diverse forms of research synthesis about three times as often as Women's Studies. These results suggest that Social Work is a better fit for RSM.

Table 1. Relative prevalence of characteristics related to research synthesis methods

	Social Work	*Women's Studies*
Evidence based practice (1992 -)	1.49%	0.15%
Reviews (S/SCI document type)	1.33%	0.12%
Year of first RSM publication	1977	1985
Diverse RS methods / all RSM	6.73%	2.33%

Approximately two-thirds of all Social Work and one-third of all Women's Studies publications identified via the search protocol were judged as directly or indirectly related to research synthesis (Table 2). These relatively low (Social Work) and low (Women's Studies) proportions were unique to the social science fields selected from the larger study for more in-depth analysis. Review of results suggests that the S/SCI "KeywordsPlus" field, which includes words extracted from titles of cited references, contributed substantially to publications retrieved but apparently not related to research synthesis. Adherence to publishing practices and guidelines that proscribe indicating a paper reports a research synthesis in the title likely increases the prevalence of "meta-analysis" and similar terms in the KeywordsPlus field.

Table 2. Extent of relationship between retrieved publications and RSM

	Social Work		*Women's Studies*	
Relationship	Count	Percent	Count	Percent
Direct	284	57.96	95	31.77
Indirect	34	6.94	16	5.35
Direct or Indirect	318	64.9	111	37.12
No apparent	172	35.1	188	62.88
Total	490		299	

3.1 Knowledge Base Interactions

Social Work and Women's Studies differ in the extent and diversity of references to the knowledge base. Though Social Work references a greater number of fields, most references are concentrated within Social Work and cognate fields. Lower Rao-Stirling diversity (Table 3) and the proximity of larger nodes to Social Work (Fig. 1, yellow node with red ring) reflect these differences. Women's Studies references are more evenly distributed across fields, as indicated by higher Shannon evenness and nodes that are more similar in size. Women's Studies seldom references work published in Women's Studies journals (Fig. 1, yellow node with red ring). The juxtaposition of a concentration of RSM publications within a few Women's Studies journals with broad referencing patterns suggests Women's Studies scholars engage with content of other fields through RSM. The pattern echoes observations that Women's Studies scholars tend to have dual allegiances: to Women's Studies and another field; and the description of feminist scholarship as one that "simultaneously challenges and is shaped by disciplinary inquiry" [17, p. 121]. Social Work, in contrast, mobilizes research in Social Work and cognitively similar fields for the benefit of the field broadly.

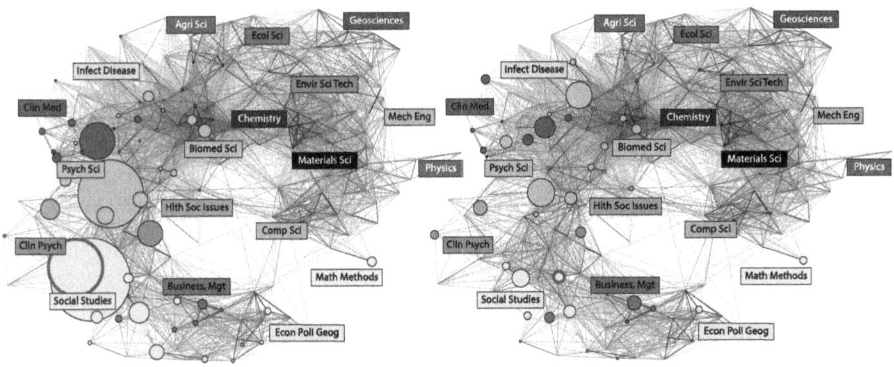

Fig. 1. Fields referenced by Social Work (left) and Women's Studies (right) RSM publications

Table 3. Diversity of fields referenced

Measure	Social Work	Women's Studies
Fields referenced	111	88
Shannon evenness	0.6361	0.7317
Rao-Stirling diversity	0.6299	0.7256

3.2 Diffusion Dynamics and Topics Associated with RSM

The temporal distribution of research synthesis publications in each field reveals a substantial lag between introduction and broader engagement with RSM. Across science broadly, the first research synthesis publications appeared in the early to mid 1970s [4,18], shortly before the first in Social Work (1977) and Women's Studies (1985). In Social Work, sustained engagement with the methods did not occur until the late 1990s, about twenty years after the first research synthesis publication in the field (Fig. 2). In Women's Studies, engagement with the methods has remained modest, but became more prevalent in 2003, and expanded again in 2009.

Topical distributions of research synthesis publications suggest broad engagement with research synthesis across Social Work; but selective and unevenly distributed engagement in Women's Studies. While each field is associated with 38 titles in the *JCR*, 36 of the Social Work titles, but only 14 of the Women's Studies titles include RSM publications. Concentration within journals is similarly uneven: 4 journals (10.5%) contain 81% of the RSM related publications in Women's Studies; and 17 journals (44.7%) contain 81% in Social Work.

Evidence-based practice, methodologies, and intervention research are salient issues in Social Work. In Fig. 2, these issues are represented by the topics "practice evidence-bas", "meta-analysis design result method", "effect size meta-analysis", "systematic search database", and "intervent treatment effect". The prevalence of topics associated with practice and EBP reflects the practice orientation of Social Work research [19]. EBP and RSM are central to discussions about research-practice divides; and have been cast in opposition to traditionally prevalent research approaches

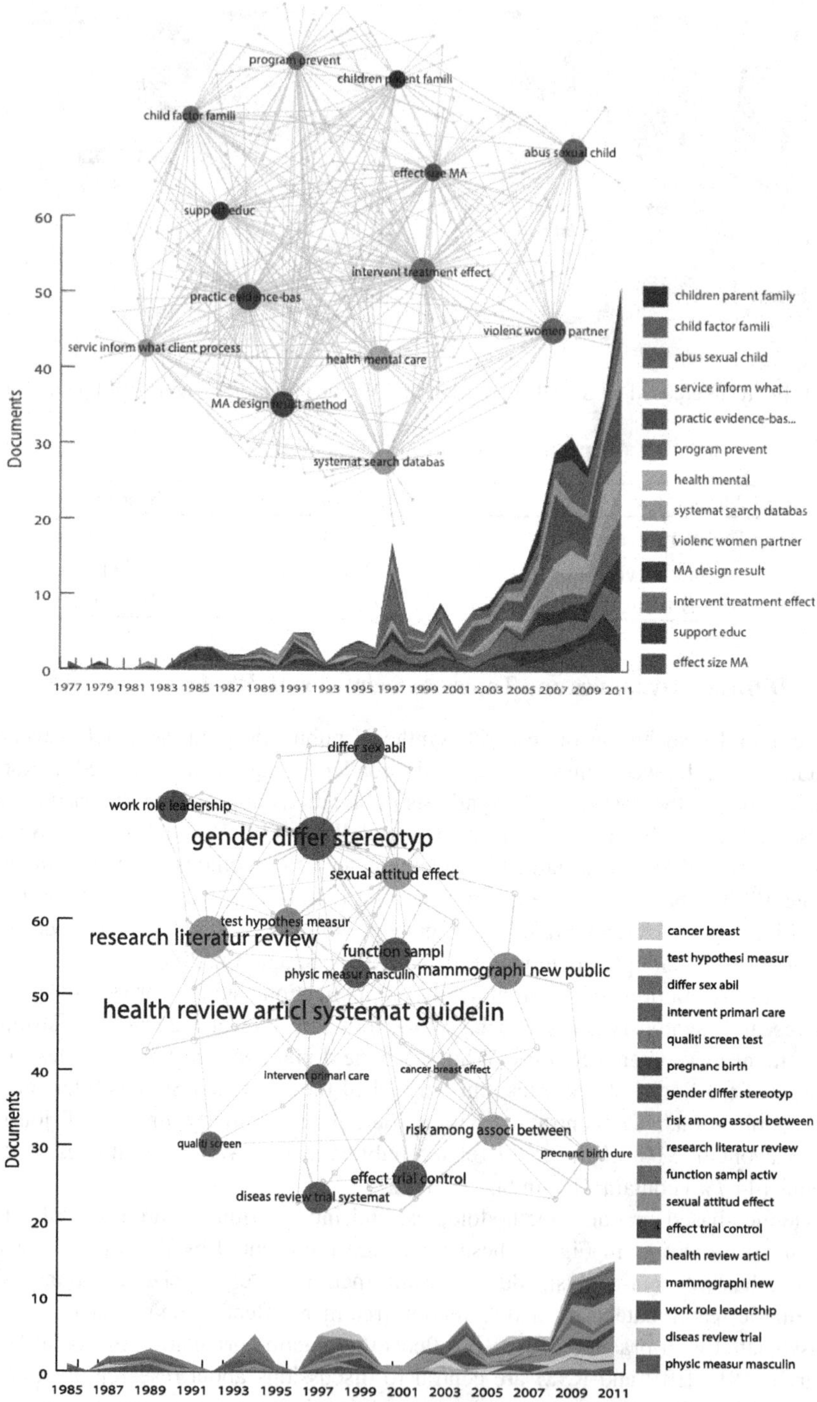

Fig. 2. Topics of Social Work (top) and Women's Studies (bottom) RSM publications

such as qualitative case studies [20], which are difficult to systematically synthesize. Since the late 1990s, interest in intervention research, which is prototypically amenable to synthesis methods, has emerged. Preventative programs ("program prevent") and client services ("service inform what client process"), which are linked to EBP, are also focal interests.

The limited range of topics and uneven distributions of content across topics provide evidence of selective engagement with RSM in Women's Studies. RSM publications are centered on research related to gender differences in psychology and related fields; women's health; and methodological issues (Fig. 2). Psychology topics include work and leadership ("work role leadership"), gender differences and stereotypes, and abilities linked with sex such as spatial navigation ("differ sex abil"). Women's health issues include breast cancer, mammography, reproductive issues, and conditions not specific to females. Topics that link health-oriented and psychological gender-oriented research include those focused on literature and synthesis ("research literature review", "health review articl systemat guidelin"), and methods (e.g., "function sampl"), though topic locations reveal emphases on different methods topics in health versus psychology fields. For example, "test hypothesi measure" is embedded in psychology topics; and interventions and guidelines are associated with health topics.

3.3 Summary: RSM in Social Work and Women's Studies

Contrasts in the cultures and approaches to research in Social Work and Women's Studies are reflected in the topics associated with research synthesis and the dynamics of the RSM diffusion process. In Social Work, engagement with EBP and a tradition that prioritizes social work practice over research has fundamentally shaped engagement. While a practice orientation provides motivation to synthesize, limitations associated with past research have moderated the applicability of prevalent research synthesis methods, and likely impeded diffusion. In Women's Studies, research synthesis methods have been used primarily in subfields associated with disciplines in which RSM are common - psychology and the health sciences. An activist stance is exemplified by engagement with research in cognate fields to communicate the value of the alternative lens Women's Studies offers; and to comment on prior research not compatible with addressing interests of diverse populations.

3.4 Conclusion and Future Research

Topic modeling and bibliometric mapping, in combination with domain expertise, can be used to visualize and analyze relationships and processes at the level of research fields. The methods were mobilized in this study to contrast dynamics and relationships associated with diffusion of research synthesis methods. Future research should focus on use of the methods to contrast differences within fields, including to examine research that is discussed prospectively versus that which is actually performed; to analyze and inform distribution and use of resources; and to provide high-level summaries of research in fields over time. Challenges include development and visualization of interpretable and comparable topic maps, accessibility of texts used to represent research, and flexible yet specifiable approaches to defining research fields.

References

1. Abbott, A.: Boundaries of Social Work or Social Work of boundaries? The Social Service Review Lecture. Soc. Serv. Rev. 69, 545–562 (1995)
2. Boxer, M.J.: For and about women: The theory and practice of Women's Studies in the United States. Signs 7, 661–695 (1982)
3. Hedges, L.V.: How hard is hard science, how soft is soft science? The empirical cumulativeness of research. Am. Psychol. 42, 443–455 (1987)
4. Toews, L.: The information infrastructure that supports evidence-based veterinary medicine: A comparison with human medicine. J. Vet. Med. Educ. 38, 123–134 (2011)
5. Bastian, H., Glasziou, P., Chalmers, I.: Seventy-five trials and eleven systematic reviews a day: How will we ever keep up? PLoS Medicine 7, e1000326 (2010)
6. Sheble, L.: Appendices B and C (2013), `http://laurasheble.web.unc.edu/dissertation/`
7. Rafols, I., Porter, A.L., Leydesdorff, L.: Science overlay maps: A new tool for research policy and library management. J. Am. Soc. Inform. Sci. Tech. 61, 1871–1887 (2010)
8. Batagelj, V., Mrvar, A.: Pajek. Program for Large Network Analysis. Connections 21, 47–57 (1998)
9. Stirling, A.: A general framework for analysing diversity in science, technology and society. J. R. Soc. Interface 4, 707–719 (2007)
10. Rafols, I., Meyer, M.: Diversity and network coherence as indicators of interdisciplinarity: case studies in bionanoscience. Scientometrics 82, 263–287 (2010)
11. Asuncion, A., Welling, M., Smyth, P., Teh, Y.W.: On smoothing and inference for topic models. In: Twenty-Fifth Conference on Uncertainty in Artificial Intelligence (UAI 2009), pp. 27–34. AUAI Press, Corvallis (2009)
12. Blei, D.M., Ng, A.Y., Jordan, M.I.: Latent Dirichlet Allocation. J. Mach. Learn. 3, 993–1022 (2003)
13. Ramage, D., Rosen, E.: Stanford TMT (2011), `http://nlp.stanford.edu/software/tmt/`
14. Bastian, M., Heymann, S., Jacomy, M.: Gephi: Open source software for exploring and manipulating networks. In: International AAAI Conference on Weblogs and Social Media (2009)
15. Blondel, V.D., Guillaume, J.-L., Lambiotte, R., Lefebvre, E.: Fast unfolding of communities in large networks. J. Stat. Mech. Theor. Exp., 10008 (2008)
16. Evidence-Based Medicine Working Group: Evidence-based medicine: A new approach to teaching the practice of medicine. J. Am. Med. Assoc. 268, 2420–2425 (1992)
17. Klein, J.T.: Crossing boundaries: Knowledge, disciplinarities, and interdisciplinaries. University Press of Virginia, Charlottesville (1996)
18. Chalmers, I., Hedges, L.V., Cooper, H.: A brief history of research synthesis. Eval. Health Prof. 25, 12–37 (2002)
19. Anastas, J.W.: Research design for Social Work and the human services, 2nd edn. Columbia University Press, New York (2000)
20. Herie, M., Martin, G.W.: Knowledge diffusion in Social Work: A new approach to bridging the gap. Social Work 47, 85–95 (2002)

Temporal Dynamics of Scale-Free Networks

Erez Shmueli, Yaniv Altshuler, and Alex "Sandy" Pentland

The Media Lab, Massacusetts Institute of Tecnology
Cambridge MA, USA
{shmueli,yanival,sandy}@media.mit.edu

Abstract. Many social, biological, and technological networks display substantial non-trivial topological features. One well-known and much studied feature of such networks is the scale-free power-law distribution of nodes' degrees.

Several works further suggest models for generating complex networks which comply with one or more of these topological features. For example, the known Barabasi-Albert "preferential attachment" model tells us how to create scale-free networks.

Since the main focus of these generative models is in capturing one or more of the static topological features of complex networks, they are very limited in capturing the temporal dynamic properties of the networks' evolvement. Therefore, when studying real-world networks, the following question arises: what is the mechanism that governs changes in the network over time?

In order to shed some light on this topic, we study two years of data that we received from *eToro*: the world's largest social financial trading company.

We discover three key findings. First, we demonstrate how the network topology may change significantly along time. More specifically, we illustrate how popular nodes may become extremely less popular, and emerging new nodes may become extremely popular, in a very short time. Then, we show that although the network may change significantly over time, the degrees of its nodes obey the power-law model at any given time. Finally, we observe that the magnitude of change between consecutive states of the network also presents a power-law effect.

1 Introduction

Many social, biological, and technological networks display substantial non-trivial topological features. One well-known and much studied feature of such networks is the scale-free power-law distribution of nodes' degrees [1]. That is, the degree of nodes is distributed according to the following formula: $P[d] = c \cdot d^{-\lambda}$. This heavy tail property of networks was found to be helpful in a wide range of domains, including robustness to failures, vulnerability to deliberate attacks, immunization strategies, advertising to opinion leaders, and many more.

As the study of complex networks has continued to grow in importance and popularity, many other features have attracted attention as well. Such features include among the rest: short path lengths and a high clustering coefficient [2, 3], assortativity or dis-assortativity among vertices [4], community structure [5] and hierarchical structure [6] for undirected networks and reciprocity [7] and triad significance profile [8] for directed networks.

W.G. Kennedy, N. Agarwal, and S.J. Yang (Eds.): SBP 2014, LNCS 8393, pp. 359–366, 2014.

Several works further suggested models for generating complex networks which comply with one or more of these topological features. For example, the known Barabasi-Albert model [1] tells us how to create scale-free networks. It incorporates two important general concepts: growth and preferential attachment. Growth means that the number of nodes in the network increases over time and preferential attachment means that the more connected a node is, the more likely it is to receive new links. More specifically, the network begins with an initial connected network of m_0 nodes. New nodes are added to the network one at a time. Each new node is connected to $m \leq m_0$ existing nodes with a probability that is proportional to the number of links that the existing nodes already have.

More sophisticated models for creating scale-free networks exist. For example, in [9], at each time step, apart of m new edges between the new node and the old nodes, m_c new edges are created between the old nodes, where the probability that a new edge is attached to existing nodes of degrees d_1 and d_2 is proportional to $d_1 \cdot d_2$. A very similar effect produces a rewiring of edges [10]. That is, instead of the creation of connections between nodes in the existing network, at each time step, m_r randomly chosen vertices lose one of their connections. In m_{rr} cases, a free end is attached to a random vertex. In the rest $m_{rp} = m_r - m_{rr}$ cases, a free end is attached to a preferentially chosen vertex.

The main focus of these generative models is in capturing one or more of the static topological features of complex networks. However, these models are very limited in capturing the temporal dynamic properties of the networks' evolvement. Therefore, when studying real-world networks, the following question arises: what is the mechanism that governs changes in the network over time?

A few recent studies (e.g., [11–16]) have started to investigate different aspects of this temporal evolution. However, their findings are limited and the above question has yet to be answered.

In order to shed some more light on this question, we studied two years of data (from 2011/07/01 to 2013/06/30) that we received from *eToro*: the worlds largest social financial trading company.

We discover three key findings. First, we demonstrate how the network topology may change significantly along time. More specifically, we illustrate how popular nodes may become extremely less popular, and emerging new nodes may become extremely popular, in a very short time. Then, we show that although the network may change significantly over time, the degrees of its nodes obey the power-law model at any given time. Finally, we observe that the magnitude of change between consecutive states of the network also presents a power-law effect.

2 Datasets

Our data come from *eToro*: the world's largest social financial trading company (See http://www.etoro.com). *eToro* is an on line discounted retail broker for foreign exchanges and commodities trading with easy-to-use buying and short selling mechanisms as well as leverages up to 400 times.

Similarly to other trading platforms, *eToro* allows users to trade between currency pairs individually. In addition, *eToro* provides a social network platform which allows users to watch the financial trading activity of other users (displayed in a number of statistical ways) and copy their trades (see Fig. 1). More specifically, users in *eToro* can place three types of trades: (1) Single trade: The user places a normal trade by himself, (2) Copy trade: The user copies one single trade of another user and (3) Mirror trade: The user picks a target user to copy, and *eToro* automatically places all trades of the target user on behalf of the user.

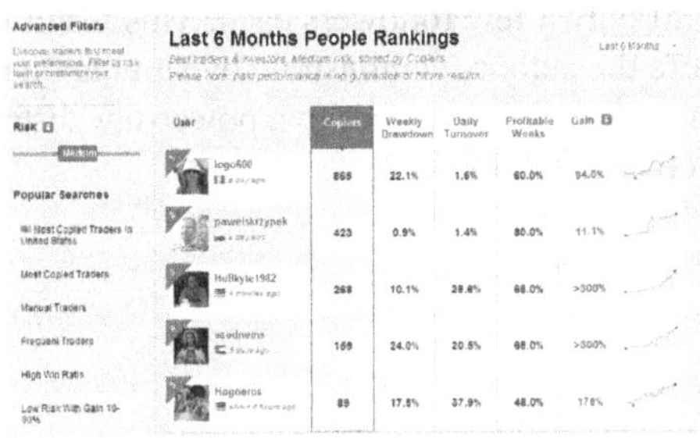

Fig. 1. The *eToro* platform. Illustrating the trading activity of the top-ranked users.

Our data contain over 67 million trades that were placed between 2011/07/01 and 2013/06/30. More than 53 million of these trades are automatically executed mirror trades, less than 250 thousand are copy trades and roughly 13 million are single trades. The total number of unique traders is roughly 275 thousands and the total number of unique mirror operations is roughly 850 thousands (one mirror operation may result in several mirror trades).

In the remainder of this paper, we use these trades to construct snapshot networks as we proceed to describe. Given a start time s and an end time e, the snapshot network's nodes consist of all users that had at least one trade open at some point in time between s and e. An edge from user u to user v exists, if and only if, user u was mirroring user v at some point in time between s and e.

Figure 2 illustrates how the size of the *eToro* network grows along time terms of both the number of nodes and the number of edges. For each day during the two years period, a snapshot network is constructed, and the number of nodes and edges for that network are counted.

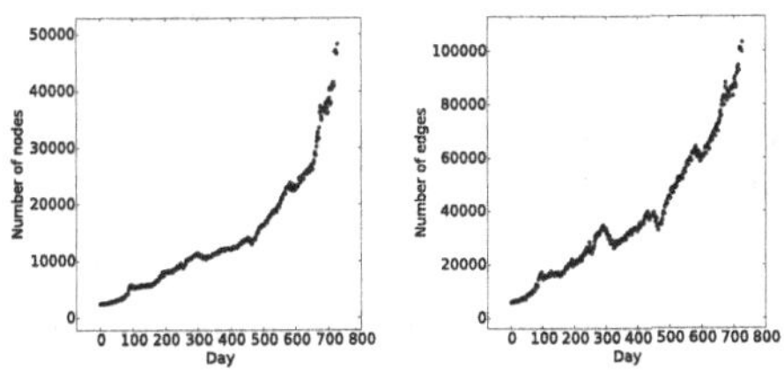

Fig. 2. The size of the *eToro* network in terms of the number of nodes (left) and the number of edges (right) along time

3 Results

First, we examined the in-degrees of nodes in the *eToro* network, over the entire period of two years. As can be seen in Figure 3, the degree distribution presents a strong power-law pattern. Although, quite expected, this result is non-trivial. One might expect to see a bunch of users that are mirrored by the others, but what we actually witness is a heavy tail of users with only a few followers each. This result is consistent with the observation in [17] where the authors demonstrate by simulation that the degree distribution of social-learning networks converges to a power-law distribution, regardless of the underlying social network topology.

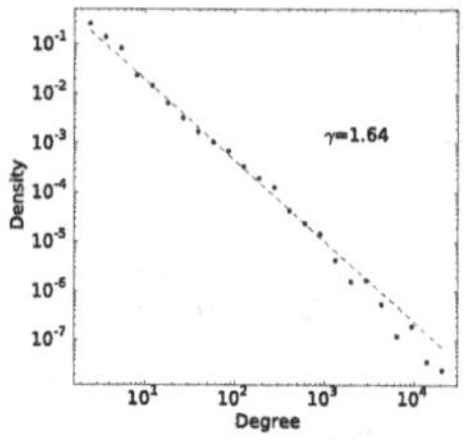

Fig. 3. In-degree distribution of nodes in the entire *eToro* network. (The in-degree of a node depicts the number of mirroring traders for the trader represented by that node).

Next, we investigated how the popularity of traders in *eToro*, in terms of the number of mirroring traders, changes along time. Fig. 4 illustrates the popularity of four traders. As can be seen in the figure, popular traders may become extremely less popular, and emerging new traders may become extremely popular, in a very short time. Note how this behavior differs significantly from the state-of-the-art "rich get richer" behavior.

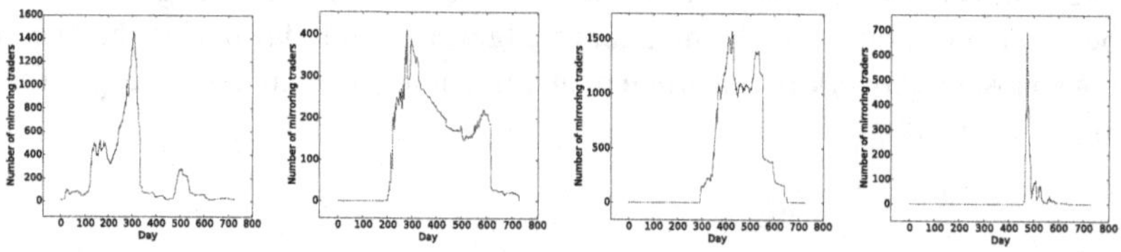

Fig. 4. The in-degree of four nodes in the evolving *eToro* network. (Depicting the popularity of the four corresponding traders along time).

To illustrate this point further we checked how similar different snapshots of the network are. Figure 5 presents the top 50 popular nodes for four different time periods: July-September 2011 (snapshot 1), January-March 2012 (snapshot 2), July-September 2012 (snapshot 3) and January-March 2013. That is four three-month snapshots with three-month gaps in between. As can be seen in the figure, only 11 nodes that were included in the top 50 popular nodes of snapshot 1 remained in the top 50 popular nodes of snapshot 2; only 17 nodes that were included in the top 50 popular nodes of snapshot 2 remained in the top 50 popular nodes of snapshot 3 and only 19 nodes that were included in the top 50 popular nodes of snapshot 3 remained in the top 50 popular nodes of snapshot 4. That is, the network may change significantly along time.

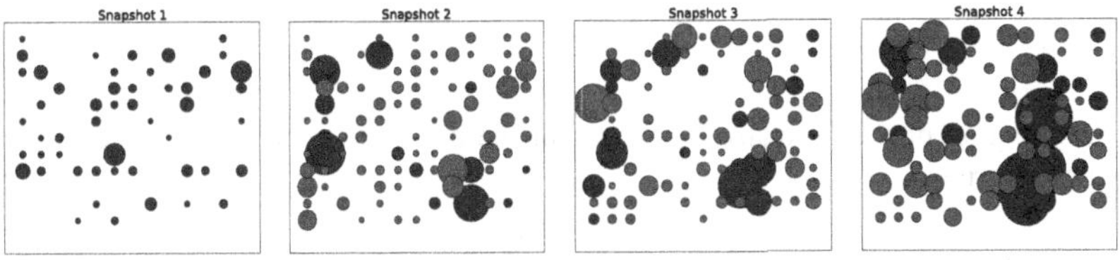

Fig. 5. The 50 most popular nodes in each one of the four snapshots. Green nodes represent nodes that are included in the 50 most popular nodes of the current snapshot but were not included in the previous one. Red nodes represent nodes that were included in the 50 most popular nodes of the previous snapshot but are not included in the current one. Blue nodes represent nodes that were included in both snapshots. The node's circle area is proportional to its popularity.

We then examined the degree distribution for each one of the four snapshots above. As can be seen in Figure 6, although the four snapshots differ significantly, the degree distribution for each one of them obey the power-law model.

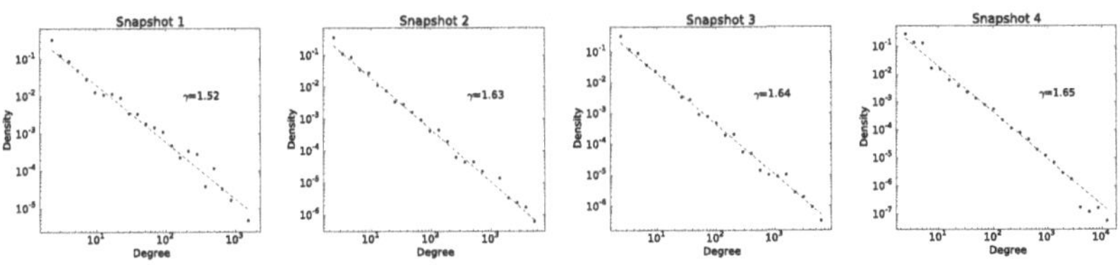

Fig. 6. Degree distribution for each one of the four snapshots that are shown in Figure 5

Next, we studied more carefully the *eToro* network changes between consecutive days. More specifically, we measured the number of added edges (i.e., edges that did not appear in the previous day and appear in the current day) and the number of removed edges (i.e., edges that appeared in the previous day and do not appear in the current day). Since the size of the *eToro* network grows over time (see Fig. 2), we normalized the above quantities by dividing them in the number of edges that were present in the previous day. We found that, the normalized magnitude of change between each two consecutive snapshots (according to each one of the two measures) follows a power-law distribution (see Figure 7).

In order to understand better this finding, we tried to break down the overall network changes into two smaller components.

First, we measured the changes by taking into account only the nodes that were added and removed between the two consecutive days. That is, we considered only users that

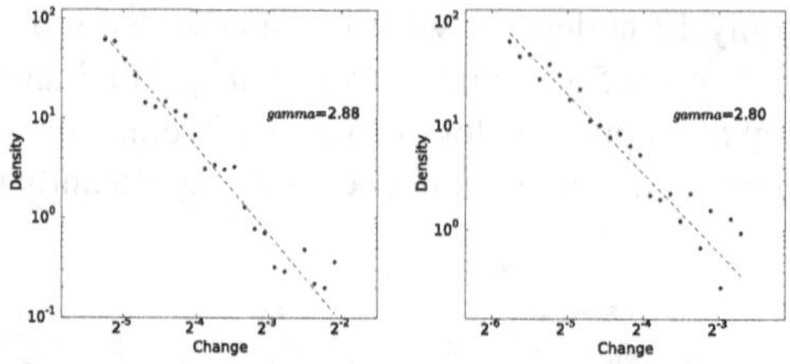

Fig. 7. Distribution of the normalized changes in the *eToro* network: added edges (left) and removed edges (right)

were not trading in the previous day but are trading in the current day and users that were trading in the previous day but are not trading in the current day. As can be seen in the top two subfigures of Figure 8, the normalized number of added and removed nodes also follows a power-law distribution. That is, in most days, only a small number of nodes are added to or removed from the network, but occasionally, a large number of nodes are added or removed. We repeated the same analysis, when taking into account only the edges that at least one of their nodes was added or removed. As can be seen in the bottom two subfigures of Figure 8, the result was again a power-law distribution.

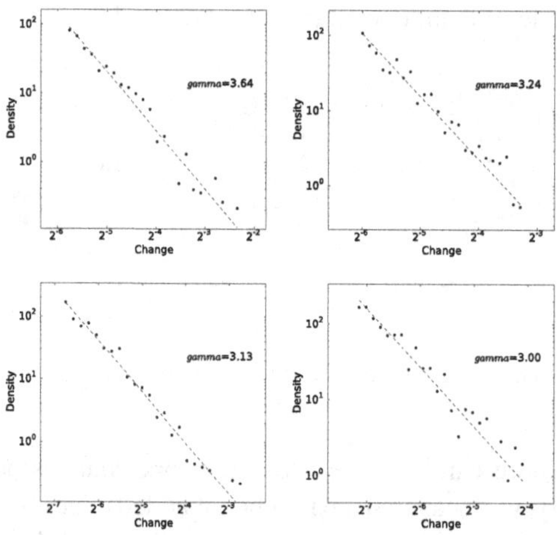

Fig. 8. Distribution of the normalized changes in the *eToro* network, as reflected by the added and removed nodes: added nodes (top left), removed nodes (top right), added edges (bottom left) and removed edges (bottom right)

Then, we measured the changes by taking into account only the nodes that existed in both of the two consecutive days. That is, we considered only users that were trading in the previous day and are also trading in the current day. As can be seen in Figure 9, even when only the common nodes are considered, the normalized number of added and removed edges follows a power-law distribution.

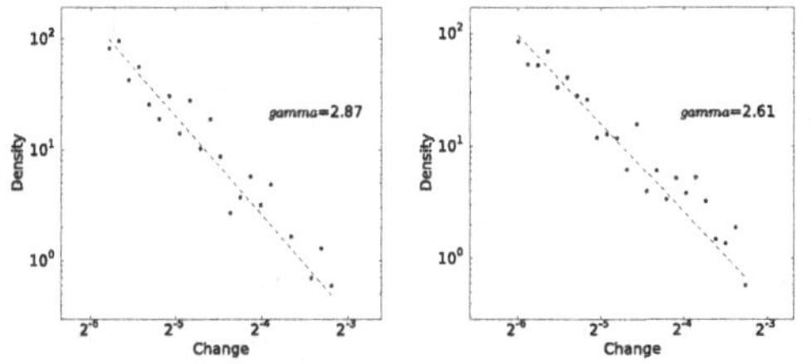

Fig. 9. Distribution of the normalized changes in the *eToro* network, as reflected by the common nodes: added edges (left) and removed edges (right)

Our results were validated using the statistical tests for power-law distributions that were suggested in [18]. First, we applied the goodness of fit test. As can be seen in Table 1, the p-values for all cases are greater than 0.1, suggesting that power-law is a plausible fit. Second, we tested alternative types of distribution. As can be seen in the table, the distribution is more likely to be truncated power-law than general power-law in all cases (the GOF value is negative), and the results are significant in three out of eight of the cases (the p-values are lower than 0.05); the distribution is more likely to be truncated power-law than exponential and the result is significant in five out of eight of the cases cases and the distribution is more likely to be truncated power-law than log-normal in all cases and the result is significant in five out of eight of the cases.

Table 1. Statistical tests for power-law distributions. The numbers in the three right columns represent the p-value and the sign of the GOF value in brackets.

Fig.	Subfigure	xmin	alpha	Goodness of Fit	Power-Law vs. Trunc. Power-Law	Trunc. Power-Law vs. Exponential	Trunc. Power-Law vs. Log-Normal
7	added eges	0.024	2.88	0.121	(-) 0.108	(+) 0.012	(+) 0.396
	removed edges	0.025	2.80	0.207	(-) 0.012	(+) 0.008	(+) 0.000
8	added nodes	0.073	3.64	0.613	(-) 0.613	(+) 0.093	(+) 0.732
	removed nodes	0.023	3.24	0.111	(-) 0.099	(+) 0.160	(+) 0.005
	added edges	0.018	3.13	0.545	(-) 0.544	(+) 0.063	(+) 0.411
	removed edges	0.012	3.00	0.110	(-) 0.108	(+) 0.159	(+) 0.006
9	added edges	0.014	2.87	0.123	(-) 0.039	(+) 0.027	(+) 0.032
	removed edges	0.014	2.61	0.131	(-) 0.009	(+) 0.014	(+) 0.000

4 Summary and Future Work

In this paper, we investigate how scale-free networks evolve over time. Studying a real-world network, we find that: (1) the network topology may change significantly along time, (2) the degree distribution of nodes in the network obeys the power-law model at any given state and (3) the magnitude of change between consecutive states of the network also presents a power-law effect.

Better understanding the temporal dynamics of scale-free networks would allow us to develop improved and more realistic algorithms for generating networks. Moreover, it would help us in better predicting future states of the network and estimating their probabilities. For example, it may help in bounding the probability that a given node remains popular over a certain period of time.

In future work we intend to check how the distribution of changes between consecutive states of the networks influences the overall networks performance. We hypothesize that in cases where the distribution of changes is closer to a power-law distribution, the overall network performance would be higher. Furthermore, we would like to investigate the mechanism that is responsible for the power-law shape of the distribution. Finally, we would like to suggest a generative model for networks based on the above findings.

References

1. Barabási, A.L., Albert, R.: Emergence of scaling in random networks. Science 286(5439), 509–512 (1999)
2. Watts, D.J., Strogatz, S.H.: Collective dynamics of 'small-world' networks. Nature 393(6684), 440–442 (1998)
3. Amaral, L.A.N., Scala, A., Barthélémy, M., Stanley, H.E.: Classes of small-world networks. Proceedings of the National Academy of Sciences 97(21), 11149–11152 (2000)
4. Newman, M.E.: Assortative mixing in networks. Physical Review Letters 89(20), 208701 (2002)
5. Girvan, M., Newman, M.E.: Community structure in social and biological networks. Proceedings of the National Academy of Sciences 99(12), 7821–7826 (2002)
6. Ravasz, E., Barabási, A.L.: Hierarchical organization in complex networks. Physical Review E 67(2), 026112 (2003)
7. Garlaschelli, D., Loffredo, M.I.: Patterns of link reciprocity in directed networks. Physical Review Letters 93(26), 268701 (2004)
8. Milo, R., Itzkovitz, S., Kashtan, N., Levitt, R., Shen-Orr, S., Ayzenshtat, I., Sheffer, M., Alon, U.: Superfamilies of evolved and designed networks. Science 303(5663), 1538–1542 (2004)
9. Dorogovtsev, S.N., Mendes, J.F.F.: Scaling behaviour of developing and decaying networks. EPL (Europhysics Letters) 52(1), 33 (2000)
10. Albert, R., Barabási, A.L.: Topology of evolving networks: local events and universality. Physical Review Letters 85(24), 5234 (2000)
11. Berger-Wolf, T.Y., Saia, J.: A framework for analysis of dynamic social networks. In: Proceedings of the 12th ACM SIGKDD International Conference on Knowledge Discovery and Data Mining, pp. 523–528. ACM (2006)
12. Carley, K.M.: Dynamic network analysis. In: Dynamic Social Network Modeling and Analysis: Workshop Summary and Papers, Comittee on Human Factors, National Research Council, pp. 133–145 (2003)
13. Chin, A., Chignell, M., Wang, H.: Tracking cohesive subgroups over time in inferred social networks. New Review of Hypermedia and Multimedia 16(1-2), 113–139 (2010)
14. Kossinets, G., Watts, D.J.: Empirical analysis of an evolving social network. Science 311(5757), 88–90 (2006)
15. Palla, G., Barabási, A.L., Vicsek, T.: Quantifying social group evolution. Nature 446(7136), 664–667 (2007)
16. Skyrms, B., Pemantle, R.: A dynamic model of social network formation. In: Adaptive Networks, pp. 231–251. Springer (2009)
17. Anghel, M., Toroczkai, Z., Bassler, K.E., Korniss, G.: Competition-driven network dynamics: Emergence of a scale-free leadership structure and collective efficiency. Physical Review Letters 92(5), 58701 (2004)
18. Clauset, A., Shalizi, C.R., Newman, M.E.: Power-law distributions in empirical data. SIAM Review 51(4), 661–703 (2009)

Big Data-Driven Marketing: How Machine Learning Outperforms Marketers' Gut-Feeling

Pål Sundsøy[1], Johannes Bjelland[1], Asif M. Iqbal[1],
Alex "Sandy" Pentland[2], and Yves-Alexandre de Montjoye[2]

[1] Telenor Research, Norway
{pal-roe.sundsoy,johannes.bjelland,asif.iqbal}@telenor.com
[2] Massachusetts Institute of Technology, The Media Laboratory, USA
{pentland,yva}@mit.edu

Abstract. This paper shows how big data can be experimentally used at large scale for marketing purposes at a mobile network operator. We present results from a large-scale experiment in a MNO in Asia where we use machine learning to segment customers for text-based marketing. This leads to conversion rates far superior to the current best marketing practices within MNOs.

Using metadata and social network analysis, we created new metrics to identify customers that are the most likely to convert into mobile internet users. These metrics falls into three categories: discretionary income, timing, and social learning. Using historical data, a machine learning prediction model is then trained, validated, and used to select a treatment group. Experimental results with 250 000 customers show a 13 times better conversion-rate compared to the control group. The control group is selected using the current best practice marketing. The model also shows very good properties in the longer term, as 98% of the converted customers in the treatment group renew their mobile internet packages after the campaign, compared to 37% in the control group. These results show that data-driven marketing can significantly improve conversion rates over current best-practice marketing strategies.

Keywords: Marketing, Big Data, Machine learning, social network analysis, Metadata, Asia, Mobile Network Operator, Carrier.

1 Introduction

For many people in Asia, mobile phones are the only gateway to the Internet. While many people have internet capable phones, they are often not aware of their capabilities. The overall penetration of internet in these countries is very small which causes a large digital discrepancy [1]. In the market of this study, internet penetration is less than 10%.

Mobile Network Operators (MNOs) commonly use texts as a way to raise customer awareness of new products and services - and to communicate with their customers. In Asian markets, MNOs typically run thousands of text campaigns a year, resulting in customers receiving several promotional texts per month. Making sure

W.G. Kennedy, N. Agarwal, and S.J. Yang (Eds.): SBP 2014, LNCS 8393, pp. 367–374, 2014.

they are not seen as spammers by customers but rather as providing useful information is a major concern for MNOs. For this particular operator, the policy is to not send more than one text per customer every 14 days. This limit is currently maxed and is preventing new campaigns from being launched.

Targeting, deciding which offer to send to which customer, often relies on the marketing team's "gut-feeling" of what the right audience is for this campaign. In a recent IBM study, 80% of marketers report making such decisions based on their "gut-feeling" [2]. A data-driven approach might lead to an increased efficiency of text-based campaigns by being one step closer to delivering "the right offer to the right customer". For example, previous research showed that data-driven approaches can reliably predict mobile phone and Facebook user's personality [3-5,18], sexual orientation [6], or romantic relationships [7].

Our data-driven approach will be evaluated against the MNO's current best-practice in a large-scale "internet data" experiment [8,9]. This experiment will compare the conversion rates of the treatment and control groups after one promotional text.

We show that this data-driven approach using machine learning and social network analysis leads to higher conversation rates than best-practice marketing approach. We also show that historical natural adoption data can be used to train models when campaign response data is unavailable.

2 Best Practice

The current best practice in MNOs relies on the marketing team's experience to decide which customers should receive a text for a specific campaign. The marketing team typically selects customers using a few simple metrics directly computed from metadata such as call sent, call received, average top-up, etc. For this particular "internet data" campaign, the marketing team recommended to use the number of text sent and received per month, a high average revenue per user (ARPU) [10], and to focus on prepaid customers. Table 1 shows the variables used to select the control group, the customers that, according to the marketing team, are the most likely to convert.

The control group is composed of 50 000 customers selected at random amongst the selected group.

Table 1. Variables used to select the control group

Sending at least four text per month
Receiving at least four text per month
Using a data-enabled handset
'Accidental data usage' (less than 50kb per month)
Customer in medium to high ARPU segment (spending at least 3.5 USD per month)

3 Data-Driven Approach

3.1 Features

For each subscriber in the experiment, we derive over 350 features from metadata, subscription data, and the use of value added services. Earlier work show the existence of influence between peers [11,12] and that social influence plays an important role in product adoption [13,14]. We thus inferred a social graph between customers to compute new features. We only considered ties where customers interacted more than 3 times every month. The strength of ties in this social graph is a weighted sum of calls and texts over a 2-month period. Using this social graph, we computed around 40 social features. These include the percentage of neighbors that have already adopted mobile internet, the number of mobile data users among your closest neighbors (ranked by tie strength), or the total and average volume of data used by neighbors.

3.2 Model

We develop and train the model using 6 months of metadata. As the outcomes of previous mobile internet campaigns were not stored, we train our model using natural adopters. We then compare these natural adopters, people who just started using mobile internet, to people who over the same period of time did not use mobile internet. Our goal is to identify the behavior of customers who 1) might be interested in using internet and who 2) would keep using mobile internet afterwards. We then select 50 000 natural adopters and 100 000 non-internet users at random out of these groups. Note that natural converters are only a way for us to extract characteristics of customers who are likely to convert. The conversion rates are likely to have been better if we had access to data about previous campaigns and previously persuaded adopters.

Table 2. Training set

Sample size	Classifier	Definition
50k	Natural adopters	Less than 50KB of data per month from December to March (accidental data usage). More than 1MB of data per month in April and May
100k	Reference users not using internet	No internet usage

We tested several modeling algorithms such as support vector machine and neural networks to classify natural converters. The final model is a bootstrap aggregated (bagging) decision tree [15] where performance is measured by accuracy and stability. The final model is a trade-off between accuracy and stability where, based on earlier experience, we put more weight on stability. The accuracy of the final model is slightly lower than other considered models. The bagging decision tree however turned out to be more stable when tested across different samples.

The final cross-validated model only relies on a few key variables. Only 20 features out of the initial 350 where selected for the final model. Table 3 shows the top 10 most useful features to classify natural converters as ranked by the IBM SPSS modeler data mining software. The features tagged as binned are handled by the software optimal binning feature.

Table 3. Top 10 most useful features to classify natural converters. Ranked by importance in the model.

Rank	Type	Description
1	Social learning	Total spending on data among close social graph neighbors
2	Discretionary income	Average monthly spending on text (binned)
3	Discretionary income	Average monthly number of text sent (binned)
4	Discretionary income	Average monthly spending on value added services over text (binned)
5	Social Learning	Average monthly spending on data among social graph neighbors
6		Data enabled handset according to IMEI (Yes/No)
7	Social Learning	Data volume among social graph neighbors
8	Social Learning	Data volume among close social graph neighbors
9	Timing	Most used handset has changed since last month
10		Amount of 'accidental' data usage

3.3 Out-of Sample Validation

Before running the experiment, we validated our model on natural adopters in a new, previously unseen, sample using other customers and another time period. The performance on historical data is measured using lift curves. Fig. 1 shows a lift of around 3 among the 20% highest scored customers. This means that if we were to select the 20% highest scored customers, the model would predict 3 times better than selecting at random from the sample.

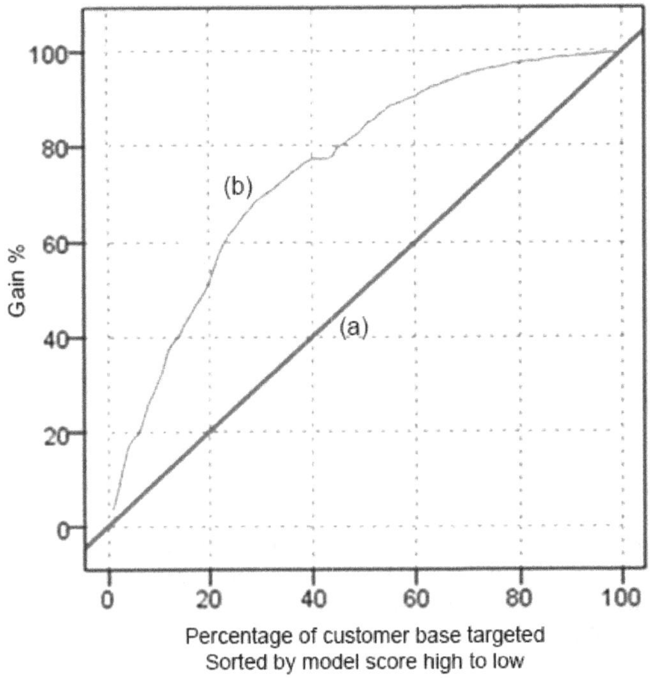

Fig. 1. Out-of-sample validation of the model using lift curves. (a) is the gain if customer were selected at random while (b) shows selection by using the model.

We then select the treatment group using our model. We let the marketing team pick their best possible (control) group first and then specifically excluded them when selecting our treatment group. The treatment group is composed of the top 200 000 customers with the highest score.

4 Experiment

A large-scale experiment is then run to compare our data driven approach to the current best-practice in MNOs. The approaches will be compared using the conversion rates of the control and treatment group.

In this experiment, the selected customers receive a text saying that they can activate a 15MB bundle of data usage for half of the usual price. The 15MB have a limited validity and are only valid for 15 days. The customer can activate this offer by sending a text with a code to a short-number. This is a common type of campaign and is often used in this market. The text contains information about the offer and instructions on how to activate it.

The conversion rates between treatment and control group were striking. The conversion rate in the treatment group selected by our model is 6,42% while the conversion rate of the control group selected using the best-practice approach is only 0.5%, as shown in Fig. 2. The difference is highly significant (p-value < 2.2e-16). Our data-driven approach leads to a conversion rate 13 times larger than the best-practice approach.

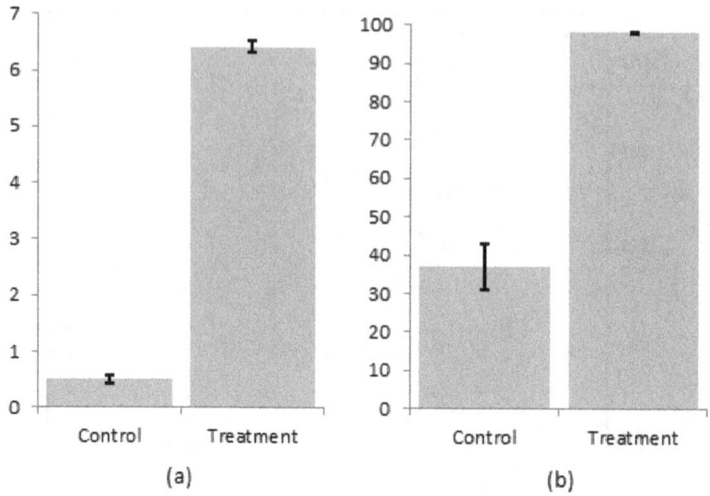

Fig. 2. (a) Conversion rate in the control (best practice) and treatment (data-driven approach) groups. (b) the percentage of converted people who renewed their data plan after using the volume included in the campaign offer. Error bars are the 95% confidence interval on the mean using the blaker method.

The goal of this campaign is not only to have customers take this offer but also to have them renew it after the trial period. We thus compare the renewal rate, customers buying another data plan after using the half-priced package, between the converted people in the two groups. Here too, the results are striking and highly significant (p-value < 2.2e-16). We find that while 98% of converted people in the treatment group buy a second, full price package, only 37% of the converted people in the control group renew their plan. This means that 6.29 % of the treatment group is converted in month two, compared to 0.19% of the control group.

5 Discussion

Although it was not a goal for our data-driven approach to be interpretable, a posteriori categorization of the features selected by our model leads to some interesting qualitative insights. Indeed, most of the features fall under three categories: discretionary income, timing, and social learning, see Table 3.

Discretionary income was expected by the marketing team to be important overall. They hypothetised that customers with a high total ARPU would be more likely to convert. The model does however not select total spending as an important variable to help predict conversion. In fact, looking at the ARPU of those who received an SMS and then adopt, we see that the low ARPU segment is slightly overrepresented. Our text and data focused discretionary spending variables are however selected as important by the model. Text and data focused spending variables seem to contain relevant information to help predict adoption more than overall spending.

Timing measured through using a new phone is our 9th most useful feature.

Finally, the social learning features we computed for this study turn out to be very helpful to help classify the natural converters. The total spending on data among the closest social graph neighbors is our most predictive feature.

Using social features in selecting customers for the offer might have improved the retention rates. We speculate that the value customers derive from mobile data increases when their neighbors are also mobile data users. In other words that we expect that a network externality effect exists in mobile internet data. This means that selecting customers whose closest neighbors are already using mobile data might have locally used this network effect to create the lasting effect we observe with very high retention rate in the second month.

The success of this pilot study triggered new technical developments and campaign data are now being recorded and stored. Future models might be refined using this experimental data and new insights might be uncovered using statistical models or interpretable machine learning algorithms. We also plan to refine the different attribute categories; social learning, discretionary income, and timing and to use them a priori in further model building.

The marketing team was impressed by the power of this method and is now looking into how this can be implemented in the MNO's operations and used more systematically for future campaigns. They see such data-driven approach to be particularly useful to help launch new products where little prior experience exists. This was the case with this campaign as the overall mobile internet penetration rate in the country is very low. The marketing team usually learns what the right segment for a new product through extensive trial-and-error.

When performing research on sensitive data, privacy is a major concern. [16,17] showed that large scale simply anonymized mobile phone dataset could be uniquely characterized using as little as four pieces of outside information. All sensitive information in this experiment was hashed and only the local marketing team had the contacts of the control and treatment groups.

We believe our findings open up exciting avenues of research within data-driven marketing and customer understanding. Using behavioral patterns, we increased the conversion rate of an internet data campaign by 13 times compared to current best-practice. We expect such an approach will greatly reduce spamming by providing the customer with more relevant offers.

Acknowledgments. This research was partially sponsored by the Army Research Laboratory under Cooperative Agreement Number W911NF-09-2-0053, and by the Media Laboratory Consortium. The conclusions in this document are those of the authors and should not be interpreted as representing the social policies, either expressed or implied, of the sponsors.

References

1. Chinn, M.D., Fairlie, R.W.: The Determinants of the Global Digital Divide: A Cross-Country Analysis of Computer and Internet Penetration. Economic Growth Center (2004)
2. IBM Center for applied insight, `http://public.dhe.ibm.com/common/ssi/ecm/en/coe12345usen/COE12345USEN.PDF`
3. de Montjoye, Y.-A., Quoidbach, J., Robic, F., Pentland, A(S.): Predicting Personality Using Novel Mobile Phone-Based Metrics. In: Greenberg, A.M., Kennedy, W.G., Bos, N.D. (eds.) SBP 2013. LNCS, vol. 7812, pp. 48–55. Springer, Heidelberg (2013)
4. Staiano, J., Lepri, B., Aharony, N., Pianesi, F., Sebe, N., Pentland, A.: Friends don't lie: inferring personality traits from social network structure. In: Proceedings of the 2012 ACM Conference on Ubiquitous Computing, pp. 321–330 (2012)
5. Schwartz, H.A., Eichstaedt, J.C., Kern, M.L., Dziurzynski, L., Ramones, S.M., et al.: Personality, Gender, and Age in the Language of Social Media: The Open-Vocabulary Approach. PLoS One 8(9), e73791 (2013), doi:10.1371/journal.pone.0073791
6. Jernigan, C., Mistree, B.F.T.: Gaydar: Facebook friendships expose sexual orientation. First Monday, [S.l.] (2009), `http://journals.uic.edu/ojs/index.php/fm/article/view/2611/2302`, ISSN 13960466
7. Backstrom, L., Kleinberg, J.: Romantic Partnerships and the Dispersion of Social Ties: A Network Analysis of Relationship Status on Facebook, arXiv preprint arXiv:1310.6753 (2013)
8. Bond, R.M., Fariss, C.J., Jones, J.J., Kramer, A.D.I., Marlow, C., Settle, J.E., Fowler, J.H.: A 61-million-person experiment in social influence and political mobilization. Nature 489(7415), 295 (2012), doi:10.1038/nature11421
9. Aral, S., Walker, D.: Identifying Influential and Susceptible Members of Social Networks. Science 337(6092), 337 (2012)
10. Wikipedia, `http://en.wikipedia.org/wiki/Arpu`
11. Turner, J.C.: Social influence. Mapping social psychology series, p. 206. Thomson Brooks/Cole Publishing Co., Belmont (1991)
12. Christakis, N.A., Fowler, J.H.: The spread of obesity in a large social network over 32 years. New England Journal of Medicine 357(4), 370–379 (2007)
13. Sundsøy, P., Bjelland, J., Engø-Monsen, K., Canright, G., Ling, R.: Comparing and visualizing the social spreading of products on a large-scale social network. In: Ozyer, T., et al. (eds.) The influence on Technology on Social Network Analysis and Mining, XXIII, 643 p. 262. Springer (2013)
14. Sundsøy, P., Bjelland, J., Canright, G., Engø-Monsen, K., Ling, R.: Product adoption networks and their growth in a large mobile phone network. IEEE Advanced in Social Network Analysis and Mining (2010)
15. Breiman: Technical Report No. 421, Department of Statistics University of California (1994)
16. de Montjoye, Y.-A., Hidalgo, C.A., Verleysen, M., Blondel, V.D.: Unique in the Crowd: The privacy bounds of human mobility. Nature Scientific Reports 3, id. 1376 (2013)
17. de Montjoye, Y.-A., Wang, S., Pentland, A.: On the Trusted Use of Large-Scale Personal Data. IEEE Data Engineering Bulletin 35(4) (2012)
18. de Montjoye, Y.-A., Quoidbach, J., Robic, F., Pentland, A(S.): Predicting personality using novel mobile phone-based metrics. In: Greenberg, A.M., Kennedy, W.G., Bos, N.D. (eds.) SBP 2013. LNCS, vol. 7812, pp. 48–55. Springer, Heidelberg (2013)

"Frogs in a Pot": An Agent-Based Model of Well-Being versus Prosperity

Chris Thron

Texas A&M University – Central Texas Department of Mathematics,
Founder's Hall, Killeen TX USA 76549
thron@ct.tamus.edu

Abstract. Surveys of rapidly-developing countries have shown that huge increases in average personal wealth are frequently accompanied by little or no increase in average (self-reported) happiness. We propose a simple agent-based model that may help to explain this phenomenon. The model shows that under certain conditions, the cumulative effect of individuals' free choices of employment that maximizes their (self-perceived) personal well-being may actually produce a continuing decrease in the population's average well-being. Like the proverbial "frog in a pot", the eventual effect is worse when the onset of the decrease is more gradual. More generally, the model indicates that there is a natural tendency in free-market societies for well-being to become defined in increasingly materialistic terms. We discuss the implications of our model on the issue of incentive pay for teachers, and argue that our model may also provide insight into other situations where individuals' free-market choices lead to progressive worsening of the population's average well-being.

Keywords: Well-being, economic development, prosperity, agent-based model.

1 Introduction

Money can't buy happiness: this adage applies to nations as well as individuals. During the decade from 1995 to 2005, the mean per-capita income in mainland China rose by 150 percent; while studies report the mean level of (self-reported) well-being or happiness dropped significantly during the same period[1,2].The example of China may be the most striking, but it is not unique. Japan from 1958 and 1987 saw a 400 percent increase in real income, with no significant increase in average self-reported happiness level [3]. Similarly, the U.S. experienced strong economic growth from 1946-1990, while many indicators show a decrease in happiness [4].

Some authors attribute such results to rising expectation levels which increase as rapidly as real income [5]. Others cite "relative deprivation", and contend that those that get wealthier still find themselves increasingly worse off relative to those they consider to be their peers [6]. But there are reasons to suppose that there may be more going on than changes in expectations or perceived relative status. Increasing incomes are usually accompanied by worsening of other social indicators such as divorce, crime, and delinquency rates: such was certainly the case in China[7,8].

W.G. Kennedy, N. Agarwal, and S.J. Yang (Eds.): SBP 2014, LNCS 8393, pp. 375–383, 2014.

It seems implausible that such changes are due to feelings of financial inferiority. Clearly many social changes are occurring during such periods of growth, and no simple explanation can account for all of the trends that are observed.

In this paper we propose a simple agent-based model that may help to explain some of these trends. We argue that under certain conditions, the cumulative effect of people's own free choices may produce decreases in the well-being of the population as a whole. These are "real" decreases, not just relative to a changing norm.

Before we present the model, some caveats are in order. In the sociological literature the terms "happiness" and "well-being" have various definitions, and are not necessarily synonymous. In particular, some research has shown that there is a significant difference between self-reported happiness and other measures of "actual" well-being [9]. We will use the term "well-being," which we do not attempt to define precisely. However, we presume certain relations between a person's well-being and various factors related to his/her employment, as we explain in the following sections.

We also emphasize that our model is *heuristic*, and as such the terminology should not be taken literally. For example, in reality there is certainly no clear line between "incentives" and "collaterals" such as we have drawn in this paper. Also, "job offers" may include a variety of factors, including the restructuring of existing jobs, employees taking on additional employment, and so on. The terms that we employ should be taken as proxies for complicated factors that exert certain types of influence on agents within the system. They are not intended to denote directly-measurable quantities.

Finally, we admit that the model is tremendously oversimplified. We have neglected many, effects in order to focus on a few important factors. The model may easily be elaborated, so there are many prospects for future research.

2 Assumptions and Specification of the Agent-Based Model

2.1 Basic Assumptions of the Model

Our model of a competitive job market is based on the following basic assumptions:

1. Potential employees decide whether or not to accept a particular position offer based on their perception of the relative benefits of the position being offered.
2. Besides evident incentives, positions possess other conditions which impact the employee's well-being. Employees may not be fully cognizant of the impact of these conditions on their well-being.
3. Employers offer incentives comparable to what they see other employers offering.
4. To improve incentives, employers make tradeoffs that tend to have a negative impact on less-obvious employment conditions. Alternatively, an employer may use high incentives to entice potential employees to consider a position despite personal costs such as relocation and termination of on-the-job friendships. These factors both indicate a negative correlation between incentives and other job conditions.

2.2 Detailed Specification of the Model

We represent the working population as a set of agents, where each agent possesses a job. In this simple model, we make the following assumptions about the agents:

- The number of agents (denoted by N) is fixed, and all agents are employed;
- All agents in the population compete for the same positions.
- All agents have the same preferences (that is, under the same conditions all make the same choice).

Every job has several factors which impact the job-holder's well-being. These factors may be broadly divided into two categories:

– *Incentives* are the "selling points" which the employer advertises to make the position attractive. First and foremost of these are salary, bonuses, and benefits. Other factors may include advancement potential, vacations, flexibility, job security, and so on. Incentives also tend to serve as rewards for good job performance.
– *Collaterals* on the other hand are less likely to be discussed during job negotiations, but nonetheless impact the employee's physical, emotional, and spiritual well-being. These may include working atmosphere, performance expectations, on-the-job relationships, implicit responsibilities, ethics, and so on. Also included are factors particular to the employee such as relocation, commuting, relationships, etc.

Our model is time-dependent, and for simplicity we use a discrete time index t. The model's evolution over time is governed by the following specifications:

1. At each time t=0,1,2,3,…,, each agent n holds a position that is characterized by two indices. The first index (denoted by $A_t(n)$) reflects the effect on well-being of the position's incentives; while the second (denoted by $B_t(n)$) reflects the effect of collaterals. The total well-being $W_t(n)$ of agent at time t is obtained as:

$$W_t(n) \equiv A_t(n) + B_t(n).$$

2. The initial index values $A_0(n)$ and $B_0(n)$ for each n are independent, mean-zero normal random variables with standard deviations s_A and s_B, respectively.
3. For each time step t=1,2,3,…, a single new position is created. This new position has incentive and collateral indices A_t^* and B_t^* respectively, which are chosen randomly according to independent normal distributions. The mean of A_t^* is the average of the existing incentive indices for all agents: mathematically, this means

$$\text{Mean}[A_t^*] = \frac{1}{N}\sum_{n=1}^{N} A_{t-1}(n).$$

The mean of B_t^* on the other hand is negatively correlated with the mean of A_t^*:

$$\text{Mean}[B_t^*] = -k \cdot \text{Mean}[A_t^*],$$

where k is called the *degradation parameter*. The standard deviations of A_t^* and B_t^* are s_A and s_B respectively (the same as for the initial index distributions).

Note that Mean[A_t^*] is determined by comparison with current market conditions, reflecting basic assumption (3) from Section 2.1. On the other hand, Mean[B_t^*] decreases with increasing Mean[A_t^*], reflecting basic assumption (4).

4. Each new position is offered sequentially to up to M randomly-chosen agents (where M is a fixed model parameter) as follows. Suppose agent n is the first to be chosen for the new position at time t. Then agent chooses to accept the offer if:

$$A_t^* + q{\cdot}B_t^* > A_{t-1}(n) + q{\cdot}B_{t-1}(n) \Rightarrow \text{Accept} \quad (1)$$

Here q is a model parameter (called the *awareness parameter*) with $0 \leq q \leq 1$. (This range of values reflects basic assumption (2) from Section 2.1.) If agent n accepts, his/her incentive and collateral indices are updated as follows:

$$\text{Accept} \Rightarrow A_t(n) = A_t^* \text{ and } B_t(n) = B_t^*. \quad (2)$$

If agent n rejects the offer, then another agent is randomly chosen, who in turn chooses whether or not to accept by comparing his current position's indices with the new position according to criterion (1). His indices are then updated according to rule (2) if he accepts the offer. The same procedure is repeated up to M times. All agents m that do not accept the new position (including those that receive no offer) retain their index values, so that $A_t(m) = A_{t-1}(m)$and $B_t(m) = B_{t-1}(m)$.

3 Simulation Results

3.1 General

In this section, we present the results of simulations that show the effect on model behavior of the collateral degradation and awareness parameters (k and q respectively). Other model parameters were set at the following values: N=1000, M=10, s_A=s_B=1. These parameters have little effect on the model's qualitative behavior.

3.2 Fully-Aware Agents (q=1)

We look first at the case where agents are fully aware of the effect of collaterals on their well-being for different values of degradation parameter k from 1 to 2. Higher values of k correspond to cases where raising incentives comes at a greater cost in collaterals: such would be the case, for instance, in a tight economy where employers need to hire, but cannot improve incentives without cutting back in other areas. An alternative high-k scenario would correspond to an agent population that is looking farther afield for higher-paying jobs, and are willing to make greater personal sacrifices to obtain them.

Figure 1 shows that for fully-aware agents, regardless of k we see a steady (but decreasing) rise in both the incentive index and total well-being. The rise is slower for larger values of k: this is to be expected, because a larger k means that raising the incentive index is (on average) more costly in terms of overall well-being. As to collaterals index, in all cases there is an initial rise and then a steady decrease. The initial "euphoric" period reflects the case where an uncompetitive job market suddenly (or rapidly) becomes competitive. Such was the case in China in the late 1990's, when the "iron rice bowl" system of guaranteed employment in state-owned enterprises rapidly gave way to a competitive job market. The freedom to choose their own positions puts agents in a win-win situation: they can improve incentives without sacrificing collaterals. But as the competitive system continues, agents find themselves making tradeoffs: the population looks increasingly to material wealth for satisfaction, even though it means sacrificing less tangible benefits. This increasingly materialistic nature of well-being is exactly what has been observed in China [6].

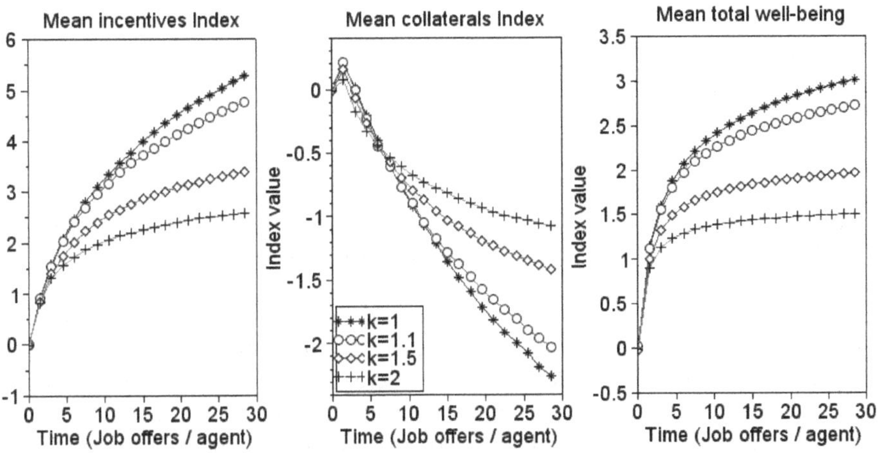

Fig. 1. Incentives, collaterals, and well-being indices when awareness parameter q=1

Figure 1 also reveals some surprising behaviors of the collaterals index. One would expect the collaterals index to be largest when the degradation parameter k is smallest, since this means that employers can improve incentives with less negative impact on collaterals. In fact, the opposite is true! Although the peak value of the collaterals index (which occurs when offers/agent is around 1) does increase for decreasing k, in the long-term the order is reversed. Another bewildering result is the continuing increase in mean incentives index even when the degradation parameter equals 2, although this means that the new job offers are becoming (on average) less and less attractive to agents in terms of overall well-being.

Both of these surprising results are related to the fact that not all positions offered are accepted. Indeed, Figure 2 (*left*) shows that the acceptance rates for new positions decrease dramatically over time. Smaller values of degradation parameter k lead to higher acceptance rates, which end up producing larger decreases in the mean collaterals index. This produces a "frog in a pot" effect, where milder negative correlations between mean collaterals and mean incentives cause heavier losses in collaterals in the long term. The mean incentives of new job offers continue to rise because employers feel obliged to match incentives of currently-held positions. Increasingly, offers are accepted based on high incentives (despite decreasing collaterals), which continues to push up the incentives index for offered positions.

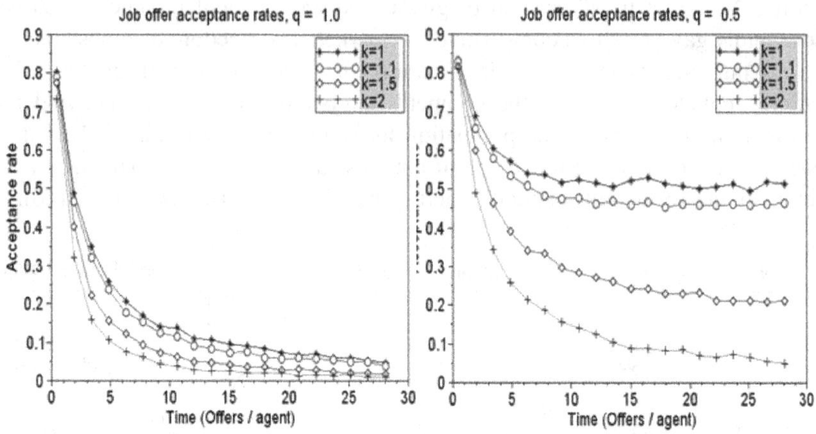

Fig. 2. Job offer acceptance rates for various values of k, in the case where awareness parameter $q = 1$ (*left*) and $q=0.5$ (*right*). Curves are smoothed using a Gaussian filter.

Many empirical studies show that wealthier individuals in a given population tend to be happier. In our model, this would correspond to a positive correlation between incentives and well-being. Indeed, initially there is a strong positive correlation ($\rho=0.5$) between the two. But over time the picture changes. When $J/N = 10$, for small degradation ($k\leq 1.1$) we find small positive correlations ($\rho\approx 0.1$) between well-being and incentives; and for larger values of k, we see negligible or very small negative correlations ($\rho\approx -0.02$). These results do not necessarily contradict the empirical studies: since all agents are competing for the same positions on equal footing, we are effectively modeling a subpopulation of peers. These results should not be expected to apply to the overall population, which is employed in a variety of professions and have varying qualifications and abilities.

3.3 Partially-Aware Agents

Next, we look first at the case where agents are only partially aware of the effect of collaterals on their well-being ($q<1$). There are many reasons for supposing that this is actually the case. Collateral factors are less obvious than incentives. Agents may also have their eyes set on "move up the ladder," and thus willfully ignore collateral deficiencies. It is likely also be that employers do not disclose collateral deficiencies to potential employees, since this brings no advantage to the employer.

Figure 3 shows model evolution for q=0.5 with the same values of k as in Figure 1: qualitatively similar results are obtained with other values of q less than 1. In contrast to the q=1 case, we now observe a *decrease* in well-being, even when k is only slightly greater than 1. Figure 2 (*right*) shows that offer acceptance rates stabilize at much higher values than the q=1 case. This ensures that the negative trends will continue, particularly when k is close to 1.

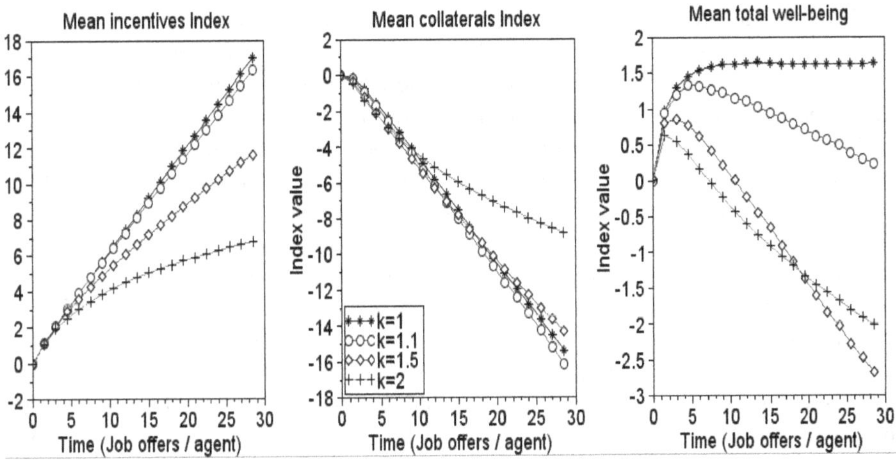

Fig. 3. Incentives, collaterals, and well-being indices when awareness q=0.5

We also find that the correlation between agents' incentives and well-being becomes definitely negative in the q<1 case, as shown in the scatterplots in Figure 4.

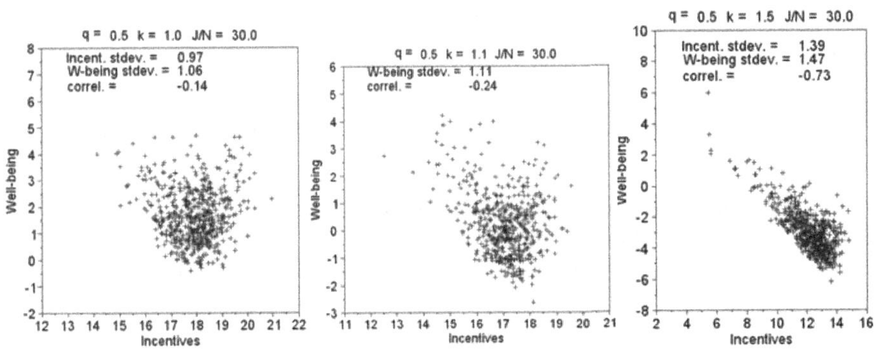

Fig. 4. Scatterplots of Incentives versus well-being for partially-aware agents (q=0.5) at time J/N=30 for the cases k=1 (*left*), k=1.1 (*middle*), and k=1.5 (*right*)

4 Discussion and Conclusions

Our model supports Lane's assertion (see [4]) that there is a natural tendency in free-market societies for well-being to become defined in increasingly materialistic terms. Competitive pressures on employers produce new job offers with greater material incentives, at the expense of less material factors. The prevalence of materially top-heavy job offers means that many are accepted, further reinforcing this trend. If employed agents are fully aware of the tradeoffs they are making, their average well-being continues to increase—though at the expense of a steady decrease in collaterals. But if they fail to take the influence of non-material factors on well-being sufficiently into account, the result is a decrease in agents' average well-being.

Due to the generality of its assumptions, the model may be applied to many different situations. Here we mention two. First, consider the situation in American public schools. In response to poor student performance (particularly in inner-city schools), some policy-makers have suggested giving incentive pay to teachers that produce "results". Our model suggests that this policy would be disastrous, both for teachers' well-being and for quality of education. Teachers tend to leave their jobs for collateral reasons such as poor student behavior, lack of input into school policy, and inadequate preparation time [8]. Pay incentives would create a natural downward pressure on these collaterals, as school budgets shift their emphasis. This would further tilt the incentives-collaterals balance for teachers, making them even worse off on average and leading to an increase in teacher attrition that is already at catastrophic levels.

Second, consider the mortgage crisis of 2007. Although the economic choices here did not involve jobs, they did involve financial transactions with evident incentives as well as less-visible collaterals. Competitive pressures led naturally to mortgages with progressively easier requirements and lower apparent interest rates, while invisibly eroding the financial stability of home-buyers. The eventual result is well-known.

References

1. Burkholder, R.: Chinese Far Wealthier Than a Decade Ago—but Are They Happier? The Gallup Organization, `http://www.gallup.com/poll/14548/chinese-far-wealthier-than-decade-ago-they-happier.aspx` (retrieved November 5, 2013)
2. Wong, C.K., et al.: Subjective well-being, societal condition and social policy—the case study of a rich Chinese society. Social Indicators Research 78, 405–428 (2006)
3. Easterlin, R.A.: Will Raising the Income of all Increase the Happiness of All? Journal of Economic Behavior and Organization 27(1), 35–47 (1995)
4. Lane, R.: The Joyless Market Economy. In: Ben-Ner, A., Putterman, L. (eds.) Economics, Values, and Organization, pp. 461–485. Cambridge University Press (1999)
5. Graham, C.: Happiness around the World: The paradox of happy peasants and miserable millionaires. Oxford University Press (2009)
6. Brockmann, H., Delhey, J., Welzel, C., Yuan, H.: The China Puzzle: Falling Happiness in a Rising Economy. Journal of Happiness Studies, doi:10.1007/s10902-008-9095-4

7. Wang, D.: The Study of Juvenile Delinquency and Juvenile Protection in the People's Republic of China. Crime and Justice International 22(94), 4–14 (2006)
8. Wang, Q., Zhou, Q.: China's divorce and remarriage rates: Trends and regional disparities. Journal of Divorce & Remarriage 51, 257–267 (2010)
9. Kahneman, D., Krueger, A.B.: Developments in the Measurement of Subjective Well-Being. Journal of Economic Perspectives 20(1), 3–24 (2006)
10. Alliance for Excellent Education, Teacher Attrition: A Costly Loss to the Nation and to the States (2005), `http://www.all4ed.org` (retrieved November 5, 2013)

Frabjous: A Declarative Domain-Specific Language for Agent-Based Modeling*

Ivan Vendrov, Christopher Dutchyn, and Nathaniel D. Osgood

University of Saskatchewan
firstname.lastname@usask.ca

Abstract. Agent-based modeling (ABM) is a powerful tool for the study of complex systems; but agent-based models are notoriously difficult to create, modify, and reason about, especially in contrast to system dynamics models. We argue that these difficulties are strongly related to the choice of specification language, and that they can be mitigated by using functional reactive programming (FRP), a paradigm for describing dynamic systems. We describe Frabjous, a new language for agent-based modeling based on FRP, and discuss its software engineering benefits and their broader implications for language choice in ABM.

Keywords: functional reactive, functional programming, simulation, dynamic model, domain-specific language, agent-based simulation, agent-based modeling.

1 Introduction

For systems that evolve continuously in space and time, the language of differential equations—honed by centuries of application to the physical sciences—has no substitute. Its syntax is extremely terse, with precise mathematical semantics that permit sophisticated analysis.

There are, however, a number of processes that are difficult to express with differential equations, such as those involving networks, history dependence, or heterogenous populations[8, 11]. The need to model these processes is addressed by agent-based modeling (ABM), a more general approach which involves specifying the behavior of individuals and allowing the global dynamics to emerge from their interactions.

This generality comes with a number of costs. With existing tools and frameworks, agent-based models are significantly harder to create, extend, and understand; and significantly more expensive to calibrate and run relative to models based on systems of differential equations [11, 12].

Although the increased cognitive and computational costs of agent-based models are to some degree unavoidable due to their increased complexity and generality, we argue that these costs have been greatly compounded by the use of imperative object-oriented languages such as Java and C++ to express

* This work was funded in part by an NSERC Undergraduate Student Research Award.

W.G. Kennedy, N. Agarwal, and S.J. Yang (Eds.): SBP 2014, LNCS 8393, pp. 385–392, 2014.

model logic. As used in modern ABM frameworks, these languages force modelers and users to think at a low level of abstraction, and fail to cleanly separate domain-level structure from implementation details such as input/output, the time-stepping mechanism, and the data structures used, which obscures the essential model logic [11].

On the other hand, the underlying language of ODE models is not imperative but declarative: rather than describing rules by which model variables change, differential equations specify relationships between model variables that hold at all times. We believe that the declarative nature of ODE models accounts for much of their success by simplifying model creation, modification, and analysis. It then stands to reason that ABM could be similarly simplified by basing it on an appropriate declarative language. To support this hypothesis, we developed Frabjous, a new declarative language for ABM. In this paper, we describe Frabjous and demonstrate its benefits on a standard example model.

2 Background

In this section, we briefly describe the existing languages and technologies we employ to create Frabjous, as well as explain why we chose them.

2.1 Haskell

Haskell is a purely functional programming language; that is to say, a Haskell program is a list of equations, each defining a value or a function.

Since Haskell lacks a mechanism for changing the value of a variable, it comes very close to the declarative ideal - specifying what things are, not how they change - and reaps the associated benefits: Haskell programs are often an order of magnitude shorter than programs written in imperative languages, are clearer to read, and are much easier to analyze mathematically. For these reasons, Haskell is the base language of Frabjous: Frabjous code is largely composed of segments of Haskell code, and compiles directly to Haskell.

2.2 Functional Reactive Programming

An apparent weakness of Haskell is the difficulty of representing systems that vary with time, since there is no mechanism for changing state. As pioneered by Elliott and Hudak, *functional reactive programming* (FRP) is a paradigm that augments functional programming with time-varying values as well as a set of primitive operations on these values [4]. Arrowized functional reactive programming (AFRP) is a version of FRP that shifts the focus onto functions between time-varying values, called *signal functions*[9].

The simplest AFRP operator is `constant`, which defines a constant signal function. So the output of the signal function `constant 1` is `1` at all times, and for all inputs. Integration over time can also be viewed as a signal function, since it operates on a function of time and produces a function of time. So

```
integral . constant 1
```

is a signal function that ignores its input and outputs the current time ('.' is the Haskell function composition operator).

Following the Netwire version of AFRP[14], we allow signal functions to sometimes not produce values. For example, `rate` is a signal function that takes a time-varying number and produces a value at a rate specified by that number, so `rate . constant 2` produces twice (on average) in a given time unit. This allows us to model, for example, the Poisson process:

```
poisson lambda = count (rate . constant lambda)
```

where `count` is an operator that counts the number of instants that its argument produces a value, and `lambda` is the rate parameter of the Poisson process.

Another common signal operator is `after`, which starts producing values after a given delay. Two signal functions can be combined in parallel using the <|> operator, which acts like its left hand side when it produces, and like its right hand side otherwise, so

```
constant 1 . after 3 <|> constant 0
```

is a signal function that produces `0` for the first three time units, then `1` forever.

We base Frabjous on FRP because its declarative nature provides the clarity and concision associated with declarative modeling [10, 11]. The utility of FRP as a specification language for complex systems has been demonstrated in a number of domains, including graphics [4], robotics [5], and games[3].

2.3 Frabjous

The generality of FRP comes at a cost, however. Understanding the syntax used in existing FRP libraries such as Netwire or Yampa [3] requires familiarity with advanced functional programming concepts such as monads [15] and arrows [6]. While these concepts allow for a great deal of conceptual elegance and generality, a domain-specific language that packages those portions relevant to ABM is desirable.

To explore this, the original version of Frabjous[13] realized concision and clarity compared to the popular AnyLogic framework. However, it placed severe restrictions on agent behavior and network structure. In this paper, we completely redesign Frabjous to yield a language that is still concise and readable, but is general enough to describe, in principle, any agent-based model.

3 The Frabjous Modeling Language

At the top level, a Frabjous model consists of a set of populations evolving in time. Each population is a dynamic collection of agents. Each agent comprises a set of time-varying values (such as income, age, or educational level) called *attributes*. Agents can be added to and removed from populations by processes such as birth, death, and migration. Any pair of populations can be linked together by a network, which represents relationships between agents.

Time-varying values are not specified in Frabjous directly, but implicitly by means of signal functions (introduced in 2.2). In particular, the dynamics of an agent attribute are specified by a signal function whose input is the entire agent.

Normally one defines signal functions by combining simpler functions with one of the provided operators. For example, we might define the attribute `age` of an agent as the amount of time elapsed since the agent was added to the model:

```
age = integral . constant 1
```

But how would we declare an attribute `isAdult`, which should be `False` during the first 18 years of the agent's life, and `True` from the 18th birthday on? This is a special case of a *functional dependency* between signal function, which is declared by appending `(t)` to all signal functions in the declaration:

```
isAdult(t) = age(t) ≥ 18
```

which makes explicit the signal functions' dependence on time.

4 An Extended Example: The SIR Model

In this section, we use Frabjous to implement an adaptation of the classic Susceptible, Infectious, Recovered (SIR) model of the spread of infectious disease, then extend it in a number of directions. Our purpose is to give examples of the clarity, concision, and flexibility of Frabjous models, paving the road for a deeper discussion in Section 5.

Our basic agent type is called `Person`, with an attribute for the agent's current infection state. In Frabjous we declare this as follows:

```
data State = Susceptible | Infectious | Recovered
agent Person { infectionState :: State}
```

where `data` is a Haskell keyword that creates a new type with a given set of named values, similar to C++ or Java `enum`, `agent` declares a new agent type with the given name and list of attributes, and :: means "is of type". We also declare, for convenience, a boolean-valued helper function that determines whether a given Person is currently infectious:

```
infectious person = (get infectionState person) == Infectious
```

To capture the structured character of human contact patterns, we introduce a *neighbor* relation between people: a network which has an edge between two people if they come into contact on a regular basis. We do this by amending the agent declaration for `Person`:

```
agent Person { infectionState :: State,
        neighbors :: Vector Person}
```

where `Vector` is a standard Haskell collection, similar to a C++ vector - so each person has a reference to a collection of other people in the population. Now the core dynamics of the model can be specified by defining `infectionState`:

```
infectionState = hold . repeatedly transition
      where
    transition person =
      case (get infectionState person) of
            Susceptible → constant Infectious . rate .
                        infectionRate
            Infectious → constant Recovered . rate .
                        constant recovery_rate
            Recovered → never
    infectionRate = per_contact_rate * numContacts
        numContacts(t) = count infectious neighbours(t)
```

The interesting part here is the `transition` function, which selects (using Haskell's `case` statement, an analogue to C++ or Java **switch**) between three possible evolution paths depending on the current state of the person.

A `Susceptible` person becomes `Infectious` with a rate determined by multiplying its count of infected neighbours by `per_contact_rate`. An `Infectious` person will recover at a constant rate of `recovery_rate`, and if the current state is `Recovered`, the person's `infectionState` will never change.

Finally, the first line defines the overall behavior of `infectionState`: an evolution path is repeatedly selected using the transition function, holding the most recently produced value (the value of the last transition taken). Both `hold` and `repeatedly` are FRP operators in the Frabjous standard library.

4.1 Adding Time-Varying Infectiousness

An implicit assumption of the SIR model is that all `Infectious` people are equally infectious at all times. In ODE models, relaxing this assumption and allowing infectiousness to vary over time has been shown to yield a more accurate model for the spread of diseases such as HIV[7]; how can we relax it in Frabjous?

The first step is to add a new attribute, `infectiousness`, to `Person`:

```
agent Person { ... , infectiousness :: Double}
```

Then we specify infectiousness after infection as an explicit function of time, perhaps linearly decreasing over three days:

```
after_infection t =  if t < 3 then 1 - t/3 else 0
```

then convert it to a time-varying value with the Frabjous operator `timeFunction`, which yields the following definition:

```
infectiousness = trigger (edge infectious)
             (timeFunction after_infection)
```

where `edge` is an FRP operator that only produces a value at the instant that its argument becomes `True`, and `trigger` produces nothing until its first argument (the "trigger") produces a value, then acts as dictated by its second argument. In this case, `infectiousness` will stay at its initial value (presumably 0) until the agent first becomes infectious, at which point it will behave like `after_infection` - jumping to 1, then declining back to 0.

Finally, we change the calculation of `infectionRate` to be the sum of the infectiousness values of the person's neighbors:

```
infectionRate(t) = sumBy (get infectiousness) neighbors(t)
```

4.2 Adding Dynamic Networks

So far we have not bound the `neighbors` attribute to any value. In fact, if all we want is a static network, we need not specify it at all, since a Frabjous model only describes change, not initial state. But we often do want the network to vary with time, whether randomly or in response to local changes in the agents.

We cannot specify the dynamics of a network by binding `neighbors` to a time-varying value as we would with any other attribute, since agents need to agree on network structure. Instead, we recognize that dynamic networks involve global interactions between agents, and specify them at global scope. For example, suppose we want `neighbors` to describe a random, dynamic network where each link has a 30% probability of existing at any point in time. We start by attaching an explicit name, `people`, to a population of `Persons`:

```
population people of Person
```

and declare the network as follows:

```
network people neighbours by randomLinks (const 0.3)
```

where `randomLinks` is a Frabjous standard library operator that creates dynamic networks by connecting two agents with a given probability. Using the standard Haskell `const` function gives equal weights to all pairs; a different function could be used to implement preferential mixing.

Networks can also be described between two different populations, which allows the specification of hierarchical models (e.g. persons within neighborhoods within cities within countries).

4.3 Usage

The Frabjous compiler currently generates, for every model, a single Haskell function that takes four arguments—the initial state (all the agent populations), the timestep, the amount of time for which to run the model, and a function that specifies the desired output (e.g. all the agent states, or the percent of agents currently infected)—and returns an array of the desired outputs.

5 Discussion

As a language for ABM, Frabjous is distinguished by two key properties: the high-level constructs it provides to hide the computational details of common ABM mechanisms (state-charts, event queues, functional dependencies), and the language's declarative nature. These properties provide a number of important benefits, which we discuss below:

Concision. The use of a high-level language allows models to be expressed more concisely, as can be seen from the example implementation of a fairly sophisticated SIR model in only 14 lines of code. The program reads like an executable specification, going to the heart of the unique, defining characteristics of the model rather than low-level implementation details.

This offers a major benefit for scientific communication. One of the great challenges of conducting research with agent-based models is that they are hard to communicate in a fully transparent and reproducible way. By contrast, models written in Frabjous are sufficiently short and free of boilerplate that many of them can be feasibly provided in complete form within papers introducing them.

Clarity. The key benefit of Frabjous' declarative nature is that models are expressed directly in terms of *processes* rather than as sequences of imperative state changes. Since Frabjous models are composed of equations linking processes, the dynamic hypotheses made about the world are laid clear.

Correctness. The encapsulation of common ABM mechanisms together with greater code clarity both help reduce the risk of error during model creation. Concision and clarity together lead to fewer places where bugs can arise, and make it easier to perceive the essential governing logic of the model, which eases developing confidence in a model's correctness. Contrast this to an approach which interleaves the model logic with implementation and visualization details, where the low-level code must be understood in order to gauge correctness.

Flexibility. Modeling is typically undertaken for the purpose of discovery, which means that the model will frequently evolve in unexpected directions. The flexibility of Frabjous means there is less inertia when adding features or changing directions. For example, adding time-varying infectiousness, a drastic change of one of the core SIR mechanics, required only 4 lines of additional code.

This flexibility is largely due to the modularity enforced by a declarative specification: all the code that can modify a particular agent attribute *must* appear in the attribute definition, easing the identification of code that needs to change and minimizing unintended side effects from modification.

6 Future Work

The primary area for future work is to make Frabjous a more complete framework, with the normal features and conveniences modelers expect, including support for collection of statistics over agent populations, parameter calibration and sensitivity analysis, and graphical visualization of model outputs.

This will make it Frabjous a useful language, at least for the purposes of pedagogy and communication; it will also pave the way for a direct quantitative and qualitative comparison to existing ABM frameworks.

To make Frabjous an industrial-strength ABM framework, performance issues must also be addressed. The declarative nature of the language provides many opportunities for optimization and parallelization. In particular, we are exploring the possibility of leveraging Data Parallel Haskell[1] as well as GPU acceleration[2] to speed up the execution of Frabjous models.

References

[1] Chakaravarty, M.M., Leschinskiy, R., Peyton-Jones, S., Keller, G., Marlow, S.: Data parallel Haskell: a status report. In: Workshop on Declarative Aspects of Multicore Programming, pp. 10–18. ACM Press (January 2007)

[2] Chakravarty, M.M., Keller, G., Lee, S., McDonell, T.L., Grover, V.: Accelerating Haskell array codes with multicore GPUs. In: Sixth Workshop on Declarative Aspects of Multicore Programming, DAMP 2011, pp. 3–14. ACM, New York (2011)

[3] Courtney, A., Nilsson, H., Peterson, J.: The yampa arcade. In: 2003 ACM SIGPLAN Workshop on Haskell, Haskell 2003, pp. 7–18. ACM, New York (2003)

[4] Elliott, C., Hudak, P.: Functional reactive animation. In: International Conference on Functional Programming, pp. 263–273 (1997)

[5] Hudak, P., Courtney, A., Nilsson, H., Peterson, J.: Arrows, robots, and functional reactive programming. In: Jeuring, J., Jones, S.L.P. (eds.) AFP 2002. LNCS, vol. 2638, pp. 159–187. Springer, Heidelberg (2003)

[6] Hughes, J.: Generalising monads to arrows. Science of Computer Programming 37, 67–111 (2000)

[7] Jacquez, J.A., Simon, C.P., Koopman, J., Sattenspiel, L., Perry, T.: Modeling and analyzing HIV transmission: the effect of contact patterns. Mathematical Biosciences 92(2), 119–199 (1988)

[8] Keeling, M.: The implications of network structure for epidemic dynamics. Theoretical Population Biology 67(1), 1–8 (2005)

[9] Nilsson, H., Courtney, A., Peterson, J.: Functional reactive progamming, continued. In: 2002 ACM SIGPLAN Workshop on Haskell, pp. 51–64 (2002)

[10] Osgood, N.: Systems dynamics and agent-based approaches: Clarifying the terminology and tradeoffs. In: First International Congress of Business Dynamics (2006)

[11] Osgood, N.: Using traditional and agent based toolsets for system dynamics: Present tradeoffs and future evolution. In: 2007 International Conference on System Dynamics (2007)

[12] Rahmandad, H., Sterman, J.: Heterogeneity and network structure in the dynamics of contagion: Comparing agent-based and differential equation models. In: 2004 International Conference on System Dynamics (2004)

[13] Schneider, O., Dutchyn, C., Osgood, N.: Towards frabjous: a two-level system for functional reactive agent-based epidemic simulation. In: 2nd ACM SIGHIT International Health Informatics Symposium, IHI 2012, pp. 785–790. ACM, New York (2012)

[14] Soeylemez, E.: Netwire (2012) (accessed August 29, 2013)

[15] Wadler, P.: Monads for functional programming. In: Jeuring, J., Meijer, E. (eds.) AFP 1995. LNCS, vol. 925, pp. 24–52. Springer, Heidelberg (1995)

Social Maintenance and Psychological Support Using Virtual Worlds

Peggy Wu[1], Jacquelyn Morie[2], J. Benton[1], Kip Haynes[2], Eric Chance[2], Tammy Ott[1], and Sonja Schmer-Galunder[1]

[1] Smart Information Flow Technologies, Minneapolis, MN
{PWu,JBenton,TOtt,SGalunder}@Siftech.com
[2] All These Worlds, LLC, Los Angeles, CA
Morie@usc.edu, {KipHaynes,Chance.Abattoir}@gmail.com

Abstract. In the space exploration domain, limitations in the Deep Space Network and the lack of real-time communication capabilities will impact various aspects of future long duration exploration such as a multi-year mission to Mars. One dimension of interest is the connection between flight crews and their Earth-based social support system, their family, friends, and colleagues. Studies in ground-based analogs of Isolated and Confined Environments (ICE) such as Antarctica have identified sensory deprivation and social monotony as threats to crew psychological well-being. Given the importance of behavioral health to mission success and the extreme conditions of space travel, new methods of maintaining psycho-social health and social connections to support systems are critical. In this paper we explore the use of Virtual Environments (VEs) and Virtual Agents (VAs) as tools to facilitate asynchronous human-human communication, and counteract behavioral health challenges associated with prolonged isolation and deep space exploration.

Keywords: Virtual Worlds, Virtual Agents, Psychological support, Communications, Psychological Health.

1 Introduction

At this very moment, there are tens of thousands of real people currently logged on, socializing, conducting commerce, and essentially living their lives in a virtual environment. Individuals unsatisfied with their real life can gain life satisfaction with their virtual life [1]. We speculate that satisfaction with social connections and interpersonal skills can be increased in real life, and environmental stressors can be alleviated through strategically designed experiences delivered through virtual environments. Currently, Virtual Environments (VEs) primarily provide entertainment value, but their predecessor, Virtual Reality, has been used to train technical skills that result in real life performance enhancements. VEs are no longer a novel concept, but their use as a construct to deliver real world benefits is just in its infancy.

In the domain of space exploration, NASA has identified the need to develop support and adaptation countermeasures to social isolation among the flight crew from

W.G. Kennedy, N. Agarwal, and S.J. Yang (Eds.): SBP 2014, LNCS 8393, pp. 393–402, 2014.

their family, friends, and the ground crew. This problem currently occurs on Earth, where geographically distributed groups desire to share their lives or work together remotely. Both real time solutions such as video, voice, and text chat, and time delayed solutions such as email, social network websites, and micro voice messaging serve to keep communities connected. However, there are unique challenges for their application to long duration space missions. Given technical restrictions such as communication delays and data transfer/bandwidth limitations, no single interaction method can accommodate all different types of interactions (e.g. short casual messages vs. long discussions). The social monotony and isolation of a long duration mission can result in limited perspective taking, increasing ruminations and likelihood for misunderstandings. Maintaining and strengthening of social skills are crucial to counteract the risk for deterioration after extended periods of isolated activity. The astronaut in-flight needs the appropriate outlet and support system in place, such as to allow for the possibility to engage in social activities and to withdraw into a more private sphere on a routine basis. The solution must maintain ties with existing friends and family, but may also accommodate developing new ties with new friends and colleagues as suggested by prior research [2].

In the past decade VEs and intelligent Virtual Agents (VAs) have come together to make possible a new range of human interaction opportunities. Virtual Worlds (VWs) have taken the paradigm of Virtual Reality and transformed it into rich, persistent, networked 3D spaces populated by tens of thousands of people using embodied avatars. Intelligent VAs are able to work as tutors, guides, conversational partners, and even as virtual therapists [3]. VWs combined with VAs have been used by ourselves and others to address human interactions in a diverse set of domains, from language and culture training [4, 5] to medical consulting [6, 7]. Morie et al. used VWs as an advanced form of tele-health care to facilitate and support soldiers in reintegration into civilian life in a project called "coming home" [8]. The authors have been developing a socio-linguistics based engine for driving intelligent VA social behavior and for detecting emotional changes in humans through their speech or written word. This document describes the use of VEs for long distance, long duration exploration class space missions. VEs have demonstrated their utility in enhancing telemedicine [9, 10, 11, 12, 13] supporting behavioral therapy [e.g. 14], and applications in language and culture training [15, 16], as well as robotics and equipment maintenance training [17, 18]. We believe its promise can extend to enhancing psychosocial health in space flight to directly address the risk of adverse behavioral conditions. We describe our process and results below.

2 Benefits Gained from Virtual Worlds Can Carry into the Real World

There is increasing evidence that behaviors learned in VWs can migrate into the real world. For example, in a meta-study of the field of VRET, Parsons and Rizzo [19] note the efficacy of Virtual Reality Exposure Therapy (VRET) in treating various anxiety disorders, such as acrophobia, agoraphobia, social anxiety and post-traumatic stress disorder. More directly, Yee and Balienson describe "The Proteus effect" [20, 21], the observed phenomenon where self-perception and skills obtained through a digital persona (i.e. through manipulations of their online avatar), including attitudes gained from virtual worlds, are maintained even when the human reenters the real world. All of this advocates that humans not only interact socially and meaningfully in VEs, they can also be primed for behavior modification and interpersonal sensitivities, and the effects of that priming can carry over into their offline interactions in live situations.

The ability to harness benefits acquired from virtual spaces translates to several applications in space. Psychological support can play an important role in risk mitigation since astronauts may experience anxieties while in flight, which then could be alleviated (or even prevented) by in vivo virtual world intervention. Given the limitations of the habitat, VEs may also be used to encourage relaxation and mindfulness: to create a spot to gather one's thoughts, to calm down, or to be alone in a manner that facilitates a restful state in the person. For instance, an astronaut may choose to sit in the VW near a quiet spot designed for meditation, or practice mindfulness therapy (as developed by John Kabat-Zinn [22]) utilizing awareness of the present moment and centering of one's thoughts. This type of work is being tested in Second Life at the University of Southern California's Institute for Creative Technologies [23]. Such quiet activities could help in providing support to people of different personality types, whereby some people gain energy by socializing (extroverted people) and others (introverted people) gain energy by isolating and recuperating alone. A tonic environment may be, for instance, a waterfall in a forest, or a mountaintop on a snow-filled valley. VEs may thus also provide a way to relax, recover and gather one's thoughts by a feeling of quiet and connection through one's environment. For many, this is an important way to "ground" oneself, even though one's physical environment is stressful, disengaging with that environment and changing it to a relaxed environment is crucial.

VWs may also provide good environments to acclimatize and acculturate crews that have been long removed from the world as a form of re-entry preparation. They can do this by providing them with updates on world events, news, new movies and cultural and historical development throughout the mission. As these long duration missions keep them out of the loop, we see a great benefit in helping astronauts get additional information through VEs that help them acclimate to the changes they have not witnessed or experienced while away from earth.

3 Method

Our goal is to create a holistic, evidence-based approach for formulating natural scenarios that address crew and Earth psychosocial health while ensuring its compatibility with currently known operational limitations such as bandwidth, communication delay as well as habitat restrictions. To design this system, we first leveraged a prior review of over 50 publications and technical reports to extract possible psycho-social dimensions of interest [24]. Using this literature review, we focused on psychosocial dimensions relevant to long duration exploration missions that can be addressed by virtual environments. We then conducted a review of current evidence-based best practices in supporting the psycho-social dimensions of interest identified above. In addition, we investigated methods of augmenting time delayed communication in VEs through VAs powered by Artificial Intelligence (AI) engines. In the last decade, researchers have begun focusing efforts on the merger of VEs and AI agents. A full survey of this work can be found in Morie, et al.'s survey on Embodied Conversational Agents (ECAs)[25]. The work of incorporating VAs into VEs is very much in its infancy, even more so than incorporating autonomous intelligent robots into the real world. Compared to other research into AI agents, the empirical data specific to VA effectiveness in either realism or helpfulness is limited. There are a few exceptions to this, however. For example, Jan D., et al. [26] note that virtual AI helpers are asked more questions than one might expect, while Weitnauer, et al. observed that when virtual world participants were unaware that they were conversing with an AI agent (named Max), they reported frustration with its repetitiveness [27], which paralleled reports about ELIZA, a simple natural language processor built in the mid-1960s which posed as a Rogerian psychotherapist[28]. Our own prior work showed that human subjects readily anthropomorphized even non-embodied agents [29], confirming Nass's findings that humans generalize patterns of conduct and expectations for human-human interaction to human-computer interaction [30]. Further, we found the social aspects of human-machine interaction affected not only subjective measures such as trust, but also objective performance metrics such as compliance [29]. More generally, we believe embodied VA will only amplify the human tendency to anthropomorphize technology, and increase the psychological investment and engagement of the human.

This work resulted in a list of broad and wide ranging strategies such as the use of plant life to combat monotony and sensory deprivation, training for marriage and family specific as well as general interpersonal skills, recalling moments of gratitude, and focusing on past and future responsibilities and expectations to reinforce sense of meaningfulness. These strategies can be implemented in a variety ways as VE realizations, ranging from novel virtual scenarios that families can experience together to the use of VAs to add social diversity and deliver training. Next, we selected a subset of strategies, reviewed commercially available VW technologies, and began to design activities surrounding those strategies within a virtual space. We describe the result of our designs below.

4 Implementation in Virtual Environment

We created a virtual space in which to implement the evidence based strategies that promote social connectedness and individual psychological well-being. The strategies we selected can be categorized as having the following objectives:

- Enhance human-human asynchronous communications
- Counter social monotony and isolation
- Combat sensory deprivation due to habitat and vehicle limitations

The resulting virtual environment consists of a main building, the Family Communication Center, or FAMCOM, and extensive outdoor spaces (see Figure 1).

Fig. 1. View of main building, FAMCOM from exterior

The lighting and weather in the virtual environment can mimic those on Earth, providing the end user with reminders of the changing seasons and providing visual and audio stimuli otherwise not available. Currently on the International Space Station,

Fig. 2. A virtual space outside of FAMCOM for enjoying nature

the flight crew has a constant and majestic view of Earth. However, Earth will be a distant dot of light during some phases of a long duration mission. This can further increase the feeling of isolation and disconnectedness. Visuals as well as audio of nature sounds such as rain and wildlife will help to combat sensory deprivation. Looking at nature for even brief periods of time has been shown to decrease stress. There are many opportunities for virtual world implementation to leverage the positive effects of nature on psychological well-being [31]. Activities can be solitary, such as watching the sunrise/sunset and the movement of clouds, or they can be layered with social activities such as enjoying a camp fire (see Figure 2).

Fig. 3. Games such as chess are asynchronous in nature thus can be implemented in this application despite time latencies

In addition to outdoor spaces, we created a variety of interior settings within FAMCOM. These include spaces for large groups, such as conference areas for conducting asynchronous educational outreach events, as well as gathering places for small groups where individuals can engage in activities such as an asynchronous game of chess with a friend on Earth (see Figure 3). Within the virtual space, end users can create and exchange virtual "care packages". This not only provides novelty to combat monotony, but gift giving boosts mood [32] and increases social connection to giver [33].

Due to the vast distance between the spacecraft and Earth, a communication delay will prevent real time network connections. As a result, flight crews and Earth-bound family and friends will not be able to interact in real time in the virtual environment. However, avatars that represent individuals can be automated and logged in to perform pre-recorded and even interactive behaviors. This activity of watching avatars in the virtual space is similar to watching a movie or a home video but with increased interactivity. Recently, the fields of neuroscience and film studies have come together in "neurocinematics". One area of interest is the role of "mirror neurons" and the effects of film watching on the viewer's emotions [34]. We believe this also applies in our application, where mirror neurons can be activated when viewing one's avatar interacting with objects and other individuals in the virtual space, causing real emotional responses. This idea is compatible with observations from both the "The Proteus effect" [20, 21], and "The Media Equation" [30], and is reported anecdotally by thousands of virtual environment users.

5 Future Work and Conclusion

The next phase of this work includes a ground based validation study to identify the effects of performing health promoting activities within the deigned virtual spaces. The specific application of supporting space exploration has a highly limited intended audience, but the ideas developed within this project can be applied to any groups for asynchronous communication, whether they are geographically distributed or not. The tremendous growth of asynchronous communication (e.g. IM, texting, email) shows that at least sometimes, the benefit of convenience outweighs the cost of coordinating with multiple parties for real time communication. While VEs are not meant to replace existing methods with which people connect, they can provide additional value. In the real world, interactions with people often go hand-in-hand with interactions with tangible things. VEs can unleash new dimensions of interaction that allows for shared manipulation of objects and shared experiences, potentially enhancing shared mental models and improving cohesion and coordination. Further, the persistence of objects within VEs makes them excellent vehicles for delivering experiential learning in the growing industry of online self-paced learning, which is estimated to become a $51.5 B worldwide market by 2016[35].

Acknowledgements. The above work was sponsored by NASA's Human Research Program. We would like to thank our NASA Program Mangers Lauren Leventon, Brandon Vessey, Diana Arias, and the BHP Element for their oversight, direction, and support.

References

1. Castronova, E., Wagner, G.G.: Virtual Life Satisfaction. Interanational Review for Social Sciences 64(3) (2011)
2. Cartreine, J.A.: Risks for Future Depression in Astronauts: A Narrative Review of the Literature, http://humanresearchroadmap.nasa.gov/Evidence/other/BMed%20Additional%20Evidence.pdf (accessed January 2014)

3. Litz, B.T., Wang, J., Williams, L., Bryant, R., Engel Jr., C.C.: A Therapist-Assisted Internet Self-Help Program for Traumatic Stress. Professional Psychology: Research and Practice 35(6), 628–634 (2004)
4. Wu, P., Miller, C.: Interactive Phrasebook - Embedding Human Social Dynamics in Language Training. In: Proceedings of Second IEEE International Conference on Social Computing (SocialCom 2010), August 20-22, Minneapolis, MN (2010)
5. Miller, C., Wu, P., Funk, H., Wilson, P., Johnson, W.L.: A Computational Approach to Etiquette and Politeness: Initial Test Cases. In: Proceedings of 2006 BRIMS Conference, Baltimore, MD, May 15-18 (2006)
6. Bickmore, T.: Relational agents: effecting change through human–computer relationships. Unpublished Ph.D. Thesis, Massachusetts Institute of Technology, Cambridge, MA, USA (2003)
7. Wu, P., Miller, C., Ott, T.: Utilizing Human Computer Etiquette to Encourage Human-Machine Therapeutic Alliance. In: Proceedings of the 13th International Conference on Human-Computer Interaction, San Diego, CA, July 19-24 (2009)
8. Morie, J.: Re-Entry: Online worlds as a healing space for veterans. In: The Engineering Reality of Virtual Reality 21st Annual IS&T/SPIE Symposium, San Jose, CA (2009), http://people.ict.usc.edu/~morie/ComingHome/Morie-2009-2SPIE-FINAL.pdf
9. Schmer-Galunder, S., Sayer, N., Wu, P., Keller, P.: Continuous Anger Level Monitoring (CALM) - A mental health smartphone application. Paper submitted to SOTICS 2013 (2013)
10. Scherer, S., Stratou, G., Mahmoud, M., Boberg, J., Gratch, J., Rizzo, A., Morency, L.-P.: User-State Sensing for Virtual Health Agents and TeleHealth Applications. In: 10th IEEE International Conference on Automatic Face and Gesture Recognition, Shanghai, China (April 2013)
11. Rizzo, A., Strickland, D., Bouchard, S.: The Challenge of Using Virtual Reality in Telerehabilition. Telemedicine Journal and e-Health 10(2) (2005)
12. Morie, J.F., Haynes, E., Chance, E., Purohit, D.: Virtual Worlds and Avatars as the New Frontier of Telehealth Care. Stud. Health Technol. Inform. 181, 27–31 (2012)
13. Johnsen, K., Dickerson, R., Raij, A., Lok, B., Jackson, J., Shin, M., Hernandez, J., Stevens, A., Lind, D.S.: Experiences in Using Immersive Virtual Characters to Educate Medical Communication Skills. To Appear in Proceedings of IEEE Virtual Reality 2005 (VR 2005), Bonn, Germany (March 2005)
14. Morie, J.F., Antonisse, J., Bouchard, S., Chance, E.: Virtual worlds as a healing modality for returning soldiers and veter-ans. In: Studies in Health Technolology and Informatics: Annual Review of Cybertherapy and Telemedicine 2009 – Advanced Technologies in the Behavioral, Social and Neurosciences, vol. 144, pp. 273–276 (2009)
15. Zheng, D., Newgarden, K.: Rethinking Language Learning: Virtual Worlds as a Catalyst for Change. International Journal of Learning and Media 3(2), 13–36 (2011)
16. Johnson, W.L., Friedland, L., Schrider, P., Valente, A., Sheridan, S.: The Virtual Cultural Awareness Trainer (VCAT): Joint Knowledge Online's (JKO's) Solution to the Individual Operational Culture and Language Training Gap. In: Proceedings of ITEC 2011. Clarion Events. Describes the instructional design and implementation of Alelo's VCAT courses, London (2011)
17. Bluemel, E., Hintze, A., Schulz, T., Schumann, M., et al.: Virtual environments for training of maintenance and service training tasks. In: Simulation Conference (2003)

18. Guidali, M., Duschau-Wicke, A., Broggi, S., Klamroth-Marganska, V., Nef, T., Riener, R.: A robotic system to train activities of daily living in the virtual environment. Medical & Biological Engineering & Computing 49(10), 1213–1223 (2011)
19. Parsons, T.D., Rizzo, A.A.: Affective Outcomes of Virtual Reality Exposure Therapy for Anxiety and Specific Phobias: A Meta-Analysis. Journal of Behavior Therapy and Experimental Psychiatry 39, 250–261 (2008)
20. Yee, N., Bailenson, J., Ducheneau, N.: The Proteus Effect: Implications of Transformed Digital Self-Representation on Online and Offline Behavior. Communication Research (2009)
21. Yee, N., Bailenson, J.N.: The Proteus Effect: The Effect of Transformed Self-Representation on Behavior. Human Communication Research 33, 271–290 (2007)
22. Kabat-Zinn, J.: Full catastrophe living: Using the wisdom of your body and mind to face stress, pain, and illness. Bantam Dell, New York (1990)
23. Morie, J.F.: Virtual Worlds as a Healing Modality for Returning Soldiers and Veterans. In: 14th Conference on CyberTherapy and CyberPsychology Proceedings. Mary Ann Liebert Publisher, Inc. (2009); CyberPsychology and Behavior 12(5)
24. Wu, P., Rye, J., Miller, C., Schmer-Galunder, S., Ott, T.: Non-Intrusive Detection of Psycho-Social Dimensions using Sociolinguistics. In: Third Workshop on Social Network Analysis in Applications (SNAA) of IEEE/ACM International Conference on Advances in Social Networks Analysis and Mining (ASONAM 2013), Niagara Falls, Canada, August 25-28 (2013)
25. Morie, J.F., Chance, E., Haynes, E., Purohit, D.: Embodied Conversational Agent Avatars in Virtual Worlds: Making Today's Immersive Environments More Responsive to Participants. In: Hingston, P. (ed.) Believable Bots: Can Computers Play Like People, pp. 99–118. Springer (2012)
26. Jan, D., Roque, A., Leuski, A., Morie, J., Traum, D.: A Virtual Tour Guide for Virtual Worlds. In: Ruttkay, Z., Kipp, M., Nijholt, A., Vilhjálmsson, H.H. (eds.) IVA 2009. LNCS, vol. 5773, pp. 372–378. Springer, Heidelberg (2009)
27. Weitnauer, E., Thomas, N.M., Rabe, F., Kopp, S.: Intelligent agents living in social virtual environments – bringing max into second life. In: Prendinger, H., Lester, J.C., Ishizuka, M. (eds.) IVA 2008. LNCS (LNAI), vol. 5208, pp. 552–553. Springer, Heidelberg (2008)
28. Weizenbaum, J.: ELIZA — A Computer Program For the Study of Natural Language Communication Between Man And Machine. Communications of the ACM 9(1), 36–45 (1966), doi:10.1145/365153.365168
29. Wu, P., Miller, C.: Can Polite Computers Produce Better Human Performance? In: Proceedings of ACM Multimedia 2010 International Conference, Firenze, Italy, October 25-29 (2010)
30. Reeves, B., Nass, C.: The Media Equation. Cambridge University Press, Cambridge (1996)
31. Valtchanov, D., Barton, K.R., Ellard, C.: Restorative effects of virtual nature settings. Cyberpsycholo. Behav. Soc. Network. 13(5), 503–512 (2010)
32. Anik, L., Aknin, L.B., Norton, M.I., Dunn, E.W.: Harvard Business School Working Paper. Feeling Good about Giving: The Benefits (and Costs) of Self-Interested Charitable Behavior (2009), `http://www.hbs.edu/faculty/Publication%20Files/10-012.pdf` (accessed July 2013)

33. Zhang, Y., Epley, N.: Exaggerated, Mispredicted, and Misplaced: When "It's the Thought that Counts" in Gift Exchanges. J. Experimental Psychology: General 141(4), 667–681 (2012), http://faculty.chicagobooth.edu/nicholas.epley/ZhangEpleyJEPG2012.pdf (accessed)
34. Konigsberg, I.: Film Theory and the New Science. Projections: The Journal for Movies and Mind 1(1), 1–24 (2007), http://www.ingentaconnect.com/content/berghahn/proj/2007/00000001/00000001/art00002?token=005310af573d257025702c492b2f2d3125765747357c4e7547543c7e386f642f466f771614f4710e488
35. http://www.ambientinsight.com/Reports/eLearning.aspx

Civil Unrest Prediction: A Tumblr-Based Exploration

Jiejun Xu, Tsai-Ching Lu, Ryan Compton, and David Allen

HRL Laboratories, LLC
{jxu,tlu,rfcompton,dlallen}@hrl.com

Abstract. This work focuses on detecting emerging civil unrest events by analyzing massive micro-blogging streams. Specifically, we propose an early detection system consisting of a novel cascade of text-based filters to identify civil unrest event posts based on their topics, times and locations. In contrast to the model-based prediction approaches, our method is purely extractive as it detects relevant posts from massive volumes of data directly. We design and implement such a system in a distributed framework for scalable processing of real world data streams. Subsequently, a large-scale experiment is carried out on our system with the entire dataset from Tumblr for three consecutive months. Experimental result indicates that the simple filter-based method provides an efficient and effective way to identify posts related to real world civil unrest events. While similar tasks have been investigated in different social media platforms (e.g., Twitter), little work has been done for Tumblr despite its popularity. Our analysis on the data also shed light on the collective micr-oblogging patterns of Tumblr.

Keywords: Early Event Detection, Social Media, Big Data.

1 Introduction

The ability to provide early warnings of impending civil unrest events has tremendous impact on both government and the general public. From the government point of view, being able to forecast violent strikes allows law enforcement to devise more effective mitigation plans, mobilize resources, and coordinate responses. Similarly, being able to identify protest events ahead of time allows the general public to be informed and stay away from unsafe or hostile regions. Traditionally, civil unrest events are predicted by intelligence services which generally need to place their agents in different regions of the world for risk assessments. This could introduce significant delays on the report of the events. Furthermore, it is incredibly difficult (if not impossible) to have human analysts to monitor all regions of the world at the same time.

On the other hand, the growing popularity of micro-blogging platforms (e.g., Twitter, Tumblr, and others) provides a convenient mechanism for users to create and share any messages online. In general, users can publish posts about what they did, what they are currently doing, and more interestingly what they are

W.G. Kennedy, N. Agarwal, and S.J. Yang (Eds.): SBP 2014, LNCS 8393, pp. 403–411, 2014.

planning to do in the (near) future. For instance, it has been shown that Twitter served an important role in terms of organizing and planning events during Arab Spring, which is a sequence of protests that swept through the Middle East in early 2011[5]. Thus it is possible to generate early warnings regarding forthcoming civil unrest events by monitoring these micro-blogging platforms.

The value of event forecasting with social media has attracted much attention from research communities across disciplines. For example, in [2], a stochastic hybrid dynamical system (S-HDS) model is proposed to generate predictions for social phenomena. In [3][6], effective keyword based approaches are proposed to detect emerging popular events using publicly available Twitter data. Similar approach is adopted in [11] for the civil unrest domain. In [4], the connection between online discussion and the diffusion of protests is studied. Through the analysis of the recruitment patterns of blogging data, connections between online network, social contagion, and collective dynamics are found. In [5] the role of modern information communication technologies in the 2011 Egyptian protests is studied. Their work shows that the sociopolitical protests were facilitated by social media networks, particularly in regard to their organizational and communication aspects, and social network played an important role in the rapid disintegration of the regime. In [7], a spatial surrogate model is proposed to forecast social mobilization and civil unrest. Specifically, a dictionary of key terms related to protests is defined and a statistical model is applied to identify future civil unrest regions and their potential magnitudes. In its subsequent work [8], a semi-supervised system is proposed to detect and interactively visualize events of a targeted type. The system utilize transfer learning and label propagation to accurately estimate event locations. Similar studies can be found in [1] [9] [10].

Motivated by the success of prior work, we propose an automated system for civil unrest prediction based on social media signals. Specially we focus on the micro-blogging platform Tumblr due to its growing popularity. According to the statistics reported on Tumblr, there are about 133 million users registered for this social networking service. Since its debut in 2007, a total of 59.4 billion posts have been generated. On average, more than 80 million posts are generated every day. The large amount of data spreading over Tumblr provides information or clues for virtually any major social movements in the world. In addition, Tumblr is most popular among younger social media users who are more likely to be the driving forces for protests and other social movements. These factors make Tumblr an ideal platform for civil unrest monitoring and forecast.

In a nutshell, our proposed system functions by continually applying multiple text-based filters to the Tumblr data stream. Our method is based on direct extraction of a set of highly relevant posts instead of a physical model describing a large-scale theory of population behavior (e.g., [12]). The rest of the paper is organized as follows: Section 2 describes the design and implementation of our system in detail. Section 3 showcases our results from a large-scale experiment. Finally, we conclude the paper and discuss future work in Section 4.

2 Early Detection Method

The overall architecture of the civil unrest prediction system is shown in Figure 1. There are three main components in the detection pipeline. Each component is responsible for different sections of a post as indicated by the arrows. Given a massive collection of Tumblr posts, the system searches each post for 1) a set of carefully selected civil unrest related keywords, 2) a set of pre-defined location terms covering the area of interest, and 3) mentions of future dates. Essentially this pipeline is a cascade of filters which is used to continually monitor and detect events of interest from a large data stream. Posts pass through all the filters are considered relating to an upcoming civil unrest events. In the current implementation, we focus on civil unrest events related to Latin America.

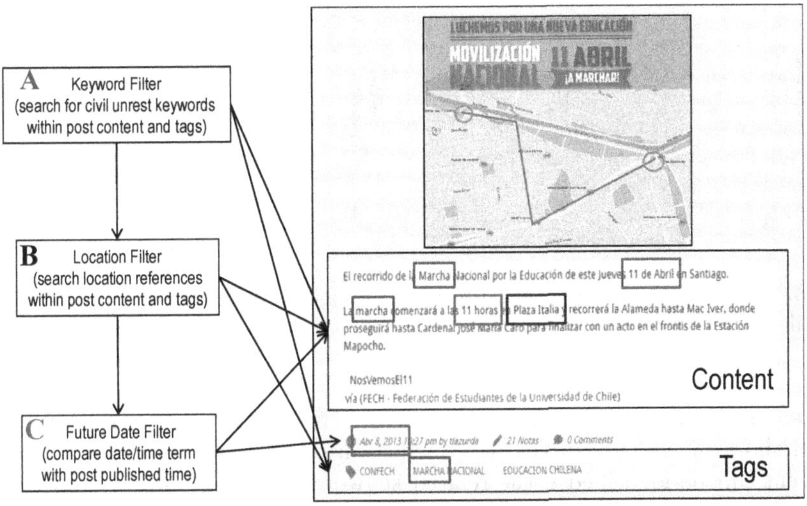

Fig. 1. A sample Tumblr post describing a future protest event. Three filters are applied to determine the nature of the post. Rectangles in RGB colors indicate matched criterias from different filters.

Keyword Filter. The first filter that a Tumblr post must pass is a simple check for mentions of civil unrest keywords. We have manually compiled a collection of 59 keywords which our domain experts believe are of substantially important and distinctive to identify civil unrest events. A few samples of the list are shown in Figure 2(a). Note that these pre-defined keywords are Spanish and Portuguese which are commonly used across Latin America regions. For each Tumblr post, the system searches its textual content for matches of any keywords. For instance, the word "Marcha" is matched from the sample post in Figure 1 (as marked in red rectangle). In addition, the system also scans through the tag section for matches. Note that tags are commonly used to boost content visibility, and it is the only search mechanism provided by Tumblr.

```
"cacerolazo","manifestación","manifestac
iónes","marcha", "marchado", "marcham",
"ocupanban","protesta", "reforma",......
```

(a)

(b)

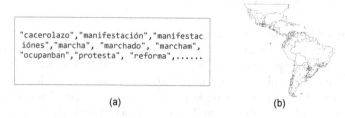

Fig. 2. (a) A list of sample civil unrest keywords chosen by domain experts; (b) Geo-distribution of the pre-defined candidate locations

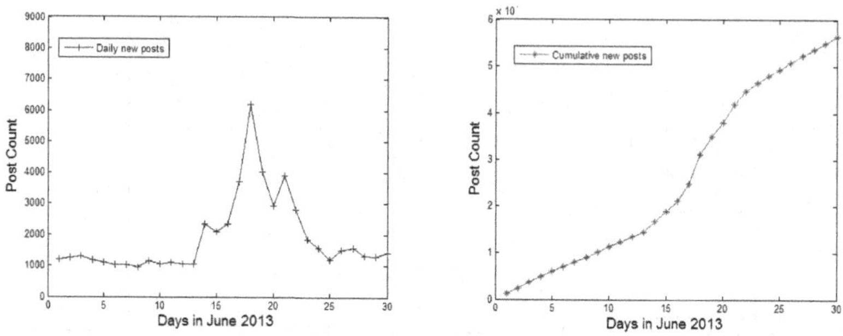

Fig. 3. (a) Daily counts of new posts with pre-defined keywords in June 2013; (b) Cumulative counts of the same month

The underlying assumption for the keyword based filter is that related words would show an increase in the usage when an event is emerging [3]. Therefore an event can be identified if the related keywords showing burst in appearance count. Figure 3 shows the daily count of new Tumblr posts which contains one or more of the pre-defined keywords in June 2013. A spike in the middle of the month corresponds to the escalation of the wide spread protests in Brazil during the time frame. The observed number of keyword mentions is 6 times more than the average keyword mentions in the earlier part of the month.

Location Filter. Location information is crucial for generating early warnings of civil unrest. This is because precautious actions can only be taken if the location of a civil unrest event is known in advance. In Twitter and several other micro-blogging platforms, geo-tags (i.e., associated GPS coordinates in tweets) are supported. Based on a small portion of geo-tagged tweets/posts, location information of majority posts can be identified through friendship analysis [13]. However, unlike Twitter, Tumblr does not provide any location-based tagging service. Thus a different tack is required to infer the location of a post where it is published.

Here we resort to the text-based search approach, in which a set of pre-defined locations are matched against each Tumblr post. The list of locations consists

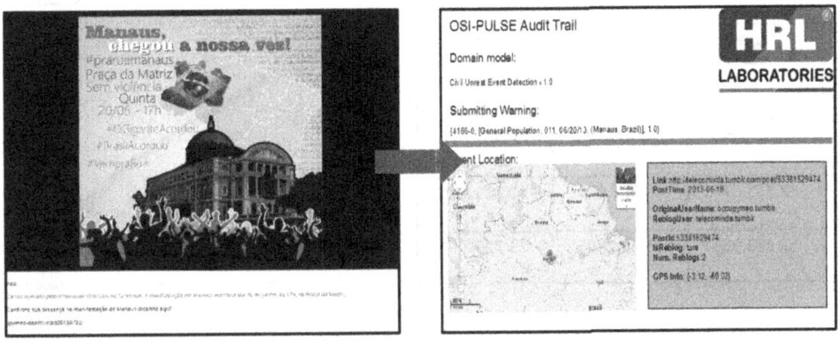

Fig. 4. A detected future event post is re-structured into an easily-interpretable audit trail displayed in the front-end interface

of major cities of Latin America countries. The intuition is that major protests and strikes are more likely to occur in capital or big cities. We have defined a total of 1022 different locations. The geographic distribution of these candidate locations is shown in Figure 2 (b). In addition to scanning for location terms within post content, our system also searches for tags for potential matches. This is because a large portion of Tumblr posts contain images. When users assign tags to images, location information is among the top selections.

Future Date Filter. Future date mentions extracted from social media posts are effective indicators of upcoming events. In addition, future date is a necessary component for any meaningful forecast of civil unrest events. In this step, our system first searches for month names and abbreviations in Spanish and Portuguese, and second for numbers less than 31 within three whitespace separated tokens from each other. Thus, an example matching date pattern would be 11 de Abril. Four-digit years are quite rare in social media posts, in order to determine the year of the mentioned date we use the year which minimizes the number of days between the mentioned date and the posts published time. In our example shown in Figure 1, the Tumblr post mentions "11 de Abril" (marked by a green box) on the date of 2013-04-08, thus we assume the user is talking about the future event on 2013-04-11 as 2013 minimize the difference in days between published time and mention time (e.g., compared to 2012). Once we have extracted dates from the text, we assert that the mentioned dates occur after the post published time. We believe that posts mentioning past dates are unlikely to be indicative of future events. After this step, for each Tumblr post which passes all three filters, we consider it as a candidate post of civil unrest event on the mentioned date.

3 Experimental Results

3.1 Dataset and System Implementation

We have been obtaining the full Tumblr firehose since 2013-04-01. Every public post (along with update, and delete request) is delivered to our system in real

	June	July	August
Number of Posts	2,345,420,462	2,602,044,120	2,682,210,840
Number of Unique Posts	169,858,637	183,672,482	194,020,188
Number of Users	12,753,874	13,399,945	13,339,285
Number of Tags	59,113,858	66,635,895	73,725,817

(a)

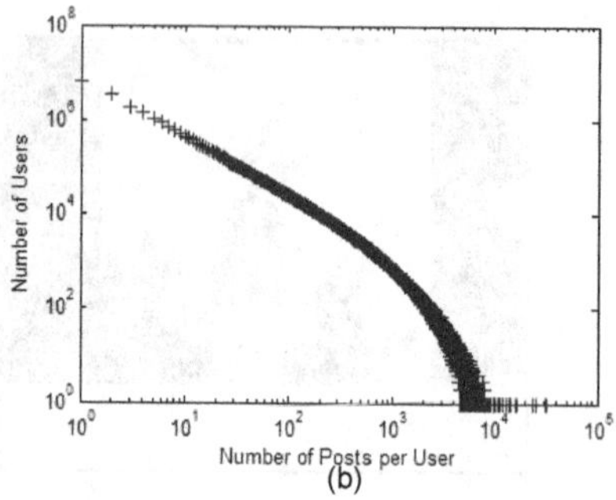

(b)

Fig. 5. Dataset statistics

time. The size of data collected between 2013-04-01 and 2013-11-04 is roughly 10.6 TB (with bzip2 compression). Data is stored via Hadoop Distributed File System[1] (version 0.20.2-cdh3u3) deployed across a multi-node multi-core cluster with combined memory in the terabyte scale. The event detection system is implemented using a combination of standard Java MapReduce code as well as Apache Pig [2]. On average, it takes less than 10 minutes to complete a daily event detection task. In addition to the core back-end algorithm, our system provides an front-end interface summarizing each detected emerging civil unrest events for easy end-user interpretation. The communication protocol between front-end and back-end is encoded in standard JSON format. Figure 4 shows an example of a detected event post and its audit trail in the front-end interface.

In order to provide quantitative evaluation for our system, we examine the full corpus of Tumblr data between 2013-06-01 and 2013-08-31. The selection is because dense ground-truths are provided by government agencies covering all major civil unrest events in Latin America during this period. Figure 5 (a) summarizes the statistic of the corpus[3]. As can be seen, our system is truly working in the "big data" regime. Figure 5 (b) shows the monthly distribution of number of posts per users. It shows that for large majority of the users, each user publishes only a handful of posts in a month, while a small portion of users publish much more than the average. This distribution mirrors a power-law distribution with α of 2.53 and mean square error of 0.80 in the loglog plot.

Recall that our system consists of a set of filters. The Venn diagrams in Figure 6 show the numbers of resulting posts from each filter. As seen, the final number of detected civil unrest events (lies in the overlapping part of the circles) is substantially smaller and manageable compared to the original size of input data. Only highly relevant posts are distilled from the data stream. Note that depending on the applications, we can customize each filter to adjust (e.g., increase/decrease the number of domain keywords or locations) the number of remaining posts for event forecast.

[1] `http://hadoop.apache.org/`

[2] `http://pig.apache.org`

[3] Unique posts indicate original content posted by Tumblr users, but not reblogs (reblog is similar to retweet in Twitter).

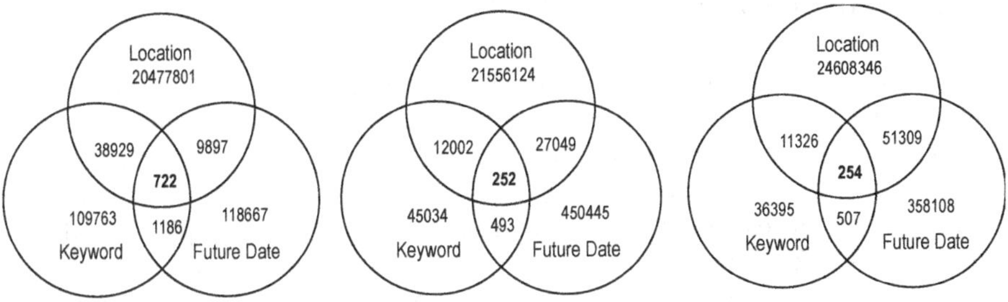

Fig. 6. Venn diagram showing the number of passed posts for each filter

	No. Detected Events	Precision	Ave. Lead Time (days)
Jun	722	97.65%	3.8
Jul	252	92.06%	6.6
Aug	254	93.70%	6.0

(a)

	No. Detected Unique Events	Precision	Ave. Lead Time (days)	No. Distinct Real world Events Covered
Jun	274	95.62%	5.1	92
Jul	113	84.96%	5.4	65
Aug	108	89.81%	6.4	63

(b)

Post Date	Lead Time (days)	Snapshot of the Post
2013-06-01	12	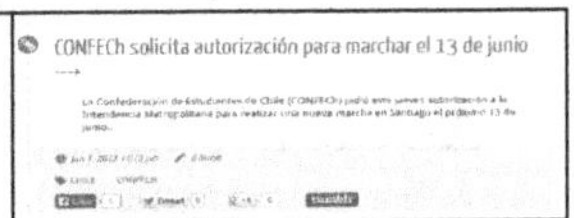
2013-06-16	9	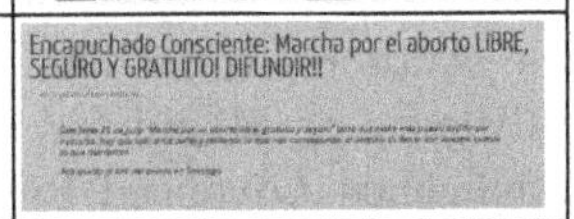
2013-06-14	8	
2013-07-31	1	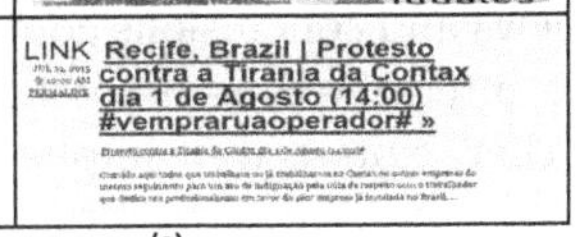

(c)

Fig. 7. (a) Monthly detections from our system and their accuracies, (b) Results only based on unique posts (no reblogs) from (a), (c) Snapshots of Tumblr posts (detected by our system) showing planned future civil unrest events

To evaluate the accuracy of detected civil unrest posts, we compare the system results with the provided ground-truths. Figure 7(a) summarizes our system performance. Let us first look at the results from June. There are a total of 722 detected posts related to future civil unrest events. Out of those, 97.65% are considered correct as they can be matched to at least one event in the ground-truths with the same date and location. The average lead time (i.e., difference between event date and the post published date) is 3.8 days. There are a total of 274 unique posts (See Figure 7(b)), in other words, the rest of 448 posts are simply reblogs of the original posts, and share identical contents. If only consider unique detections, the average lead time is increased to 5.1 days. Finally, these posts can be matched to 92 distinct real-world civil unrest events. Similar interpretations can be made for the results in July and August. Overall, our results are quite satisfactory given the simplicity of the detection system. Visual samples of detected posts can be seen in Figure 7(c).

4 Conclusion and Future Work

To summarize, social media has become a powerful tool for the organization of mass gatherings of all types. However, the massive volume of data makes it difficult to automatically extract valuable information in a timely fashion. In this work, we have provided a straightforward approach for the detection of upcoming civil unrest events in Latin America based on a simple cascade of textual filters. We conducted our initial experiment on three months of Tumblr data and showed promising result from our approach.

Immediate future work includes more advanced geo-coding mechanism for Tumblr posts. Preliminary analysis shows that it is possible to align users in different social media platforms, and then propagate their geo-information across. We also plan to analyze different types of social movements in the global scale.

5 Acknowledgments and Disclaimer

The Authors would like to thank Joe Palacios and our collaborators at the University of Miami for their contributions in the project. Supported by the Intelligence Advanced Research Projects Activity (IARPA) via Department of Interior National Business Center (DoI / NBC) Contract Number D12PC00285. The U.S. Government is authorized to reproduce and distribute reprints for Governmental purposes notwithstanding any copyright annotation thereon. The views and conclusions contained herein are those of the author(s) and should not be interpreted as necessarily representing the official policies or endorsements, either expressed or implied, of IARPA, DoI/NBC, or the U.S. Government.

References

1. Petrovic, S., Osborne, M., Lavrenko, V.: Streaming First Story Detection with application to Twitter. In: Annual Conference of the North American Chapter of the Association for Computational Linguistics (2010)
2. Colbaugh, R., Glass, K.: Early Warning Analysis for Social Diffusion Events. In: International Conference on Intelligence and Security Informatics (2010)
3. Weng, J., Yao, Y., Leonardi, E., Lee, F.: Event Detection in Twitter. In: International Conference on Weblogs and Social Media (2011)
4. Gonzalez-Bailon, S., Borge-Holthoefer, J., Rivero, A., Mareno, Y.: The Dynamics of Protest Recruitment through an Online Network. Nature Scientific Reports 1, Article 197 (2011)
5. Stepanova, E.: The Role of Information Communication Technologies in the "Arab Sping". PONARS Eurasia Policy Memo No. 159 (2011)
6. Weerkamp, W., De Rijke, M.: Activity Prediction: A Twitter-based Exploration. In: SIGIR Workshop on Time-Aware Information Access (2012)
7. Chen, F., Arredondo, J., Khandpur, R.P., Lu, C.T., Mares, D., Gupta, D., Ramakrishnan, N.: Spatial Surrogates to Forecast Social Mobilization and Civil unrests. In: CCC Workshop on "From GPS and Virtual Globes to Spatial Computing" (2012)

8. Hua, T., Cheng, F., Zhao, L., Lu, C.T., Ramakrishnan, N.: STED: Semi-Supervised Targeted Event Detection in Twitter. In: Knowledge Discovery And Data Mining (2013)
9. Unankard, S., Li, X., Sharaf, M.A.: Location-Based Emerging Event Detection in Social Networks. In: Ishikawa, Y., Li, J., Wang, W., Zhang, R., Zhang, W. (eds.) APWeb 2013. LNCS, vol. 7808, pp. 280–291. Springer, Heidelberg (2013)
10. Manrique, P., Qi, H., Morgenstern, A., Velasquez, N., Lu, T.C., Johnson, N.: Context matters: Improving the uses of big data for forecasting civil unrest. In: International Conference on Intelligence and Security Informatics (2013)
11. Compton, R., Lee, C., Lu, T.C., De Silva, L., Macy, M.: Detecting Future Social Unrest in Unprocessed Twitter Data. In: International Conference on Intelligence and Security Informatics (2013)
12. Braha, D.: Global Civil Unrest: Contagion, Self-organization, and Prediction. PLOS One (2012)
13. Jurgens, D.: That's What Friends Are For: Inferring Location in Online Social Media Platforms Based on Social Relationships. In: International Conference on Weblogs and Social Media (2013)

Studying the Evolution of Online Collective Action: Saudi Arabian Women's 'Oct26Driving' Twitter Campaign

Serpil Yuce[1], Nitin Agarwal[1], Rolf T. Wigand[1], Merlyna Lim[2], and Rebecca S. Robinson[2]

[1] Department of Information Science, University of Arkansas at Little Rock, USA
{sxtokdemir,nxagarwal,rtwigand}@ualr.edu
[2] School of Social Transformation - Justice and Social Inquiry, Arizona State University, USA
{Merlyna.Lim,Rebecca.Robinson}@asu.edu

Abstract. Social media have played a substantial role in supporting collective actions. Reports state that protesters use blogs, Facebook, Twitter, YouTube and other online communication media and environments to mobilize and spread awareness. In this research, we focus on studying the process of formation of online collective action (OCA) by analyzing the diffusion of hashtags. We examine the recently organized Saudi Arabian women's right to drive campaign, called 'Oct26Driving' and collected the Twitter data, starting from September 25, 2013 to the present. Given the definitive nature of hashtags, we investigate the co-evolution of hashtag usage and the campaign network. The study considers the dominant hashtags dedicated to the Oct26Driving campaign, viz., '#oct26driving' and '#اكتوبر26_قيادة'. Morteover, it identifies cross-cultural aspects with individual hashtag networks, with Arabic hashtags relating to local factors and English hashtags contributing to transnational support from other organizations, such as those related to human rights and women's rights. Despite the wide news media coverage of social movements, there is a lack of systematic methodologies to analytically model such phenomena in complex online environments. The research aims to develop models that help advance the understanding of interconnected collective actions conducted through modern social and information systems.

Keywords: Online collective action, *Oct26Driving*, social movement, hashtag, diffusion, transnational, interorganization, cross-cultural, Twitter, Saudi Arabia.

1 Introduction

The prevalence of contemporary forms of information and communication technologies (ICTs), such as social media, have transformed the ways people interact, communicate, and share information. Such transformation has fundamentally altered how people coordinate and mobilize, leading to manifestations of collective actions in various forms: whether they are social movements for sociopolitical transformation, campaigns for better governance through citizen journalism and engagement, or flash mobs for promoting a cause or simply entertainment. As evident in the mass protests during the Arab Spring, the Occupy campaign and other recent movements, social

W.G. Kennedy, N. Agarwal, and S.J. Yang (Eds.): SBP 2014, LNCS 8393, pp. 413–420, 2014.

media platforms helped protesters to instantly spread messages, organize, and mobilize support for their campaigns. People are able to self-report incidents of crime and harassment, even potholes, on crowdsourcing platforms elevating citizen journalism and engagement to a new level. There are many other examples of similar Internet-driven collective actions. Despite the wide news media coverage of such collective action efforts, there is a lack of systematic methodologies to understand such phenomena in complex online environments.

As collective action theories were mostly developed in the pre-Internet era [1], it is imperative to reassess conventional theories of collective action. The emergence of Internet-driven collective actions has prompted us to examine and explore some fundamental aspects of collective action that remain theoretically underdeveloped [1, 2] and call for innovative foundational research that can provide insights in re-conceptualizing collective action theories in an online environment. A number of studies attempt to understand collective action processes in online environments [2]. Such efforts, however, have not answered many questions related to the emergence of various forms of Internet-driven collective actions. Here we focus on the process of formation of online collective action (OCA) by studying the diffusion of 'hashtags'. Given the definitive nature of hashtags, we investigate the co-evolution of hashtag usage and campaign network growth. We examine the recently organized Saudi Arabian women's right to drive campaign, called 'Oct26Driving' and collected corresponding Twitter data, starting from September 25, 2013 to the present. Considering several hashtags, the study analyzes the dominant ones dedicated to the Oct26Driving campaign, viz., '#oct26driving' and '#قيادة_26اكتوبر'. The study identifies cross-cultural aspects with individual hashtag networks, where the Arabic hashtags relate to local factors and English hashtags help in bringing transnational support from various organizations such as human rights and women's rights. Our research aims to develop models to advance the understanding of interconnected collective actions conducted through modern social and information systems.

2 The *Oct26Driving* Campaign

Saudi women face some of the most inequitable laws and practices when compared to international standards, including the prohibition of driving motorized vehicles. Until recently, Saudi women have only been granted the right to ride bicycles. In order to create awareness about these inequitable laws and practices, Saudi women have organized several campaigns as part of a bigger movement. Our earlier studies [3, 4, 9, 10] have analyzed these campaigns to understand various aspects of online collective action. Recently, a group of Saudi women activists had set October 26th, 2013, as a day for defying the state ban on women driving and launched an online petition website (www.oct26driving.com) on September 25, 2013. Called "The 26th October Campaign", it quickly gained momentum with its online petition garnering more than 16,000 signatures (according to the official campaign website) despite the Kingdom's restrictions on protests. The campaign website was hacked on October 9, 2013 that led to a surge in Twitter activity, as demonstrated later in Fig. 2. This campaign follows a series of previous campaigns, the most notable of which was on June 17, 2011 (Women2Drive). The October 26th campaign is a grassroots campaign with the participation of the women and men of Saudi Arabia and aims to revive the demand to lift the ban on women driving. Although King Abdullah bin Abdulaziz left this matter to society, the government's reaction makes it very clear that this is not a societal decision but a political decision. Supporters shared videos of previous campaigns and uploaded new videos and pictures of themselves driving to protest the driving ban. The online initiative was boosted by the fact that residents of Saudi Arabia are highly active on social media, especially Twitter and YouTube. The campaign discussed above demonstrates the important role of Twitter in facilitating online collective actions. They further afford studying the formation and growth of this online collective action via the diffusion of hashtags within the Twitter network. The overarching research question studied here is: How can we track the process of online collective action using Twitter hashtags as markers? Further, through the co-occurrence of the dominant hashtags, the study attempts to understand the implications of cross-cultural nature of the campaign over the transnational and interorganizational aspects of the observed online collective action. To follow up on this and related questions, next we provide a literature background in Section 3.

3 Literature Review

Many studies have recently examined the influence of Twitter in political contestations in the Middle East and North Africa (MENA) region, starting with #iranelection in 2009 that attempted to position popular demonstrations as the "Twitter Revolution" [5]. More recent literature has turned to the "Arab Spring" uprisings. According to Skinner, "Hashtags are used both to coordinate planning for certain events or to tie one's tweets with a larger discussion on the subject" [6]. The Saudi women's driving campaign used the Twitter network in both capacities. Hashtag diffusion in different languages has been linked to drawing local collective action into the transnational Twittersphere networks through bridging mechanisms [5], which contribute to transnational support for localized movements. The shifts in popularity of hashtags, in the case of the Egyptian revolution, was related to the growth of an elite group of Twitter users, elsewhere referred to as "key actor types" [7], who were most influential in orienting discourses on the uprisings [5]. These shifts are important in gauging not only key actors but also how movements spread on Twitter. Lerman and Ghosh (2010) [8] outline the importance of users in information cascades, which rely on a triad of network-retweets-followers in the Twittersphere. Our research is interested in identifying key actors, information cascades, and transnational bridges on Twitter that supported the Saudi women's driving campaign.

While "Arab Spring" on-the-ground tweets were largely unstructured, real-time event reporting, leading to the dissemination of information and mobilization efforts under various hashtags, our research investigates loosely organized grassroots examples of collective action, manifested on Twitter as @Oct26driving, so that there were fewer possibilities of sub-clusters formulating under divergent hashtags. In other words, since @Oct26driving was established as a Twitter profile, Twits interested in finding out about the movement and sharing information had a central platform to engage with the campaign. However, since the Saudi women's driving movement was disseminated by two hashtags in Arabic and English, their development was locally and transnationally positioned to mobilize linked clusters within the larger movement.

4 Methodology

Twitter data can be mined to track information and data about emerging trends and behaviors. Moreover, such data may also demonstrate and reveal information about precisely how ideas diffuse and how trends develop and take hold. In our research, we focus on the formation process of online collective action (OCA) by studying the diffusion of 'hashtags'. Given the definitive nature of hashtags, we investigate the co-evolution of the hashtag usage and the campaign network. We examine the recently organized Saudi Arabian women's right to drive campaign, called 'Oct26Driving' and collected the associated Twitter data, starting from September 25, 2013 to November 14, 2013. The study considered several hashtags associated with the campaign and analyzed the dominant ones, viz., '#oct26driving' and '#قيادة_26اكتوبر'. Our research identifies cross-cultural aspects with individual hashtag networks, with the Arabic hashtags relating to local factors and English hashtags helping in bringing transnational support from organizations such as human rights and women's rights.

Data Collection for the *Oct26Driving* Campaign

The content of about 70,000 tweets from 116 different countries was collected. 152 unique hashtags were created by users and associated with the campaign. Two most dominant hashtags, i.e. '#oct26driving' and '#قيادة_26اكتوبر', were selected. Other available tweet information, such as language, user name, retweeted user name (RT), and other hashtags, was parsed. Since these tweets are updated with frequencies varying between hundred to thousands of tweets per day, a crawler (viz., ScraperWiki, www.scraperwiki.com/) was configured to run continuously to collect, parse, and index the data.

Tweet Crawling with ScraperWiki

The crawler allows us to store the extracted data in an option of XLS or CSV formats. Collected data includes the unique ID of the tweet, the URL of the tweet, tweet content, the timestamp when the tweet was created or retweeted, the language of the tweet, screen name of the tweeter, name of any mentioned user in the tweet, the hashtags included, and the number of retweets. From the crawled tweets, as mentioned above about 70,000 tweets were posted for the *Oct26Driving* campaign at different time periods (from September 25, 2013 to November 14, 2013). However, to address the goal of this study, i.e. to track the formation of online collective action, our analysis focuses on the October 9, 2013 to October 30, 2013 time period.

Tweet Classification and Overlap Detection

The indexed tweets were filtered based on their hashtag usage, particularly the two dominant ones, i.e. '#oct26driving' and '#قيادة_26اكتوبر'. The tweets were further grouped based on their inclusion of either or both of the hashtags. Our first aim was to focus on the cumulative traffic of the tweets and tracking the development of online collective action. Tweet-retweet networks were created during the time periods corresponding to the formation of collective action. These networks helped us track the hashtag diffusion among the supporters and compare with network growth. This co-evolution of the hashtag usage and the network further elucidated the formation of the online collective action. To study the role of each of the dominant hashtags in the evolution of online collective action, as a second filtering process, we created a hashtag network for each of these hashtags. The networks and their respective growths were compared to identify the overlap(s) between these two networks. The overlaps were further examined to study the cross-cultural significance of the Arabic and English hashtags in the evolution of the online collective action.

Network Construction and Visualization

The classification result sets were analyzed by Gephi to study their structures and visualize the relationship of members within the respective movement. We measured the modularity of the network to detect and study the compartmentalized classifications of the network. The method consists of two phases. First, it looks for "small" communities by performing local optimization of the modularity. Second, it accumulates nodes of the same community and builds a new network across the communities. These steps are repeated iteratively until maximum modularity is achieved.

5 Analysis and Results

For our analysis, we focused on the time period between October 9, 2013 and October 30, 2013. Figure 1 shows the total activity of tweets related to the *Oct26Driving* campaign from September 25, 2013 to November 14, 2013.

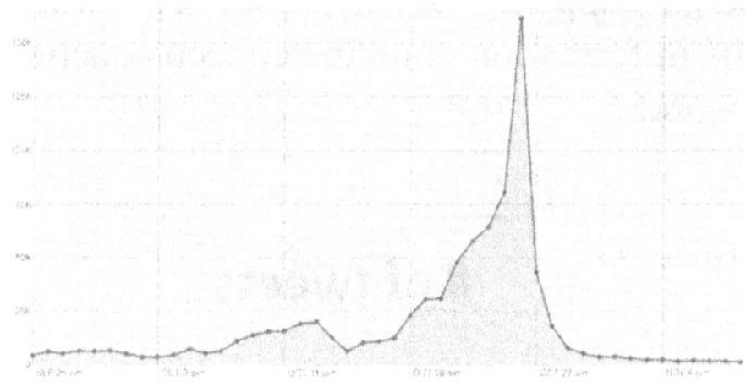

Fig. 1. (On **Left**) Total tweet activity for *Oct26Driving* campaign

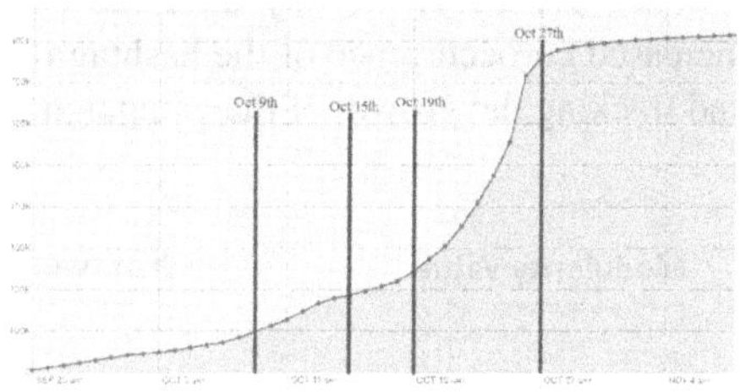

Fig. 2. (On **Right**) Cumulative tweet activity for *Oct26Driving* campaign

5.1 Diffusion of Dominant Hashtags

Using the two dominant hashtags as markers, we study the co-evolution of the hashtag usage and the campaign network. As we can see from Figures 1 and 2 above, the movement started gaining online awareness by October 9th, 2013 when the official Twitter account was created for the movement. The online initiative was boosted by the fact that residents of Saudi Arabia are highly active on social media. The fact that the campaign website was hacked and shut down on October 9, 2013 also contributed toward consolidating the web traffic to other social media platforms, primarily Twitter. To be able to track the formation of online collective action we picked the dates

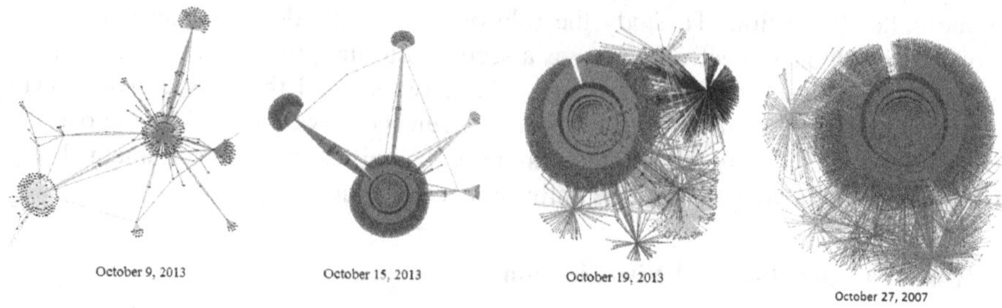

Fig. 3. Tweet-Retweet network for Oct26Driving campaign at different dates with considerable web traffic. Green cluster denotes the @oct26driving Twitter account.

when there was a spike in the web traffic (noted as peaks in Figure 1) and created tweet-retweet networks for each date as shown in Figure 3.

Although at the beginning (Oct. 9) we have very scattered and visible clusters, the clusters are no longer very distinct as the network activity increases. With the diffusion of dominant and other related hashtags, the clusters become increasingly connected. As the cumulative activity in the tweet-retweet network increases, the modularity of the network decreases from 0.607 (on October 9, 2013) to 0.225 (on October 27, 2013), reflecting connectivity between nodes increases and different clusters to start uniting. Figure 3 shows the snapshot of the tweet-retweet network for October 27, 2013, the day after the campaign happened. As hypothesized earlier, increased co-occurrence of the hashtag results in formation of online collective action and tracking the growth of the campaign network.

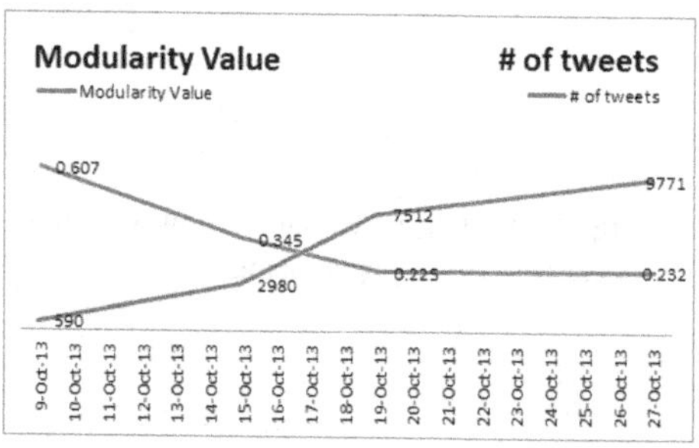

Fig. 4. Modularity distribution

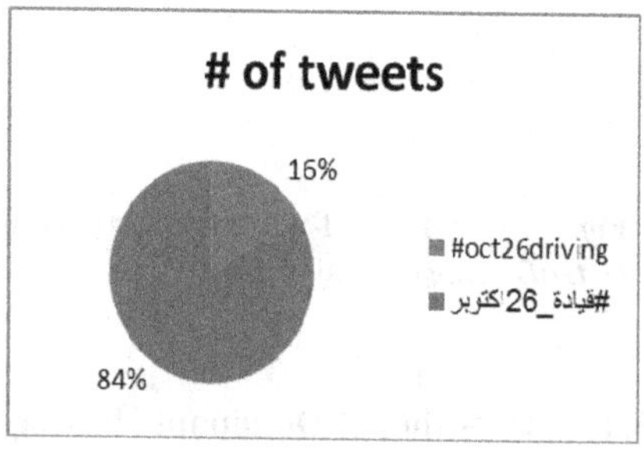

Fig. 5. Distribution of English and Arabic Hashtags

5.2 Identifying Transnational Support

We created a hashtag network for each dominant hashtag (‘#oct26driving’ and ‘#قيادة_26اكتوبر’) and compared their growth with each other. First, we filtered our tweet data and grouped the English hashtag and the Arabic hashtag tweets separately. When we compare the volume of tweets, the Arabic hashtag is used more frequently than the English hashtag, meaning more tweets are associated with the Arabic hashtag. Figure 5 shows the distribution of tweets using the two dominant hashtags. We tracked the behaviors of users and discovered that, if the user is tweeting in English, the person is 80% more likely to use/pair another English hashtag within the same tweet. On the other hand if the user is tweeting in Arabic, that user is 96% more likely to use/pair another Arabic hashtag within the same tweet. When we compared the overlap between hashtag networks, the results show that 60% of the common hashtags were in English. English hashtags are more likely to bridge different clusters. While the Arabic hashtags are more likely to relate to local factors, such as affects of driving on women's ovaries, English hashtags are helping to promote transnational and interorganizational support from various organizations such as human rights and women's rights, viz., Women2Drive Campaign. Figure 6 shows the network of both English and Arabic hashtags illustrating the interconnected nature of online collective action observed in this campaign.

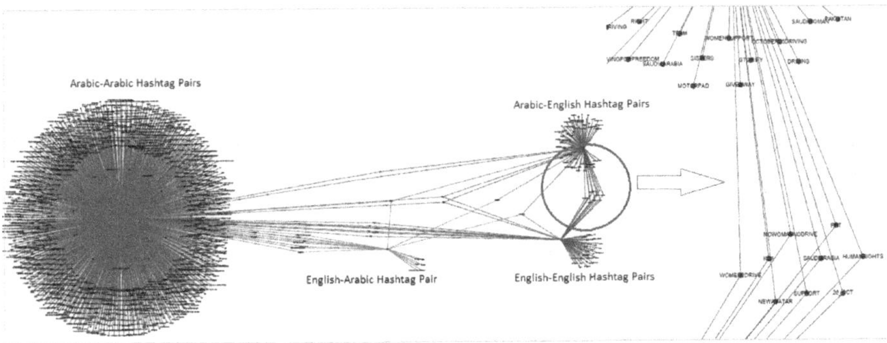

Fig. 6. Distribution of English and Arabic Hashtag. Arabic hashtags are depicted in red color on the left and English hashtags are depicted in blue and green colors on the right.

6 Conclusion

The prevalence of contemporary forms of information and communication technologies (ICTs), such as social media, have fundamentally altered how people coordinate and mobilize leading to manifestations of collective actions in various forms. Here we focus on the process of formation of online collective action by studying the diffusion of hashtags. We examine the recently organized Saudi Arabian women's right to drive campaign, called 'Oct26Driving' and collected Twitter data, starting from September 25, 2013 to November 14, 2013. Through the definitive nature of hashtags, we investigated the co-evolution of hashtag usage and campaign network growth, helping us to track the formation of collective action. Considering the dominant hashtags dedicated to the Oct26Driving campaign, viz., '#oct26driving' and '#اكتوبر26_قيادة', the study identifies cross-cultural aspects with individual hashtag networks. While Arabic hashtags are relating more to the local factors, English hashtags are helping in bringing transnational and interorganizational support from various organizations such as human rights and women's rights. Findings reveal interesting hashtag usage behavior among Twitter users. If a user is tweeting in English, that person is 80% more likely to use or pair another English hashtag within the same tweet. On the other hand, if the user is tweeting in Arabic, that user is 96% more likely to use or pair another Arabic hashtag within the same tweet. Furthermore, English hashtags are more likely to serve as transnational brokers bridging culturally diverse communities.

The research enables model development to help advance the understanding of interconnected collective actions conducted through modern social and information systems. At a broader level, the research helps to examine the role of information technology mediated communications in the formation of emergent organizations with implications to business, marketing (viral behaviors) and many other settings.

Acknowledgments. This research is funded in part by the National Science Foundation's Social Computational Systems (SoCS) research program (Award Numbers: IIS-1110868 and IIS-1110649) and the US Office of Naval Research (Grant number: N000141010091).

References

1. Lupia, A., Sin, G.: Which public goods are endangered? How evolving technologies affect the logic of collective action. Public Choice 117, 315–331 (2003)
2. Bimber, B., Flanagin, A.J., Stohl, C.: Reconceptualizing collective action in the contemporary media environment. Communication Theory 15, 389–413 (2005)
3. Agarwal, N., Lim, M., Wigand, R.T.: Finding her master's voice: The power of collective action among female Muslim bloggers. In: Proceedings of the 19th European Conference on Information Systems (ECIS), Helsinki, Finland, June 9-11, Paper 74 (2011a)
4. Agarwal, N., Lim, M., Wigand, R.T.: Collective action theory meets the blogosphere: A new methodology. In: Fong, S. (ed.) NDT 2011. CCIS, vol. 136, pp. 224–239. Springer, Heidelberg (2011)
5. Bruns, A., Highfiel, T., Burgess, J.: The Arab Spring and Social Media Audiences: English and Arabic Twitter Users and Their Networks. American Behavioral Scientist 57(7), 871–898 (2013)
6. Skinner, J.: Social Media and Revolution: The Arab Spring and the Occupy Movement as Seen through Three Information Studies Paradigms. Sprouts: Working Papers on Information Systems 11(169) (2011), `http://sprouts.aisnet.org/11-169`
7. Lotan, G., Graeff, E., Ananny, M., Gaffney, D., Pearce, I., Boyd, D.: The revolutions were tweeted: Information flows during the 2011 Tunisian and Egyptian Revolutions. International Journal of Communication 5, 1375–1405 (2011)
8. Lerman, K., Ghosh, R.: Information Contagion: An Empirical Study of the Spread of News on Digg and Twitter Social Networks. In: ICWSM 2010, pp. 90–97 (2010)
9. Yuce, S., Agarwal, N., Wigand, R.T.: Mapping Cyber-Collective Action among Female Muslim Bloggers for the *Women to Drive* Movement. In: Greenberg, A.M., Kennedy, W.G., Bos, N.D. (eds.) SBP 2013. LNCS, vol. 7812, pp. 331–340. Springer, Heidelberg (2013)
10. Yuce, S., Agarwal, N., Wigand, R.: Cooperative Networks in Interor-ganizational Settings: Analyzing Cyber-Collective Action. In: Proceedings of the 17th Pacific Asia Conference on Information Systems (PACIS), Jeju Island, Korea, June 18-22 (2013)

A Comparative Study of Smoking Cessation Programs on Web Forum and Facebook

Mi Zhang, Christopher C. Yang, and Katherine Chuang

College of Computing and Informatics, Drexel University
Philadelphia, PA 19104, United States
{Mi.Zhang,Chris.Yang}@drexel.edu

Abstract. Without time and geographical limitations, online smoking cessation programs attract a lot of users to help them quit smoking. This study compares two online smoking cessation support groups, QuitNet Forum and QuitNet Facebook, to evaluate the influence of software features that affect communication. We collect data from posts and comments of these two communities respectively, and compare user behavior from aspects of response immediacy and social network analysis. The associations between user behavior and quit stages were investigated. It is found that users of QuitNet Forum participate in communications more actively than users of QuitNet Facebook. The user behavior of QuitNet Facebook has a wider spectrum than that of QuitNet Forum.

Keywords: Smoking Cessation, Social Media, Social Network Analysis, Health Informatics, QuitNet.

1 Introduction

Smoking and tobacco use are acquired behaviors and the most preventable cause of death in the U.S. [1]. Many intervention programs of smoking cessation are developed to help people quit smoking, and face-to-face intervention methods are traditionally very popular in achieving satisfactory outcomes [2]. However, because the communication for these traditional intervention programs are limited by space and time constraints, health professionals use online means to reach their patient population. The Internet provides a widely accessible communication channel, which could reach a large number of people by easily overcoming geographical or time limitations. With this advantage, more and more intervention programs of smoking cessation are developed on the Internet, which are effective to increase the ratio of users who quit in a target group [3-6].

In fact, with the plethora of social media websites available, many of them are adapted to healthcare peer-to-peer usage, which introduces the term "Health 2.0" [7]. Communication among peers is an important feature of Health 2.0 to promote public health [8, 9]. The technical development of these online communities are geared for healthcare communications, but it is not clear from related literature what makes for a better website in terms of social support in smoking cessation. Generally, health support communities can be divided into two categories, some of which are

W.G. Kennedy, N. Agarwal, and S.J. Yang (Eds.): SBP 2014, LNCS 8393, pp. 421–429, 2014.

health focused and other that are general audience. Health specific online communities include MedHelp and PatientsLikeMe, where user group forums are developed on these sites for specific health conditions or problems. Other online health communities are built on popular social networking sites, including Facebook, Twitter, Linkedln, Blogs, etc [10]. These social network sites are developed for broader social uses, not necessarily limited to health interventions.

QuitNet (http://www.quitnet.com/qnhomepage.aspx) is a popular online intervention program of smoking cessation. It provides different services to help users quit smoking [6]. There are 11 forums on QuitNet website, which are traditional communities for discussions among smokers, smoking quitters, medical professionals and researchers. QuitNet has also created a public page on Facebook (https://www.facebook.com/QuitNet) to attract broader population. The admin of QuitNet Facebook posts messages on the public page every day, inviting discussions. Facebook users could browse QuitNet Facebook and participate in the discussions. Users who "like" the public page could receive messages in their Facebook "news feeds" and get updated information in a timely manner.

QuitNet Forum and QuitNet Facebook represent two different types of online communities for healthcare topics. Although both of them are supervised by the same organization – QuitNet, we do not know whether they attract users with similar characteristics, or whether users behave similarly in the two different communities. Quit stage is an important user characteristic that is analyzed in many studies of online intervention programs of smoking cessation, especially as an indicator of intervention outcome [4, 11-13]. In our previous study [14], we found that user distribution across quit stages were different on QuitNet Forum and QuitNet Facebook. We also built social networks to compare user activities and behaviors between the two communities.

Besides social network analysis in [14], we also analyze response time of users on QuitNet Forum and QuitNet Facebook in this study. User behaviors are explored based on the two types of analysis. Two research questions are proposed in this study:

RQ1: What are the differences of user behavior between QuitNet Forum and QuitNet Facebook?

RQ2: Is user behavior associated with quit stages on QuitNet Forum and QuitNet Facebook?

2 Methods

2.1 Data Source

We collected discussion thread data from QuitNet Forum and QuitNet Facebook, each has different accessibility features. All data collected are publicly and freely accessible. On the QuitNet Forum website, any registered user is allowed to initiate new threads (posts) and make comments. However, only the creator of the public page could start new threads on QuitNet Facebook. Only Facebook users who subscribed to the page by "liking" it can comment on posts. As a result, there are substantially more threads on QuitNet Forum than those on QuitNet Facebook during the same time period. In order to keep the sample sizes relatively similar from each website, we collected one-month data from a QuitNet Forum (May 1, 2011 – May 31, 2011), and three-month data of QuitNet Facebook (April 1, 2011 - June 30, 2011). This extraction process resulted in 3,017 posts and 24,713 comments made by 1,169 users from QuitNet Forum, and 111 posts and 2,574 comments made by 664 users from QuitNet Facebook.

User quit statuses were acquired from both QuitNet Forum and QuitNet Facebook. On QuitNet Forum, users voluntarily post their quit dates on their profile pages and their quit statuses could be calculated. On QuitNet Facebook, a weekly post of "shout your quit status" is created on every Friday. Users reply to this post and disclose their quit statuses. By analyzing the comments, the days of abstinence of these users can be calculated. From our collected data set, we extract 534 out of 1,169 users (45.6%) of QuitNet Forum and 394 out of 664 users (59.3%) of QuitNet Facebook whose days of abstinence can be identified.

2.2 Data Analysis

Measurements of User behavior (RQ1)

To solve RQ1, user behavior is studied from two aspects: user response time and social network analysis. User response time reflects the user's immediacy to respond to a post. Social networks are built from user interactions in these two communities, and a series of analysis are implemented to analyze user behavior.

Analysis of Response Time
After a thread is initiated and a post is published, it takes some time for people to respond to the post and participate in the discussion. The response time to a post could reflect a user's interests in the discussion topic as well as his/her activeness in the community. In this study, for each user of QuitNet Forum and QuitNet Facebook, we extract all posts that he/she comments on and calculate the amount of time (in seconds) that he/she takes to comment on each post since the post is published. The average value of response time to all posts that a user responds to is defined as his/her average response time, which reflects the response immediacy of that user. User average response time could indicate how fast a user responds to a post on average if participating in the thread. In our study, we calculate the average response time of all users on QuitNet Forum and QuitNet Facebook, and a T test is used to compare the means of average response time between users of the two communities. In addition, for each of the community, users are divided into seven groups according to their average response time. In each group, users have average response times of 10 minutes, 10 to 30 minutes, 30 minutes to 1 hour, 1 hour to 2 hours, 2 hours to 5 hours, 5 hours to 1 day, and more than 1 day, respectively.

Social Network Analysis
Two undirected graphs are constructed as social networks of QuitNet Forum and QuitNet Facebook. For each of the social network, actors (nodes) represent users of the community, and a tie (link) connects two actors who have participated in the same thread of discussions. Each tie carries a weight indicating the number of threads that

the corresponding users participate in together. For the social networks of QuitNet Forum and QuitNet Facebook, the network size and average weighted degree centrality are analyzed and compared. The network size of a social network is defined as the number of actors involved in linkages of the network [15]. The weighted degree centrality of an actor is defined as the sum of weights of all ties linked to it. This indicator reflects the frequency and activeness of a user's participation in discussions.

Measurement of Quit Stages (RQ2)

People usually go through multiple stages in the process of smoking cessation. Velicer et al. [16] extracted five stages, which are precontemplation, contemplation, preparation, action and maintenance. Once a smoking quitter stops smoking, he/she enters the action stage, which is composed by two periods: an early action period and a late action period [16]. In the early action period, the user has been abstinent for 0 to 3 months, and in the late action period, the user has been abstinent for 3 to 6 months. After being abstinent for 6 months, the smoking quitter moves into maintenance stage, which is suggested to be 6 to 60 months after stopping smoking.

In our previous study [14], for each user of QuitNet Forum and QuitNet Facebook, we calculated the days of abstinence from the date that he/she quits smoking to the date that he/she posts the last message on QuitNet Forum or QuitNet Facebook in our dataset. We divided users into five groups of different quit stages according to their days of abstinence. The first group is composed by users at the early action stage who have been abstinent for 0 to 3 months. The second group is composed by users in the late action stage who have been abstinent for 3 to 6 months. Users in the third and fourth groups are in the maintenance stage abstinent for 6 months to 5 years. And the fifth group is composed by users who have been abstinent for more than 5 years. In [14], we compared the distributions of user frequency with five different quite stages between QuitNet Forum and QuitNet Facebook, and found a significant difference on user frequency distributions between these two communities. Most users of QuitNet Forum are in the first group with 0-90 days of abstinence, which indicates that they have been abstinent for less than three months and are at the early action stage of smoking cessation. But for QuitNet Facebook, the numbers of users in different groups are not substantially different from each other except the last group.

In this study, we adopt ANOVA tests to compare user average response time and average weighted degree centralities between five the groups of quit stages of QuitNet Forum and QuitNet Facebook.

3 Results

3.1 User Behavior Analysis (RQ1)

Analysis of Response Time

The average response time is calculated for each user of QuitNet Forum and QuitNet Facebook. Among 1,169 users of QuitNet Forum, 1,040 of them have made comments in threads. Others only initiate new threads and do not respond to any posts. For the average response time of 1,040 users of QuitNet Forum and 664 users of

QuitNet Facebook, two outliers are detected for each community, which are removed for analysis. For QuitNet Forum, the mean of average response time is 10768.24 seconds (nearly 2 hours), and that for QuitNet Facebook is 19765.38 seconds (more than 5 hours). A T test is conducted, which indicates significant difference ($p<0.001$) on average response time between users of these two comminties. In general, users of QuitNet Forum respond faster than Facebook users.

Users of QuitNet Forum and QuitNet Facebook are divided into seven groups according to their average response time. Figure 1 presents the user distribution in each group for QuitNet Forum and QuitNet Facebook. There are 1024 out of 1038 Forum users (98.7%) and 625 out of 662 Facebook users (94.4%) with the average response time of less than 24 hours. For QuitNet Facebook, the numbers of Facebook users in the first six groups (different groups within 24 hours) are close. However, the Forum users in the group with average response time of 1 hour to 2 hours have the highest percentage.

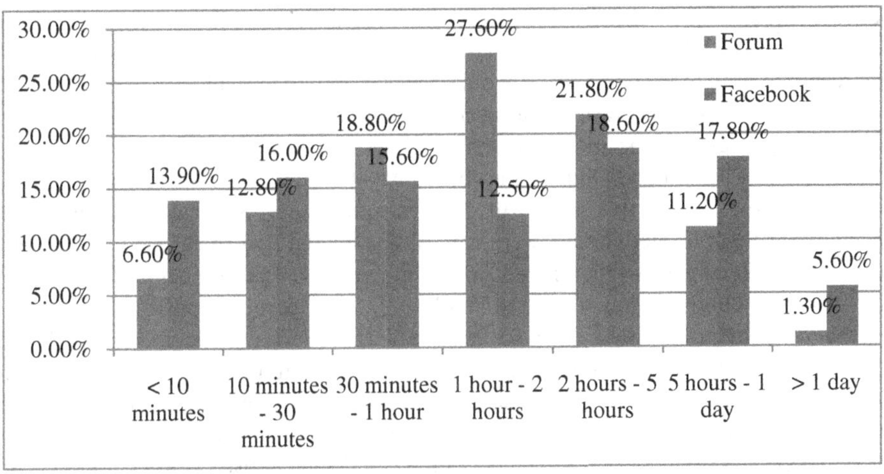

Fig. 1. User Distribution Across Time Groups of QuitNet Forum and QuitNet Facebook

Social Network Analysis (RQ)

Two social networks are developed for QuitNet Forum and QuitNet Facebook respectively. The social network of QuitNet Forum has 1,169 actors and 40,540 ties, and that of QuitNet Facebook has 664 actors and 19,904 ties. In the social networks of QuitNet Forum and QuitNet Facebook respectively, one and two actors do not involve in any linkages with other actors. So the network sizes of Forum and Facebook are 1168 and 662.

The weighted degree centralities of all actors are calculated for the two social networks to understand the distribution of user activity in the network. The average weighted degree centralities of QuitNet Forum and Quit Facebook are 213.223 and 78.374 respectively. The former is nearly as three times as that of the later. However, the network size of QuitNet Forum (1168) is about twice as that of QuitNet

Facebook (662), and the Forum data is collected in a shorter time period. On average, every two Forum users contact with each other 0.322 (213.223/1168) times in one month, but every two Facebook users contact with each other 0.118 (78.374/662) times in three months. So generally, users of QuitNet Forum behave more actively than users of QuitNet Facebook. They participate in discussions more frequently on average, and closely interacted with others.

3.2 User Behavior of Different Quit Stages (RQ2)

For QuitNet Forum and QuitNet Facebook, we compare user average response time between the five quit stages using ANOVA with LSD post-hoc tests. For both of the communities, the P values of ANOVA tests were above 0.1 (P = .47 for QuitNet Forum and P = .25 for QuitNet Facebook). For QuitNet Forum, LSD post-hoc tests showed no significant difference on user average response time between any groups at the level of P =.1. However for QuitNet Facebook, there was significant difference between Group 1 (0-90 days) and Group 3 (181-720 days) with P =.096, and between Group 3 (181-720 days) and Group 5 (>1800 days) with P =.094. The mean of average response time of users in different quit stages is shown in Table 1. Users of Group 3 were likely to respond faster than users of most other groups. The mean of their average response time was about 4 hours.

For QuitNet Forum and QuitNet Facebook, we also compare user weighted degree centralities between the five groups of different quit stages. Kruskal-Wallis test and Independent Sample Median test were carried out for the two social media channels respectively. Results show that there is no significant difference on degrees across the five groups of QuitNet Froum (P = .14 and P = .07), but there is significant difference for that of QuitNet Facebook (P = .03 and P = .02). The average degree of Facebook users in each group is shown in Table 1. To further explore the differences, ANOVA with LSD post-hoc test is implemented for QuitNet Facebook. Although P =.19 for ANOVA test, the LSD post-hoc tests showed significant differences between Group 1 (0-90 days) and Group 3 (181-720 days) with P =.035, and between Group 3 (181-720 days) and Group 5 (>1800 days) with P =.1. It indicates that the actors of Group 3 (181-720 days) are likely to have higher degree centrality than those of other groups, which is similar to the difference on user response time across five quit stages.

So, on QuitNet Forum, users at different quit stages have similar behaviors. But on QuitNet Facebook, users at early maintenance stage behave more actively.

Table 1. Degree Centrality and Average Response Time in Each Group of Quit Status of QuitNet Facebook

Quit Status (days)	0-90	91-180	181-720	721-1800	>1800
Mean of Degree Centrality	88.12	116.87	127.96	117.78	84
Mean of Average Response Time	25919.69	15830.18	15650.40	18348.51	29198.32

4 Discussions

Summarizing results in [14] and this work, there are differences of the two samples for average response time, social network structures, and distribution of quit stages. Generally, users of QuitNet Forum perform more actively than Facebook users, with the evidences that Forum users respond faster and more frequently. Different users of QuitNet Forum behave similarly to each other, because most users respond to post in 1 to 2 hours, and there is no difference of user behavior between different quit stages. Users of QuitNet Facebook behave differently from each other, with the evidences that Facebook users have variant response time, and users at early maintenance stage perform more actively than others. The features of the two social media channels may explain user differences between the two communities.

Active participation of QuitNet Forum requires each user to have an account registered. Only recent smoking quitters who have great enthusiasm and determination are likely to log on QuitNet Forum frequently to participate in discussions. So, most forum users are active and behave similarly to each other. QuitNet Forum is a community of practice (CoP), which is a group of people who "share a concern or a passion for something they do and learn how to do it better as they interact regularly" [17]. In QuitNet Forum, people share the same purpose of smoking cessation and practice together to achieve this purpose. With high average weighted degree centrality, the social network of QuitNet Forum has strong ties between different actors, which indicate strong and deep relations between users in this community [18].

QuitNet Facebook is a community page on a popular and open platform for social communications, requiring Facebook accounts. Because their main account is their personal account with QuitNet Facebook as an ancillary mailing list, they can passively receive messages from QuitNet in their "news feeds". This allows for discussions of smoking cessation while interacting with other friends on Facebook. As a result, people from a wider range are attracted to QuitNet Facebook. Their main purposes of logging on Facebook may not be smoking cessation. They participate in the discussions casually and frequently, as evidenced by the longer average response time and more scattered network shape. QuitNet Facebook is a community of interest (CoI) that is regarded as a "community of communities" [19]. It brings together people with different purposes and the same interest. The ties in the social network of QuitNet Facebook are weak, with the evidence that it has low weighted degree centrality on average. This finding is consistent with a general finding that Facebook emerges a publicly open structure with loose behavioral norms of participants [20]. The weak ties play an important role of "bridge", which connects to different networks and expands user relations [18].

The response patterns of users on QuitNet Forum are similar. A relatively higher percentage of messages have an average response time of 1 hour to 2 hours. The reason for this quicker response may be that QuitNet Forum has a unified and static interface for all users. The latest posts and comments always appear at the top of the Forum page. So the interface and content are consistent for all users when they log on QuitNet Forum at the same time. Old posts are replaced by new posts and removed from the first page of the forum. Users need to make additional efforts to go to the second or later pages to read and respond to old posts. As a result, all users tend to reply more posts at the top of the first page, which might be published in a few hours. So their average response time is very similar.

However, the response time of QuitNet Facebook users varies substantially from each other. Users' average response times are distributed widely along a spectrum. This may be because the users of QuitNet Facebook are notified of new posts automatically in their main page, which contains a "news feeds". It is not necessary for them to navigate to the public page of QuitNet Facebook for the ability to interact with others of the community. From their homepage of Facebook through a computer device, they can read messages from their subscribed "news feeds". During the same periods, different Facebook users receive different numbers of messages in their "news feeds". Some users may receive a lot of messages from variant sources on Facebook, so posts of QuitNet Facebook could not stay for a long time at the top of their "news feeds". But some users may receive fewer messages, so posts of QuitNet Facebook could stay much longer at the top of their "news feeds". So, different users of QuitNet Facebook make different efforts to read old posts of QuitNet Facebook. As a result, the response immediacy of QuitNet Facebook users varies greatly from each other.

5 Conclusion

QuitNet Forum and QuitNet Facebook represent two general types of communities for healthcare topics as mentioned above. Our study initially compares specific examples of the two types of communities with user behavior investigation. In the future, similar studies could be implemented for other health topics, and the two types of communities could be explored from different aspects. Understanding features of different communities could help us provide better services to achieve the best effects of health interventions.

References

1. Agaku, I., et al.: Current Cigarette Smoking Among Adults — United States, 2011, Centers for Disease Control and Prevention (2012)
2. Russell, M.A.H., et al.: Effect of General Practitioners' Advice Against Smoking. British Medical Journal, 2231–2235 (1979)
3. Graham, A.L., et al.: Effectiveness of An Internet-Based Worksite Smoking Cessation Intervention at 12 Months. Journal of Occupational and Environmental Medicine 49, 821–828 (2007)
4. Cobb, N.K., et al.: Initial Evaluation of A Real-World Internet Smoking Cessation System. Nicotine Tob. Res. 7, 207–216 (2005)
5. Shahab, L., McEwen, A.: Online Support for Smoking Cessation: A Systematic Review of the Literature. Addiction 104, 1792–1804 (2009)
6. An, L.C., et al.: Utilization of Smoking Cessation Informational, Interactive, and Online Community researches as Predictors of Abstinence: Cohort Study. J. Med. Internet Res. 10, e55 (2008)

7. Hughes, B., et al.: Health 2.0 and Medicine 2.0: Tensions and Controversies in the Field. Journal of Medical Internet Research 10(3), e23 (2008)
8. Van De Belt, T.H., et al.: Definition of health 2.0 and medicine 2.0: a systematic review. Journal of Medical Internet Research 12(2), e18 (2012)
9. Eysenbach, G.: Medicine 2.0: social networking, collaboration, participation, apomediation, and openness. Journal of Medical Internet Research 10(3), e22 (2008)
10. Backman, C., et al.: Social Media and Healthcare, `http://library.ahima.org/xpedio/groups/public/documents/ahima/bok1_048693.hcsp?dDocName=bok1_048693`
11. Cobb, N.K., Graham, A.L.: Characterizing Internet Searchers of Smoking Cessation Information. J. Med. Internet Res., e17 (2006)
12. Selby, P., et al.: Online Social and Professional Support for Smokers Trying to Quit: An Exploration of First Time Posts from 2562 Members. J. Med. Internet Res. 10, e55 (2010)
13. Cobb, N.K., et al.: Social Network Structure of a Large Online Community for Smoking Cessation. American Journal of Public Health 100, 1282–1289 (2010)
14. Zhang, M., Yang, C.C., Li, J.: A comparative study of smoking cessation intervention programs on social media. In: Yang, S.J., Greenberg, A.M., Endsley, M. (eds.) SBP 2012. LNCS, vol. 7227, pp. 87–96. Springer, Heidelberg (2012)
15. Chang, H.-J.: Online supportive interactions: Using a network approach to examine communication patterns within a psychosis social support group in Taiwan. Journal of the American Society for Information Science and Technology 60(7), 1504–1517 (2009), doi:10.1002/asi.21070
16. Velicer, W.F., et al.: Assessing Outcome in Smoking Cessation Studies. Psychological Bulletin 111(1), 23–41 (1992), doi:10.1037/0033-2909.111.1.23
17. Wenger, E.: Communities of practice: A brief introduction (2005), `http://www.ewenger.com/theory/index.htm`
18. Granovetter, M.S.: The Strength of Weak Ties. American Journal of Sociology 78(6), 1360–1380 (1973)
19. Fischer, G.: Communities of Interest: Learning through the Interaction of Multiple Knowledge Systems. In: Proc. 24th Annual Information Systems Research Seminar In Scandinavia, 2001 of Conference, pp. 1–14 (2001)
20. Papacharissi, Z.: The virtual geographies of social networks: a comparative analysis of Facebook, LinkedIn and ASmallWorld. New Media & Society 11(1-2), 199–220 (2009), doi:10.1177/1461444808099577

[illegible] and Cues [illegible] in the [illegible]
[illegible] 10(4), [illegible]
[illegible]
[illegible] Medicine [illegible] information [illegible]
[illegible] Health [illegible]
[illegible] Consumer [illegible]
[illegible] Social [illegible] to Con[illegible]
[illegible] 10, [illegible] (2010)
[illegible] Community for Smoking [illegible]
[illegible] 2, [illegible]
[illegible] intervention pro[illegible]
[illegible] SIG A[illegible]
[illegible] to examine [illegible]
[illegible] Journal of [illegible]
[illegible] 1504-1[illegible] (2[illegible])
[illegible]
[illegible] Technological [illegible]
[illegible] (2007) [illegible]
[illegible]
[illegible]
[illegible] Journal of Sociology 78(6) [illegible]
[illegible]
[illegible] Information [illegible] Know[illegible]
[illegible] in Scandina[illegible]
[illegible]
[illegible] Comparative analysis of [illegible]
[illegible] 11(1 2), [illegible] (2010)
[illegible]

Author Index